Lecture Notes in Control and Information Sciences

Volume 449

For further volumes:
http://www.springer.com/series/642

Danielle C. Tarraf
Editor

Control of Cyber-Physical Systems

Workshop Held at The Johns Hopkins University, March 2013

Springer

Editor
Danielle C. Tarraf
Department of Electrical and Computer
Engineering
The Johns Hopkins University
Baltimore, Maryland
USA

ISSN 0170-8643 ISSN 1610-7411 (electronic)
ISBN 978-3-319-01158-5 ISBN 978-3-319-01159-2 (eBook)
DOI 10.1007/978-3-319-01159-2
Springer Cham Heidelberg New York Dordrecht London

Library of Congress Control Number: 2013941662

Printed on acid-free paper

Springer is part of Springer Science+Business Media (www.springer.com)

Preface

This volume brings together 19 invited papers presented at the Workshop on Control of Cyber-Physical Systems, held at The Johns Hopkins University in Baltimore, USA on March 20–21, 2013 (www.ece.jhu.edu/~dtarraf/CPSWorkshop/). The workshop was organized as a single track event spanning two days, punctuated by the plenary lecture "Towards a Theory of Resilient Cyber-Physical Systems" delivered by S. Shankar Sastry as part of the 2013 CISS program. The workshop, which brought together 40 researchers from 16 national and 5 international universities, provided a forum for exchanging ideas, identifying fundamental scientific challenges, debating solution approaches, and ultimately, for paving the way forward in Cyber-Physical Systems (CPS) research.

Cyber-physical systems consist of deeply integrated, tightly coupled computational and physical components, equipped with communication capabilities. These systems are highly complex, spanning multiple scientific and technological domains. They thus pose several fundamental challenges. They are also critically important to society's advancement and security. The design and deployment of the adaptable, reliable CPS of tomorrow requires the development of a basic science foundation, synergistically drawing on various branches of engineering, mathematics, and computer science, coupled with domain specific knowledge.

The present volume, thematically organized into seven sections, highlights the central roles of control theory and systems thinking in developing the theoretical foundations of CPS, in addressing the challenges of cyber-trust and cyber-security, and in advancing emerging cyber-physical application domains ranging from smart grids to smart buildings, cars and robotic systems. While each invited paper was assigned to a single section for organizational purposes, each of the papers is equally at home in a second or even third section, underlining the inherently broad and multifaceted nature of CPS research problems.

I would like to address a word of thanks to the Department of Electrical & Computer Engineering at The Johns Hopkins University for hosting this event, and to all the participants for making the workshop a success!

The Johns Hopkins University
Baltimore, MD

Danielle C. Tarraf
Organizer & Program Chair

Organization

The invited papers were subject to a thorough peer review process. We wish to thank our referees, listed below, for their efforts.

Referees

Tansu Alpcan	*The University of Melbourne*
Carolyn L. Beck	*University of Illinois at Urbana-Champaign*
Lijun Chen	*University of Colorado at Boulder*
Noah J. Cowan	*The Johns Hopkins University*
Jean-Charles Delvenne	*Université Catholique de Louvain*
Geir Dullerud	*University of Illinois at Urbana-Champaign*
Magnus Egerstedt	*Georgia Institute of Technology*
Adriano Fagiolini	*University of Palermo*
Makan Fardad	*Syracuse University*
Mazen Farhood	*Virginia Polytechnic Institute and State University*
Laura Giarré	*University of Palermo*
Jessy W. Grizzle	*University of Michigan*
Christoforos Hadjicostis	*University of Cyprus*
Rabih Jabr	*American University of Beirut*
Maryam Kamgarpour	*ETH Zürich*
Scott Kuindersma	*Massachusetts Institute of Technology*
Andrew Lamperski	*University of Cambridge*
Cédric Langbort	*University of Illinois at Urbana-Champaign*
Eugene Lavretsky	*The Boeing Company*
Michael D. Lemmon	*University of Notre Dame*
Jerome Le Ny	*École Polytechnique de Montréal*
Laurent Lessard	*University of California, Berkeley*
Na Li	*California Institute of Technology*
Daniel Liberzon	*University of Illinois at Urbana-Champaign*
Enrico Lovisari	*Lund University*
Manuel Mazo Jr.	*Delft University of Technology*

Nader Motee	*Lehigh University*
Nils Napp	*Harvard University*
Angelia Nedich	*University of Illinois at Urbana-Champaign*
Mitra Osqui	*Massachusetts Institute of Technology*
Fabio Pasqualetti	*University of California, Santa Barbara*
Ketan Savla	*University of Southern California*
Iman Shames	*University of Melbourne*
Bruno Sinopoli	*Carnegie Mellon University*
Roy Smith	*ETH Zürich*
Eduardo Sontag	*Rutgers University*
Dušan Stipanović	*University of Illinois at Urbana-Champaign*
Xiaofeng Wang	*University of South Carolina*
Tichakorn Wongpiromsarn	*Ministry of Science and Technology, Thailand and Singapore-MIT Alliance for Research and Technology, Singapore*
Tansel Yucelen	*Georgia Institute of Technology*

Contents

Part IV: Fault Diagnosis and Sensor Fusion in CPS

Part V: Application Domains: Robotics and Power Systems

Part VI: Control of Multi-agent Systems

Part VII: Networked Control Systems

Part I
Paradigms for CPS Analysis and Design

Robust Global Almost Sure Synchronization on a Circle via Stochastic Hybrid Control*

Matthew Hartman, Anantharaman Subbaraman, and Andrew R. Teel

Electrical and Computer Engineering Department
University of California
Santa Barbara, CA 93106-9560 USA
{mhartman,anantharaman}@umail.ucsb.edu,
teel@ece.ucsb.edu

Abstract. This chapter describes some recent advances in modeling and stability analysis for stochastic hybrid systems from the viewpoint of applications to cyber-physical systems. As an illustration, it discusses synthesizing an algorithm for robust, global, almost sure synchronization of a large number of agents evolving on a circle under all-to-all communication. The robustness includes achieving near synchronization even in the presence of adversarial perturbations. This behavior is not something that is achieved via non-stochastic, non-hybrid almost global synchronization algorithms.

Keywords: Stochastic hybrid systems, stochastic stability, Lyapunov methods, synchronization.

1 Introduction

Cyber-physical systems refer to systems with integrated computational and physical capabilities [1]. Examples include next generation power grids, automobiles, air transportation systems, embedded medical devices and smart prosthetics. Further applications of cyber-physical systems are mentioned in [19] along with future directions for research.

Security in cyber-physical systems is an aspect that has received a lot of attention recently. In [5] various types of attacks that can be expected in a cyber-physical system are summarized. The shortcomings of the existing results in information security and networked control systems to ensure security of cyber-physical systems along with new directions and challenges in improving security are discussed. The results in [17], [18] develop a framework for detecting and identifying measurement errors induced by adversaries in power networks.

We consider a security problem, the solution to which requires a stochastic, hybrid systems approach. The problem is to achieve global almost sure synchronization for a large number of agents evolving on a circle in the presence

* Research supported in part by NSF under grant ECCS-1232035 and AFOSR under grant AFOSR FA9550-12-1-0127.

D.C. Tarraf (ed.), *Control of Cyber-Physical Systems*,
Lecture Notes in Control and Information Sciences 449,
DOI: 10.1007/978-3-319-01159-2_1,

adversarial perturbations that attempt to prevent synchronization. We show how a non-stochastic almost global synchronization algorithm is susceptible to faulty behavior in the presence of such perturbations; on the other hand, the stochastic hybrid algorithm is (almost surely) not susceptible to faulty behavior. We emphasize that both the stochastic aspect and the hybrid aspect of the algorithm are crucial for this robustness, each in a different region of the configuration space.

The paper is organized as follows: In Section 2 we briefly review various frameworks proposed for non-stochastic hybrid systems in the literature, but focus on the general framework established in [11]. Stability concepts and tools for Lyapunov analysis are also reviewed. In Section 3 we present a framework recently proposed in [28] for stochastic hybrid systems with stochastic elements affecting the discrete-time dynamics. Stability notions such as asymptotic stability in probability are described for stochastic hybrid systems with non-unique solutions. An overview of the global synchronization problem statement and the shortcomings of the existing non-stochastic algorithms with respect to robustness are discussed in Section 4. In Section 5 we present an algorithm for robust synchronization of agents on a circle using the tools developed for stochastic hybrid systems in [28]. Simulation results are also provided to illustrate how the algorithm achieves the synchronization task. We finish with concluding comments in Section 6.

2 A Hybrid Systems Modeling Framework

There are many ways to model systems having states capable of both continuous and discrete change. Options include hybrid automata [14] and impulsive differential equations [16]. The motivation for considering an alternative to these models is to have a general framework that extends useful results from nonlinear systems; in particular, results on invariance principles [21], converse theorems [4], robustness of stability [11], singular perturbation theory [22] and averaging theory [30] are easily accessible through the framework we use. Moreover, systems modeled as hybrid automata can be reformulated into this model [10].

We consider hybrid systems with state $x \in \mathbb{R}^n$ written formally as

$$\begin{aligned} \dot{x} &\in F(x), \quad x \in C \\ x^+ &\in G(x), \quad x \in D \end{aligned} \tag{1}$$

where C, D represent the flow and jump sets and F, G represent the flow and jump maps. Generally speaking, solutions to (1) are functions that map a time domain to vectors in Euclidean space, while satisfying (1). Note that in the more specific case, one can replace the set-valued mappings F and G with functions f and g, thus giving us differential and difference equations rather than inclusions.

We require the notion of a *hybrid time domain*: a subset E of $\mathbb{R}_{\geq 0} \times \mathbb{Z}_{\geq 0}$, which is the union of infinitely many intervals of the form $[t_j, t_{j+1}] \times j$, where $0 = t_0 \leq t_1 \leq t_2 \leq ...$, or finitely many of such intervals, with the last one possibly of the form $[t_j, t_{j+1}] \times \{j\}$, $[t_j, t_{j+1}) \times \{j\}$, or $[t_j, \infty) \times \{j\}$. A function that maps a hybrid time domain to the real numbers is called a *hybrid arc*: a function $\phi : E \to \mathbb{R}^n$ such that E is a hybrid time domain and $t \to \phi(t, j)$ is locally absolutely continuous for fixed j.

A hybrid arc is a *solution* to (1) if $\phi(0, 0) \in C \cup D$ and:

1) for all $j \in \mathbb{Z}_{\geq 0}$ and almost all t such that $(t, j) \in \text{dom } \phi$: $\phi(t, j) \in C$, $\dot{\phi}(t, j) \in F(\phi(t, j))$
2) for all $(t, j) \in \text{dom } \phi$ such that $(t, j + 1) \in \text{dom } \phi$: $\phi(t, j) \in D$, $\phi(t, j + 1) \in G(\phi(t, j))$.

A solution ϕ to a hybrid system is *complete* if dom ϕ is unbounded (in either the t or j direction). A solution ϕ is *maximal* if it cannot be extended, i.e., it is not a truncation of another solution ϕ' to some proper subset of dom ϕ'. We denote the set of solutions to a hybrid system starting from a set $\mathcal{X} \subset \mathbb{R}^n$ as $\mathcal{S}(\mathcal{X})$. We define the graph of a solution as $\text{graph}_{\geq\tau}(x) := \{(t, j, z) \in (\text{dom } x) \times \mathbb{R}^n : z = x(t, j),\ t + j \geq \tau\}$ and $\text{graph}(x) := \text{graph}_{\geq 0}(x)$.

There is a set of mild regularity conditions on (1) that are typically assumed in papers using this framework. They are stated as follows:

Assumption 1. *The sets C and D are closed; the mappings F and G are outer semicontinuous*[1] *and locally bounded*[2]*; $F(x)$ is non-empty and convex for all $x \in C$; $G(x)$ is non-empty for all $x \in D$.*

This assumption ensures that the set of solutions is sequentially compact and semicontinuous with respect to initial conditions.

Often, we wish to consider the stability of sets rather than just equilibrium points. Referring to the statement[3]

$$\text{graph}_{\geq\tau}(x) \subset (\mathbb{R}^2 \times (\mathcal{A} + \epsilon\mathbb{B})), \quad \forall \zeta \in \mathcal{A} + \delta\mathbb{B}, \quad x \in \mathcal{S}(\zeta), \tag{2}$$

a compact set $\mathcal{A} \subset \mathbb{R}^n$ for a hybrid system is

- *Lyapunov stable* if for $\tau = 0$ and each $\epsilon > 0$ there exists $\delta > 0$ such that (2) holds,
- *Lagrange stable* if for $\tau = 0$ and each $\delta > 0$ there exists $\epsilon > 0$ such that (2) holds,
- *Uniformly attractive* if for each $\epsilon > 0$ and $\delta > 0$, there exists $\tau > 0$ such that (2) holds,
- *Globally asymptotically stable* (GAS) if it is Lyapunov stable, Lagrange stable, and uniformly attractive.

Note that the empty set is a subset of any set by convention.

---

[1] A set-valued mapping $F : \mathbb{R}^n \rightrightarrows \mathbb{R}^n$ is outer semicontinuous if its graph $\{(x, y) : x \in \mathbb{R}^n, y \in F(x)\} \subset \mathbb{R}^{2n}$ is closed.

[2] The mapping F is locally bounded if for each compact set K, the set $F(K) := \bigcup_{x \in K} F(x)$ is bounded.

[3] The sets $\mathbb{B}$ and $\mathbb{B}^o$ indicate the closed and open unit ball, respectively, in the Euclidean norm of appropriate dimension.

For hybrid systems that satisfy Assumption 1, for a compact set $\mathcal{A}$ to have the property of GAS is equivalent to it having the properties of Lyapunov stability and attractivity, the latter being that each solution is bounded, and each complete solution converges to $\mathcal{A}$.

Lyapunov analysis is a useful tool for establishing asymptotic stability. Given a hybrid system $\mathcal{H} = \{C, D, F, G\}$, compact set $\mathcal{A}$, a function $V : \operatorname{dom} V \to \mathbb{R}$ is a *Lyapunov function candidate* for $(\mathcal{H}, \mathcal{A})$ if i) V is continuous and non-negative on $(C \cup D) \backslash \mathcal{A} \subset \operatorname{dom} V$, ii) V is continuously differentiable on an open set $\mathcal{O}$ satisfying $C \backslash \mathcal{A} \subset \mathcal{O} \subset \operatorname{dom} V$, and iii)

$$\lim_{\{x \to \mathcal{A}, x \in \operatorname{dom} V \cap (C \cup D)\}} V(x) = 0.$$

The following invariance principle is adapted from [21, Thm. 7.6][4].

Theorem 1. *Consider the system $\mathcal{H} = \{C, D, F, G\}$ satisfying Assumption 1, and the compact set $\mathcal{A} \subset \mathbb{R}^n$ satisfying $G(\mathcal{A} \cap D) \subset \mathcal{A}$. If $C \cup D$ is compact and there exists a Lyapunov function candidate V for $(\mathcal{H}, \mathcal{A})$ that is positive on $(C \cup D) \backslash \mathcal{A}$ and satisfies*

$$\begin{aligned} \langle \nabla V(x), f \rangle &\le 0 \qquad \forall x \in C \backslash \mathcal{A},\ f \in F(x) \\ V(g) - V(x) &\le 0 \qquad \forall x \in D \backslash \mathcal{A},\ g \in G(x) \backslash \mathcal{A} \end{aligned}$$

then the set $\mathcal{A}$ is Lyapunov stable. If furthermore, for every compact neighborhood K of $\mathcal{A}$ and each $\mu > 0$, no complete solution to $\mathcal{H}$ remains in $L_V(\mu) \cap K$, then the set $\mathcal{A}$ is globally asymptotically stable.

The following is a simple version of a converse Lyapunov theorem found in [4][5,6].

Theorem 2. *For system $\mathcal{H} = \{C, D, F, G\}$ satisfying Assumption 1, if the compact set $\mathcal{A}$ is globally asymptotically stable, then there exist a C^∞ function $V : \mathbb{R}^n \to \mathbb{R}_{\ge 0}$ and $\alpha_1, \alpha_2 \in \mathcal{K}_\infty$ such that*

$$\begin{aligned} \alpha_1(|x|_\mathcal{A}) \le V(x) &\le \alpha_2(|x|_\mathcal{A}) && \forall x \in \mathbb{R}^n, \\ \langle \nabla V(x), f \rangle &\le -V(x) && \forall x \in C,\ f \in F(x) \\ V(g) - V(x) &\le \frac{V(x)}{2} && \forall x \in D,\ g \in G(x). \end{aligned}$$

In anticipation of the framework for stochastic hybrid systems, we describe a hybrid system with inputs, as shown in (3).

$$\begin{aligned} \dot{x} &\in F(x), \quad x \in C \\ x^+ &\in G(x, v^+), \quad x \in D. \end{aligned} \tag{3}$$

[4] For $\mu \in \mathbb{R}$ and a function $V : \operatorname{dom} V \to \mathbb{R}$, $L_V(\mu) := \{x \in \operatorname{dom} V : V(x) = \mu\}$.

[5] A function $\alpha : \mathbb{R}_{\ge 0} \to \mathbb{R}_{\ge 0}$ is of class-$\mathcal{K}_\infty$ if it is continuous, zero at zero, strictly increasing and unbounded.

[6] For a closed set $S \subset \mathbb{R}^n$ and $x \in \mathbb{R}^n$, $|x|_S := inf_{y \in S}|x - y|$ is the Euclidean distance to S.

A pair of hybrid arcs (ϕ, v) is said to be a *standard solution* to (3) starting at x if $\phi(0,0) = x$ and the following conditions hold:

(1) $\phi(t,j) \in C$ and $\dot{\phi}(t,j) \in F(\phi(t,j))$ for each $j \in \mathbb{Z}_{\geq 0}$ and almost all $t \in I_j := \{s : (s,j) \in \operatorname{dom} \phi\}$;
(2) $\phi(t,j) \in D$ and $\phi(t,j+1) \in G(\phi(t,j), v(t,j+1))$ for all $(t,j) \in \operatorname{dom} \phi$ such that $(t,j+1) \in \operatorname{dom} \phi$.

3 A Stochastic Hybrid Systems Modeling Framework

Stochastic hybrid systems have been extensively studied in the literature and several frameworks have been proposed. One of the important distinguishing factors of these frameworks relates to how the randomness affects the dynamics of the system. A summary of the various ways by which stochastic elements can be introduced in hybrid systems is listed in [6, Ch. 1].

In [8], [7] piecewise deterministic Markov process are modeled, in which the continuous-time dynamics are deterministic, discrete-time dynamics are random and jumps of the state occur either at random times or when the state exits an open domain. In [31],[13] hybrid switching diffusions are analyzed where the continuous-valued states are driven by a stochastic differential equation associated with a certain discrete-valued state and jumps occurring at random time lead to changes in this discrete-valued state. A general framework for stochastic hybrid systems is proposed in [15],[3] that can model hybrid systems with randomness affecting both the continuous and discrete dynamics while allowing for random jump times.

Most of the existing frameworks in the literature for modeling stochastic hybrid systems do not encompass systems that permit non-unique solutions. Non-unique solutions arise in the case of stochastic systems when analyzing robustness of stability properties [29],[27] and while defining notions of generalized random solutions [12]. In [28] the framework is extended to hybrid systems with non-unique solutions while allowing for stochastic elements in the jump map. This class of systems can be used to model systems with spontaneous transitions as illustrated in [28]. We note that this model does not allow for stochasticity in the flow map; such a generalization would rely on further progress in the theory of stochastic differential inclusions.

3.1 Modeling Framework

For the rest of this paper we will adopt the mathematical framework in [28]. We consider stochastic hybrid systems with a state $x \in \mathbb{R}^n$ and random variable $v \in \mathbb{R}^m$ written formally as

$$x \in C \qquad \dot{x} \in F(x) \tag{4a}$$

$$x \in D \qquad x^+ \in G(x, v^+) \tag{4b}$$

$$\mu(\cdot). \tag{4c}$$

As before, C and D denote the flow and jump sets while F and G denote the flow and jump maps. The distribution function μ is derived from the probability space $(\Omega, \mathcal{F}, \mathbb{P})$ and a sequence of independent, identically distributed (i.i.d.) input random variables defined on $(\Omega, \mathcal{F}, \mathbb{P})$. Let $\mathbf{v}_i : \Omega \to \mathbb{R}^m$, $i \in \mathbb{Z}_{\geq 1}$ denote a sequence of i.i.d random variables and $\mathbf{B}(\mathbb{R}^m)$ denote the Borel-σ field of $\mathbb{R}^m$. Then μ is defined as $\mu(A) := \mathbb{P}(\omega \in \Omega : \mathbf{v}_i(\omega) \in A)$ for every $A \in \mathbf{B}(\mathbb{R}^m)$ and is independent of i because the sequence of random variables $\{\mathbf{v}\}_{i=1}^{\infty}$ are i.i.d. We denote by $\mathcal{F}_i$ the collection of sets $\{\omega : (\mathbf{v}_1(\omega), ..., \mathbf{v}_i(\omega)) \in A\}$, $A \in \mathbf{B}((\mathbb{R}^m)^i)$ which are the sub-σ fields of $\mathcal{F}$ that form the natural filtration of $\mathbf{v} = \{\mathbf{v}_i\}_{i=1}^{\infty}$. The data of the system (4) will be represented as (C, F, D, G, μ) for simplicity. The definition of random solution to (4) requires concepts of measurability for set-valued mappings. For a measurable space (T, Γ), a mapping $M : T \rightrightarrows \mathbb{R}^n$ is *measurable* [20, Def. 14.1], if for each open set $\mathcal{O} \subset \mathbb{R}^n$, the set $M^{-1}(\mathcal{O}) := \{t \in T : M(t) \cap \mathcal{O} \neq \varnothing\} \in \Gamma$. When the values of M are closed, measurability is equivalent to $M^{-1}(\mathcal{C})$ being measurable for each closed set $C \subseteq \mathbb{R}^n$ [20, Thm. 14.3].

We now define the notion of random solution to the stochastic hybrid systems. A mapping $\mathbf{x}$ from Ω to the set of hybrid arcs is a *random solution* of (4) starting at x, denoted by $\mathbf{x} \in \mathcal{S}_{\mathbf{r}}(x)$, if it satisfies the following properties.

1. (Feasibility) For every $\omega \in \Omega$, the pair $(\mathbf{x}_\omega, \mathbf{u}_\omega)$ with $\mathbf{x}_\omega := \mathbf{x}(\omega)$ and $\mathbf{u}_\omega$ a hybrid arc with dom $\mathbf{u}_\omega =$ dom $\mathbf{x}_\omega$ and $\mathbf{u}_\omega(t,j) := \mathbf{v}_j(\omega)$ for all $(t,j) \in$ dom $\mathbf{x}(\omega) \cap (\mathbb{R} \times \mathbb{Z}_{\geq 1})$ is a standard solution staring at x.
2. (Causal measurability) For each $i \in \mathbb{Z}_{\geq 0}$, the mapping $\omega \mapsto \text{graph}_{\leq i}(\mathbf{x}(\omega)) := \text{graph}(\mathbf{x}(\omega)) \cap (\mathbb{R}_{\geq 0} \times \mathbb{Z}_{\leq i} \times \mathbb{R}^n)$ has closed values and is $\mathcal{F}_i$- measurable with $\mathcal{F}_0 = \{\varnothing, \Omega\}$, and $(\mathcal{F}_1, \mathcal{F}_2, ...)$ the natural filtration of $\mathbf{v}$.

To guarantee the existence of random solutions defined above, we impose the following regularity property on the data (C, F, D, G, μ) as in [28].

Assumption 2. *(Hybrid Basic conditions)*
1. The sets $C \subset \mathbb{R}^n$ and $D \subset \mathbb{R}^n$ are closed.
2. F is outer semicontinuous, locally bounded and for each $x \in C$, $F(x)$ is nonempty and convex.
3. G is locally bounded and for each $v \in \mathbb{R}^m$, the mapping $x \mapsto G(x,v)$ is outer semicontinuous.

Assumption 3. *(Stochastic Hybrid Basic condition) The set-valued mapping $v \mapsto$ graph$(G(\cdot, v)) := \{(x,y) \in \mathbb{R}^n \times \mathbb{R}^n : y \in G(x,v)\}$ is measurable.*

The existence of random solutions to the stochastic hybrid system (4) under Assumptions 2-3 is established in [28, Thm 3.1].

3.2 Stochastic Stability Notions

In this section we review some of the stochastic stability concepts explored in [28]. The definition of a random solution to (4) states that the graphs of the solutions are measurable and so we express the stability notions in terms of probabilities on the solution graphs. The dependence of random solutions on ω is suppressed and we write "$\mathbf{x}(t,j) \in S$ for $(t,j) \in \text{dom } \mathbf{x}$" in place of "$\mathbf{x}_\omega(t,j) \in S$ for $(t,j) \in \text{dom } \mathbf{x}_\omega$" where $\mathbf{x}_\omega := \mathbf{x}(\omega)$ to save on notation. The stability notions defined in this section are "strong" stability notions, meaning they hold for every random process $\mathbf{x}$ generated by the stochastic hybrid system (4) from a particular initial condition.

Referring to the statement

$$\mathbb{P}(\text{graph}_{\geq\tau}(\mathbf{x}) \subset (\mathbb{R}^2 \times (\mathcal{A} + \epsilon\mathbb{B}^o))) \geq 1 - \rho, \quad \forall \zeta \in \mathcal{A} + \delta\mathbb{B}^o, \quad \mathbf{x} \in \mathcal{S}_{\mathbf{r}}(\zeta), \tag{5}$$

a compact set $\mathcal{A} \subset \mathbb{R}^n$ for a stochastic hybrid system is

- *Uniformly Lyapunov stable in probability* if for $\tau = 0$ and each $\epsilon > 0$ and $\rho > 0$ there exists $\delta > 0$ such that (5) holds,
- *Uniformly Lagrange stable in probability* if for $\tau = 0$ and each $\delta > 0$ and $\rho > 0$ there exists $\epsilon > 0$ such that (5) holds,
- *Uniformly attractive in probability* if for each $\epsilon > 0$, $\delta > 0$, and $\rho > 0$, there exists $\tau > 0$ such that (5) holds,
- *Uniformly globally asymptotically stable (UGAS) in probability* if it is Uniformly Lyapunov stable in probability, Uniformly Lagrange stable in probability, and uniformly attractive in probability.

Next we present sufficient Lyapunov conditions as established in [28] to certify uniform global asymptotic stability in probability for stochastic hybrid systems.

Let $\mathcal{V} := \bigcup_{\omega \in \Omega, i \in \mathbb{Z}_{\geq 0}} \mathbf{v}_{i+1}(\omega)$. A function $V : \text{dom } V \to \mathbb{R}$ is a certification candidate for (C, D, G, μ) if

C1. $C \cup D \cup G(D \times \mathcal{V}) \subset \text{dom } V$,
C2. $0 \leq V(x)$ for all $x \in C \cup D \cup G(D \times \mathcal{V})$, and
C3. $\int_{\mathbb{R}^m} \sup_{g \in G(x,v)} V(g)\mu(dv)$ is well defined for each $x \in D$, with the convention that $\sup_{g \in G(x,v)} V(g) = 0$ when $G(x,v) = \varnothing$.

Under sufficient regularity assumptions (Assumptions 2-3) on the data of the stochastic hybrid systems it is established in [28, Lemma 4.1] if $V : \text{dom } V \to \mathbb{R}$ is upper semicontinuous[7] and satisfies conditions C1-C2 then it satisfies the condition C3.

Let $\mathcal{A} \subset \mathbb{R}^n$ be compact. A continuously differentiable certification candidate for (C, D, G, μ) is a *Lyapunov function* for $\mathcal{A}$ if there exists $\alpha_1, \alpha_2 \in \mathcal{K}_\infty$ and a continuous positive definite function $\rho : \mathbb{R}_{\geq 0} \to \mathbb{R}_{\geq 0}$ such that

$$\begin{aligned} &\alpha_1(|x|_{\mathcal{A}}) \leq V(x) \leq \alpha_2(|x|_{\mathcal{A}}) && \forall x \in C \cup D \cup G(D \times \mathcal{V}) \\ &\langle \nabla V(x), f(x) \rangle \leq -\rho(|x|_{\mathcal{A}}) && \forall x \in C, f \in F(x) \\ &\int_{\mathbb{R}^m} \sup_{g \in G(x,v)} V(g)\mu(dv) \leq V(x) - \rho(|x|_{\mathcal{A}}) && \forall x \in D. \end{aligned} \tag{6}$$

The next result then follows from [28, Thm 4.4].

---

[7] A function $\phi : \mathbb{R}^n \to \mathbb{R}$ is *upper semicontinuous* if, for each sequence x_i converging to x, $\limsup_{i \to \infty} \phi(x_i) \leq \phi(x)$.

Theorem 3. *Let $\mathcal{A} \subset \mathbb{R}^n$ be compact. If Assumptions 2-3 hold for the system (4), then the existence of a Lyapunov function for $\mathcal{A}$ implies uniform global asymptotic stability in probability of $\mathcal{A}$ for (4).*

The results developed in [28] also pertain to establishing certification candidates for Lyapunov stability and Lagrange stability in probability. Weakened sufficient conditions for uniform global asymptotic stability in probability and uniform global recurrence using nested Matrosov functions are also developed in [28].

4 The Robust Synchronization Problem on the Circle

4.1 Overview

We now illustrate the utility of a stochastic hybrid systems modeling framework. In particular, we consider the development of an algorithm for *robust, global, almost sure* synchronization of N agents evolving on the circle. The position of an agent on the circle might correspond to angular orientation of an agent in the plane or the relative phase of a clock. As we describe later, non-stochastic, non-hybrid algorithms have been proposed in the literature to achieve *almost global* synchronization on the circle. Such algorithms do not have the robustness properties that we seek. For example, we aim to design an algorithm that brings every agent to a very similar orientation, regardless of initial conditions, even in the presence of small perturbations that try to prevent synchronization. The algorithm that we consider has a stochastic mode and a non-stochastic mode; due to the former, we settle for *almost sure* robust global synchronization. An alternative "fully stochastic" algorithm for global almost sure synchronization has been proposed in [24] based on gossip algorithms that have appeared in the literature [2]. Since our algorithm uses randomness only when the average position has a very small norm, empirically it appears to converge faster than gossip-based algorithms.

The qualifier "robust" has been used in the context of synchronization on Euclidean spaces in [9]. There it means that the value to which agents synchronize should be independent of initial conditions and instead should depend on (constant) inputs. This notion is somewhat problematic for synchronization on the circle. The reason is that the projection to the circle of the average of these inputs might be very sensitive to small measurement errors. Our algorithm does not attempt to synchronize to a position that is independent of initial conditions. We use the qualifier "robust" to indicate that a small disturbances cannot prevent the agents from becoming very close to one another.

4.2 Basic Mathematical Modeling

Let $N \in \mathbb{Z}_{\geq 2}$. We consider the interaction of N cooperative agents $\xi_i \in \mathbb{S}^1 \subset \mathbb{R}^2$, where $\mathbb{S}^1$ denotes the unit circle. The ith agent has the dynamics

$$\dot{\xi}_i = u_i J \xi_i \qquad J = \begin{bmatrix} 0 & 1 \\ -1 & 0 \end{bmatrix} \tag{7}$$

where u_i is a control to be designed for synchronization. We assume all-to-all communication, that is, where every agent has access to the position of every other agent.

We also consider the presence of $M \in \mathbb{Z}_{\geq 0}$ non-cooperative agents (henceforth "rogue agents"), with $M \ll N$. The rogue agents have knowledge of the control laws u_i, are allowed to coordinate with each other and will attempt to prevent synchronization by moving to appropriate points on the circle.

4.3 An Algorithm for Local Synchronization

A rich overview of analysis results for synchronization algorithms on the circle appear in [25] and [23]. For example, the algorithm

$$u_i = \sum_{j=1}^{N} a_{ij}(t)\xi_j^T J \xi_i \tag{8}$$

where $a_{ij}(t) = 1$ if there is a connection between node i and node j at time t, and $a_{ij}(t) = 0$ otherwise, is considered extensively. It is established that for some communication graphs, including all-to-all communication, i.e., $a_{ij}(t) = 1$ for all t and all pairs $i \neq j$, this algorithm achieves almost global synchronization, i.e., synchronization from all initial conditions except for those of Lebesgue measure zero on $\mathbb{S}^1$.

4.4 Robustness Issues

Our subsequent stochastic, hybrid algorithm is motivated by the fact that small disturbances have the ability to prevent the algorithm (8) from reaching near synchronization. To illustrate this, we add rogue agents to the model by changing the summation from one to N in (8) to be summations from one to $N + M$ where the extra variables corresponding to the indices $j \in \{N+1, \ldots, N+M\}$ are exogenous inputs constrained to the unit circle or disk.

Rogue agents can cause problems by shifting the average of the agents' positions so that a particular agent chooses to go the "wrong way" around the circle. This phenomenon can be seen in Figure 1 where ten agents each have the control algorithm (8) with all-to-all communication, and are joined by a single rogue agent which moves according to

$$\xi_{rogue} = R(sat_{[-\pi/2,\pi/2]}(\kappa\xi_{lone}^{\top} J\xi_{avp}))\xi_{lone} \tag{9}$$

where $R(\cdot)$ is the rotation matrix, $sat_{[-\pi/2,\pi/2]}(\cdot)$ is the saturation function, $\kappa > 0$, ξ_{avp} is the average position of the non-rogue agents projected onto $\mathbb{S}^1$, and ξ_{lone} is the position of the agent that is initially closest to $-\xi_{avp}$. Starting from random initial conditions, we see in Figure 1 that the rogue agent, represented by a red dotted line, singles out an agent ξ_{lone} starting near $(-1, 0)$, and shifts the average of the agents' positions just enough to keep ξ_{lone} on the opposite side of the circle from the majority of the agents. This is the kind of behavior to which we want to be robust.

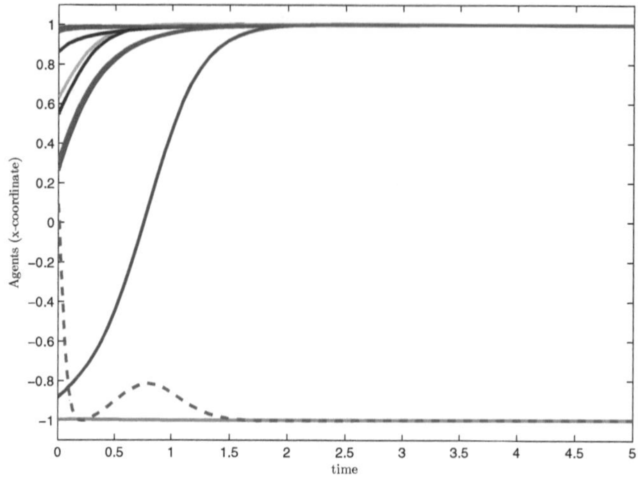

Fig. 1. Rogue agent acting on group of agents that use a standard almost global synchronization algorithm

5 Robust Global Almost Sure Synchronization on the Circle

In this section we present a hybrid algorithm that is robust to adversarial perturbations. It is composed of two parts: the first part is a stochastic algorithm that ensures that the average of all of the agents' positions, ξ_{avg}, reaches a certain distance from the origin. The second part of the algorithm is non-stochastic, and ensures robust synchronization of the agents from initial conditions where $|\xi_{avg}|$ greater than a given threshold[8].

5.1 Stochastic Hybrid Algorithm for Small Averages

In this part of algorithm (see Appendix A), each agent has a state $x_i := (\xi_i, w_i, \tau_i)$ where ξ_i indicates position on the circle, $w_i \in \{-1, 1\}$ indicates the angular velocity, and $\tau_i \in [0, T]$ is a timer state that keeps track of when the agent is allowed to jump. Overall, the algorithm functions as a random walk: each agent moves at a constant velocity either clockwise or counter clockwise, and after an interval of flowing, changes direction with probability 0.5, where the period of flow is randomly distributed in $[0, T]$, $T > 0$ (see Assumption 4 in the appendix). The algorithm runs as long as the mode label is "stochastic" and the state is in $\Theta(c) := \{x \in \mathbb{R}^{4N} : |\xi_{avg}| \leq c\}$ where $c > 0$.

[8] $|\cdot|$ refers to the Euclidean norm.

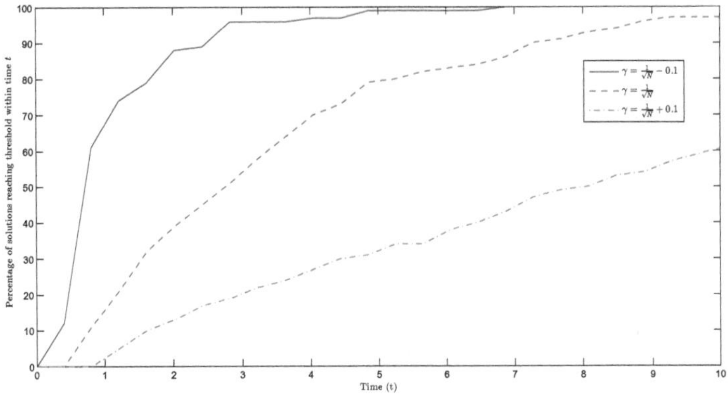

Fig. 2. Histogram : Stochastic Algorithm

It is possible to show that the set $\mathcal{A} := \{x, \in \mathbb{R}^{4N} : \xi_i \in \mathbb{S}^1, |\xi_{avg}| = 1, w_i \in \{-1, 1\}, \tau_i \in [0, T], \forall i \in \{1, \ldots, N\}\}$ is UGAS in probability. This result can be seen with the the Lyapunov function candidate $V(x)$ defined as:

$$\Psi(x) := \frac{1}{N} \sum_{i=1}^{N} \exp(J w_i \tau_i) \xi_i \tag{10}$$

$$V(x) := 1 - |\Psi(x)|^2, \tag{11}$$

It turns out that, for c sufficiently small, $V(x)$ is constant during flows, and decreasing on average during jumps when $x \in \Theta(c)$. It is easy to modify $V(x)$ to make it decrease both during flows and on average during jumps (see [26]). According to Theorem 3, this decrease implies that $\mathcal{A}$ is UGAS in probability.

The behavior of the algorithm is illustrated in Figure 2, where a histogram is generated from 100 simulations for the case of $N = 20$ agents, $T = 0.5$, and an initial condition such that $\xi_{avg} = 0$. Three different thresholds $\gamma = 1/\sqrt{N} - 0.1, 1/\sqrt{N}, 1/\sqrt{N} + 0.1$ are plotted, indicating the percentage of solutions for which the $|\xi_{avg}|$ reaches a threshold within a certain amount of time.

5.2 Non-stochastic Hybrid Algorithm for Non-small Averages

This part of the algorithm (see Appendix B) is responsible for robustly synchronizing the agents from initial conditions where $|\xi_{avg}|$ is above a certain threshold. The hybrid nature of the algorithm allow us to introduce hysteresis to the behavior of the agents, thereby avoiding the problems shown in Figure 1.

The way the algorithm works is that each agent has a state $x_i := (\xi_i, \beta_i)$ where ξ_i indicates the agent's position on the circle, and $\beta_i \in \{a_{cw}, a_{ccw}, b\} \subset \mathbb{R}$

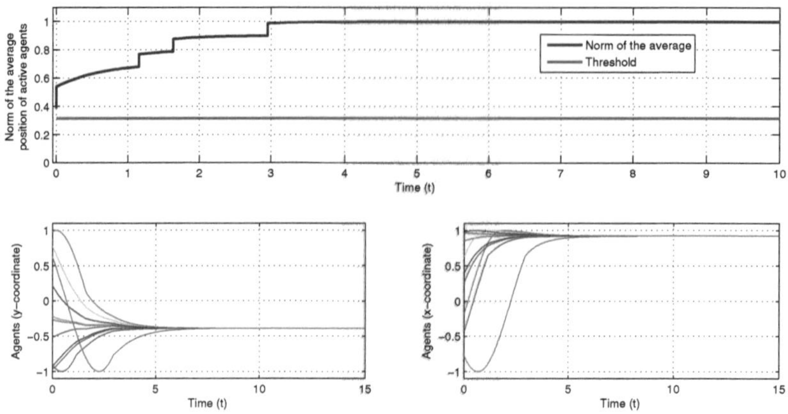

Fig. 3. Synchronization in the non-stochastic algorithm

indicates the mode of the agent. The variable $\tilde{w}(x)$ indicates a modified version of ξ_{avg} where the positions of the non-mode b agents are set to zero, and $w(x)$ indicates $\tilde{w}(x)$ projected onto the unit circle. At initialization, all agents that are in mode b and not in the semicircle centered on $w(x)$ change their mode to either a_{cw} or a_{ccw}. This process repeats itself for each new $w(x)$ until agents are no longer required to change their mode. At this point agents are allowed to flow continuously for some period of time: in mode a_{cw}, agents flow clockwise, in a_{ccw} agents flow counter-clockwise, and in mode b, agents flow according to a synchronization algorithm that looks at only other agents in mode b.

As agents in mode a_{cw} or a_{ccw} make their way around the circle into a cone centered at $w(x)$, they are allowed to change their mode back to b, and thus synchronize with other b mode agents, until all agents are once more in mode b.

This behavior is illustrated in Figure 3, which simulates the algorithm for $N = 10$ agents, starting from an initial condition such that $|\xi_{avg}| = 0.3850$. One can see in the top graph, that $|\tilde{w}(x)|$ is non-decreasing, with discontinuous increases happening when new agents join mode b. From the other graphs, it's clear that the agents achieve consensus.

In fact, by taking the Lyapunov function candidate $V(x) = 1 - |\tilde{w}(x)|_2^2$, it can be shown that $V(x)$ is non-increasing during both flows and jumps. By exploiting an invariance principle and a converse theorem such as Theorems 1 and 2, it is possible to get a Lyapunov function that decreases during both flows and jumps outside of the set $\mathcal{A}_1 := \{x \in \mathbb{R}^{3N} : \xi_i \in \mathbb{S}^1, |\tilde{w}(x)| = 1, \beta_i = b, |\hat{\xi}_i| = 0, \forall i \in \{1, \ldots, N\}\}$, where $\hat{\xi}_i$ denotes the velocity of the i'th agent in mode b, implying that the algorithm leads to synchronization of the agents.

An important point here is that hysteresis is used for transitions between sub-modes, as illustrated by Figure 4. The hysteresis ensures that flows eventually follow jumps, which guarantees that the invariance principle gives the correct convergence result.

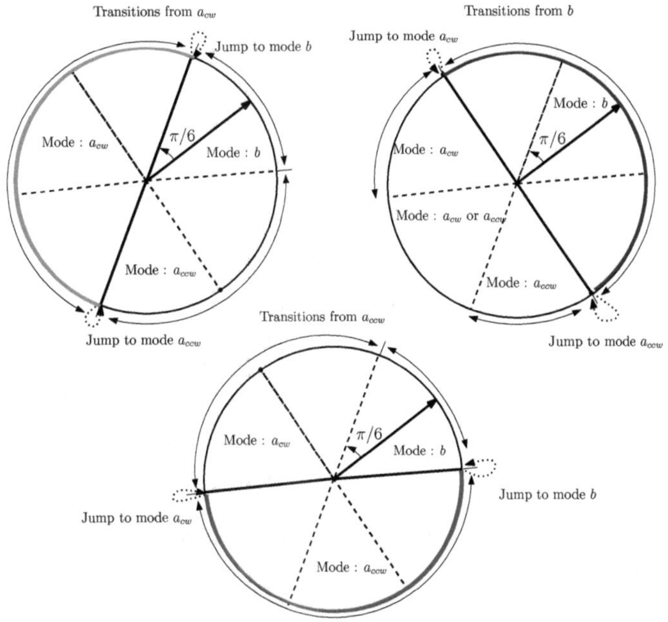

Fig. 4. Transitions between modes

5.3 Combination with Hysteresis

When the two modes are combined, each agent has a state $x_i := (\xi_i, \beta_i, w_i, \tau_i)$. We define $\mathcal{Q} := \{r, s\}$, where r and s indicate the non-stochastic mode and stochastic mode respectively, and note that due to all-to-all communication, it will be the case that all agents are in either stochastic mode or non-stochastic mode.

In switching between the two modes, we again use hysteresis. Using $|\xi_{avg}|$ and $|\tilde{w}(x)|$ as indicators, jumps from stochastic mode to non-stochastic mode occur when $|\xi_{avg}|$ becomes greater than the upper threshold θ_u. But the algorithm will switch from non-stochastic to stochastic mode only when $|\tilde{w}(x)|$ becomes lower than θ_l where $\theta_u > \theta_l > 0$.

It turns out (see [26]) that a Lyapunov function can be constructed from the Lyapunov functions of the previous two subsections to show that the set $\mathcal{A}_2 := \{(x,q) \in \mathbb{R}^{5N} \times \mathcal{Q} : \xi_i \in \mathbb{S}^1, |\xi_{avg}| = 1, q = r, \beta_i = b, |\hat{\xi}_i| = 0, w_i \in \{-1,1\}, \tau_i \in [0,T], \forall i \in \{1,\ldots,N\}\}$ is UGAS in probability. The existence of this Lyapunov function implies robust synchronization for the overall algorithm.

In choosing θ_u and θ_l, one encounters a tradeoff between robust stability and nominal performance. In particular, choosing the hysteresis widths θ_u and $\theta_u - \theta_l$ to be large makes the algorithm more robust to perturbations, but in the nominal case, requires the algorithm to spend more time in stochastic mode before

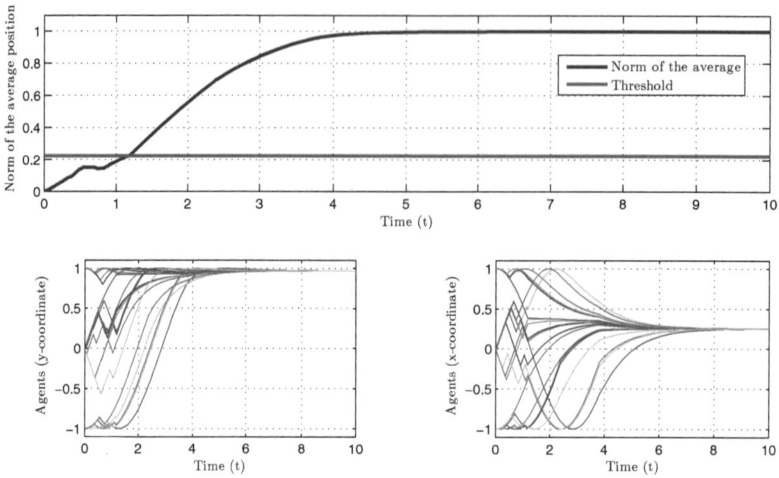

Fig. 5. Synchronization of Agents

synchronizing. This may harm nominal performance because the convergence rate is slower in stochastic mode than in non-stochastic mode.

The simulation results for the combination of the stochastic and non-stochastic algorithms are presented in Figure 5 for $N = 20$ agents, $T = 0.5$, initial conditions such that $|\xi_{avg}| = 0$, and $\theta_u = 1/\sqrt{N} = 0.2236$. The synchronization behavior is captured by the plot in which $|\xi_{avg}|$ converges to one.

6 Conclusion

We have presented a stochastic hybrid systems framework for modeling cyber-physical systems. Stochastic stability concepts and tools to certify stability properties are presented for stochastic hybrid systems with non-unique solutions. The advantages of this framework are illustrated using a synchronization problem. We have shown that an existing non-stochastic algorithm exhibits non-robust behavior in the presence of rogue agents. On the other hand stochastic algorithms like the "gossip algorithm" tend to have slow convergence rates. We have attempted to overcome both these issues using the stochastic hybrid systems framework. An algorithm for robust almost sure synchronization is presented which exploits the tools developed recently in [28]. The non-stochastic part of the algorithm is robust to adversarial perturbations, and the stochastic part of the algorithm is used only when the norm of the average position of the agents is small, leading to better convergence rates. Simulation results are used to illustrate the efficiency of this approach.

A Stochastic Part of the Algorithm

From the starting point of (7), we extend the state of the ith agent with dynamical states (w_i, τ_i) taking values in $\{-1,1\} \times [0,T] \subset \mathbb{R}^2$ where $T > 0$.

We define

$$x_i := (\xi_i, w_i, \tau_i) \tag{12a}$$

$$\widehat{C} := \mathbb{S}^1 \times \{-1,1\} \times [0,T] \tag{12b}$$

$$\widehat{f}(x_i) := \begin{bmatrix} w_i J \xi_i \\ 0 \\ -1 \end{bmatrix} \tag{12c}$$

$$\widehat{D} := \mathbb{S}^1 \times \{-1,1\} \times \{0\} \tag{13a}$$

$$\widehat{g}(x_i, v_i) := \begin{bmatrix} \xi_i \\ v_{1,i} \\ v_{2,i} \end{bmatrix}. \tag{13b}$$

We make the following assumption:

Assumption 4. *The inputs $v_{1,i}$ and $v_{2,i}$ are generated from i.i.d. random variables, the values of the random variable generating $v_{1,i}$ being uniformly distributed on $\{-1,1\}$ and the values of the random variable generating $v_{2,i}$ has a distribution with support on $[0,T]$ not all concentrated at the origin.* □

This leads to a formulation of the N-agent system. Defining

$$x := (x_1, \ldots, x_N) \in \mathbb{R}^{4N} \tag{14a}$$

$$C := \widehat{C} \times \cdots \times \widehat{C} \tag{14b}$$

$$f(x) := \begin{bmatrix} \widehat{f}(x_1) \\ \vdots \\ \widehat{f}(x_N) \end{bmatrix} \tag{14c}$$

$$D := \left\{ x \in C : x_i \in \widehat{D} \text{ for some } i \in \{1, \ldots, N\} \right\} \tag{14d}$$

$$g_i(x, v) := \begin{bmatrix} x_1 \\ \vdots \\ x_{i-1} \\ \widehat{g}(x_i, v_i) \\ x_{i+1} \\ \vdots \\ x_N \end{bmatrix} \qquad \forall x \in \left\{ x \in C : x_i \in \widehat{D} \right\} \tag{14e}$$

$$G(x, v) := \bigcup_{i \in \{1, \cdots, N\} : x_i \in \widehat{D}} g_i(x, v) \qquad \forall x \in D. \tag{14f}$$

We let $c \in (0,1)$ and we define

$$\Theta(c) := \left\{ x \in \mathbb{R}^{4N} : \left| \frac{1}{N} \sum_{i=1}^{N} \xi_i \right| \le c \right\} . \tag{15}$$

B Non-stochastic Part of Algorithm

Here we extend the state of the ith agent with dynamical state $\beta_i \in \{a_{cw}, a_{ccw}, b\} \subset \mathbb{R}$ where a_{cw}, a_{ccw}, and b are distinct integers and we provide it with an exogenous input w that will be common to all agents. We specify agent dynamics

$$(w, x_i) \in \widehat{C} \qquad \dot{x}_i = f(w, x_i) \tag{16a}$$

$$(w, x_i) \in \widehat{D} \qquad x_i^+ \in G(w, x_i) \tag{16b}$$

via the definitions

$$x_i := (\xi_i, \beta_i) \tag{17a}$$

$$\widehat{C}_{a_{cw}} := \left\{ (w, \xi) \in \mathbb{S}^1 \times \mathbb{S}^1 : \xi^T R\left(-\frac{\pi}{3}\right) w \le 0 \right\} \tag{17b}$$

$$\widehat{C}_{a_{ccw}} := \left\{ (w, \xi) \in \mathbb{S}^1 \times \mathbb{S}^1 : \xi^T R\left(\frac{\pi}{3}\right) w \le 0 \right\} \tag{17c}$$

$$\widehat{C}_b := \{ (w, \xi) \in \mathbb{S}^1 \times \mathbb{S}^1 : \xi^T w \ge 0 \} \tag{17d}$$

$$\widehat{C} := \bigcup_{\beta_i \in \{a_{cw}, a_{ccw}, b\}} \{\beta_i\} \times \widehat{C}_{\beta_i} \tag{17e}$$

$$\widehat{f}(w, x_i) := \begin{bmatrix} \left(\mathbb{I}_{a_{cw}}(\beta_i) - \mathbb{I}_{a_{ccw}}(\beta_i) + \mathbb{I}_b(\beta_i) w^T J \xi_i\right) J \xi_i \\ 0 \end{bmatrix} \tag{17f}$$

$$\widehat{D}_{a_{cw}} := \left\{ (w, \xi) \in \mathbb{S}^1 \times \mathbb{S}^1 : \xi^T R\left(-\frac{\pi}{3}\right) w \ge 0 \right\} \tag{18a}$$

$$\widehat{D}_{a_{ccw}} := \left\{ (w, \xi) \in \mathbb{S}^1 \times \mathbb{S}^1 : \xi^T R\left(\frac{\pi}{3}\right) w \ge 0 \right\} \tag{18b}$$

$$\widehat{D}_b := \{ (w, \xi) \in \mathbb{S}^1 \times \mathbb{S}^1 : \xi^T w \le 0 \} \tag{18c}$$

$$\widehat{D} := \bigcup_{\beta_i \in \{a_{cw}, a_{ccw}, b\}} \{\beta_i\} \times \widehat{D}_{\beta_i} \tag{18d}$$

$$\widehat{G}(w, x_i) := \begin{bmatrix} \xi_i \\ \alpha \in \{a_{cw}, a_{ccw}, b\} \setminus \{\beta_i\} : (w, \xi_i) \in \widehat{C}_\alpha \end{bmatrix} \tag{18e}$$

$$\mathbb{I}_b(\beta) := \begin{cases} 1, & \beta = b \\ 0, & \text{otherwise.} \end{cases} \tag{19}$$

$$\widetilde{w}(x) := \frac{1}{N} \sum_{i=1}^{N} \mathbb{I}_b(\beta_i) \xi_i , \tag{20}$$

$$\Pi(\theta_r) := \left\{x \in \mathbb{R}^{3N} : |\widetilde{w}(x)|_2 \geq \theta_r\right\}, \tag{21}$$

$$w(x) := \frac{\widetilde{w}(x)}{|\widetilde{w}(x)|} \in \mathbb{S}^1, \tag{22}$$

where $\theta_r > 0$ and $w(x)$ is defined on the set $\Pi(\theta_r)$. Finally, we define

$$x := (x_1, \ldots, x_N) \in \mathbb{R}^{3N} \tag{23a}$$

$$C := \left\{x \in \Pi(\theta_r) : (w(x), x_1), \ldots, (w(x), x_N) \in \widehat{C} \times \cdots \times \widehat{C}\right\} \tag{23b}$$

$$f(x) := \begin{bmatrix} f(w(x), x_1) \\ \vdots \\ f(w(x), x_N) \end{bmatrix} \tag{23c}$$

$$D := \left\{x \in C : (w(x), x_i) \in \widehat{D} \text{ for some } i \in \{1, \ldots, N\}\right\} \tag{24a}$$

$$G_i(x, v) := \begin{bmatrix} x_1 \\ \vdots \\ x_{i-1} \\ \widehat{G}(w(x), x_i) \\ x_{i+1} \\ \vdots \\ x_N \end{bmatrix} \quad \forall x \in \left\{x \in C : (w(x), x_i) \in \widehat{D}\right\} \tag{24b}$$

$$G(x, v) := \bigcup_{i \in \{1, \cdots, N\} : x_i \in \widehat{D}} G_i(x, v) \qquad \forall x \in D. \tag{24c}$$

References

1. Baheti, R., Gill, H.: Cyber-physical systems. The Impact of Control Technology, pp. 161–166 (2011)
2. Boyd, S., Ghosh, A., Prabhakar, B., Shah, D.: Randomized gossip algorithms. IEEE Transactions on Information Theory 52(6), 2508–2530 (2006)
3. Bujorianu, M., Lygeros, J.: Toward a general theory of stochastic hybrid systems. In: Stochastic Hybrid Systems, pp. 3–30. Springer, Berlin (2006)
4. Cai, C., Goebel, R., Teel, A.R.: Smooth Lyapunov functions for hybrid systems part ii:(pre) asymptotically stable compact sets. IEEE Transactions on Automatic Control 53(3), 734–748 (2008)
5. Cardenas, A.A., Amin, S., Sastry, S.: Secure control: Towards survivable cyber-physical systems. In: 28th International Conference on Distributed Computing Systems Workshops, pp. 495–500. IEEE (2008)
6. Cassandras, C.G., Lygeros, J.: Stochastic hybrid systems, vol. 24. CRC (2006)
7. Davis, M.H.A.: Markov Models & Optimization, vol. 49. Chapman & Hall/CRC (1993)

8. Davis, M.H.A.: Piecewise-deterministic markov processes: A general class of non-diffusion stochastic models. Journal of the Royal Statistical Society. Series B (Methodological), 353–388 (1984)
9. Freeman, R.A., Nelson, T.R., Lynch, K.M.: A complete characterization of a class of robust linear average consensus protocols. In: Proc. American Control Conference, pp. 3198–3203 (2010)
10. Goebel, R., Sanfelice, R.G., Teel, A.R.: Hybrid dynamical systems. IEEE Control Systems 29(2), 28–93 (2009)
11. Goebel, R., Teel, A.R.: Solutions to hybrid inclusions via set and graphical convergence with stability theory applications. Automatica 42(4), 573–587 (2006)
12. Grammatico, S., Subbaraman, A., Teel, A.R.: Discrete-time stochastic control systems: a continuous Lyapunov function implies robustness to strictly causal perturbations. Automatica (submitted, 2013)
13. Hanson, F.B.: Applied stochastic processes and control for jump diffusions: modeling, analysis, and computation. Society for Industrial and Applied Mathematics (2007)
14. Henzinger, T.A.: The theory of hybrid automata. Springer (2000)
15. Hu, J., Lygeros, J., Sastry, S.: Towards a theory of stochastic hybrid systems. In: Hybrid Systems: Computation and Control, pp. 160–173 (2000)
16. Lakshmikantham, V., Bainov, D., Simeonov, P.S.: Theory of impulsive differential equations, vol. 6. World Scientific Publishing Company Incorporated (1989)
17. Pasqualetti, F., Dorfler, F., Bullo, F.: Cyber-physical attacks in power networks: Models, fundamental limitations and monitor design. In: Proc. 50th IEEE Conference on Decision and Control and European Control Conference, pp. 2195–2201 (2011)
18. Pasqualetti, F., Dorfler, F., Bullo, F.: Cyber-physical security via geometric control: Distributed monitoring and malicious attacks. In: Proc. 51st IEEE Conference on Decision and Control, pp. 3418–3425 (2012)
19. Poovendran, R.: Cyber–physical systems: Close encounters between two parallel worlds [point of view]. Proceedings of the IEEE 98(8), 1363–1366 (2010)
20. Rockafellar, R.T., Wets, R.J.B.: Variational Analysis. Springer, Berlin (1998)
21. Sanfelice, R.G., Goebel, R., Teel, A.R.: Invariance principles for hybrid systems with connections to detectability and asymptotic stability. IEEE Transactions on Automatic Control 52(12), 2282–2297 (2007)
22. Sanfelice, R.G., Teel, A.R.: On singular perturbations due to fast actuators in hybrid control systems. Automatica 47(4), 692–701 (2011)
23. Sarlette, A., Sepulchre, R.: Synchronization on the circle. arXiv preprint arXiv:0901.2408 (2009)
24. Sarlette, A., Tuna, S.E., Blondel, V., Sepulchre, R.: Global synchronization on the circle. In: Proc. 17th IFAC World Congress (2008)
25. Sepulchre, R.: Consensus on nonlinear spaces. Annual Reviews in Control 35(1), 56–64 (2011)
26. Subbaraman, A., Hartman, M., Teel, A.R.: A stochastic hybrid algorithm for robust global almost sure synchronization on the circle: all-to-all communication. In: Proc. IEEE Conference on Decision and Control (submitted, 2013)
27. Subbaraman, A., Teel, A.R.: A converse Lyapunov theorem for strong global recurrence. Automatica (2012) (submitted) username: guest, password: ABCD1234, `http://www.ece.ucsb.edu/%7Eteel/submitted/converse2.pdf`
28. Teel, A.R.: Lyapunov conditions certifying stability and recurrence for a class of stochastic hybrid systems. Annual Reviews in Control (2013), `http://dx.doi.org/10.1016/j.arcontrol.2013.02.001`

29. Teel, A.R., Hespanha, J.P., Subbaraman, A.: A converse theorem for global asymptotic stability in probability. IEEE Transactions on Automatic Control (2012) (submitted) username: guest, password: ABCD1234, http://www.ece.ucsb.edu/%7Eteel/submitted/converse1.pdf
30. Wang, W., Teel, A.R., Nešić, D.: Analysis for a class of singularly perturbed hybrid systems via averaging. Automatica 48(6), 1057–1068 (2012)
31. Yin, G., Zhu, C.: Hybrid switching diffusions: properties and applications, vol. 63. Springer (2009)

CPS Modeling Integration Hub and Design Space Exploration with Application to Microrobotics

Yuchen Zhou and John S. Baras

The Institute for Systems Research
and Electrical and Computer Engineering Department,
University of Maryland, College Park, Maryland, USA
{yzh89,baras}@umd.edu

Abstract. We describe a new methodology and environment for Cyber Physical Systems (CPS) synthesis and demonstrate it in the design of microrobots viewed as CPS. Various types of microrobots have been developed in recent years for applications related to collaborative motion such as, sensor networks, exploration and search-rescue in hazardous environments and medical drug delivery. However, control algorithms for these prototypes are very limited. Our new approach for modeling and simulation of the complete microrobotics system allows the robots to complete more complex tasks as per specifications. Since the microrobots tend to have small features, complex micro-structures and hierarchy, the control laws cannot be designed separately from the physical layer of the robots. Such a type of microrobot is indeed a CPS, as control in the cyber side, and the material properties and geometric structure in the physical side, are tightly interrelated. This design approach is important for microrobots, capable of collaborating and completing complex tasks.

Keywords: Modelica, Microrobot, CPS, System model.

1 Introduction: Synthesis Environment for CPS

The rapid development of information technology (in terms of processing power, embedded hardware and software systems, comprehensive IT management systems, networking and Internet growth, system design and integration environments) is producing an increasing number of applications and opening new doors. In addition, over the last decade, we entered a new era where systems complexity has increased dramatically. Complexity is increased both by the number of components that are included in each system as well as by the dependencies between those components. Cyber-Physical Systems (CPS) are *engineered systems* constructed as networked interactions of physical and computational (cyber) components. In CPS, computations and communications are deeply embedded in and interacting with physical processes, and add new capabilities to physical systems. The challenge in CPS is to incorporate the inputs (and their characteristics and constraints) from the physical components in the logic of the

D.C. Tarraf (ed.), *Control of Cyber-Physical Systems*,
Lecture Notes in Control and Information Sciences 449,
DOI: 10.1007/978-3-319-01159-2_2,

cyber components (hardware and software). Whole industrial sectors are transformed by new product lines that are CPS-based. Modern CPS are not simply the connection of two different kinds of components engineered by means of distinct design technology, but rather, a new system category that is both physical and computational. Current industrial experience tells us that, in fact, we have reached the limits of our knowledge of how to combine computers and physical systems. The shortcomings range from technical limitations in the foundations of cyber-physical systems to the way we organize our industries and educate engineers and scientists that support cyber-physical system design. If we continue to build systems using our very limited methods and tools but lack the science and technology foundations, we will create significant risks, produce failures and lead to loss of market.

If a successful contribution is to be made in shaping this change, the revolutionary potential of CPS must be recognized and incorporated into internal development processes at an early stage. For that *Interoperability and Integratability* of CPS is critical. In our recent research [1], [2], [3], [4], [5], [6] we have initiated the development of a framework to facilitate interoperability and integratability of CPS. Currently there is a lack of well-defined tools and synthesis environments for CPS. CPS synthesis requires cross-domain concepts for architecture, communication and compatibility at all levels. The effects of these factors on existing or yet undeveloped systems and architectures represent a major challenge. The aim of our recent research is precisely to clarify these objectives and systematically develop detailed recommendations and synthesis environments for CPS. We have focused our efforts in two essential problems:
(i) A framework for developing cross-domain *integrated modeling hubs* for CPS.
(ii) The creation and demonstration of an initial framework for linking the integrated CPS modeling hub of (i) with powerful and diverse *tradeoff analysis methods* and tools for *design space exploration* for CPS.

1.1 Model-Based Systems Engineering

MBSE [7] has emerged as a promising methodology for the systematic design, performance evaluation and validation of complex engineering systems. "Model-Based Systems Engineering (MBSE) is the formalized application of modeling to support system requirements, design, analysis, verification and validation activities beginning in the conceptual design phase and continuing throughout development and later life cycle phases" [7]. MBSE facilitates the flow of requirements through models, a methodology that is at the same time compact and enforces consistency between data and requirements (through the models). Figure 1 describes the basic steps of the MBSE process that we have developed, and have been teaching at the University of Maryland (UMD) for several years. A most recent development of particular importance is the development and teaching of a new hands-on undergraduate course at UMD, *ENES489P "Hands-on Systems Engineering Projects"*. This MBSE process has the following steps (phases): Requirements Collection, Construction of System Structure Model (what the system consists of), Construction of System Behavior Model (what the system does), Mapping of Behavior onto Structure (what structure components will perform parts of behavior), Allocation of Requirements to Structure and Behavior Components, Trade-Off Analysis, Validation and Verification. As illustrated in Figure 1, the process moves between these steps in an iterative manner, until satisfactory alternative system designs are developed. The process is executed at different levels of granularity (detail/aggregation). As the MBSE process executes a *system architecture* is developed through the creation of behavior and structure components, their interrelationships and the allocation of behavior components to structure components.

1.2 Systems Modeling Language (SysML)

SysML [8] is a general purpose graphical modeling language that was developed based on UML and is a key enabler for the MBSE process by providing ways for the representation and analysis of complex engineering systems. SysML supports the specification, analysis, design, verification, and validation of systems that include hardware, software, data, personnel, procedures, and facilities. SysML supports model and data interchange via XML Metadata Interchange (XMI) and the AP233 standard. Recent research has demonstrated the use of SysML [8] as a centerpiece abstraction for team-based system development, with a variety of interfaces and relationship types (e.g., parametric, logical and dependency) providing linkages to detailed discipline-specific analyses and orchestration of system engineering activities. The four fundamental pillars of SysML are the support of models for the *structure* of the system, models of the *behavior* of the system, models for capturing the requirements for the system via the new *requirements diagram* of the system, and the new and innovative *parametric diagram* of the system, which ties design variables and metric parametric representations to the structure and behavior models (a kind of annotation of these models). Parametric diagrams are the key to linking SysML-based system models to analysis models, including trade-off analysis models such as multi-metric optimization (e.g. IBM-ILOG CPLEX) and constraint based reasoning tools (e.g. IBM-ILOG Solver). SysML, as a language for describing the system architecture, is a catalyst for the integration of various modeling environments, as well as analysis/design environments, for complex systems, while allowing multiple disciplinary views of the system and its components, as illustrated in Figure 2, where the System Architecture Model

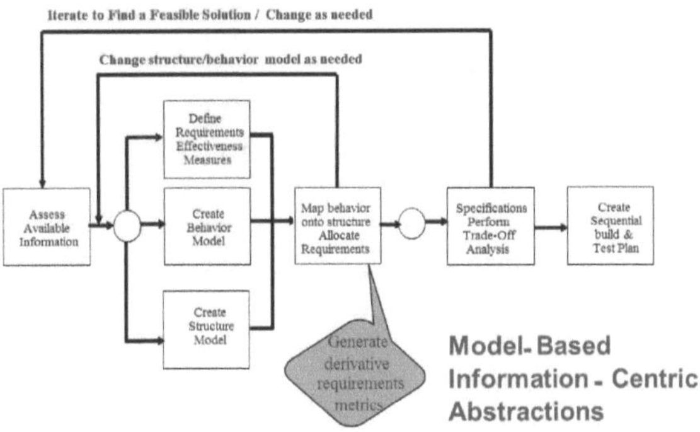

Fig. 1. Model-Based Systems Engineering Process [9]

is described via SysML. Our research has taken several key steps towards the development of new foundations for this model integration framework we call **CPS modeling integration hubs.** We have recently developed [3], [6] such hubs for power grids, microrobotics, energy efficient buildings, vehicle management systems for next generation all-electric aircraft, sensor networks, robotics and collaborative swarms.

1.3 CPS Modeling Integration Hub Architecture

A major challenge in MBSE for CPS is to have models that are consistent with each other. However, besides having consistent data there is a need for the models to work together in order to offer a holistic Systems Engineering approach to the designer of CPS. SysML is used in the core of our modeling integration hub (Fig. 2 and Fig. 3). The main aim is to integrate this core module with external tools, each one used in a different phase of the Systems Engineering process [10]. The resulting MBSE environment can be thought of as a "virtual" product line management (PLM) environment for CPS, across discipline tools. To achieve this integration a three-layer approach needs to be followed. Initially, for the tool we need to integrate, a domain specific profile is created in SysML. Then a model transformation is defined, followed by the implementation of tool adapters that are used as a middleware for exchanging information between the model transformation layer and the other components of the hub. Fig. 3 presents these layers as well as the areas for which we need to integrate tools with the core module to realize the MBSE vision of a system design experience for CPS.

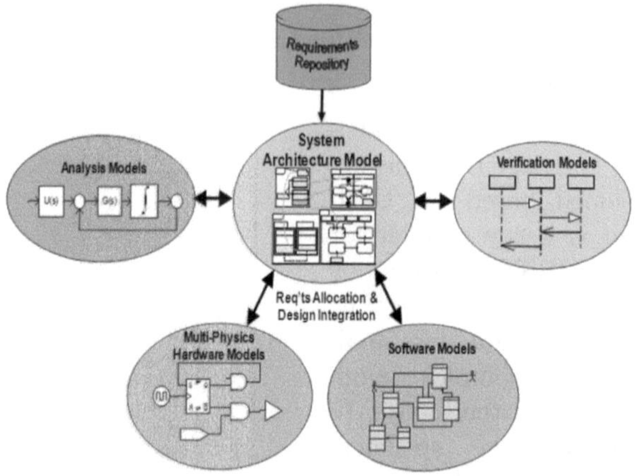

Fig. 2. Multi-domain model integration via system architecture model (SysML)

A key component of the emerging framework is a metamodeling environment with its associated languages and its semantics based on sophisticated versions of annotated block diagrams and bond graphs [6]. A metamodeling layer stands one abstraction layer above the actual design implementation in a modeling language. A metamodel consists of the constructs of a modeling language together with the rules that specify the allowable relationships between these constructs. It can be considered as the grammar of that modeling language. At the metamodeling layer model transformations take place. There are many alternatives in terms of model transformation tools, like ATL, GME, eMoflon, QVT. In our research the eMoflon model transformation tool was used [11], [12]. Finally, tool adapters work as the "glue" between the different pieces of software. Their role is to access/change information inside a model and call the appropriate Java functions generated by the eMoflon tool to perform model transformations [6], [13].

1.4 Tradeoff Analysis and Design Space Exploration

Although progress to date in MBSE facilitates the integration of system component models from different domains, we still need an integrated environment to optimize system architecture, manage the analysis and optimization of diverse measures of effectiveness (MoE), manage the various acceptable designs and most than anything else perform tradeoff analysis. Tradeoff is an essential part of system design, as it implements design space exploration. SysML does not provide a way for engineers to formally evaluate and rank design criteria, conduct sensitivity analysis, search design spaces for better design solutions, and conduct trade studies. To address this challenge we have introduced [6] the concept that SysML needs to be integrated with industrial-strength multi-objective algorithms, constraint-based reasoning algorithms, with appropriate linkages to modeling/simulation environments. An integration of SysML with a tradeoff tool will allow the designer to make decisions faster and with more confidence.

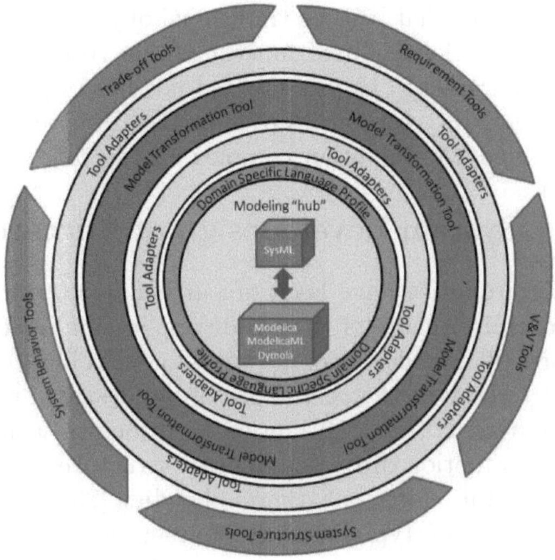

Fig. 3. The Modeling Integration Hub for CPS

We have recently developed and demonstrated [6] the first ever integration of a powerful tradeoff analysis tool (and methodology), Consol-Optcad, which is a sophisticated multi-criteria optimization tool developed at the University of Maryland, with our SysML-based modeling integration hubs for CPS. Consol-Optcad is a multi-objective optimization tool that allows interaction between the model and the user. It can handle non-linear objective functions and constraints with continuous values. Another version of Consol-Optcad has been developed to handle also logical variables, via integer and constraint programming [14]. In systems development and after the system structure is defined there is a need to calculate the design parameters that best meet the objectives and constraints. Usually when we deal with complex systems and optimization is under consideration, this is not a trivial task. The support of an interactive tool, like Consol-Optcad, to help the designer resolve the emerging trade-offs is necessary.

A major advantage of Consol-Optcad is that it allows the user to interact with the tool, while the optimization is under way. The designer might not know or might not be in a position at the beginning to specify what preferred design means. Therefore such interaction with the tool could be of great benefit [15], [16]. Another key feature of Consol-Optcad is the use of the Feasible Sequential Quadratic Programming (FSQP) algorithm for the solver. FSQP's advantage is that as soon as we get an iteration solution that is inside the feasible region, feasibility is guaranteed for the following iterations as well. Moreover, very interesting is the fact that besides traditional objectives and constraints Consol-Optcad allows the definition of functional constraints and objectives that depend on a free parameter. Consol-Optcad has been applied to the design of flight control systems [17], rotorcraft systems [18], integrated product process design (IPPD) systems [14] and other complex engineering systems.

2 System Level Design of Microrobots as CPS

Microrobotics have been of particular interest to researchers in Robotics and Control, because of their wide application in collaborative control, medical sensors [19], mobile sensor networks for surveillance [20] and microrobot self-assembly [21]. Many of the recent prototype designs are based on Microelectromechanical systems (MEMS) fabrication processes using specific mechanisms to create planar motion through miniature structures. These include using force from electrical static force [22], thermo bending [23] and chemical reactions [24]. However, current prototype-based design methodology for microrobots is not systematic. In the design process, control policies are normally designed completely separately from the structure after the prototype is manufactured [21], [23]. Such a process either makes it impossible for the robot to accomplish complicated tasks autonomously, or require external force to control the motion of the robot [19]. Modeling the robot requires a very precise description to the physical layers in the process. Material constraints and material properties are critical for the microrobotic design [25]. Complex control laws, which are the cyber side, will not perform well if the physical robot is not well modeled. Moreover, the simulation and design process of the cyber part become increasingly dependent on the physical model and will be directly influenced by any changes made in the physical design. On the other hand, the cyber part affects the stability and controllability of the physical model as well. This makes the microrobot a complicated CPS. Therefore, a model-based systems engineering approach including simulation and validation is needed for this process.

In this paper we follow the methodology described in Sections 1.1-1.4, for the system level modeling and design of microrobots viewed (properly) as CPS. To model the cyber and physical layers, we chose the Modelica language, due to its well-known capability for modeling complex physical systems problems [26].

In this paper we focus on a type of walking robot that uses six legs to alternatively support the structure and moving forward similar to that of an insect. Instead of designing a whole new robot, we demonstrate the possible design exploration enabled by our methodology on a particular prototype of a walking microrobot described in [27]. The subject robot [27] utilizes flexible joints to damp the impact with the ground so as to stabilize the walking motion. Although the overall structure is not complex as shown in Fig. 4a, the model can be easily made unstable due to the ground collisions even with the flexible joints. In order to create a more stable model and further explore the design space, a system level model for this particular robot is created to investigate its stability, structure alternatives and efficiency.

The rest of the paper is organized as follows. Section 3 gives the analytical approach for the physical model of the subject walking robot [27]. Section 4 includes Modelica simulation and results related to the stability and planar motion of the robot. Section 5 presents possible material choices and design exploration suggested by the simulation and also its effects on the control laws. Section 6 gives suggestions on tradeoff analysis and validation before prototype fabrication using a model-based system engineering approach.

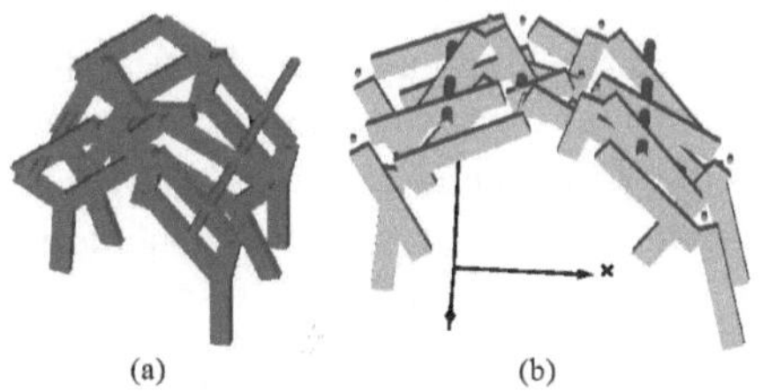

Fig. 4. The original millirobot is on the left and the modified millirobot is on the right

3 The Physical Model

The particular microrobots we are interested in are small robots with micro features, more specifically with flexible joints which make them more stable. In the first part of this section, the mechanism of flexible joints will be discussed and approximated by a torsion spring according to beam theory. In the second part, the rigid part of the microrobot will be analyzed using multibody dynamics and kinematics derived from a Lagragian formulation. The ground interaction is discussed at the end of this section. The physical model creates constraints for the controller, since the flexible part will break if large force or torque is applied, and instability can easily arise from the poorly designed structure.

3.1 Flexible Joint Model

Assuming small bending of the joint, all flexible joints of the robot are modeled as torsion springs derived from beam theory. According to beam theory, the local curvature ρ of the bending beam is

$$\frac{1}{\rho} = \frac{M}{EI_z}. \tag{1}$$

where M is the applied moment, E is the Young Modulus of the beam and I_z is the inertia about the rotational axis. In the case of a rectangular beam, $I_z = 1/12bh^3$, where b is the beam height and h is the beam width.

For small values of the angle, we obtain a linear relation between the angle and the torque. The spring constant is therefore,

$$k = \frac{EI_z}{l},$$

where l is the beam length.

3.2 Kinematics Model

Consider the kinematics model described in Fig. 5. Let $\mathcal{R}_0$, $\mathcal{R}_1$, $\mathcal{R}_2$ be the coordinate frames attached to the joints as shown. Denote $(V_{i/j}^P)_{\mathcal{R}_k}$ as the velocity of point P attached to body i (B_i) relative to body j (B_j) expressed in the coordinate frame $\mathcal{R}_k$. The twist, including rotational and translational velocities, of P can be described with respect to $\mathcal{R}_0$ using $\hat{V} = [w\ v]^T$ [28],

$$\left(\hat{V}_{3/0}^P\right)_{\mathcal{R}_3} = \left(\hat{V}_{3/2}^P\right)_{\mathcal{R}_3} + \left(\hat{V}_{2/1}^P\right)_{\mathcal{R}_3} + \left(\hat{V}_{1/0}^P\right)_{\mathcal{R}_3}.$$

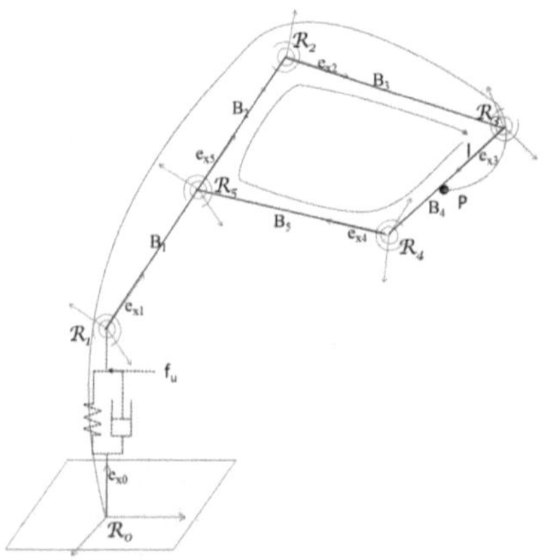

Fig. 5. Mechanical model of one single leg

Let l_1 be the length of B_1 and B_2 as shown in Fig. 5 (B_1 and B_2 constitute one rigid segment), l_2 be the length of B_3 and l be the distance between the origin of $\mathcal{R}_3$ and the point of interest P. Then the relative twists are,

$$\left(\hat{V}_{3/2}^P\right)_{\mathcal{R}_3} = \begin{pmatrix} \dot{\theta}_3 e_z^3 \\ l e_x^3 \times \dot{\theta}_3 e_z^3 \end{pmatrix}$$

$$\left(\hat{V}_{2/1}^P\right)_{\mathcal{R}_3} = \begin{pmatrix} \dot{\theta}_2 e_z^2 \\ (l_2 e_x^2 + l e_x^3) \times \dot{\theta}_2 e_z^2 \end{pmatrix}$$

$$\left(\hat{V}_{1/0}^P\right)_{\mathcal{R}_3} = \begin{pmatrix} \dot{\theta}_1 e_z^1 \\ (l_1 e_x^1 + l_2 e_x^2 + l e_x^3) \times \dot{\theta}_1 e_z^1 \end{pmatrix},$$

where $e_{x,y,z}^i$ are unit vectors codirectional with the axes of frame $\mathcal{R}_i$, for $i = 0, 1, \ldots 5$. All unit vectors of different frames can be expressed in terms of $e_{x,y,z}^3$ in $\mathcal{R}_3$ alone using coordinate transformations. The Jacobian J of the robot is

$$\left(\hat{V}_{3/0}^P\right)_{\mathcal{R}_3} = J \begin{pmatrix} \dot{\theta}_1 \\ \dot{\theta}_2 \\ \dot{\theta}_3 \end{pmatrix}$$

Then, we can express J at the point P on B_4 as,

$$\begin{pmatrix} 0 & 0 & 0 \\ 0 & 0 & 0 \\ 1 & 1 & 1 \\ -l_2 \sin\theta_3 - l_1 \sin\theta_2 + \theta_3 & -l_2 \sin\theta_3 & 0 \\ -l - l_2 \cos\theta_3 - l_1 \cos\theta_2 + \theta_3 & -l - l_2 \cos\theta_3 & -l \\ 0 & 0 & 0 \end{pmatrix}$$

The inner loop of the leg in Fig. 5 represents the additional constraint of the system dynamics, which reduces the system degrees of freedom (DOF). Using the same method as above, we can compute J_c which has constraint $0 = J_c\dot{q}$, where q is the vector of the generalized coordinates of the robot with entries θ_i, $i = 1, 2, \ldots, 5$. In the leg model this constraint reduces the necessary coordinates from 5 to 4.

To solve the dynamics using Euler-Lagrange equation, the kinetic energy has to be expressed in terms of the generalized coordinates. By using the method above, we can compute the velocity of every interest point P in terms of generalized coordinates θ_1, θ_2, θ_3, and θ_4. For instance, the kinetic energy of B_1 and B_2 relative to the lab frame is,

$$(T_{1/0})_{\mathcal{R}_0} = \frac{1}{2}(\dot{\theta}_1 e_z^1) I (\dot{\theta}_1 e_z^1) + \frac{1}{2} m_1 \left(\dot{\theta}_1 e_z^1 \times \frac{l_1}{2} e_x^1\right)^2,$$

which can be expressed in terms of θ_1 and $\dot{\theta}_1$ after performing coordinate transformation on e_x^1. We apply the same process to express the kinetic energy of B_3, B_4, B_5 but with an extra term to incorporate motion of the center of mass.

The conservative forces are gravity force, ground interaction force, and torsion spring tension. Using the same coordinate transformation method, we can express them in generalized coordinates as well.

We can invoke the Euler-Lagrange equation for every leg segment, with Q being the sum of all conservative forces on that segment,

$$\frac{d}{dt}\frac{\partial T}{\partial \dot{q}} - \frac{\partial T}{\partial q} = Q \qquad (2)$$

3.3 Ground Interaction

The ground interaction as shown in Fig. 5 is modeled as spring and damper with kinetic friction whenever contact is present. The static friction is not included because the contact time is short and the tangential speed is always not zero. First the state is augmented with a relative vertical distance between the origin of $\mathcal{R}_1$ and the origin of $\mathcal{R}_0$. The kinetic friction and normal force is shown in Equation (2) only when the relative distance is less than zero. There is no rotation in the normal direction of collision, thus no torque.

The normal force f_v and horizontal force f_h is formulated as follows, [29]

$$f_v = \min(ky + dv_n, 0)$$
$$f_h = -\mu_m \|f_v\| \begin{cases} 1 & \text{if } \|v_h\| > v_{min} \\ \|v_h\|/v_{min} & \text{else} \end{cases}$$

where k is the spring constant, d is the damping term, y is the deformation in the vertical direction, v_n is the approaching velocity in the vertical direction, v_h is the relative velocity in the horizontal direction, v_{min} is an adjustment term to avoid abrupt changes in the friction force through sign changes in v_n.

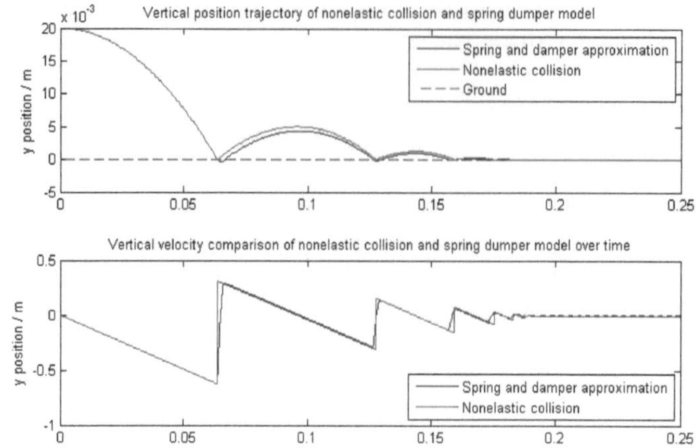

Fig. 6. Simple example of a bouncing ball using spring damper model and nonelastic collision. Upper plot shows y position while bottom one gives the velocity comparison.

This contact model is a model that is linear in distortion and linear in approaching velocity. The coefficient is tuned to achieve similar energy loss with nonelastic collision with coefficient of restitution 0.5. As shown in Fig. 6, the result of the spring and damper model is well fit with the nonelastic collision model with the chosen spring and damper constants. A more precise static model is the Hertz model. In this case, the contact force is proportional to the distortion to the power of 2/3. The Hertz model is a precise static model and it requires a more precise understanding of the contact point. Because the contact point of the millirobot has different shapes over the simulation time, the simplified spring and damper models are more suitable for ground modeling.

4 Simulation Results and Discussion

The Modelica millirobot model is pieced together according to a Pro Engineering model used for the initial structure design of the subject robot [27]. The leg, as shown in Fig. 5, is modeled in Modelica as in Fig. 7(a). The joints have specific details such as joint length, specified by the designer, and spring constant determined by Equation (1) according to the material properties and geometry of the flexible joint. The leg model is then linked together with other pieces as shown in Fig. 7(b).

In simulation animation, the robot is seen as in Fig. 4. The overall Modelica model of the robot is close to the physical model, with modifiable parameters for geometry and material properties of the joints and rigid body parts. The simulation results show that the model behaves close to the experiment (Fig. 8).

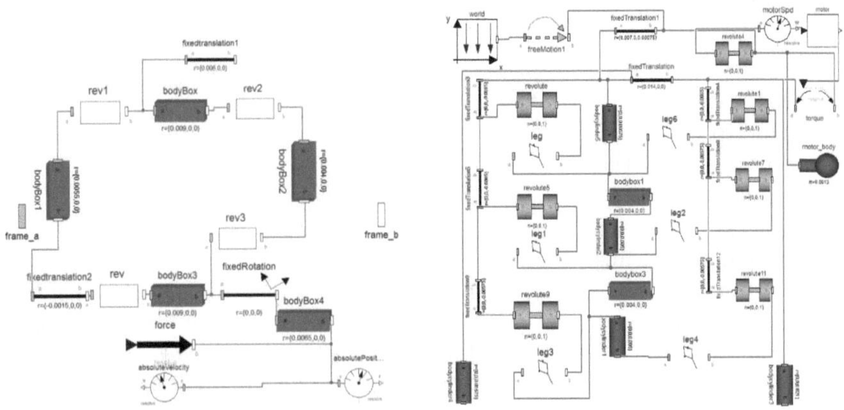

(a) Modelica leg model (b) Overall model

Fig. 7. Fig. (a) describes the structure of the leg model in Modelica block diagram. The joints rev, rev1, rev2 and rev3 are the joints with flexible material. Fig. (b) gives a simplified structure of the robot using the leg submodel. The shaft mechanism is in the middle with linkages to six legs on either side. The top right portion depicts the motor.

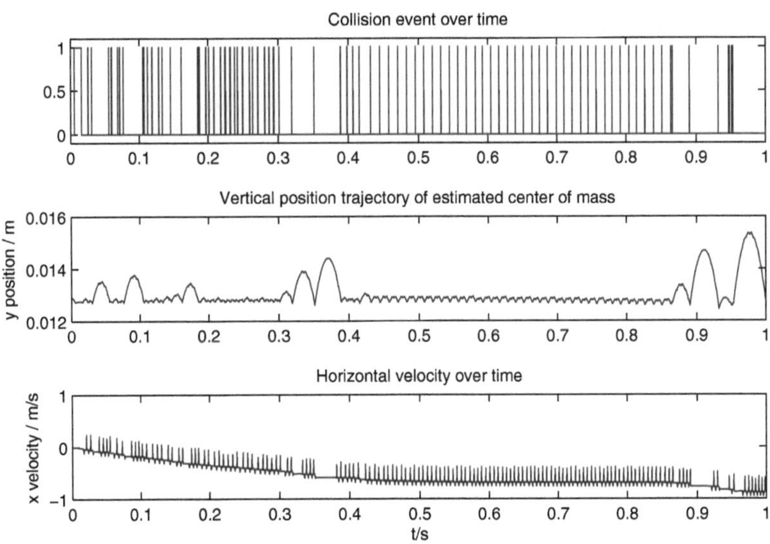

Fig. 8. Modelica simulation results for the millirobot. The top plot depicts the ground collision events over time of all six legs. The second plot shows the relative y position of the motor to the ground. When the simulated robot flips, unstable behavior will be shown in this plot. For this particular setup, it is stable and close to the stable behavior in the experiment. The last plot on the bottom depicts the horizontal velocity.

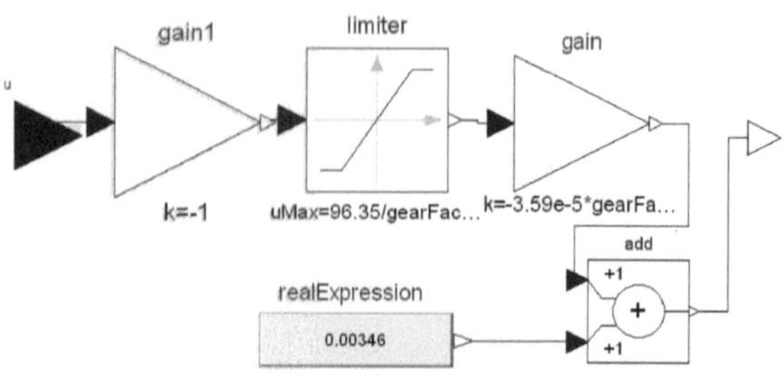

Fig. 9. Modelica motor model. Because of physical nature of electrical motors, the output torque is affine to rotational speed. This model described the particular motor used in the prototype design using its datasheet.

The motor block, as shown in Fig. 9, is the feedback control of the robot, i.e. the cyber component. Due to the physical properties of the electrical motor, the output torque is affine in the rotational speed input of the motor. This model is directly obtained using the physical properties of the electrical motor used in the prototype. This motor can be controlled using Pulse Width Modulation (PWM), so that the power is reduced and stability is improved. The PWM will also reduce the torque to prevent joints from breaking. Other improvements are possible. For example, in [27], the authors propose that future models can use additional weight to create a complete 2D planar motion instead of only back and forward motion. The controller for the weight will be more complicated and will require model-based design instead of experimental only methods. Therefore, if more features are to be added to the original prototype, the cyber component will be more complicated and will need to be modified accordingly.

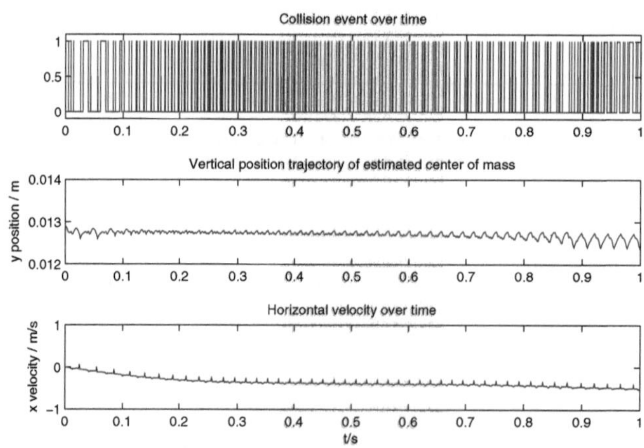

Fig. 10. The stability is improved if the torque is controlled using the sensor input related to the ground contact.

From the second plot in Fig. 8, we note that the robot often bounces away from the ground. The friction force, which is used to move the robot forward, is not present, so such design is not efficient. More importantly, this causes instability in the long term simulation. The subject robot has similar jumping instability in real experiments, but no solutions have been proposed to improve stability [27]. Now consider a very simple modification of the model that has a PWM motor control unit included so that the power output of the motor is reduced, when the legs of the robot are not in contact with the ground. This makes the model more stable as shown in Fig. 10. To understand further how the PWM motor control helps the stability, one can observe the changes in the associated limit cycle. Fig. 11(a) gives the initial trajectory of θ_l, $\dot{\theta}_l$, θ_r and $\dot{\theta}_r$ within the robot model, where θ_l and θ_r are the generalized coordinates θ_1 of the left and right legs, as shown in Fig. 5. From the trajectory, one deduces the limit cycle of the hybrid system, and reset points due to collision as shown. After adding the PWM motor control, the trajectory takes similar shapes (Fig. 11(b)). The major change is that the swinging speed decreases by about 67%. The converging speed from the reset point towards the limit cycle is faster as well.

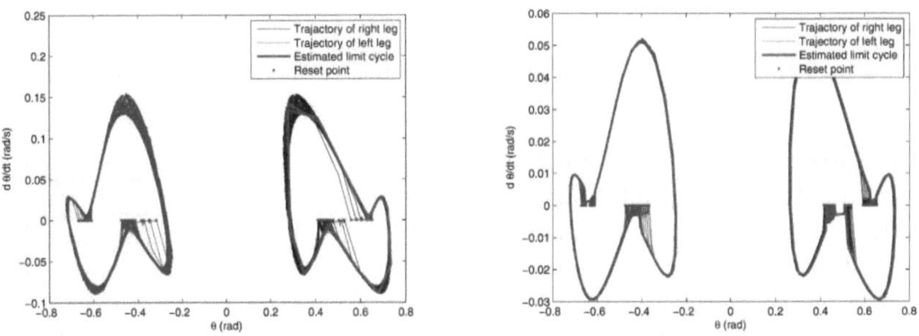

(a) The original design. (b) After adding PWM motor control.

Fig. 11. Fig. (a) (b) describe the trajectories of θ_l, $\dot{\theta}_l$, θ_r and $\dot{\theta}_r$ before and after adding the PWM motor control unit. Compared to the original design, the limit cycle with PWM control takes similar shape, but swinging speeds, $\dot{\theta}_r$ and $\dot{\theta}_l$, decrease by about 67%.

Simulation and system modeling lead to a new design that improves the efficiency of the cyber side of the CPS. The new design also induces the cyber and physical layers to cooperatively behave in a more stable manner.

5 Material Choice and Geometry Exploration

The material selection of the joints is very limited in [27], but in [30] the authors proposed a way of constructing microstructures so that the overall performance of the structure reflects the properties of different material layers. Though this method was mainly used and implemented for thermal bending purposes, this approach can be used in other areas. For microrobots, this means that material selection can consider a much wider range. In the design process, one can design a joint or segments with materials that are unknown but have properties within some reasonable range. In the last step, one can design the microstructure so as to fit the desired properties (specifications).

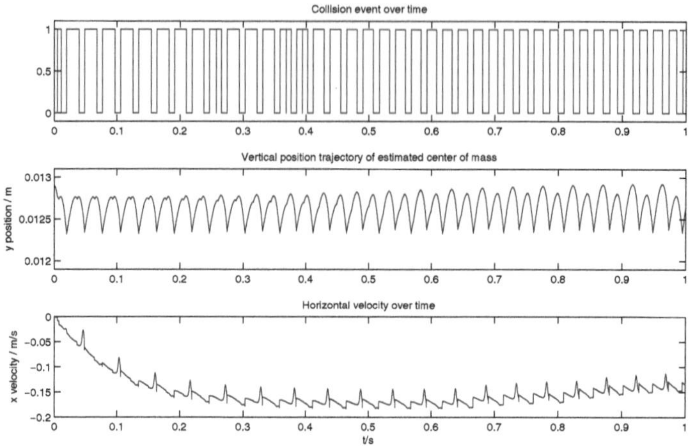

Fig. 12. The stability is improved by reducing the spring constant

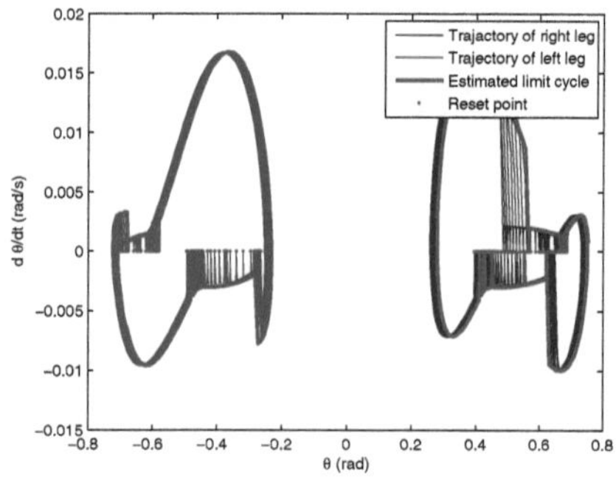

Fig. 13. The trajectories of θ_l, $\dot{\theta}_l$, θ_r and $\dot{\theta}_r$ after using modified joint width. Comparing to the original design, the limit cycle takes different shape, and swinging speed decrease by about 90%.

In the subject robot [27], the key design parameter is the joint spring constant, which is affected linearly by the elasticity modulus of the material, and is proportional to h^3. Therefore, the internal torque between the joints can be made 8 times larger by doubling the joint width. Initially the spring constant is chosen so that the internal torque between joints has about the same magnitude as the maximum motor torque. This may induce instability. Fig. 12 shows the result for the same shape of robot structure but with half the joint width, which is still within the reasonable range of joint width in [27]. This design change seems to make the robot more stable. Further exploration using limit cycle methods gives different results. As shown in Fig. 13, the trajectory shows that the swinging speed decreases by 90%, and the shape of the limit cycle changes. However the limit cycle may be unstable since the trajectory keeps on shifting to the right with no sign of converging.

The material choice can increase the range of possible values for the joint spring constant and even make the joint sustainable under large tension when required in the design. However, changes in the material and geometry of the joints add constraints to the controller, and in particular to the maximum torque output of the motor.

Now suppose we modify the geometry significantly and the new model takes the shape of Fig. 4b. The shape of the legs is modified to emulate the legs of a crab. The new design is obtained through trial and error to achieve a more regulated walking behavior, i.e. bouncing forward but with similar height.

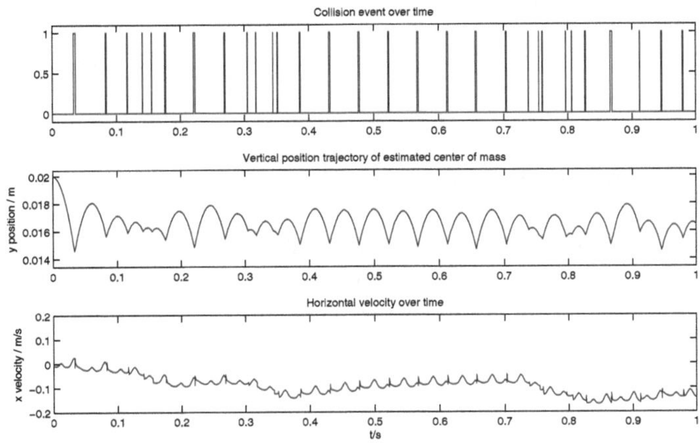

Fig. 14. Collision and motion behavior are different due to different geometry

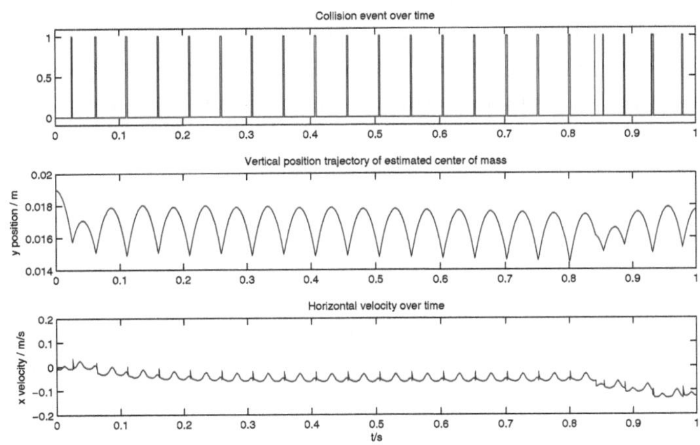

Fig. 15. By adding motor control, as can be seen in the second plot, the jumping behavior is regulated and more stable compared to Fig. 14. The magnitude of the speed is hardly increasing because the robot is constantly lifting off the ground.

As shown in Fig. 14, the robot bounces frequently but it is more regulated and more stable compared to the previous design. Although the joint spring constant has about the same magnitude as in Fig. 12, the behavior of the robot is different. If explored further using the limit cycle, one concludes that the collision for this design actually happened more irregularly, and the limit cycle

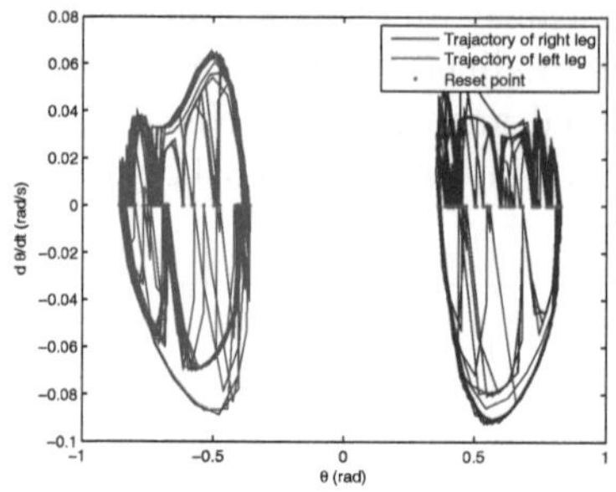

(a) Microrobot with modified geometry.

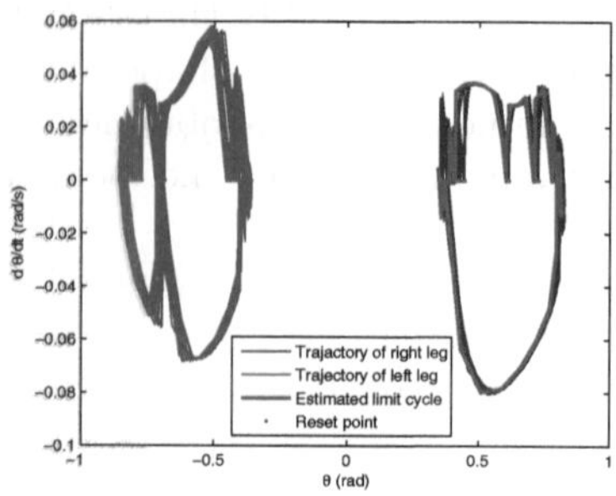

(b) After adding PWM motor control.

Fig. 16. Fig. (a) (b) describe the trajectories of θ_l, $\dot{\theta}_l$, θ_r and $\dot{\theta}_r$ of the modified microrobot before and after adding the PWM motor control unit. Compared to the original design, the limit cycle takes different shape. For (a), the collision happened so irregularly that the limit cycle is hardly visible. In (b), the collision is more regular and converges to the limit cycle faster.

is hardly seen as shown in Fig. 16(a). One can use similar control to reduce the instability as discussed in the previous section. The result is shown in Fig. 15 with its associated limit cycle analysis in Fig. 16(b). The jumping behavior is regulated to be more stable but it cannot be removed completely in the case of the new leg shape, which also shows the close relationship between the cyber and the physical components of the robot. Thus the cyber components have to be completely reconsidered as a result of changes in the physical part.

6 Tradeoff and Model Based System Design

As discussed in the previous section, the cyber components have to be adjusted or even redesigned because of the changes in the physical components. A systematic way of jointly considering physical modeling and control design is required. We propose the framework described in Sections 1.2, 1.3, 1.4 for designing millirobots as CPS (Fig. 3). SysML is used as a language for the structure description of the robot, and also used as a linkage with trade-off tools so that material properties trade-off can be performed based on efficiency and stability matrices.

So far the system level design is done in ModelicaML, which is a Java and XML based metamodel which bidirectionally transforms between the UML model and Modelica [31]. As can be seen in Fig. 17, the designer of the robots can use ModelicaML tool (in Eclipse) to modify and simulate design using class definition diagram with underlying simulation engine OpenModelica. One can design control algorithms and have control parameters tuned together with material properties selection and associated trade-offs using the framework described earlier. The model based approach makes the control algorithm easily modifiable, so that different controller designs can be tested and verified.

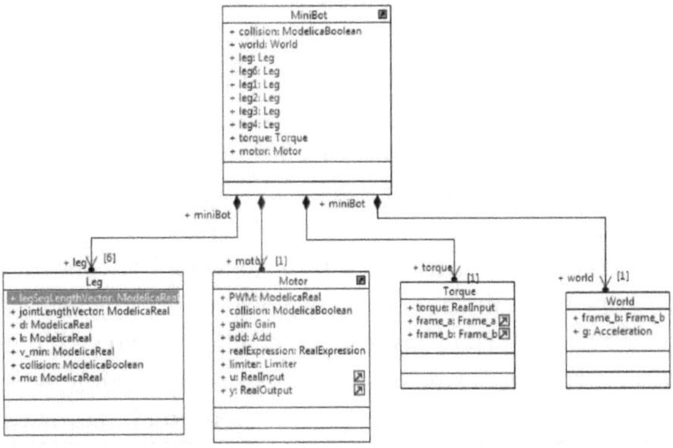

Fig. 17. This is the class diagram created in ModelicaML implementation in Eclipse [31]. The structure model is abstracted from the synchronized Modelica model. The model based approach in developing the Modelica model gives the designer of the robots convenient ways to modify the key physical model parameters (joint width and geometry specified by leg segments length) and the cyber components parameters like PWM. The underline OpenModelica compiler is able to simulate the model at the same time to perform verification tasks for material constraints.

Finite Element Analysis tools like COMSOL for material oriented simulation can be integrated with Modelica to provide a more detailed model of the robot. In this paper a joint is modeled as a torsion spring but clearly it is a simplification of the linkage. One can use COMSOL to provide a more detailed nonlinear model of the joint. The Modelica library FlexBody [32] can be used to solve this problem as well. This library uses the output from FEA tools, such as Nastran, Genesis, to reduce complex finite element models to models which consist of only few boundary nodes, or attachment points.

For this particular prototype, the PWM modification comes from the fact that decreasing torque applied by the motor directly increases the stability. In general this insight can be drawn from the trade-off tools directly. Given material constraints, in particular the deformation condition of the joints, one can formulate this problem as an optimization problem with objective to maximize efficiency and stability matrices such as maximizing forward moving speed and minimizing jumping heights. The tool we implemented in MagicDraw-SysML as shown in Sections 1.2, 1.3, 1.4 can then be used to direct such modifications in the original design. Geometrical modeling and design exploration need to take a different approach from our view. Few modifications of the geometry can dramatically alter the problem. We suggest the combination of design and optimization in earlier stages so that the overall geometry is fairly fixed with only minor changes. This requires the CAD design to be integrated into the design as well. CATIA [33], the 3D CAD tool which can interface with Dymola simulation, might be the best approach for this. We suggest the following design process steps.

1. The designer of the robots defines system structure and high level system design according to the requirement in SysML (System level design) and detailed structure model in CATIA (geometric modeling), alternatively the user can generate SysML/UML from Modelica/CATIA modeling using SysML4Modelica transformation [34] or ModelicaML [31]. The tool will generate model structure as well as Modelica code through model transformation.
2. To do design space exploration such as material trade-off, designer can use FEA tool to generate joint model for several materials so that it can be linked with Modelica through the FlexBody Modelica library. Then the designer has to specify objectives in terms of stability matrices such as jumping height or limit cycle criteria, and performance matrices such as forward motion speed and energy transformation ratio. Constraints, such as the maximum torques the link can sustain, also need to be specified. The trade-off is then done using Consol-Optcad with Modelica simulation.
3. Based on the Consol-Optcad suggestions, the designer has to modify the initial design and go back to system level design and verify that all the requirements are met. If not, the constraints of the previous step have to be refined and the designer will go through the process again until all the requirements are satisfied.

7 Conclusions

To conclude, microrobots are complex CPS, and their cyber part cannot be designed separately from the physical part. In this paper, we described a model-based systems engineering methodology and framework for the design of microrobots as CPS. The physical model and associated control of a particular prototype were examined to demonstrate the close relationship between the physical and cyber parts. We also proposed improvements of the control laws so that the system is more stable and demonstrated these improvements by modeling and simulation. The control laws for the design of this particular type of robot, such as those shown in Fig. 4a or 4b, can be designed via tradeoff with material properties as tunable parameters. They may also need to be redesigned when the geometrical shape changes are significant.

Acknowledgements. We would like to thanks Dana E. Vogtmann for providing experimental demonstrations so we could compare with the models we built. We would also like to thank members of the OpenModelica Association who have kindly helped us solve Modelica related problems.

Research supported in part by the National Science Foundation (NSF) under grant award CNS-1035655 and by the National Institute of Standards and Technology (NIST) grant award 70NANB11H148

References

1. Austin, M.A., Baras, J.S., Kositsyna, N.I.: Combined Research and Curriculum Development in Information-Centric Systems Engineering. In: Proc. of the 12th Annual Intern. Council on Systems Engineering (INCOSE) Symposium (2002)
2. Yang, S., Baras, J.S.: Factor Join Trees in Systems Exploration. In: Proceedings of the 23rd International Conference on Software and Systems Engineering and their Applications (ICSSEA 2011), Paris, France (2011)
3. Wang, B., Baras, J.S.: Integrated Modeling and Simulation Framework for Wireless Sensor Networks. In: Proceedings of the 21st IEEE International Conference on Collaboration Technologies and Infrastructures (WETICE 2012- CoMetS track), Toulouse, France, pp. 268–273 (2012)
4. Yang, S., Zhou, Y., Baras, J.S.: Compositional Analysis of Dynamic Bayesian Networks and Applications to Complex Dynamic System Decomposition. In: Proc. of the Conf. on Systems Engineering Research, CSER 2013 (2013)
5. Yang, S., Wang, B., Baras, J.S.: Interactive Tree Decomposition Tool for Reducing System Analysis Complexity. In: Proc. of the Conf. on Systems Engineering Research, CSER 2013 (2013)
6. Spyropoulos, D., Baras, J.S.: Extending Design Capabilities of SysML with Tradeoff Analysis: Electrical Microgrid Case Study. In: Proc. of the Conference on Systems Engineering Research, CSER 2013 (2013)
7. International Council on Systems Engineering (INCOSE): Systems Engineering Vision 2020. Version 2.03, TP-2004-004-02 (2007)
8. Friedenthal, S., Moore, A., Steiner, R.: A Practical Guide to SysML. The MK/OMG Press (2009)
9. Baras, J.S.: Lecture Notes for MSSE class, ENSE 621 (2002)
10. Haskins, C., Forsberg, K., Krueger, M., Walden, D., Hamelin, D.: Systems Engineering Handbook. INCOSE, San Diego (2011)
11. The eMoflon team: An Introduction to Metamodelling and Graph Transformations with eMoflon, V 1.4. TU Darmsadt (2011)
12. Anjorin, A., Lauder, M., Patzina, S., Schurr, A.: eMoflon: Leveraging EMF and Professional CASE Tools. In: INFORMATIK 2011, Bonn (2011)
13. No Magic, Inc.: Open API-User Guide. Version 17.0.1 (2011)
14. Meyer, J., Ball, M., Baras, J.S., Chowdhury, A., Lin, E., Nau, D., Rajamani, R., Trichur, V.: Process Planning in Microwave Module Production. In: Proc. SIGMAN: AI and Manufacturing: State of the Art and State of Practice (1998)
15. Fan, M.K.H., Tits, A.L., Zhou, J., Wang, L.-S., Koninckx, J.: CONSOLE-User's Manual. Technical report, Un. of Maryland, Vers. 1.1 (1990)
16. Fan, M.K.H., Wang, L.-S., Koninckx, J., Tits, A.L.: Software Package for Optimization-Based Design with User-Supplied Simulators. IEEE Control Systems Magazine 9(1), 66–71 (1989)
17. Tischler, M.B., Colbourne, J.D., Morel, M.R., Biezad, D.J.: A Multidisciplinary Flight Control Development Environment and its Application to a Helicopter. IEEE Control Systems Magazine 19(4), 22–33 (1999)
18. Potter, P.J.: Parametrically Optimal Control for the UH-60A (Black Hawk) Rotorcraft in Forward Flight. MS Thesis, Un. of Maryland (1995)
19. Nagy, Z., Ergeneman, O., Abbott, J., Hutter, M., Hirt, A., Nelson, B.: Modeling assembled-mems microrobots for wireless magnetic control. In: Proc. of IEEE Intern. Conf. on Robotics and Automation, ICRA 2008, pp. 874–879. IEEE (2008)

20. Mohebbi, M.H., Terry, M.L., Böhringer, K.F., Kovacs, G.T.A., Suh, J.W.: Omnidirectional walking microrobot realized by thermal microactuator arrays. In: Proc. of 2001 ASME Intern. Mech. Engin. Congress and Exposition, pp. 1–7 (2001)
21. Donald, B., Levey, C., McGray, C., Paprotny, I., Rus, D.: An untethered, electrostatic, globally controllable mems micro-robot. Journal of Microelectromechanical Systems 15, 1–15 (2006)
22. Bergbreiter, S., Pister, K.: Cotsbots: an off-the-shelf platform for distributed robotics. In: Proceedings IEEE/RSJ International Conference on Intelligent Robots and Systems (IROS 2003), vol. 2, pp. 1632–1637. IEEE (2003)
23. Erdem, E., Chen, Y.M., Mohebbi, M., Suh, J., Kovacs, G., Darling, R., Öandhringer, K.B.: Thermally actuated omnidirectional walking microrobot. Journal of Microelectromechanical Systems 19, 433–442 (2010)
24. Churaman, W., Currano, L., Morris, C., Rajkowski, J., Bergbreiter, S.: The first launch of an autonomous thrust-driven microrobot using nanoporous energetic silicon. Journal of Microelectromechanical Systems 21, 198–205 (2012)
25. Hiller, J., Lipson, H.: Automatic design and manufacture of soft robots. IEEE Transactions on Robotics 28, 457–466 (2012)
26. Fritzson, P.: Introduction to Modeling and Simulation of Technical and Physical Systems. Wiley-IEEE Press (2011)
27. Vogtmann, D.E., Gupta, S.K., Bergbreiter, S.: Multi-material compliant mechanisms for mobile millirobots. In: Proceedings 2011 IEEE International Conference on Robotics and Automation (ICRA), pp. 3169–3174 (2011)
28. Bellouard, Y.: Microrobotics: Methods and Applications. CRC Press (2010)
29. Otter, M., Elmqvist, H., Dí López, J.: Collision Handling for the Modelica Multi-Body Library. In: Proc. of the 4th Intern. Modelica Conf., pp. 45–53 (2005)
30. Shin, M., Gerratt, A.P., Metallo, C., Brindle, A., Kierstead, B.P., White, R.D.: Characterization of a micromachined parylene-based thermal c-shape actuator. Journal of Micromechanics and Microengineering 21(9), 095028 (2011)
31. Schamai, W.: Modelica Modeling Language (ModelicaML): A UML Profile for Modelica. Tech. Rep. 2009:5, Linköping University, Department of Computer and Information Science (2009)
32. Dymola libraries flexbody | claytex, `http://www.claytex.com/products/dymola/model-libraries/flexbody-library`
33. CATIA systems engineering, `http://www.3ds.com/products/catia/solutions/catia-systems-engineering`
34. SysML-Modelica Transformation (SyM), `http://www.omg.org/spec/SyM/1.0`

Cloud Conveyors System: A Versatile Application for Exploring Cyber-Physical Systems

Shivakumar Sastry[1], Michael S. Branicky[2], and P.S. Sastry[3]

[1] Department of Electrical and Computer Engineering, The University of Akron
ssastry@uakron.edu
[2] Department of Electrical Engineering and Computer Science, Case Western Reserve University
mb@case.edu
[3] Department of Electrical Engineering, Indian Institute of Science
sastry@ee.iisc.ernet.in

Abstract. We present a versatile application for cyber-physical systems (CPS), called the *Cloud Conveyors System* (CCS). This system comprises a collection of mobile conveyor units with simple periodic behavior; the units move back and forth along fixed tracks. The system-level objective is to transport entities from some input port to an output port when each entity has its own target output port, deadline, and end-to-end QoS constraints. Entities ride on the mobile units to physically move from one location to another. Entities may transfer instantaneously between two units — or when the unit is at an input or an output. We refer to these transfers as cyber transfers because they involve decision-making and the entities do not have to transfer at every possible opportunity. We view the transport of each entity in CCS as a *CPS-Task* that evolves both in space and in time; more precisely, a CPS-Task is an alternating sequence of cyber transfers and physical moves that starts at an input and ends at the output of the entity. This novel model for a CPS-Task allows one to explore solutions to some of the principal CPS challenges namely, Composition, Control Strategies, Computational Abstractions, Model-driven Engineering, and Verification & Validation. Further, this abstract and well-defined problem is useful in CPS Education and Training because it has a rich structure with intertwined cyber and physical dynamics; also, the scale and complexity of the problem can be increased by adding more units or changing the configuration of the system without increasing the implementation burden, which is critical to validating CPS solution techniques on physical testbeds.

Keywords: Cloud Conveyors System, Cyber-Physical Systems, spatio-temporal dynamics, abstraction, tasks.

1 Introduction

Across a broad spectrum of application domains including advanced manufacturing, intelligent transportation, and healthcare, there is an urgent need for

D.C. Tarraf (ed.), *Control of Cyber-Physical Systems*,
Lecture Notes in Control and Information Sciences 449,
DOI: 10.1007/978-3-319-01159-2_3,

systems-engineering formalisms that can transform the state-of-the-practice from ad hoc methods to assured methods. Such methods are necessary to design, engineer and maintain a class of systems called cyber-physical systems (CPS) [1]. These systems typically have a collection of networked embedded devices that interact at multiple temporal and spatial scales. The system structure can change dynamically; many subsystems operate autonomously and may have control loops that close at multiple scales. In certain domains, the operations of the system must be dependable and certifiable. A well-developed theoretical foundation for CPS is essential to develop architectures, techniques, processes, and tools to design, maintain, and operate such systems. A critical first-step toward such a foundation is an abstraction for a *unit of activity in a CPS*, i.e., a CPS-Task, that evolves both in space and in time, exhibits complex behaviors, and is yet amenable for theoretical analysis. Identifying such a model or representation for a generic CPS is a challenging and interesting problem.

Any model for a CPS-Task must (a) represent the intertwined dynamics of the cyber and physical activities in the system; (b) embody rich possibilities for spatio-temporal dynamics to accurately reflect the operational environment of systems; (c) allow for rigorous reasoning about the Quality of Service (QoS); (d) enable selective examination of interesting dynamic behaviors; and (e) be intuitive and simple so that application-domain specialists, systems engineers, and theoreticians can use the model to develop techniques for scheduling, planning, verification, etc. that are important in a variety of CPS sectors. The model for a CPS-Task must be implementable to facilitate the validation of the theory, techniques, and tools either through simulations and/or on testbeds. We present a well-structured example of a CPS that we call the *Cloud Conveyors System* (CCS); this system allows us to formulate a model for a CPS-Task. By adjusting the configuration of the CCS, either statically or dynamically, we can alter the structure and dynamics of system and, hence, the complexity of the CPS-Tasks.

The Cloud Conveyors System (CCS) comprises of a collection of conveying units that each move periodically between two ends of a fixed track at a constant velocity. Figure 1 presents an instance of a CCS that has two sets of tracks — the horizontal and the vertical. The system-level objective is to transport physical entities (e.g., material or people) from some input port to an output port, where each entity has its own target output port, deadline, and Quality-of-Service (QoS) constraints. As illustrated, entities enter the system via an input that is located at the left end of a horizontal track, ride on one or more conveying units in the CCS, and exit via an output at the right end of a horizontal track. In this paper, we allow each mobile unit to carry only one entity; further, each unit moves one unit along its track in each time step.

The interesting cases of transport in a CCS involve entities that enter on horizontal track H_i and exit on H_j. In such a case, the entity must first ride on the horizontal track H_i, get transfered to a unit moving along some vertical track and thus reach the unit on H_j. To enable such transfers, one or more units that move along the vertical tracks must rendezvous with the two units that move on H_i and H_j. We stipulate that an entity may transfer from one unit

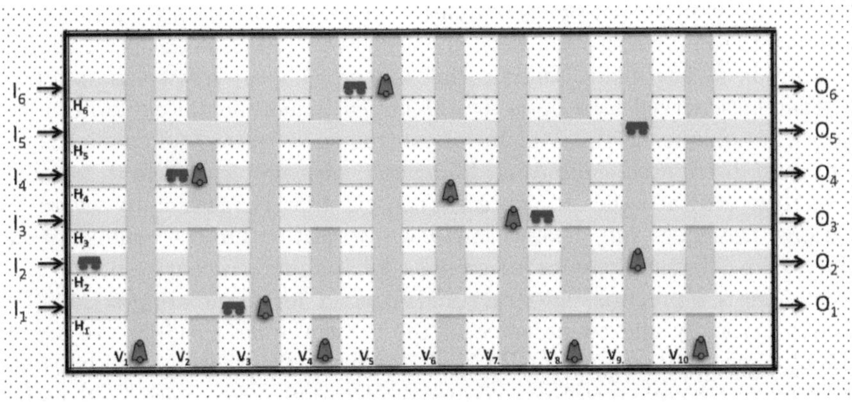

Fig. 1. *Cloud Conveyors System.* Conveying units move periodically between the two ends of fixed tracks, one grid unit each time tick. Inputs and outputs are located at the ends of the horizontal tracks. Entities that arrive at an input *physically move* toward their output by riding on the units; a *cyber transfer* can occur when two units rendezvous — here, an entity can instantaneously transfer from one unit to another.

to the other at any rendezvous, as long as the other unit can carry the entity. *We view the transport of each entity through the CCS as a "task" performed by the system*, i.e., a CPS-Task. The design and operation of such a system has to include principled methods to analyze patterns of rendezvous possible for different configurations of the system, methods to design a configuration to allow for sufficiently rich patterns of rendezvous, algorithms to decide when an entity should move from one conveyor unit to another, and so on.

When an entity moves (changes its location) by riding on a conveyor unit, we call it a *physical move*. When one unit rendezvous with another unit, a *cyber transfer* can occur; here, an entity can instantaneously transfer to the other unit. A cyber transfer need not occur at every rendezvous because the decision on such a transfer would depend on many factors, including the target output of the entity, deadline, and remaining time before which it must be delivered, QoS constraints of the entity, the current load on the downstream units in the system, etc. In order to account for these dynamic factors, every cyber transfer in CCS involves a significant amount of cyber activities (i.e., decision-making, computation, communication, sensing, actuation, and control). Thus, our abstract CPS-Task is a sequence of physical moves and cyber transfers that represent the evolution of an entity in space and time through the CCS. We present a more precise system model for CCS in Section 2.

CCS offers a precise and well-defined context to explore new solutions to key CPS challenges such as Composability, Control Strategies, Computational Abstractions, Model-driven Engineering, and Verification & Validation. In addition, CCS is a unique platform for CPS Education and Training. It is relatively easy to implement in a simulator or a physical testbed. The scale and complexity of the problem can be increased or decreased to validate solution techniques without increasing the implementation burden. The Inverted Pendulum [2], Prisoner's Dilemma [3], and Drinking/Dining Philosophers [4,5] are a few example problems that have served as a basis for theoretical/experimental exploration of a variety of techniques in the domains of control, cooperation and conflict among agents, and distributed computing. In general, such problems help to formulate and analyze critical issues in a precise context that is relevant to several applications. Designers and theoreticians have worked within the context of such problems in different domains to gain useful insights. Similarly, the CCS we present here is a precise, well-structured, context for formulating a CPS-Task and systematically addressing some of the critical issues in CPS. We are not aware of a general model for a CPS-Task and hence feel that a useful beginning can be made through this abstract model. We would like to see CCS used in such a spirit.

The planning of the sequence of physical and cyber transfers for transporting entities in CCS has many issues that are similar to real-time scheduling problems. The classical literature on real-time systems [6,7,8] is well-known. Recent investigations have focused on handling sporadic tasks [9,10], preemption overheads [11,12], resources [13,14,15], rhythmic tasks [16] and structured tasks [17,18,19]. The interplay between rendezvous patterns and scheduling techniques is an important for the design and operation of CCS. Here, the scheduling must be addressed in the context of the spatio-temporal structure of the system because the number of vertical conveyor units to deploy can also be a design decision. We believe that our model of CPS-Task is relevant for a class of CPS applications and has the potential to lead us to a spatio-temporal (CPS) analog of the classical result of Liu and Layland [6] for real-time scheduling.

Many problems related to CCS can also be approached from a hybrid systems standpoint [20]. Aspects of CCS are related to the recent framework of sampling-based methods [21,22,23,24,25,26]. The vast literature on robotic motion planning [27,28] can inform path selection and routing techniques for CCS. The most related work here is that on the optimization and coordination of multiple agents [29,30,31,32,33,34,35,36,37].

The remainder of this paper is organized as follows. We present a system-model for CCS in Section 2. To motivate the utility of CCS, we present on qualitative discussion on how specific CPS problems can be mapped to CCS in Section 3. In Section 4, we present one of the fundamental problems in CCS, i.e., the precise characterization of the pattern of rendezvous events and some initial solutions. In Section 5, we discuss how the CPS-Task model we propose is useful to address many of the CPS challenges in a systematic manner. After discussing the versatility of CCS in Section 6, we conclude in Section 7.

2 System Model

The CCS instance illustrated in Figure 1 has 10 vertical units and 6 horizontal units and the underlying grid has 8×12 lattice points that provide a discrete coordinate space for locations in CCS. Given, M horizontal units and N vertical units, and an underlying grid of $(M+2) \times (N+2)$, the horizontal tracks $H_1, H_2, \cdots, H_M$ are on rows $1, 2, \cdots, M$; similarly the vertical tracks $V_1, V_2, \cdots, V_N$ are on columns $1, 2, \cdots, N$. CCS may have a maximum of M inputs and M outputs; the jth input, $1 \leq j \leq M$ is located at $(0, j)$ and the jth output is located at $(N+1, j)$.

2.1 Periodic Movement of Mobile Units

We represent the horizontal units as $h_1, h_2, \cdots, h_M$ where unit h_j moves along the track H_j. Similarly, the vertical units $v_1, v_2, \cdots, v_N$ move along the vertical tracks[1]. System time evolves in discrete ticks and in each tick, every unit moves one unit. Initially, the horizontal unit h_j is at the grid location $(0, j)$ and the vertical unit v_i is at location $(i, 0)$. The horizontal units, h_j, move from $(0, j)$ to $(N+1, j)$ and return toward $(0, j)$. Similarly, the vertical units v_i move from $(i, 0)$ to $(i, M+1)$ and return to $(i, 0)$. All these units periodically move from one end of the track to the other at a constant velocity. Each mobile unit can be started with an initial delay (offset). We use δ^r_j to represent the offset of the horizontal unit h_j and δ^c_i to represent the offset of the vertical unit v_i.

2.2 Rendezvous Events

A rendezvous event (or rendezvous) is an important spatio-temporal event in CCS and is essential for transporting entities across horizontal tracks. Cyber transfers are necessary to improve the connectivity between inputs and outputs in CCS, and the QoS. We refer to the portion of the grid that is bounded by the locations $(1, 1), (1, M), (N, M)$ and $(N, 1)$ as the *Rendezvous Zone* because any rendezvous in CCS, and hence all interesting behaviors in CCS, occur only in this part of the grid. By physically adjusting the actual positions of the units on the tracks, we can assure that the two units will not collide when a rendezvous event occurs. Several rendezvous events can occur simultaneously at different locations of the underlying grid. We refer to a temporally ordered sequence of rendezvous, each involving either a single or multiple rendezvous, as a patterns of rendezvous events (see more details in Section 4).

2.3 Configuring CCS

The pattern of rendezvous in CCS is constrained by its configuration. From a control perspective, all the parameters of the configuration can be changed.

Definition 1. *A configuration of CCS is a specification of values for the number of horizontal units M, the number of vertical units N, the velocity of each unit, and the initial offset of each unit.*

[1] Relaxations and variants are discussed in Section 6

The number of tracks cannot be changed dynamically without causing the system to reset. However, the velocities of the units can be changed (including setting the velocity of a specific unit to zero). While it is technically possible to introduce arbitrary delays (i.e., stalls) during the execution of units, it raises schedulability issues that are similar in spirit to clairvoyant scheduling. As long as the velocities of the units can be expressed as integers or rational values, the pattern of rendezvous events is periodic; this periodic pattern determines the structure of paths in the system from the inputs to the outputs and hence affects the QoS that can be achieved.

2.4 Transporting Entities

The system-level objective in the abstract CCS is to transport entities from inputs to their outputs; the user must specify the QoS attributes, such as the maximum End-to-End Latency (Deadline) and Throughput that must be satisfied when delivering an entity to its output.

For example, suppose an entity e_s arrives at input I_j and must be transported to the output O_j. In this case, the CPS-Task for e_s begins with a cyber transfer from I_j to unit h_j, a physical move on h_j from location $(0, j)$ to $(N+1, j)$ and finally another cyber transfer from h_j to O_j. Suppose another entity e_x arrives at input I_k and its target output is O_j, $k \neq j$. The CPS-Task for e_x would begin with a cyber transfer from I_k to unit h_k, a physical move on h_k from $(0, k)$ to (p, k), a cyber transfer from h_k to unit v_p, a physical move on unit v_p from (p, k) to (p, j) where another rendezvous must occur with unit h_j. A cyber transfer must occur at (p, j) to transfer e_x from v_p to h_j. Finally, e_x must physically move on h_j to O_j to complete the transport of e_x with a cyber transfer to output O_j. This example illustrates that a CPS-Task always begins and ends with a cyber transfer. A physical move is always preceded by and succeeded by cyber transfer. When a unit reaches the end of the track and does not transfer to the output, it remains on the unit until it is involved in a cyber transfer.

We can represent each CPS-Task as τ_i; we say that this task is released when a new entity, e_i, arrives at its input. Tasks can have a relative deadline D_i before which e_i must be delivered to its output. A *periodic task* is a sequence of entities that arrive at the same input with an inter-arrival time of T_i. We can consider the jth instance of task τ_i as the jth entity, e_i^j in such a sequence. We can consider periodic and aperiodic models for a CPS-Task in a natural manner as used in the real-time systems literature [7,8]. The horizontal and vertical units that are involved in the transfer of entity e_i are the resources required for the task τ_i. Fixed priorities in CCS can be assigned to each input. Thus if input I_i has priority $prio(I_i)$, every entity that arrives via I_i will be assigned a fixed priority $prio(I_i)$ as long as it is being transported in CCS. On the other hand, a dynamic priority scheme can be used by assigning the priorities to the entities instead of to the inputs. In this approach, the priority of an entity can change when it is enroute; for example, the dynamic priority assigned to an entity can be based on the remaining time left before which its relative deadline would be violated.

There are two complementary problems that must be addressed in CCS. First, given a configuration of CCS, it is necessary to understand what QoS can be achieved for the CPS-Tasks. Second, given a collection of CPS-Tasks and the desired QoS for these tasks, it is necessary to determine what CCS configuration(s) can achieve the target QoS. To address both these problems, understanding where the rendezvous events occur and the temporal pattern of these events is critical for achieving the QoS.

3 Mapping CPS Problems to the Cloud Conveyors System

In this section, we illustrate the benefits of CCS. A modular approach to conveyor systems that is described in [38,39] uses instances of two kinds of conveyor unit called Segments and Turns to realize composable conveyor systems. Each of these units have a local microcontroller that is integrated with a low power radio transceiver. The microcontrollers interact with each other to coordinate the transfer of entities from one unit to the other, and to achieve system-level objectives. While these systems have cyber-enabled components that interact at multiple temporal and spatial scales, they do not have all the characteristics of CPS; in particular, these systems are not dynamically reconfigurable and the cyber tasks are not closely intertwined with the physics or the electromechanical aspects that impact the transport of entities through the system.

Consider a scenario in which composable conveyors are used in a warehouse or packaging application. There are a number of input docks where entities arrive at different time, perhaps with different distributions. These entities must be transported to specific output docks while respecting QoS constraints such as end-to-end latency, throughput and jitter. The structure of the conveying system must change dynamically to adapt to the changing input arrivals of entities or the QoS requirements of different kinds of entities. While predictable real-time performance in such a scenario is feasible [38], the burden of implementing the protocols and/or middleware to achieve this performance lies on the users. Thus, the user of the conveyor system is saddled with the burden of deploying and maintaining a system with dynamically changing topologies either because of the need to cope with failures or because of changing demand.

CCS offers a simpler approach to achieving reconfigurable and dynamically changing topologies. The mobile units that move periodically on their tracks constitutes the physical infrastructure on which virtual (composable and reconfigurable) conveyor systems can be realized. Since the design, operation and maintenance of these units is relatively easy, these burdens are transfered from the user of the conveyor system to the infrastructure provider. The configuration of the CCS can be changed statically or dynamically to cope with failures and changing demands. As explained further in Section 5, several virtualization and model-transformation issues must be addressed to fully realize this mapping.

Disaster recovery or resource mobilization to a disaster zone is another popular CPS problem. In such a scenario, we can consider the horizontal tracks as the primary tracks on which resources or people can move. The vertical tracks are necessary to improve the QoS in the system. By mapping such an emerging situation to CCS, we can utilize the theoretical insights and foundations for CCS to improve the QoS for the disaster recovery situation.

In a similar manner, it is possible to map other problems such as traffic, noise or pollution monitoring in urban environments to CCS. While it is intuitively clear that many CPS problems involving material transport can be mapped to CCS, we do not yet have a precise characterization of the class of CPS problems that can be mapped in this manner.

4 Patterns of Rendezvous Events

Characterizing the patterns of rendezvous events that can occur in CCS is an interesting problem. Figure 2 illustrates such a pattern in a CCS where $N = M = 4$, $\delta_j^r = \delta_i^c = N - i$, and every unit moves one step in each tick. Here, the offsets for each unit are $\delta_1^r = \delta_1^c = 3$, $\delta_2^r = \delta_2^c = 2$, $\delta_3^r = \delta_3^c = 1$, and $\delta_4^r = \delta_4^c = 0$. Consequently, at $t = 1$, since only units v_4 and h_4 can move to the locations $(4, 1)$ and $(1, 4)$, respectively. At $t = 2$, v_4 and h_4 will advance one more step and units v_3 and h_3 will move one step away from their starting positions. Thus, the first rendezvous event occurs at time $t = 4$ when all the four pairs of units rendezvous at locations (i, i) as illustrated in Figure 2(b). No rendezvous event can occur at $t = 5$; notice the rendezvous events that occur at $t = 6, 7, \cdots, 12$. At $t = 14$, the same rendezvous pattern as the one that started at $t = 4$ repeats as long as the units do not fail.

Definition 2. *A* <u>*rendezvous event*</u> *(or rendezvous) is a spatio-temporal event that occurs when two units v_i and h_j arrive at the grid location (i, j) at the same time, t. It can be represented as a binary valued function:*

$$r_{i,j}^t = f(i, j, \delta_j^r, \delta_i^c, t) \qquad (1)$$

where, $1 \leq i \leq N$ and $1 \leq j \leq M$.

$r_{i,j}^t = 1$ signifies that a rendezvous event occurred at location (i, j) at time t and a value 0 denotes that no rendezvous event occurred at that location and time.

4.1 Period of the Pattern of Rendezvous Events

Note that because there are $(M + 2)$ lattice points in each row of the grid, the number of steps that a horizontal unit can take is $M + 1$; similarly, $N + 1$ is the number of steps a vertical unit can take.

Observation 1. *Given a CCS with M horizontal units N vertical units, the pattern of rendezvous events repeat with a period that is a factor of $2 \times lcm(M + 1, N + 1)$.*

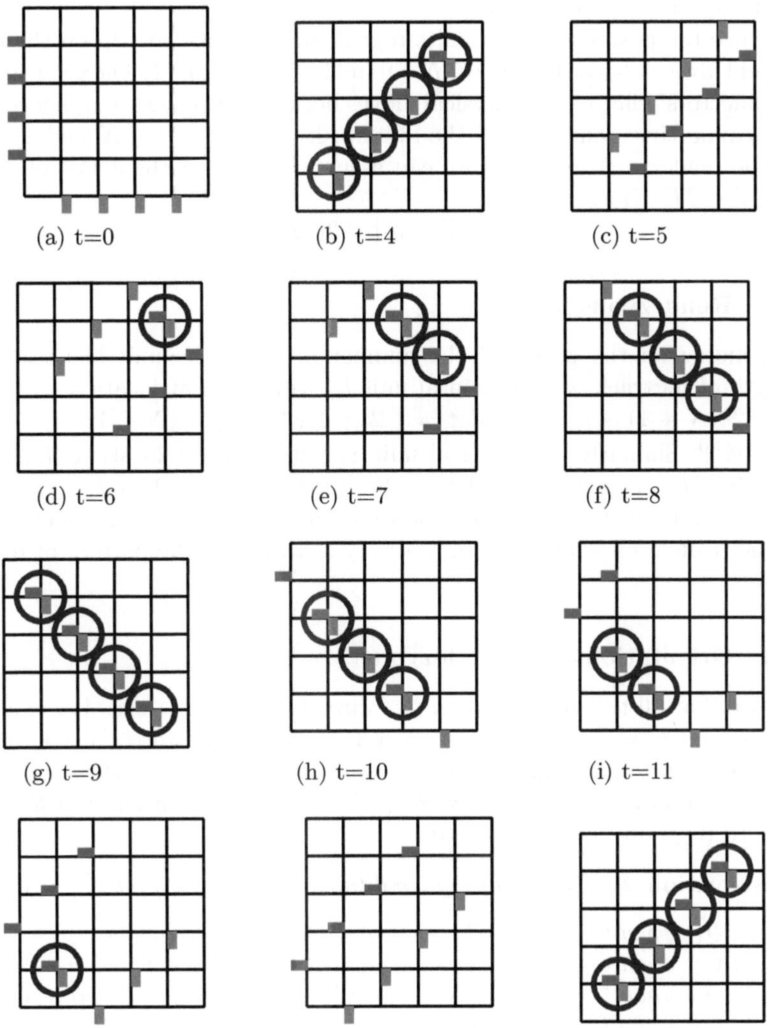

(a) t=0 (b) t=4 (c) t=5

(d) t=6 (e) t=7 (f) t=8

(g) t=9 (h) t=10 (i) t=11

(j) t=12 (k) t=13 (l) t=14

Fig. 2. Rendezvous Pattern for $N = M = 4$ and $\delta_j^r = \delta_i^c = N - i$

After $2\times lcm(M+1,N+1)$ time ticks have elapsed, all the horizontal and vertical units would have returned to their initial positions, after accounting for each of their offsets. For this reason, any rendezvous patterns that occurred in the time interval $[0, 2 \times lcm(M+1, N+1)]$ will repeat. Depending on the relationship between M and N the pattern of rendezvous events may have interesting sub patterns with shorter periods.

Since the positions of the units repeat after one round trip on the track, we should always choose $0 \le \delta_j^r \le 2N+1$ and $0 \le \delta_i^c \le 2M+1$. The vertical units are "connectors" that enable physical moves between the tracks and, hence, improve the connectivity and QoS of the system. Because $1 \le i \le N$ and $1 \le j \le M$ in the rendezvous zone, and we can assume $N < M$ without loss of generality, $|i-j| < M$. Thus,

$$|\delta_j^r - \delta_i^c| < 2(M+1). \tag{2}$$

4.2 Rendezvous Times

Depending on the parity of the number of times the unit has traversed the length of its track, the horizontal unit h_j will arrive at location (i,j) at times $t = 2k(N+1) + i + \delta_j^r$ and $t = (2k+1)(N+1) + (N+1-i) + \delta_j^r$ where $k \in \mathbb{N}^{\ge 0}$. Similarly, the vertical unit v_i will arrive at location (i,j) at times $t = 2k'(M+1) + j + \delta_i^c$ and $t = (2k'+1)(M+1) + (M+1-j) + \delta_i^c$ where $k' \in \mathbb{N}^{\ge 0}$. These constraints on the times at which rendezvous events can occur is the scaffold on which we can precisely characterize the pattern of rendezvous events.

4.3 Orientations for Rendezvous

Notice that all the units move in two directions on the tracks. For example, (1) in $\mathcal{O}_1$: unit h_j is moving toward $(N+1,j)$ and v_i toward $(i,M+1)$, (2) in $\mathcal{O}_2$: unit h_j is moving toward $(N+1,j)$ and v_i toward $(i,0)$, (3) in $\mathcal{O}_3$: unit h_j is moving toward $(0,j)$ and v_i toward $(i,M+1)$, and (4) in $\mathcal{O}_4$: unit h_j is moving toward $(0,j)$ and v_i toward $(i,0)$.

Observation 2. *The following conditions are necessary for* $r_{i,j}^t = 1$*:*

1. *In* $\mathcal{O}_1$*: when* $t = 2k(N+1) + i + \delta_j^r = 2k'(M+1) + j + \delta_i^c$.
2. *In* $\mathcal{O}_2$*: when* $t = 2k(N+1) + i + \delta_j^r = (2k'+1)(M+1) + (M+1-j) + \delta_i^c$.
3. *In* $\mathcal{O}_3$*: when* $t = (2k+1)(N+1) + (N+1-i) + \delta_j^r = 2k'(M+1) + j + \delta_i^c$.
4. *In* $\mathcal{O}_4$*: when* $t = (2k+1)(N+1) + (N+1-i) + \delta_j^r = (2k'+1)(M+1) + (M+1-j) + \delta_i^c$.

4.4 Observations When $N = M$

Characterizing when $r_{i,j}^t = 1$ for the case when $N = M$ is important because by adjusting the velocities of the physical units, we can effectively realize the cases for $N = zM$, where $z \in \mathbb{N}^{\ge 0}$. A few observations without proofs are listed below:

Observation 3

1. *In orientation* $\mathcal{O}_1$, $r^t_{i,j} = 1$ *if and only if* $\delta^c_i - \delta^r_j = i - j$.
2. *If* $\delta^c_i - \delta^r_j = i - j$, *then* $r^t_{i,j} = 1$ *whenever* $t = 2k(N+1) + j + \delta^c_i = 2k(N+1) + i + \delta^r_j \; \forall k \in \mathbb{N}^{\geq 0}$.
3. *If* $\delta^c_i - \delta^r_j = i - j$ *for all* $1 \leq i, j \leq N$, *then* $r^t_{i,j} = 1$ *for some* (i, j) *at all* t.

Determining the time ticks at which rendezvous events can occur is necessary to design algorithms that plan the needed cyber transfers. In general terms, the objective is to find a temporal order of cyber transfers that are necessary to successfully transport entities from their inputs to their outputs while satisfying their QoS constraints.

4.5 Input-Output Connectivity for Transport

The pattern of rendezvous events can be represented as a bipartite multigraph in which the two partite sets are the horizontal units and the vertical units [40] as shown in Figure 3. To simplify the illustration, multiple edges between a pair of nodes are not illustrated in the figure; such edges can be interpreted from the accompanying adjacency matrices. Since $N = M = 4$, the period $P = 10$. The weights of the edges shown in the adjacency matrices are the times in which the corresponding units rendezvous. In configuration 1, $\delta^c_i = \delta^r_j = N - i$; for

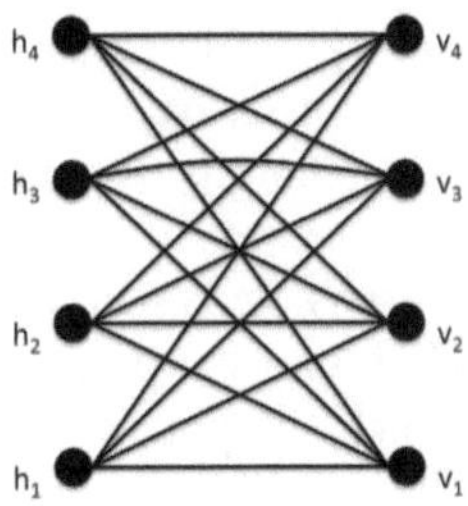

	v_1	v_2	v_3	v_4
h_4	9	8	7	4,6
h_3	10	9	4,8	7
h_2	11	4,10	9	8
h_1	4,12	11	10	9

(a) Configuration 1

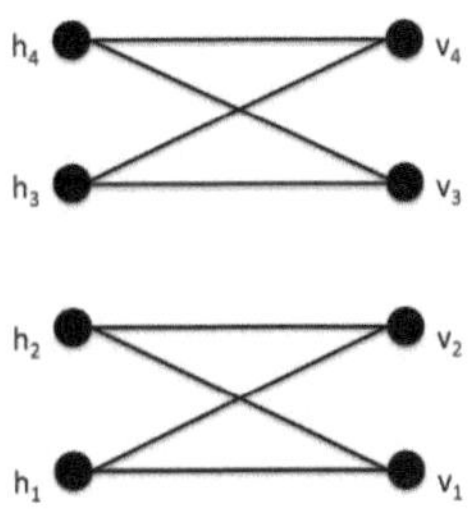

	v_1	v_2	v_3	v_4
h_4			4	5,7
h_3			3,7	4
h_2	9	2,8		
h_1	2,10	9		

(b) Configuration 2

Fig. 3. Rendezvous Graphs for two configurations of the CCS in Figure 2

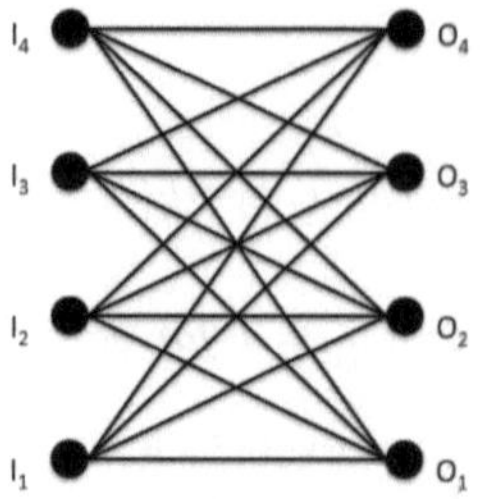

	O1	O2	O3	O4
I4	18	17	16	5
I3	17	16	5	14
I2	16	5	14	13
I1	5	14	13	12

(a) Configuration 1

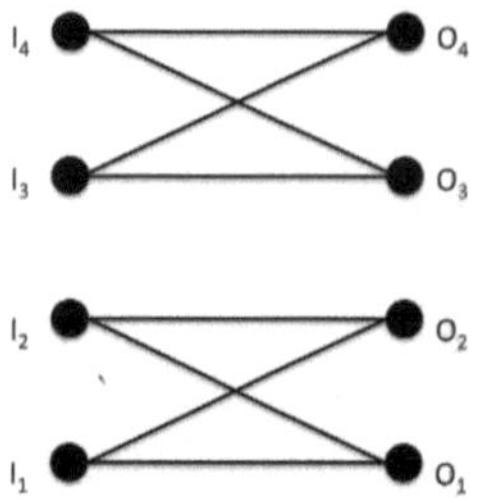

	O1	O2	O3	O4
I4	Inf	Inf	14	5
I3	Inf	Inf	5	6
I2	16	5	Inf	inf
I1	5	14	Inf	inf

(b) Configuration 2

Fig. 4. IO Connectivity Graphs for two configurations of the CCS in Figure 2

this reason, units h_1 and v_1 have an initial delay of 3. To account for this initial offset, the entry (1, 1) implies that $r^{4}_{1,1} = 1$ and $r^{12}_{1,1} = 1$. The other entries can be interpreted in a similar manner. In configuration 2, $\delta^r_2 = \delta^c_2 = \delta^r_3 = \delta^c_3 = 0$. The four other units have initial offset 1. Notice that in this configuration, all the units do not rendezvous with each other.

In the approach described fully in [40], we show how the adjacency matrix of the rendezvous graph can be obtained as a Hadamard product of the position matrices of the conveying units in the CCS. Since the position matrices can be obtained using a simple linear time (in P), the rendezvous matrix can also be obtained algorithmically. This pattern of rendezvous events affects the end-to-end transport that can be achieved in the CCS.

Figure 4 illustrates two Input-Output Connectivity graphs for the two configurations of CCS discussed above. This is a simple bipartite graph in which the two partite sets are the inputs and the outputs of the CCS. An edge in this graph means that entities can be transported from the corresponding input to the output. The weight of the edge (shown in the accompanying adjacency matrix) is the minimum latency through the system for this end-to-end transport. As described in [40], the end-to-end latencies can be obtained by executing a standard shortest-path algorithm on a graph that is derived from the line graph of the rendezvous graph illustrated in Figure 3.

5 Relevance of CCS for Addressing Some CPS Challenges

In this section, we highlight how techniques for model-based engineering, resource virtualization, composition and abstraction can be investigated in the context of CCS. In addition, we briefly outline a suite of real-time protocols that are necessary to fully realize a CCS testbed to facilitate CPS investigations and education [41].

5.1 Model-Based Engineering and Virtualization

Model-based approaches are important for realizing systems that are correct by construction. The approach in [42] offers a Domain-Specific Modeling Language to specify conveyor systems at a high-level. These specifications are transformed to code that executes on a simulator and, thus, enables users to explore performance of the system in various operational scenarios. CCS is a new target platform for such a toolset. Instead of simulating the desired conveyor systems, the tools can now map the (virtual) conveyor systems specified at a high level to the "cloud" of mobile units in a CCS.

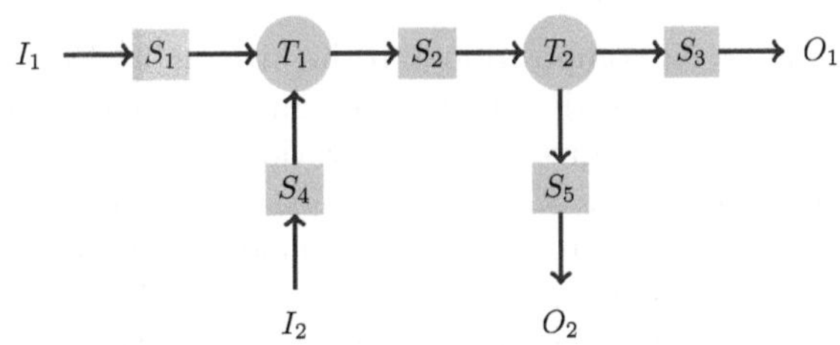

Fig. 5. Reconfigurable and Composable Conveyor System Topology

Figure 5 illustrates a virtual conveyor system that has two inputs and two outputs. Such a conveyor system may be used for package or material sorting; in such a case, the turn T_2 must incorporate an appropriate sensor (e.g., weight, color, or size) to detect the kind of entity and make a decision whether it should leave the system via O_1 or O_2. Such a conveyor system can be mapped to a CCS as follows: first, we extract the CPS-Tasks that are performed in the virtual conveyor system, including their QoS constraints. Next, we must bind the inputs and outputs of the virtual conveyor system to the inputs and outputs of the CCS. Finally, we must establish a configuration of the CCS so that the pattern of rendezvous events that can occur will enable us to schedule the CPS-Tasks extracted from the virtual conveyor system. A technique such as the one

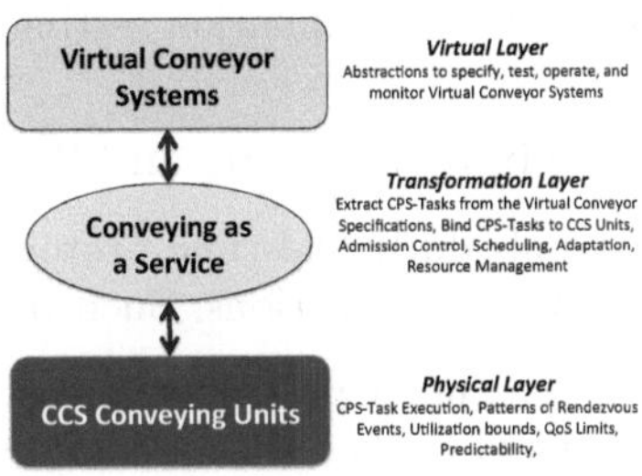

Fig. 6. Resource Virtualization in CCS

described in [40] can be used to identify the physical moves and cyber transfers that are necessary to achieve the transport requirement of each CPS-Task.

Thus, to map virtual conveyor systems to CCS, a natural three-tier architecture illustrated in Figure 6 emerges. CCS provides elastic conveying resources that can be used as a service. The top layer is a virtual conveyor systems layer, the middle layer is a transformation layer, and the bottom layer is the physical resources as illustrated in Figure 6.

Questions such as which virtual conveyor system can and cannot be realized in a given CCS configuration, how to support multiple virtual conveyor systems simultaneously, and what QoS can be achieved for a new virtual conveyor system request given the current load on CCS are all interesting problems that can be addressed in the well-defined context of CCS.

5.2 CPS Composition and Abstraction

Techniques for composition and the abstractions that are important in a variety of CPS to enable dynamic reconfiguration and improve the responsiveness to failures. In CCS, given a configuration and a collection of CPS-Tasks that are already scheduled, accepting the execution of new CPS-Tasks is one form of composition. Dynamically adding/deleting units, or initiating/stopping one or more of the mobile units in response to emerging needs or failures is another form of composition. When the configuration of the system is altered, the pattern of rendezvous events and, hence, the QoS that can be achieved will change. The graph theoretic formalism in [40] is one framework in which techniques for composition can be investigated. Other techniques based on number theory, game theory, or timed/hybrid automata[2] can also be investigated.

New abstractions can be designed to consider the physical characteristics and constraints of the units and synthesize interaction behaviors i.e., patterns of rendezvous events. For example, given a fixed capacity for the motors that move the units, users can develop simple models to account for the weight of the entities being moved; in such a scenario, the analysis of the patterns of rendezvous events must account for the continuous dynamics of the mobile units and the finite-state behaviors of entity transfers between units. CCS provides a precise context with well-defined behaviors to investigate composition and mechanisms for abstraction. Moreover, the configuration of CCS can be systematically changed to increase the scale and complexity of the problem in order to fully validate the techniques both in a simulation setting and in a testbed.

5.3 Real-Time Protocols and Techniques

Networked embedded real-time protocols and services are critical for realizing a variety of CPS applications. Such protocols must be implemented in simula-

[2] The periodicity of the pattern of rendezvous events outlined in Section 4 holds when the velocities of the units are either integers or rational values. When the velocities are arbitrary real values, we will need to define a rendezvous event as occurring in an open set, e.g, an ϵ neighborhood around the lattice points in the rendezvous zone.

tors and prototype systems to obtain an operational instance of CCS. While at the system-level the conveyor systems topologies supported are virtual, these virtual systems must be realized by coordinating the local operations of the mobile conveyor units. In addition to utilizing novel operating systems such as nano-RK [43], there is a need to design robust protocols for collision avoidance, rendezvous detection, global snapshots, alarm and exception condition reporting, data gathering, and in-network decision-making. The QoS targets for such techniques can be scaled by increasing or decreasing the velocity of the units and their mobility patterns.

From a networked systems standpoint, CCS represents a contemporary network problem that involves multiple timescales i.e., electronic messaging vs. entity transport, local cyber transfers vs. end-to-end transport of entities, intertwined cyber-physical dynamics, and dynamic adaptation. The the QoS metrics in CCS depend on the total delay in transporting entities which in turn are affected by the current traffic along various paths in CCS. Hence, as in any such networked transportation problem, issues such as admission control, congestion control etc. become important. Given periodic or stochastic input arrivals, ideally one wants design techniques that would decide on the number of vertical units and their offsets and the routing techniques so that the performance requirements can be met. In addition, or because such a design is difficult we may also have an admission controller which can decide the rate at which entities are allowed to enter the system at the input ports so as to avoid congestion. There are many interesting possibilities of simple adaptive controllers in such conveyor applications [39]. There are also other interesting ways of formulating congestion control and routing. In CCS, the vertical units are there to create sufficiently many rendezvous so that all units can transfer to the required horizontal units. In this sense, this represents an ad hoc network where the formation of links is time-dependent and is controlled at a gross level by choosing specific offsets for the vertical units. The entities can use some protocols of opportunistic transfers when rendezvous occur. Now the performance of the system is essentially dependent on choosing the offsets needed. For each vertical unit, there are only finitely many offsets possible. Thus for each vertical unit, we can have a controller which has to learn to choose one of the possible offsets based on some noisy performance evaluation in terms of some measured indices of congestion or delays etc. Such learning problems are classical and are receiving attention today because of their relevance in many machine learning problems [44,45]. The CCS will provide a challenging and useful application problem to explore such learning techniques.

In addition to addressing the above challenges, CCS may be useful for exploring verification and validation techniques, planning techniques, and techniques for cyber trust and security.

6 Versatility of the Cloud Conveyor System

The following are some of the variations of CCS that alter the cyber-physical dynamics. This versatility of CCS, which allows one to change the scale and complexity of the problem without increasing the implementation burden, is useful to improve CPS Education and Training.

1. **Easy Variants:** Some easy variants of CCS are (a) limiting the orientations of rendezvous, (b) velocities of the units — i.e., different velocities for the mobile units, and time-variant velocity profiles, (c) allowing arbitrary locations for inputs and outputs, (d) allowing units to carry multiple entities, (e) allowing multiple units per track, and (f) using typed entities.
2. **Stop and Go Model:** Suppose we allow the units to stop and wait at a location for a desired rendezvous, how will it affect the QoS of the system and the utilization? This is a fundamental change to the dynamics of the system and is likely to have important ramifications because the units must balance their current workload with a predicted load in the future in making cyber transfers.
3. **Effects of Failures:** What happens to the patterns of rendezvous events when one or more units fail? If a new unit on a different track can be started, what initial offset should we use? What are some configuration changes that one can make to cope with failures? Techniques for adaptations necessary to restore QoS and effects of dependent failures can also be explored.

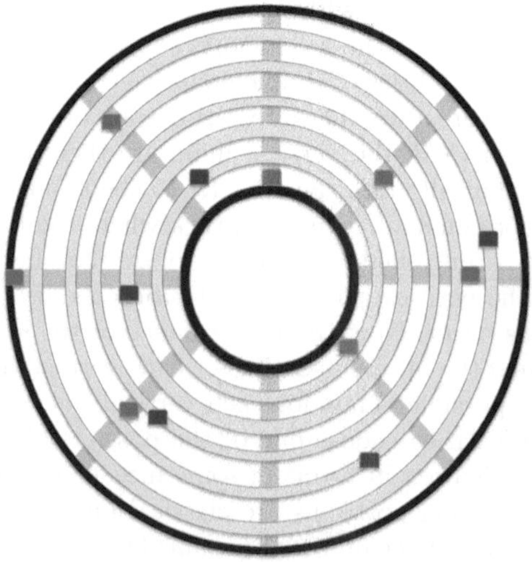

Fig. 7. CCS with Circular and Radial Tracks

4. **Changing Lattice Points and Tracks:** How would the patterns of rendezvous events change when the lattice points and tracks change? Figure 7 shows a CCS with circular and radial tracks. The inputs may be in the center of the concentric tracks and the outputs can be located at the other end of the radial tracks. In such a case, CCS can be used to capture the transport requirements in a disaster recovery scenario, alternatively if the inputs

and outputs are reversed, this CCS may now represent the scenario when resources are being mobilized for an affected region (i.e., in the center) from multiple sources. The characterization of the patterns of rendezvous events in this and other variants of CCS where the lattice points and tracks change is likely to offer new insights into a class of real-world systems.

7 Conclusions

We introduced a versatile reference application CPS called the *Cloud Conveyor System* (CCS). This system comprises a collection of mobile units that have simple periodic behavior. The objective for the system is to transport entities from some Inputs to a target Output when each entity has its own target Output, deadline and QoS constraints. Entities move through the system by riding on the mobile units and transferring between units to reach their destination.

The CCS allows us to formulate a model for a CPS-Task that evolves both in space and in time. The intertwined dynamics of the cyber and physical activities and the rich structure of interactions between the mobile units makes this model of a CPS-Task attractive. The well-defined structure of the problem enables one to systematically understand critical issues such as scheduling, resource virtualization, composition and abstractions.

CCS can be realized in simulators and testbeds in a straightforward manner. When realized in testbeds, CCS is a unique platform for research and education in networked embedded real-time systems. By adjusting the configuration of the CCS, one can alter the structure of the networked system dynamically; however, the basic model of the CPS-Task — as a sequence of cyber transfers and physical moves — remains consistent through such changes and adaptations. The scale and complexity of the simulation and the testbed can be increased without increasing the implementation burden and, hence, CCS is well-suited to rigorously validate new techniques and protocols for CPS.

References

1. Rajkumar, R., Lee, I., Sha, L., Stankovic, J.: Cyber-physical systems: The next computing revolution. In: ACM/IEEE Design Automation Conference (2010)
2. Anderson, C.W.: Learning to control an inverted pendulum using neural networks. IEEE Control Systems Magazine 9(3), 31–37 (1989)
3. Rapoport, A., Chammah, A.M.: Prisoner's dilemma. University of Michigan Press (1965)
4. Chandy, K., Misra, J.: The drinking philosophers problem. ACM Transactions on Programming Languages and Systems 6(4), 632–646 (1984)
5. Herescu, O., Palamidessi, C.: On the generalized dining philosophers problem. In: Proceedings of the Twentieth Annual ACM Symposium on the Principles of Distributed Computing, pp. 81–89 (2001)
6. Liu, C., Layland, J.: Scheduling algorithms for multiprogramming in a hard-real-time environment. ACM Journal 20(1), 46–61 (1973)

7. Buttazzo, G.: Hard Real-time Computing Systems: Predictable Scheduling Algorithms and Applications. Springer (2005)
8. Lee, I., Leung, J.T., Son, S.: Handbook of Real-Time and Embedded Systems. CRC Press (2007)
9. Li, H., Baruah, S.: An algorithm for scheduling certifiable mixed-criticality sporadic task systems. In: IEEE Real-Time Systems Symposium, pp. 183–192 (2010)
10. Bertogna, M., Buttazzo, G., Yao, G.: Improving feasibility of fixed priority tasks using non-preemptive regions. In: IEEE Real-Time Systems Symposium, pp. 251–260 (2011)
11. Bertogna, M., Baruah, S.: Limited preemption edf scheduling of sporadic task systems. IEEE Transactions on Industrial Informatics 6(4), 579–591 (2010)
12. Bertogna, M., Xhani, O., Marinoni, M., Esposito, F., Buttazzo, G.: Optimal selection of preemption points to minimize preemption overhead. In: Euromicro Conference on Real-Time Systems, pp. 217–227 (2011)
13. Niemeier, M., Wiese, A., Baruah, S.: Partitioned real-time scheduling on heterogeneous shared-memory multiprocessors. In: Euromicro Conference on Real-Time Systems, pp. 115–124 (2011)
14. Hou, I., Kumar, P.: Scheduling periodic real-time tasks with heterogeneous reward requirements. In: IEEE Real-Time Systems Symposium, pp. 282–291 (2011)
15. Kim, T., Poor, H.: Scheduling power consumption with price uncertainty. IEEE Transactions on Smart Grid 2(3), 519–527 (2011)
16. Junsung, K., Karthik, L., Raghunathan, R.: Rhythmic Tasks: A New Task Model with Continually Varying Periods for Cyber-Physical Systems. In: 2012 IEEE/ACM Third International Conference on Cyber-Physical Systems, pp. 55–64. IEEE (April 2012)
17. Saifullah, A., Agrawal, K., Lu, C., Gill, C.: Multi-core real-time scheduling for generalized parallel task models. In: IEEE Real-Time Systems Symposium, pp. 217–226 (2011)
18. Baruah, S.: The partitioned edf scheduling of sporadic task systems. In: IEEE Real-Time Systems Symposium, pp. 116–125 (2011)
19. Bastoni, A., Brandenburg, B., Anderson, J.: Is semi-partitioned scheduling practical? In: Euromicro Conference on Real-Time Systems, pp. 125–135 (2011)
20. Branicky, M.: Introduction to hybrid systems. In: Hristu-Varsakelis, D., Levine, W. (eds.) Handbook of Networked and Embedded Control Systems, pp. 91–116. Birkhauser, Boston (2005)
21. Kavraki, L., Švestka, P., Latombe, J., Overmars, M.: Probabilistic roadmaps for path planning in high-dimensional configuration spaces. IEEE Transactions on Robotics and Automation 12(4), 566–580 (1996)
22. LaValle, S., Branicky, M., Lindemann, S.: On the relationship between classical grid search and probabilistic roadmaps. International Journal of Robotics Research 23(7-8), 673–692 (2004)
23. Karaman, S., Frazzoli, E.: Sampling-based algorithms for optimal motion planning. International Journal of Robotics Research 30(7), 846–894 (2011)
24. Bhatia, A., Maly, M., Kavraki, L., Vardi, M.: Motion planning with complex goals. IEEE Robotics & Automation Magazine 18(3), 55–64 (2011)
25. Branicky, M., Curtiss, M., Levine, J., Morgan, S.: RRTs for nonlinear, discrete, and hybrid planning and control. In: Proc. IEEE Conference on Decision and Control, Lahaina, HI (December 2003)

26. Branicky, M., Curtiss, M., Levine, J., Morgan, S.: Sampling-based planning, control, and verification of hybrid systems. IEE Proceedings on Control Theory and Applications 153, 575–590 (2006)
27. Choset, H., Lynch, K., Hutchinson, S., Kantor, G., Burgard, W., Kavraki, L., Thrun, S.: Principles of Robot Motion: Theory, Algorithms, and Implementations. MIT Press (2005)
28. LaValle, S.M.: Planning Algorithms. Cambridge University Press (2006)
29. Roozbehani, H., D'Andrea, R.: Adaptive highways on a grid. In: Pradalier, C., Siegwart, R., Hirzinger, G. (eds.) Robotics Research. STAR, vol. 70, pp. 661–680. Springer, Heidelberg (2011)
30. Enright, J.J., Wurman, P.R.: Optimization and coordinated autonomy in mobile fulfillment systems. In: Workshop on Automated Action Planning for Autonomous Mobile Robots, San Francisco, Workshops at the Twenty-Fifth AAAI Conference on Artificial Intelligence (2011)
31. Coltin, B., Manuela Veloso, R.V.: Dynamic user task scheduling for mobile robots. In: Workshop on Automated Action Planning for Autonomous Mobile Robots, San Francisco, Workshops at the Twenty-Fifth AAAI Conference on Artificial Intelligence (2011)
32. Klavins, E.: Automatic synthesis of controllers for distributed assembly and formation forming. In: Proc. IEEE International Conference on Robotics and Automation (2002)
33. Arsie, A., Frazzoli, E.: Efficient routing of multiple vehicles with no explicit communications. International Journal of Robust and Nonlinear Control 18(2), 154–164 (2007)
34. Bullo, F., Frazzoli, E., Pavone, M., Savla, K., Smith, S.: Dynamic vehicle routing for robotic systems. Proceeedings of the IEEE (2010)
35. Como, G., Savla, K., Acemoglu, D., Dahleh, M., Frazzoli, E.: Distributed robust routing policies for dynamical flow networks. IEEE Trans. Automatic Control (2010)
36. Pavone, M., Frazzoli, E., Bullo, F.: Adaptive and distributed algorithms for vehicle routing in a stochastic and dynamic environment. IEEE Transactions on Automatic Control (2010)
37. Smith, S., Pavone, M., Bullo, F., Frazzoli, E.: Dynamic vehicle routing with priority classes of stochastic demands. SIAM Journal Control and Optimization 48(5), 3224–3245 (2010)
38. Archer, B., Sastry, S., Rowe, A., Rajkumar, R.: Profiling primitives of networked embedded automation. In: IEEE Conference on Automation Science and Engineering (2009)
39. Cheerala, C., Sastry, S., Sastry, P.S.: Learning automata based online input-rate control for composable conveyor systems. In: Centenary Conference of the Electrical Engineering Department, Indian Institute of Science (2011)
40. Morris, T., Ghosh, A., Ferrara, M., Gethner, E., Branicky, M.S., Sastry, P.S., Sastry, S.: Characterizing spatio-temporal pattern of events in a cloud conveyor system. Unpublished Manuscript, The University of Akron (2013)
41. Gokhale, A., Biswas, G., Sarkar, N., Sastry, S., Branicky, M.S.: CPS-Laboratory-as-a-Service: Enabling technology for readily accessible and scalable CPS education. In: First Workshop on Cyber Physical Systems Education, CPSWeek (2013)

42. An, K., Trewyn, A., Gokhale, A., Sastry, S.: Model-driven Performance Analysis of Reconfigurable Conveyor Systems used in Material Handling Applications. In: Second IEEE/ACM International Conference on Cyber Physical Systems (ICCPS 2011), pp. 141–150. IEEE, Chicago (2011)
43. Eswaran, A., Rowe, A., Rajkumar, R.: nano-RK: An energy aware resource-centric operating system for sensor networks. In: IEEE Real - Time Systems Symposium (December 2005)
44. Thathachar, M.A.L., Sastry, P.S.: Networks of Learning Automata: Techniques for Online Stochastic Optimization. Kluwer Academic Press (2004)
45. Bubeck, S., Cesa-Bianchi, N.: Regret analysis of stochastic and nonstochastic multi-armed bandit problems (2012), `http://arXiv:1204.5721`

Part II

Foundations of Cyber-Security

Cyber-Attack Forecast Modeling and Complexity Reduction Using a Game-Theoretic Framework*

Malachi Jones, Georgios Kotsalis, and Jeff S. Shamma

Georgia Institute of Technology, Atlanta, GA 30332, USA
{kye4u,gkotsalis3,shamma}@gatech.edu

Abstract. The security community has placed a significant emphasis on developing tools and techniques to address known security issues. Some examples of this emphasis include security tools such as anti-virus software and Intrusion Detection Systems (IDS). This reactive approach to security is effective against novice adversaries (i.e. script kiddies) because they typically use off-the-shelf tools and popular techniques to conduct their attacks. In contrast, the innovative adversaries often devise novel attack vectors and methodologies that can render reactive measures inadequate. These pioneering adversaries have continually pushed the security frontier forward and motivate a need for proactive security approaches. A proactive approach that we pursue in this research is actionable cyber-attack forecasting. The objectives of actionable cyber-attack forecasting are to learn an attacker's behavioral model, to predict future attacks, and to select appropriate countermeasures. The computational complexity of analyzing attacker models has been an impediment to the realization of reliable cyber-attack forecasting. We address this complexity issue by developing adversary models and corresponding complexity reduction techniques. We then introduce a heuristic for learning behavioral models of potentially deceptive adversaries online. Last, we consider a capture-the-flag problem, formulate the problem as a cyber-security game with asymmetric information, and demonstrate how the models and techniques developed in this paper can be used to forecast a cyber-attack and recommend appropriate countermeasures.

Keywords: game theory, asymmetric information games, cyber-attack forecasting, belief compression.

1 Introduction

A reactive mindset is often the status-quo in the security community. Consider popular security products that include intrusion detecion systems (i.e. Snort) and anti-virus software as examples.[1] In the case of anti-virus (AV) software, the typical scenario is the following. First, new malware is developed by cyber hackers and then tested against popular versions of AV software tools.[2] *Testing malware against the AV tools virtually*

* This research was supported by ARO/MURI Project W911NF-09-1-0553.

[1] A consequence of reactive security tools is that once a threat has been detected, damage has been done and costs have been incurred.

[2] Websites such as `http://www.virustotal.com` can test malware against over 44 different AV tools. Although these sites are intended for security professionals, they can also be used by hackers.

D.C. Tarraf (ed.), *Control of Cyber-Physical Systems*,
Lecture Notes in Control and Information Sciences 449,
DOI: 10.1007/978-3-319-01159-2_4,

guarantees that new malware will initially go undetected. Next, the malware infects computing devices, and after some period of time has elapsed, which can range from days to years, AV signatures are developed by the AV vendors (i.e Symantec, Kaspersky, and McAfee).[3] This cat-and-mouse pattern then repeats itself and can be observed within and across the security community.

Reactive security methodologies are effective against novice adversaries because these adversaries typically use off-the-shelf tools and implement popular hacking techniques. In contrast, the pioneering adversaries are able to push both the cyber-hacking and the security frontiers forward by developing new malware, designing advanced hacking techniques and methodologies, and challenging security researchers and practitioners to match their ingenuity.[4] In order to combat these adept adversaries and also address advanced persistent threats (APTs), measures and approaches that are forward looking and predictive are needed. This is why we think proactive security is the next frontier in cyber security.

A proactive security approach that we consider in this research is actionable cyber-attack forecasting. The objectives of actionable cyber-attack forecasting are to learn an attacker's behavioral model, to predict future attacks, and to select appropriate countermeasures to prevent future attacks. Impediments that have prevented the realization of reliable cyber forecasting include, but are not limitied to, difficulty in modeling the adversary in an analytical framework and the computational complexity of analyzing the model to forecast attacks. We will address the computational complexity issues in this research.

Previous work to address forecasting challenges, where uncertainty exists about the capabilities of an attacker, include the work of Alpcan et al [1]. In this work, they model the interaction between attackers and an Intrusion Detection System (IDS) using a stochastic (Markov) game. The defender who operates the IDS has uncertainty about the attacker's intent. Tools that include value iteration are used to solve Markov Decision Processes. In [2], You et al. describe how to model cyber-security problems that consider the interactions between an attacker and a defender in a two player, zero-sum game. They illustrate how the Nash and Bayesian Equilibria can be used to predict the behavior of an attacker and to analyze the interactions between attacker and defender. You et al. suggest that linear programs could be used to solve these problems.

Similar to You et al., we also model cyber-security problems as zero-sum games in a Bayesian framework. We introduce computational methods to approximate solutions to these problems. A key feature of these methods is that the solutions can be computed by solving a linear program whose complexity is invariant with respect to the number of stages of the zero-sum game. These methods also have tight lower bounds on their performance that converge asymptotically to optimal with respect to the number of stages of the game.

[3] Flame malware that was discovered in May 2012 had been operating in the wild since February 2010.

[4] Malware developers have introduced polymorphic viruses that mutate the machine code of the virus after each execution. This polymorphism feature is designed to defeat AV tools that look for patterns in new viruses that match older versions of the viruses.

The outline of this paper is as follows. In Section 2, we will discuss iCTF2010, which is a cyber-security challenge problem that we will later formulate into a strategic game. We will then introduce zero-sum games with asymmetric information and discuss basic concepts and definitions in Section 3.[5] In Section 4, we will develop capture-the-flag (CTF) adversarial models and explore methods and techniques to reduce the complexity of the models. We will then formulate the CTF problem as a security game with asymmetric information in Section 5. Last, we will use simulations to demonstrate how the models and techniques developed in this paper can be used to learn the behavioral model of an adversary, to predict future attacks, and to launch appropriate countermeasures.

2 iCTF2010

2.1 Overview

Security researchers at the University of California Santa Barbara (UCSB) host a live capture-the-flag tournament each year [3]. *We will consider the 2010 version, called iCTF2010 in this research.* There are typically over 900 participants in this tournament, and the participants are hackers from the international community. The purpose of the tournament is to observe and analyze strategies and techniques of real hackers and collect datasets that can be used in security research projects. The iCTF is designed as an abstraction of real-word cyber-security scenarios. For instance consider as a target the security system at Georgia Tech. The various departments, (i.e. ECE, ME, etc.) are subsystems that have to run certain services such as ssh, smtp, citrix, in order to facilitate the computing needs of faculty, staff and students.[6] The attacker's objective is to successfully disrupt critical services based on partial information it receives about the operational state of the system.[7]

The iCTF challenge problem is abstracted as a controlled partially observable stochastic system and each subsystem is modelled as a Markov chain. The attacker however does not have full knowledge of the operational state and is forced to estimate it based on a subset of emitted signals that are correlated with the state transitions. Given those estimates it then chooses an appropriate action that corresponds to the critical services it disrupts. The objective of the attacker is to cause maximal disruption to the overall system.

In the proceeding section, we will present a formal description of our version of the iCTF2010 capture-the-flag problem. We have made some modifications to the original iCTF2010 problem and will present these modifications as well. *The original CTF formulation captures aspects of real-world cyber-security problems that include a dynamic*

[5] This game theoretic framework will be used later in the paper to model the capture-the-flag (CTF) problem as a security game.

[6] An operational state of each subsystem in this example can be defined as the services that are not offline due to maintenance. Maintenance could consist of security patch updates and/or server upgrades that effect a particular service.

[7] It will be assumed that each subsystem hosts its own services and is therefore not impacted by service disruptions in other subsystems.

security system whose behavior is at least partially observable and an adversary that can potentially learn and predict the system's behavior. Our modifications incorporate an additional characteristic of real-world cyber-security problems. This characteristic is a defender who initially has some uncertainty about the capabilities and behavior of an attacker, but who can potentially learn an attacker's capabilities and predict its behavior by using previous observations.

2.2 Model Description

2.3 Target System

The target system is abstracted as a discrete-time finite state, finite output Hidden Markov Model (HMM). We refer to the target system as T composed of N subsystems $T_{(1)}, T_{(2)}, ..., T_{(N)}$. The operational states of subsystem T_i are denoted by

$$\mathbb{A}_{(i)} = \{a_{(i)_1}, a_{(i)_2} \dots, a_{(i)_{n_i}}\}.$$

The state space of the target system is

$$\mathbb{A} = \mathbb{A}_{(1)} \times \mathbb{A}_{(2)} \times \dots \times \mathbb{A}_{(N)}.$$

The set of observation signals of subsystem T_i is denoted by

$$\mathbb{B}_{(i)} = \{b_{(i)_1}, b_{(i)_2} \dots, b_{(i)_{m_i}}\}.$$

The observation signal of the attacker at every instant is an unordered N-tuple of output symbols generated by the subsystems. Given the observation signal $\bar{B} = \{\bar{b}_1, \dots, \bar{b}_N\}$ at some instant, the attacker knows that there there exists an ordering of its elements, say $(\bar{b}_{\sigma(1)}, \dots, \bar{b}_{\sigma(N)})$, where $\sigma : \{1, \dots, N\} \to \{1, \dots, N\}$ is a bijection, such that

$$(\bar{b}_{\sigma(1)}, \dots, \bar{b}_{\sigma(N)}) \in \mathbb{B}_{(1)} \times \mathbb{B}_{(2)} \times \dots \times \mathbb{B}_{(N)}.$$

He may not know that particular ordering, but can always perform some probabilistic inference given the HMM abstraction. The output space of the target system is denoted by $\mathbb{B}$. After relabelling of the states and outputs we denote them by $\mathbb{A} = \{a_1, \dots, a_n\}$, and $\mathbb{B} = \{b_1, \dots, b_m\}$, where $n = \prod_{i=1}^{N} n_i$ and m is the number of unordered N-tuples of output symbols generated by the subsystems.

It is assumed that the state transitions of each subsystem are independent from each other. Let $\{(X_t, Y_t)\}_{t \in \mathbb{Z}_+}$ denote the state and output process of the given HMM, then the statistical description of the model is given by an initial distribution vector π, where

$$\pi_i = \mathbf{Pr}[X_t = a_i], \;\; i \in \{1, \dots, n\},$$

and a set of m transition matrices $\{M[y_1], \dots, M[y_m]\}$ where

$$M[y_k]_{ij} = \mathbf{Pr}[Y_{t+1} = b_k, X_{t+1} = a_i \mid a_t = x_j],$$

where $k \in \{1, \dots, m\}, i, j \in \{1, \dots, n\}$.

2.4 Attacker

The action set of the attacker is denoted by

$$\mathbb{S} = \mathbb{S}_{(1)} \times \ldots \times \mathbb{S}_{(N)}.$$

It corresponds to the set of services he disrupts in each subsystem. Associated with state action pairs is a payoff structure, $r_{\mu,\lambda} : \mathbb{X} \times \mathbb{S} \to \mathbb{R}$. This payoff structure reflects whether an attacker based on his/her information chose to disrupt a service that is relevant to the current operational state of the system. The subscript μ reflects the skill level of the attacker and λ is associated with the resources that the defender allocates to the system. The notion of probing[8] is relevant to the attacker's problem. The payoff incurred at each time step provides the attacker with additional information in estimating the current operational state. As such the attacker is faced with the problem of leveraging short term payoff versus obtaining more accurate information about the current state that will prove beneficial in the long run.

2.5 Defender

The defender's objective in iCTF is to minimize the cost of an attack by an adversary on the target system. Since an attacker's goal is to target critical services, the defender can allocate resources to protect these critical services from being disrupted. Let λ_j represent the amount of resources that the defender allocates to protecting service s_j. The likelihood of a successful attack on s_j decreases as λ_j increases. Deciding on how to allocate resources among the services can be challenging for the defender because of its uncertainty about an attacker's type. We discuss this in detail in Section 5.2 and offer an approach to address the issue.

3 Asymmetric Information Games

3.1 Overview

In a repeated zero-sum game, two players (defender and attacker) repeatedly play the same zero-sum game over several stages. We assume that while both players can observe the actions of the other, only the attacker knows the specific opponent he is playing against. Although the defender has uncertainty on the type of attacker it faces, it has a probability distribution of attacker types and can use its observation of the attacker's actions during game play to eventually learn the attacker's type. The dilemma faced by an attacker is how should he trade off the short-term reward by exploiting his private information versus long-term consequences resulting from revelation of his type.

Classic work by Aumann and Maschler [4] derives a recursive formula for the value of the game, which quantifies the exploitation tradeoff, and also derives the optimal policy for the attacker. Using Aumann and Maschlers' model for explicit computations of optimal policies is prohibitive for games with multiple stages. In [5], this computational

[8] The idea of optimal probing is an interesting topic that we would like to consider in future research.

issue is addressed by introducing methods to compute suboptimal strategies by solving linear programs whose complexity is constant with respect to the number of stages. The methods from [5] are discussed in Section 5.2.

3.2 Game Setup

Game Play. Two players repeatedly play a zero-sum matrix game over l stages. The attacker is the row player and maximizer, and the defender is the column player and minimizer. There are a finite set $\mathbb{K}$ of possible attacker types that the defender can face. A specific attacker is chosen from this set to play against the defender. Let S be the set of pure strategies of an attacker and similarly define J to be the set of pure strategies of the defender. The payoff matrix for an attacker of type k will be denote as $M^k \in \mathbb{R}^{|S| \times |J|}$. Before the initial stage $m = 1$, nature selects an attacker type according to a probability distribution $p \in \Delta(K)$, which is common knowledge. The outcome of this selection is not revealed to the defender. Once selected, the attacker's type remains fixed over all stages of the game.

Strategies. Mixed strategies correspond to distributions over pure strategies. Let $x_m^k \in \Delta(S)$ denote the mixed strategy of an attacker of type k at stage m. In repeated play, this strategy can be a function of the actions of both players during stages $1, ..., N$. Likewise, let $y_m \in \Delta(J)$ denote the mixed strategy of the defender at stage m, which again can depend on player actions over stages m=1,...,N. Let $x_m = \{x_m^1, ..., x_m^K\}$ denote the collection of mixed strategies for all attacker types and for all states at stage m, and $x = \{x_1, ..., x_N\}$ denote mixed strategies over all states and stages. Likewise, let $y = \{y_1, .., y_N\}$ denote the defender's mixed strategies over all stages.

Payoffs. Let $\gamma_m^p(x, y) = \sum_{k \in K} p^k x_m^k M^k y_m$ denote the expected payoff for the pair of mixed strategies (x, y) at stage m. The payoff for the n-stage game is then defined as

$$\bar{\gamma}_n^p(x, y) = \frac{1}{n} \sum_{m=1}^{n} \gamma_m^p(x, y) \tag{1}$$

3.3 Concepts and Definitions

Beliefs Since the defender is not informed of the attacker's type k, it can build beliefs on the type. These beliefs are a function of the initial distribution p of attacker types and the observed moves of an attacker. An attacker must therefore carefully consider its actions at each stage as they could potentially reveal its type to the defender. In order to get a worse case estimate of how much information an attacker transmits about its type through its actions, it models the defender as a Bayesian player and assumes that the defender knows its mixed strategy. The updated belief p^+ is computed as

$$p^+(p, x, s) = \frac{p^k x^k(s)}{\bar{x}(p, s)} \tag{2}$$

where $\bar{x}(p, s) := \sum_{k \in K} p^k x^k(s)$ and $x^k(s)$ is the probability that an attacker of type k plays pure action s.

Non-revealing Strategies. Revealing information is defined as an attacker selecting a mixed strategy that is dependent on his type k. From (2), it follows that a mixed strategy x_m at stage m does not change the current beliefs of the defender if $x_m^k = x_m^{k'}\ \forall\ k, k'$. As a consequences, an attacker who plays as if it is oblivious of its type, ensures that its opponents beliefs about its type do not change.[9]

An optimal non-revealing strategy can be computed by solving

$$u(p) = \max_{x \in \mathrm{NR}} \min_{y} \sum p^k x^k M^k y, \tag{3}$$

where $\mathrm{NR} = \{x_m \mid x_m^k = x_m^{k'}\ \forall k, k' \in K\}$ is the set of non-revealing strategies [11]. By playing an optimal non-revealing strategy at each stage of the game, an attacker can guarantee a game payoff of $u(p)$.[10]

Definition 1. *Let Cav$[u(p)]$ denote the point-wise smallest concave function g on $\Delta(K)$ satisfying $g(p) \geq u(p)\ \forall p \in \Delta(K)$*

Short-Term vs. Long-Term Payoffs. The dynamic programming recursive formula

$$v_{n+1}(p) = \max_{x_1} \min_{y_1} \left[\frac{1}{n+1} \sum_{k \in K} p^k x_1^k M^k y_1 + n \sum_{s \in S} \bar{x}_s v_n \Big(p^+(p, x_1, s) \Big) \right], \tag{4}$$

introduced by Aumann and Maschler [4], characterizes the value of repeated zero-sum games with asymmetric information. Note that n is a non-negative integer. When $n = 0$, the problem reduces to

$$v_1(p) = \max_{x_1} \min_{y_1} \sum_{k \in K} p^k x_1^k M^k y_1, \tag{5}$$

which is the value of the 1-shot zero-sum game.

A key interpretation of this formulation is that it also serves as a model of the tradeoff between short-term gains and the long-term informational advantage. For each decision x_1 of an attacker , the model evaluates the payoff for the current stage, which is represented by the expression $\sum_{k \in K} p^k x_1^k M^k y_1$, and the long-term cost for decision x_1, which is represented by $n \sum_{s \in S} \bar{x}_s v_n \Big(p^+(p, x_1, s) \Big)$.

It is worth pointing out that the computational complexity of finding the optimal decision x_1 can be attributed to the cost of calculating the long-term payoff. Since the long-term payoff is a recursive optimization problem that grows with respect to the game length, it can be difficult to find optimal strategies for games of arbitrary length. This difficulty is because the number of decision variables in the recursive optimization problem grows exponentially with respect to the game length. A revised formulation

$$\hat{v}_n(p) = \max_{x_1} \min_{y_1} \left[\frac{1}{n} \sum p^k x_1^k M^k y_1 + (n-1) \sum_{s \in S} \bar{x}_s u\Big(p^+(p, x_1, s) \Big) \right] \tag{6}$$

[9] In stochastic games, it is possible for the defender's beliefs about an attacker's type to change even if an attacker plays as if it is oblivious of its type.

[10] In [5], this idea of non-revelation was exploited to reduce the complexity of Aumann and Maschler's formulation.

was introduced in [5] to address a complexity issue of the recursive formulation of the value of the game. In this formulation, it is assumed that the informed player uses optimal non-revealing strategies (i.e. $u(p)$) for all future stages. Therefore the cost-to-go function $v_n(p)$ in (4) can be expressed as $u(p)$ in (6). As a consequence of the non-revealing assumption, the computational complexity remains constant with respect to the number of stages of the game.

Theorem 1. *[5] A perpetual policy improvement strategy can be computed by solving a linear program online at each stage of the game, and the computational complexity of the linear program is constant with respect to the number of stages of the game.*

In [5], lower bounds on $\hat{v}_n(p)$ were established. It was also shown that the lower bounds were tight, and it was proved that $\hat{v}_n(p)$ has asymptotic convergence to optimality with respect to the number of stages n.

Theorem 2. *[5] One-time policy improvement and perpetual policy improvement achieve Cav[u(p)] and the optimality bounds are*

$$Cav[u(p)] \leq \hat{v}_n(p) \leq v_n(p) \leq Cav[u(p)] + \frac{C}{\sqrt{n}} \sum_{k \in K} \sqrt{p^k(1-p^k)} \tag{7}$$

We will use the $\hat{v}_n(p)$ formulation in Section 5.2 to reduce the computational complexity of attacker models.

4 Attacker Modeling and Complexity Reduction

A basic adversarial model should addresses the following questions about a specific adversary:

1. What are its skills? **(Skillset/Capabilites)**
   Since each critical service may require specific technical skills to disrupt, what is the probability that an attacker can successfully disrupt service s_j?
2. What is its intent? **(Intent)**
   Is the ultimate goal of an adversary to prevent the success of the security system under consideration, or just to create general disruption?
3. How patient is the adversary? **(Patience)**
   Is the attacker greedy, or is it patient and willing to forgo an immediate gain in order to maximize its long-term payoff?
4. How does it build beliefs about the system's current state? **(Beliefs)**
   Is computing the system's belief function computationally prohibitive? If so, what technique does the adversary use to approximate the belief function?
5. How does it make decisions based on its state estimates? **(Strategies)**
   Given an estimate of the system, will an adversary disrupt critical services of the most likely operational state or disrupt services that maximize its expected payoff?

In the adversarial models we develop in this section, we will make assumptions that allow us to convey the main ideas of attacker modeling and complexity reduction techniques in a clear and accessible manner. These assumptions serve as intermediate steps that allow us to explore the prominent issues in modeling an adversary and also to consider possible approaches to address those issues. We will assume the following: 1) Available actions of the attacker and the probability distribution of attacker skill types are public knowledge. 2) Intents of the attacker are zero-sum. 3) The Attacker is greedy 4) Consider the worst case with respect to the attacker's computational ability (i.e. only prohibitive for the defender to computer the belief function of the system).

4.1 Capabilities

An attacker's type will be defined as his skill level at disrupting a set of services. The skill level will be represented as a vector, where the jth component of the vector represents the attackers ability to disrupt service j. The values of the skill vector are in the range between 0 and 1, where 1 is expert skill and 0 is no skill at disrupting service s_j.

4.2 Intent

The ultimate objective of the adversary may be unknown. Although many cyber hackers aim to profit from their attacks, other hacker groups such as Annonymous employ denial of service attacks to make political statements and seek publicity. Therefore, an approach that we use to address the uncertainty of the adversaries objective is to consider the worst case with respect to the defender. In particular, we model the problem as zero-sum (i.e. a reward α for the attacker is a corresponding cost α to the defender). This zero-sum assumption allows us to make performance guarantees on security policies.

4.3 Patience

One can model an adversary as having a discount factor λ on its future payoff. A discount factor of $\lambda \approx 0$ would then indicate a greedy adversary that heavily discounts the future, while $\lambda \approx 1$ would be indicative of an adversary that is a long-term player that heavily discounts the present. An alternative interpretation of the discount factor is the patience of the adversary. Modeling a patient adversary introduces a complication that is absent in the greedy models we use. This complication is probing.[11] Since a patient attacker may be willing to defer an immediate reward, it can consider choosing actions that may provide it will a better estimate of the current state of the security system. This improved estimate along with the adversaries knowledge of the HMM can then be used to make a better prediction of the future behavior of the system. The question that follows is when should it probe and when should it attack. This is an interesting question that we would like to consider in future research.

[11] We briefly discussed probing in Section 2.4.

4.4 Beliefs and Strategies

An adversaries decision to disrsupt a particular service s_j can be dependent on his ability to disrupt service s_j (i.e. his skill set), his payoff for disrupting service s_j, and his beliefs about the current operational state of each subsystem T_i. Computing the belief function of the current state of the system T is prohibitive. Consider the following example as an illustration. Suppose that a system T' is composed of N subsystems that each have 10 operational states. The size of the state space of the system T' is the product of the individual subsystems and is equal to 10^N, and the beliefs are probabilities of state combinations of the subsystems, e.g. $\Delta(10^N)$. We assume the worst case about the adversary, which is that he can compute the belief function of the system, while the defender may only be able to compute an estimate. To address the computational challenges of the defender, we introduce two techniques, quasi-beliefs and belief compression, that can be used to calculate estimates of the belief function of the system T.

Quasi-Beliefs. The main idea of quasi-beliefs (QB) is the following. Instead of computing the true beliefs of system T (e.g. $\Delta(10^N)$), we consider computing an estimate of the beliefs of each subsystem T_i independently of the other subsystems T_{-i}. An issue that arrises with computing independent beliefs of each subsystem is signal assignment. *Recall that the attacker only observes the collection of signals emitted from the subsystems. Therefore, he has uncertainty about the mapping between each signal and the subsystem that emitted the signal.* We will first introduce a method described in Algorithm 1 that provides us with an estimate of the likely mapping between signals and subsystems.

Algorithm 1. Signal assignment

1: **procedure** SIGNALASSIGN
2: **initialize matrix** P_S
3: **while** size$(P_S) > 0$ **do**
4: find the maximum element of P_S
5: let position of max element be denoted as (i^*, m^*)
6: assign signal y_{m^*} to subsystem T_{i^*}
7: remove row i^* and column m^* from matrix P_S
8: **end while**
9: **end procedure**

Note that for the matrix P_S at step 2 of Algorithm 1, each column corresponds to a signal y_m, each row corresponds to a subsystem T_i, and element (i, m) represents the probability that signal y_m was emitted from subsystem T_i. Also note that if there are more than one maximum element at step 4, we break the tie by randomly selecting a maximum element. We then use this signal assignement method in the quasi-belief greedy (QBG) algorithm (Algorithm 2) that is described below.

Algorithm 2. QBG Strategy

1: **procedure** QBGSTRATEGY
2: start with set of individual subsystem beliefs
3: **run** procedure SIGNALASSIGN
4: update individual beliefs using assignment
5: ***attack services with highest expected reward***
6: renormalize beliefs given success/failure of attack
7: **end procedure**

Belief Compression. The point of departure is the statistical description of the HMM abstraction of the target system. Let $\mathbb{A} = \{a_1, \ldots, a_n\}$, $\mathbb{B} = \{b_1, \ldots, b_m\}$ denote the state and output space respectively. The statistics of the joint state and output process $\{(X_t, Y_t)\}_{t \in \mathbb{Z}_+}$ are encoded by the initial distribution vector π, where

$$\pi_i = \mathbf{Pr}[X_t = a_i],\ i \in \{1, \ldots, n\},$$

and a set of m transition matrices $\{M[y_1], \ldots, M[y_m]\}$ where

$$M[y_k]_{ij} = \mathbf{Pr}[Y_{t+1} = b_k, X_{t+1} = a_i \mid X_t = a_j],$$

for $k \in \{1, \ldots, m\}$ and $i, j \in \{1, \ldots, n\}$. Let $\mathbb{B}^*$ denote the set of all emitted finite strings of observation signals including the empty sting $\varnothing$. Let $v = v_k \ldots v_1$ stand for a string of length k. Let $1_n \in \mathbb{R}^n$ denote the vector whose entries are all 1. Introduce the function

$$p : \mathbb{B}^* \times \mathbb{R}^n_+ \to [0, 1],$$

where

$$p[(v, \pi)] = 1_n^T\, M[v_k]\ \ldots\ M[v_1]\pi.$$

The function p is referred to as the probability function. It is used to compute the probability of observing a particular under initial distribution π, i.e.

$$p[(v, \pi)] = \sum_{i \in \{1, \ldots, n\}} \mathbf{Pr}[Y_k = v_k, \ldots, Y_1 = v_1 \mid X_o = a_i]\, \pi_i.$$

The value $p[(v, \pi)]$ is computed recursively using the rule

$$p[(v, \pi)] = 1_n^T\, H_k,$$

where $H_t = M[v_t]\, H_{t-1}$, $t \in \{1, \ldots, k\}$, and $H_0 = \pi$. Consider also the functions

$$p_{co} : \mathbb{Y} \times \mathbb{Y}^* \times \mathbb{R}^n_+ \to [0, 1], \quad p_{cs} : \mathbb{X} \times \mathbb{Y}^* \times \mathbb{R}^n_+ \to [0, 1],$$

referred to as the conditional output probability and conditional state probability function. The value $p_{co}[(b, v, \pi)]$ corresponds to the conditional probability of emitting the signal b given that the signal v has been observed under the initial distribution π, i.e.

$$p_{co}[(b, v, \pi)] = \sum_{i \in \{1, \ldots, n\}} \mathbf{Pr}[Y_{k+1} = b \mid Y_k = v_k, \ldots, Y_1 = v_1, X_0 = a_i]\, \pi_i.$$

Similarly the value $p_{cs}[(a, v, \pi)]$ corresponds to the conditional probability of being at state a given that the signal v has been observed under the initial distribution π, i.e.

$$p_{cs}[(a, v, \pi)] = \sum_{i \in \{1,\ldots,n\}} \mathbf{Pr}[X_k = a \mid Y_k = v_k, \ldots, Y_1 = v_1, X_0 = a_i] \, \pi_i.$$

The belief function is

$$\Pi : \mathbb{Y}^* \times \mathbb{R}^n_+ \to \mathbb{R}^n_+,$$

where

$$\Pi[v, \pi]_i = p_{cs}[(a_i, v, \pi)], \quad i \in \{1, \ldots, n\}.$$

The value of the belief function is computed recursively.

$$\Pi[v, \pi] = \frac{H_k}{1_n^T \, H_k}.$$

At every time step the attacker chooses an action to maximize his/hers instantaneous expected reward. For $s \in \mathbb{S}$ let $g[s] \in \mathbb{R}^n$ where $g[s]_i = r[s, a_i]$, $i \in \{1, \ldots, n\}$. Having observed the signal v and following a greedy strategy the attacker is faced with the optimization problem

$$\max_{s \in \mathbb{S}} < g[s], \Pi[v, \pi] > .$$

The notion of belief compression is associated with projecting the dynamics of the given HMM onto a lower dimensional manifold. Let $\hat{n} < n$, $V \in \mathbb{R}^{n \times \hat{n}}$, $U \in \mathbb{R}^{\hat{n} \times n}$ with $U\,V = I_{\hat{n}}$, so that $V\,U$ is a projection matrix. The parameters of a reduced complexity model are given by

$$\hat{c}^T = 1_n^T \, V, \quad \hat{b} = U \, \pi,$$
$$\hat{A}[y] = U \, M[y] \, V, \quad y \in \mathbb{Y}.$$

Using the reduced complexity model one can determine a greedy strategy while performing the relevant calculations on a $\hat{n}$ dimensional space with obvious computational and storage advantages. In particular consider the function

$$\hat{p} : \mathbb{Y}^* \times \mathbb{R}^n \to \mathbb{R},$$

where

$$\hat{p}[(v, \hat{b})] = \hat{c}^T \, \hat{A}[v_k] \, \ldots \, \hat{A}[v_1] \hat{b}.$$

The value $\hat{p}[(v, \hat{b})]$ is computed recursively using the rule

$$\hat{p}[(v, \hat{b})] = \hat{c}^T \, \hat{H}_k,$$

where $\hat{H}_t = \hat{A}[v_t] \, \hat{H}_{t-1}$, $t \in \{1, \ldots, k\}$, and $\hat{H}_0 = \hat{b}$. The function $\hat{p}$ is a low complexity surrogate for the probability function of the given HMM. Similarly consider the function

$$\hat{\Pi} : \mathbb{Y}^* \times \mathbb{R}^{\hat{n}} \to \mathbb{R}^{\hat{n}},$$

where

$$\hat{\Pi}[v, \hat{b}] = \frac{\hat{H}_k}{1_n^T \hat{H}_k}.$$

Let $\hat{g}[s] = V^T g[s]$, when employing the low complexity model the attacker is faced with the optimization problem

$$\max_{s \in \mathbb{S}} < \hat{g}[s], \hat{\Pi}[v, \pi] > .$$

In order to compute the compression matrix U and dilation matrix V we will employ the balanced truncation algorithm developed for HMM's in [7]. The reduction method is based on stable numerical linear algebra tools employing the singular value decomposition and is accompanied by an a priori bound to the approximation error. In other words it leverages the favourable features of Hankel norm based reduction techniques for linear time invariant systems.

First one solves linear algebraic equations to obtain "gramian like" quantities W_c, $Wo \in \mathbb{R}^{n \times n}$ where $W_c, W_o \succeq 0$,

$$W_o = \sum_{y \in \mathbb{Y}} M[y]^T \, W_o \, M[y] + 1_n^T \, 1_n, \; W_c = \sum_{y \in \mathbb{Y}} M[y] \, W_c \, M[y]^T + \pi \, \pi^T. \qquad (8)$$

Denote by L_o, L_c the Cholesky factors of $W_o = L_o^T \, L_o$ and $W_c = L_c \, L_c^T$ and consider the SVD of $L_c^T \, L_o^T$,

$$L_c^T \, L_o^T = \left[\Psi^{(1)} \, \Psi^{(2)} \right] \begin{bmatrix} \Sigma^{(1)} & 0 \\ 0 & \Sigma^{(2)} \end{bmatrix} \begin{bmatrix} \Xi^{(1)^T} \\ \Xi^{(2)^T} \end{bmatrix}.$$

where $\Sigma^{(1)} = \text{diag}[\sigma_1, \ldots, \sigma_{\hat{n}}]$, $\Sigma^{(2)} = \text{diag}[\sigma_{\hat{n}+1}, \ldots, \sigma_n]$ and $\sigma_1 \geq \ldots \geq \sigma_{\hat{n}} > \sigma_{\hat{n}+1} \geq \ldots \geq \sigma_n > 0$. In the above notation

$$V = L_c[\psi_1^{(1)} \frac{1}{\sqrt{\sigma_1}}, \ldots, \psi_{\hat{n}}^{(1)} \frac{1}{\sqrt{\sigma_{\hat{n}}}}], \; U = \begin{bmatrix} \frac{1}{\sqrt{\sigma_1}} \xi_1^{(1)^T} \\ \vdots \\ \frac{1}{\sqrt{\sigma_{\hat{n}}}} \xi_{\hat{n}}^{(1)^T} \end{bmatrix} L_o.$$

The following error bound controls the approximation of the given probability function:

$$\sqrt{\sum_{v \in \mathbb{Y}^*} (p[v, \pi] - \hat{p}[v, \hat{b}])^2} \leq 2(\sigma_{\hat{n}+1} + \ldots + \sigma_n).$$

The algorithm is demonstrated on a target system comprised from 3 subsystems used in the iCTF, with $|\mathbb{A}| = 1331$ and $|\mathbb{B}| = 680$. The following figure depicts the singular values $\sigma_1, \ldots, \sigma_{1331}$ that control the error bound. There is a clear cut-off behavior indicating that a choice of a reduced complexity model with 286 states is appropriate.

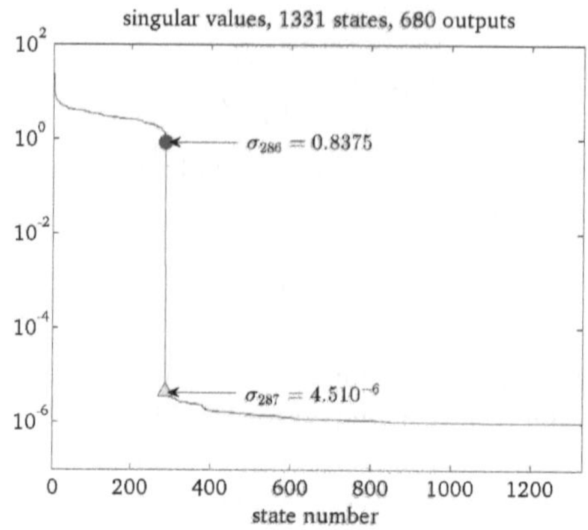

Next it is demonstrated that the reduced complexity model delivers a very accurate approximation to the conditional output probability function. The following figure depicts the vector $(p_{co}[b_1, v_k \dots v_1, \pi], \dots, p_{co}[b_{680}, v_k \dots v_1, \pi])$ conditioned on two trajectories $v_k \dots v_1$ of length 1000. The bottom row corresponds to the exact model and the rows above it correspond to approximations computed using the reduced order model. In both cases a reduced order model of at most 286 states approximates this conditional probability within 0.1%.

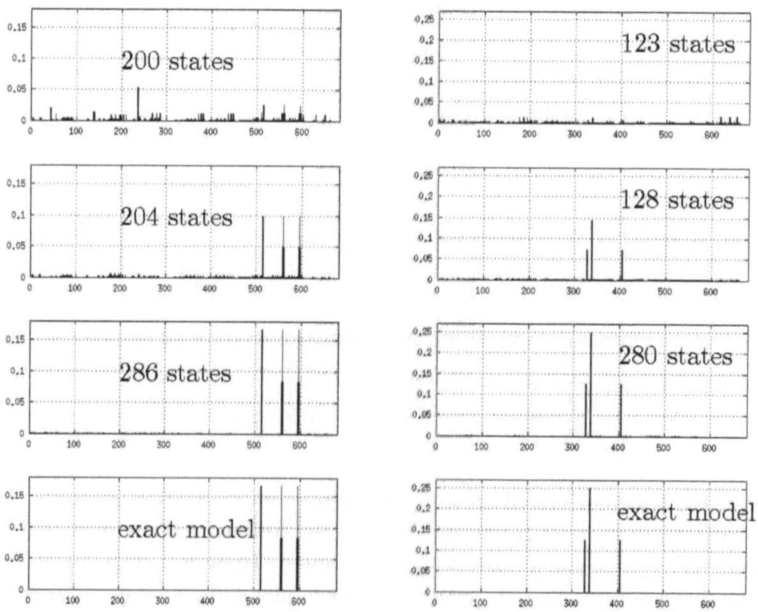

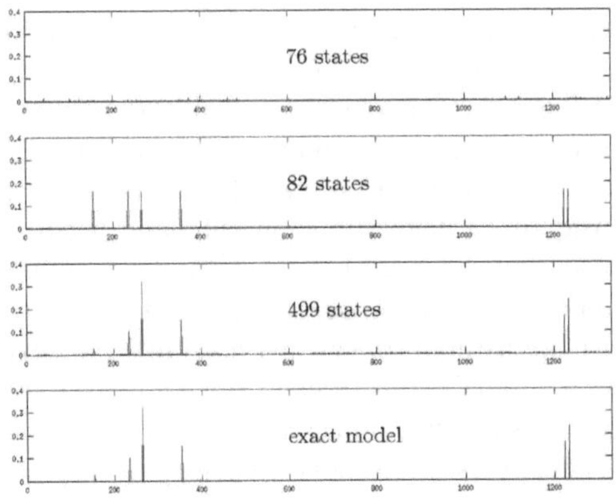

One can use the reduced order model also to compute an approximation to the belief function of the system and subsequently solve the attackers greedy optimization problem. As demonstrated in the next figure, a reduced order model of 499 states delivers a very accurate approximation to the belief function within 0.1% error. The belief function was computed for a trajectory $v_k \dots v_1$ of length 1000. The bottom row corresponds to the exact model and the rows above it correspond to approximations computed using the reduced order model. For the same trajectory greedy optimization using the reduced order model let to the same choice of action in 96% of the instances.

5 CTF Security Game Formulation

In this section, we will formulate iCTF as a security game with asymmetric information. Recall that these games were discussed extensively in Section 3. In our formulation of iCTF, we assume that only the attacker knows its opponents type. Therefore, the information asymmetry of the iCTF security game is on the side of the attacker. We will proceed by first discussing the game play. Next, we will discuss the strategies available to the players. Last, we will cover prominent issues that each player must consider in this security game and also discuss approaches to address those issues.

5.1 Game Play

One-Shot Game Formulation. The one-shot game formulation of iCTF consists of two players, an attacker and the defender. An attacker of type k is selected by nature before the start of the game from a known distribution p of attacker types.[12] The attacker's objective in this game is to maximize its reward for attacking the security system T that was defined in Section 2.3.[13] The attacker maximizes its reward by disrupting critical

[12] Attackers differ in their ability to disrupt services, and an attacker's type is defined by its skill vector.

[13] Conversely, the defender's objective is to minimize its costs incurred by the attacker.

services that are needed by the individual subsystems T_i of T. The attacker knows the specific services that are critical for each state of the subsystem. However it has uncertainty on the current state of each subsystem T_i.

An attacker can improve its estimate of each subsystem's state by computing the belief function of the system T. Since computing the belief function is computationally prohibitive for systems with large state spaces, the defender can approximate the true belief function by considering a state estimation technique discussed in Section 4.4. We assume that the defender uses the quasi-beliefs in this paper, although it could also use an alternative technique (i.e. belief compression). A particular greedy strategy that an attacker can consider is to disrupt services that maximizes the expected payoff for an attacker of its type; we will refer to this strategy as the honest strategy. Alternatively, an attacker can disrupt services that maximizes the payoffs of an attacker of another type; we will refer to this strategy as the dishonest strategy. It may seem irrational for an attacker to select a dishonest strategy in the one-shot game because the honest strategy gives him the better expected payoff. However, we will see in the next section that choosing a dishonest strategy can not only be rational, it can be optimal.

The defender has finite resources that it can allocate to maintain the availability of critical services. Dedicating more resources to a particular service s_j decreases the likelihood that service s_j will be disrupted in the event that an attacker targets that service. At the start of iCTF, it selects how these resources are allocated, and the resource allocation decision remains fixed until completion of iCTF. This resource allocation decision can be a function of the defender's belief about an attacker's type. In particular, it can chose to do a best response allocation with respect to an attacker's type or best respond to an attacker's type with some probability.

Repeated Game Formulation. We assumed in the one-shot formulation that once a QBG policy is chosen by the attacker and a resource allocation technique is chosen by the defender at the beginning of the game, the selections by both players remain fixed throughout the duration of iCTF. In the repeated game formulation, however, both players can reevaluate their decisions at specific intervals. We will refer to these intervals as stages. At each stage, the security system S is reset to its initial state. What isn't reset is each player's knowledge about the other player's past actions. In fact, each player's knowledge changes over time, affects its beliefs, and introduces a dynamic aspect to this security game.

5.2 Player Concerns

As we have discussed in the previous section, each player's knowledge is dynamic and changes in time. An important objective of the attacker is to control the beliefs of the defender about its type because knowledge of the attacker's type can allow the defender to make decisions that cost the attacker. Controlling the defender's beliefs often involves deceptive play by the attacker.[14] Because of the potential for deception,

[14] An attacker of type k can elect to select a QBG strategy of an attacker of a different type. (i.e. $QBG_{k'}$ where $k \neq k'$)

it can be difficult for the defender to learn the attacker's true type. In the preceding sections, we will discuss approaches for addressing each players concerns.

Attacker. Recall that in Section 3.3, we discussed a formulation introduced by Aumann and Maschler that allows us to model an attacker that optimally controls the beliefs of the defender and therefore attains the optimal game payoff. As part of the discussion, we mentioned complexity issues that arose with using this formulation when considering games with multiple stages. To address this complexity issue, we will use policy-improvement strategies presented in [5] to approximate the strategy selection of an optimal attacker. These policy-improvement strategies have error bounds on their performance with respect to optimal strategies, and the performance of the policy improvemement methods converge asymptotically to optimal with respect to n, the number of stages in the game. We will first discuss the one-time policy-improvement method and then preceed to discuss the perpetual policy improvement method.

In one-time policy improvement, an attacker strategizes for the *first stage* of the iCTF game while assuming that it will play in a non-revealing manner in all future stages. Perpetual policy improvement is an extension of one-time policy improvement. In the perpetual policy improvement method, the attacker strategizes for the current stage while assuming a non-revealing strategy in all future stages at *every stage*e. The perpetual policy improvement method involves solving an LP online, and the computational complexity of the LP is constant with respect tho the number of stages of the game. Below is the perpetual policy improvement algorithm.

Algorithm 3. Perpetual policy improvement

procedure PERPPOLICYIMPROVE
  initialize: set $p_1 = p$
  for $m = 1 \to N$ **do** ▷ Note that m is the current stage
    compute $\hat{x}_m$ by solving one-time policy improvement LP with p_m
    select a move s for attacker type k using mixed strategy $\hat{x}^k_m$
    update beliefs vector (i.e.$p_{m+1} = p^+(p, \hat{x}, s)$)
  end for
end procedure

Defender. Learning an attacker's true type can be challenging because it can play in a deceptive manner [8]. If the defender knew the mixed strategy x^k for each type of attacker k, then learning an attacker's type would be straightforward because the defender could follow the standard Bayesian update approach. Unfortunately, in the actual play of the game, the defender does not know each attackers mixed strategy as this information is private. Another approach that the defender can consider is solving a linear program to compute an optimal defensive strategy.[15] However, the complexity of the LP is exponential with respect to n, the number of stages of the game. We will introduce a payoff-based heuristic for learning an attacker's type that is computationally

[15] Ponssard and Sorin [13] showed that zero-sum repeated games of incomplete information can be formulated as linear programing problems to compute optimal strategies.

tractable for arbitrary n and only depends on the information that the defender has. This information is namely the history of the attacker's actions.

The main idea behind the payoff-based heuristic is as follows. The defender's belief of an attacker of type k will be correlated with the actual game payoff of the attacker. After each stage, the defender keeps track of what the overall game payoff would be for each type of attacker. The game payoff for an attacker of type k at stage n, given history h_n will be denoted $\tilde{\gamma}_n^k(h_n)$. The game payoff $\tilde{\gamma}_n^k(h_n)$ for an attacker of type k will be compared to the best possible payoff that a type k attacker can achieve. We will denote the best possible payoff by $|M^k|$, where $|M^k| := \max_{i,j} M_{i,j}^k$.[16] Let

$$\xi^k(h_n) = \frac{\tilde{\gamma}_n^k(h_n)}{|M^k|} \tag{9}$$

be a measure of the likelihood that an attacker is of type k given history h_n. The belief update procedure is then

$$\tilde{p}_{n+1}^k(h_n) = \tilde{p}_n^k \frac{\xi^k(h_n)}{\bar{\xi}(h_n)} \tag{10}$$

where $\bar{\xi}(h_n) = \sum_{k\in K} \tilde{p}_n^k \xi^k(h_n)$. To compute a best response strategy $\tilde{y}^*$ for the defender given the approximate belief $\tilde{p}_n$ at stage m, solve the optimization equation

$$\tilde{y}_m^* = \arg\min_{y_m} \max_{x_m} \sum_{k\in K} \tilde{p}_m^k x_m^k M^k y_m. \tag{11}$$

6 Simulation

6.1 Game Setup

As usual, this game consists of two players, an attacker and a defender. In this example, we assume that their are two types of attackers (i.e. type I and type II) and each attacker is uniquely defined by its skill vector. The probability distrubution of attacker types is uniform (i.e. $p^k = \frac{1}{2}$ for $k = 1, 2$), and there are two stages in this game. Matrix payoffs for the attacker types are

Type I

	BR_1	BR_2
QBG_1	23	375
QBG_2	−92	69

Type II

	BR_1	BR_2
QBG_1	−6	−28
QBG_2	128	−20

(12)

Note that an attacker of type I has the option of playing as his type by selecting QBG_I or playing deceptively by selecting QBG_{II}. Similarly, a type II attacker can opt to play either honestly or deceptively. The defenders available actions are to play a best-response resource-allocation strategy for a specific attacker type (i.e BR_I or BR_{II}.

[16] Without loss of generality, we will assume in this section that each matrix M^k is scaled with values ranging from 0 to 1.

6.2 Discussion

We will discuss the performance of four attacker strategies in this section. These strategies are dominant strategy, non-revealing strategy, one-time policy improvement, and perpetual policly improvement. In the one-shot game, the optimal strategy for an attacker is to behave as its type by selecting his domiant strategy. However, for games where $n > 1$, this can be a suboptimal strategy because it can reveal the attackers true type to the defender and cost the attacker the informational advantage. Specifically, in the two stage game, the attacker can achieve a better payoff by selecting the perpetual policy improvement strategy. For games where N is large, the dominant strategy has the worst performance out of the four strategies and the policy improvement strategies have the best performance.

Two-Stage Game. An optimal non-revealing strategy requires the attacker, regardless of his type, to play as a type I attacker with probability .70 and to play as a type II attacker with probability .30 at each stage. This strategy rewards the attacker with a payoff of 18 and has the worst performance of the four strategies in the two-stage game.[17] One-time policy improvement performs better by guranteeing an expected payoff of 27. The two strategies differ conceptually only at the first stage as how he plays when using a one-time policy improvement strategy is dependent on his type. A type I attacker plays deceptively at stage one with probability .92, while a type II attacker plays honestly with probability 1. At stage two, an attacker plays as a type II attacker with probability .96, which is independent of his type. An attacker that chooses to use his domimant strategy, which requires him to play honestly, at each stage of the game yields the attacker an expected payoff of 39, which outperforms the two previously mentioned strategies. Perpetual policy improvement yields the attacker the highest reward, 53, of the four strategies in consideration. At the first stage of perpetual policy improvement, an attacker plays the same way it would have played had it chosen one-time policy improvement. However, the key difference is at the second stage. Instead of playing non-revealing as with the former strategy, the attacker behaves as his type in the second stage of perpetual policy improvement.

***N*-Stage Game.** We discussed the performance of the four strategies in the two-stage case in the previous section and will now examine their performance as N grows large. The expected payoff for the dominant strategy converges asymptotically to 2 and has the worst performance of the four strategies for large N. This is because the defender can readily learn the attacker's type as the attacker does not play deceptively. The defender can then use this knowledge to select a resource allocation scheme that is a best response to his type. An optimal non-revealing strategy performs better than the dominant strategy because the defender is unable to learn any additional information about the attacker after observing his action at each stage. As a consequence, the defender has uncertainty on which resource alloction scheme will perform best against the attacker. An attacker who chooses this strategy can therefore guarantee a payoff of 18 at every stage. An immediate consequence of this guarantee is that an attacker can achieve a

[17] There are games where playing non-revealing is optimal for all n.

game payoff of 18 for games of any length. Policy improvement methods have the best performance of the four strategies for large N. Both methods have identical behavior and converge asymptotically to optimal and yields a payoff of 23. At stage one of the policy improvement methods, the attacker behaves deceptively with some probability that is type dependent. For all stages thereafter, the attacker plays a non-revealing strategy that is independent of its type.

References

1. Alpcan, T., Basar, T.: An intrusion detection game with limited observations. In: Proc. of the 43rd IEEE Conference on Decision and Control (2004)
2. You, X., Shiyong, Z.: A kind of network security behavior model based on game theory. In: Proceedings of the Fourth International Conference on Parallel and Distributed Computing, Applications and Technologies (2003)
3. Doupé, A., Egele, M., Caillat, B., Stringhini, G., Yakin, G., Zand, A., Cavedon, L., Vigna, G.: Hit 'em where it hurts: a live security exercise on cyber situational awareness. In: Proceedings of the 27th Annual Computer Security Applications Conference, ACSAC 2011, pp. 51–61. ACM, New York (2011)
4. Aumann, R.J., Maschler, M.: Repeated Games with Incomplete Information. MIT Press (1995)
5. Jones, M., Shamma, J.S.: Policy improvement for repeated zero-sum games with asymmetric information. In: 51st IEEE Conference on Decision and Control (December 2012)
6. Domansky, V.C., Kreps, V.L.: Eventually revealing repeated games of incomplete information. International Journal of Game Theory 23, 89–109 (1994)
7. Kotsalis, G., Megretski, A., Dahleh, M.: Balanced Truncation for a Class of Stochastic Jump Linear Systems and Model Reduction of Hidden Markov Models. IEEE Transactions on Automatic Control 53 (2008)
8. Heur, M.: Optimal strategies for the uninformed player. International Journal of Game Theory 20, 33–51
9. Zamir, S.: On the relation between finitely and infinitely repeated games with incomplete information. International Journal of Game Theory 23, 179–198
10. Gilpin, A., Sandholm, T.: Solving two-person zero-sum repeated games of incomplete information. In: International Joint Conference on Autonomous Agents and Multiagent Systems, vol. 2, pp. 903–910 (2008)
11. Zamir, S.: Repeated games of incomplete information: Zero-sum. In: Handbook of Game Theory, vol. 1, pp. 109–154 (1999)
12. Aumann, R.: Mixed and behavior strategies in infinite extensive games. Princeton University (1961)
13. Ponssard, J., Sorin, S.: The l-p formulation of finite zero-sum games with incomplete information. International Journal of Game Theory 9, 99–105 (1999)
14. Blackwell, D.: An analog of the minimax theorem for vector payoffs. Pacific Journal of Mathematics 1956(1), 1–8 (1956)
15. Freund, Y., Schapire, R.E.: Game theory, on-line prediction and boosting. In: Proceedings of the Ninth Annual Conference on Computational Learning Theory, COLT 1996, pp. 325–332. ACM, New York (1996), http://doi.acm.org/10.1145/238061.238163
16. Rosenberg, D., Solan, E., Vieille, N.: Stochastic games with a single controller and incomplete information. Northwestern University, Center for Mathematical Studies in Economics and Management Science, Discussion Papers 1346 (May 2002)

Formulating Cyber-Security as Convex Optimization Problems*

Kyriakos G. Vamvoudakis[1], João P. Hespanha[1], Richard A. Kemmerer[2], and Giovanni Vigna[2]

[1] Center for Control, Dynamical-systems and Computation (CCDC), University of California, Santa Barbara, CA 93106-9560 USA
[2] Computer Security Lab, University of California, Santa Barbara, CA 93106-9560 USA
{kyriakos,hespanha}@ece.ucsb.edu,
{kemm,vigna}@cs.ucsb.edu

Abstract. Mission-centric cyber-security analysts require a complete overview and understanding of the state of a mission and any potential threats to their completion. To facilitate this, we propose optimization-based algorithms that can be used to predict in real-time how an attacker may try to compromise a cyber-mission with a limited amount of resources, based on a model that takes into account potential damage to the mission and probabilistic uncertainty. Two different optimization schemes are considered: one where all the mission data is known a priori to the attacker and another where system identification and a moving horizon optimization is used to produce the estimates based on historical data. Our schemes are compared with real attacks carried our by human players in the 2011 international Capture The Flag (iCTF) hacking competition.

Keywords: Cyber-Security, Convex Optimization, System Identification, iCTF.

1 Introduction

Guaranteeing the security of cyber-missions is a complex, multi-dimensional challenge that demands a multi-faceted, strategic solution. The terminology *cyber-mission* refers to a set of computer transactions aimed at accomplishing a specific purpose or task, such as placing an online shopping order, submitting a paper to a conference through an online submission system, or printing a bank statement at an ATM machine. Cyber-missions typically require a large number of computer services, including encryption services, authentication servers, database engines, web servers. We are especially interested in cyber-missions

* This material is based upon work supported by ARO MURI Grant number W911NF0910553.

DOI: 10.1007/978-3-319-01159-2_5,

that go through several states, each of which may require one or more computer services. Cyber-missions are especially vulnerable to attacks because it may be possible to prevent the mission's completion by compromising just one of the multiple services required by the mission, provided that the right service is compromised at the right time.

Cyber-missions are pervasive and can be found in trading, banking, power systems management, road traffic managements, healthcare, online shopping, business-to-business transactions, etc. The disruption to cyber-missions can thus result in cyber or physical consequences that threaten National and economic security, critical infrastructure, public health, and welfare. Moreover, stealthy cyber-attackers can lay a hidden foundation for future exploitation or attack, which they can later execute at a time of greatest advantage. Securing cyberspace requires a layered security approach across the public and private sectors.

In the cyber-mission security domain, the security analyst is interested in making decisions based on the potential damage that attacks can inflict to the mission and also on the probability that the potential damage is realized. To focus their attention and coordinate defensive actions, security professionals must be able to determine which attacks presents the biggest threat and prioritize which services to defend, a problem often referred to as *cyber situation awareness*. Situation awareness [3] is a common feature of many cyber-security solutions but most of them are fragmented. In this paper, we present a model that can be used to predict how an attacker may try to compromise a cyber-mission with a limited amount of resources, based on a model that takes into account potential damage to the mission and probabilistic uncertainty.

This approach followed here motivated by the need to avoid flooding the security analyst with raw data about complex missions and detailed logs from intrusion detection systems (IDSs). Instead, an automated or semi-automated system should process this data and present the analyst with high-level information about the computer services that are currently most crucial for mission completion and thus most likely to be the target of attacks, based on the current state of the mission and its future expected evolution. To achieve this we propose a relatively general model to describe the damage to a cyber-mission caused by potential attacks. This model can be utilized in optimization schemes to discover optimal policies to distribute attack resources over time and over the different computer services relevant to the mission so as to maximize damage to the cyber mission. The models proposed, need mission parameters that typically vary with time according to complex dynamics, which are difficult to determine in an analytic fashion. To avoid this difficulty, we learn such parameters using system identification of low-order state-space models that are used to make predictions of the parameter evolution for a reasonable future time horizon.

Security competitions are exceptional venues for researchers to discover and validate novel security solutions. The international Capture The Flag (iCTF) [5] is a distributed wide-area security exercise whose goal is to test the security skills of the participants. The iCTF contest is organized by the Security Lab of the Department of Computer Science at UCSB and is held once a year. The Capture the Flag contest is a multi-site, multi-team hacking contest in which a number of teams compete independently against each other. The 2011 edition of iCTF was aimed at Cyber-Situation Awareness and, to our knowledge, produced the first experimental dataset that includes mission descriptions as well as attack logs and the statuses of computer services required by missions [2,5]. We have used this data to validate the algorithms presented in this paper and show their efficacy in predicting attacks to cyber missions by the human participants in the exercise.

The results presented in this paper were also used in the design of a high-level visualization tool to help security analysts to protect the computer systems under attack in the 2011 iCTF competition [4]. We are in the process of developing human subject experiments to demonstrate the benefits of using the predictions generated by the methodology proposed in this paper, instead of searching through mission traces and security logs.

The remainder of the paper is structured as follows. Section 2 develops the general mathematical framework for cyber-security and then describes how one can use classical system identification techniques to identify the completely unknown or partially known time-varying processes. Section 3 describes an optimization problem to discover how an attacker would optimally allocate her resources through all the services as time evolves for two different scenarios. The first assumes that the all the mission data is known to the attacker, whereas the second one uses a moving horizon optimization scheme that estimates this data online to predict when and where to attack. In Section 4, the algorithms proposed are applied to data from the 2011 iCTF competition. Comparison results between how the teams in the competition attacked and the results obtained by the optimization schemes are presented in Section 5. Finally, Section refse:conclusions concludes and discusses about future work.

2 General Framework for Cyber-Security

Our goal is to capture complex behaviors with a relatively simple model and incorporate that in a cyber-security advisory system to show its effectiveness. This section presents a general framework to model mission-critical cyber-security scenarios.

2.1 Cyber-Mission Damage Model

Suppose that the *(potential) damage* that an attacker can inflict to a cyber mission is quantified by a scalar $x_{\text{PD}} \geqslant 0$ that is a function of the level of *attack resources* $u_{\text{AR}} \geqslant 0$ devoted to the attack. The mapping from attack resources to potential damage is expressed by the so called *potential damage equation* that we approximate by a linear map:

$$x_{\text{PD}} = f(u_{\text{AR}}) := a + b\, u_{\text{AR}}, \tag{1}$$

where $a \in \mathbb{R}^+$ can be viewed as the zero-resource damage level (damage achieved without an intended attack), and $b \in \mathbb{R}^+$ the marginal damage per unit of attack resources.

Whether or not the potential damage to the mission x_{PD} is realized is assumed to be a stochastic event that occurs with a given probability $\rho \in [0, 1]$ that also depends on the attack resources $u_{\text{AR}} \in \mathbb{R}^+$, according to the so-called *uncertainty equation* that we approximate by a linear map projected to the interval $[0, 1]$:

$$\rho = g(u_{\text{AR}}) := \Pi_{[0,1]}(c - d\,u_{\text{AR}}) \tag{2}$$

where $\Pi_{[0,1]} : \mathbb{R} \to \mathbb{R}$ denotes the projection function

$$\Pi_{[0,1]}(x) = \begin{cases} 0 & x < 0 \\ x & x \in [0, 1] \\ 1 & x > 1, \end{cases}$$

the scalar $c \geqslant 0$ corresponds to a zero-resource probability of damage, and the scalar $d \geqslant 0$ to the marginal *decrease* in the probability of damage per unit of attack resources. We note that an increase in attack resources u_{AR} leads to an increase in the potential damage x_{PD} [expressed by the + sign before the b term in (1)], but may actually decrease the probability that the potential damage will actually be realized [expressed by the − sign before the d term in (2)], which is motivated by the fact that a large-scale attack is more likely to trigger defense mechanisms that can prevent the potential damage from being realized.

The total expected damage y_{TD} to the mission can be found by multiplying equations (1) and (2), leading to the *expected damage equation*

$$y_{\text{TD}} = f(u_{\text{AR}})g(u_{\text{AR}}). \tag{3}$$

In the context of cyber-missions that evolve over time and require multiple computer services, the potential damage equation (1) and the uncertainty equation (2) need to be augmented with an index $t \in \{1, 2, \ldots, T\}$ that parameterizes mission time and an index $s \in \{1, 2, \ldots, S\}$ that parameterizes the required computer services, as in

$$x_{\text{PD}}{}_t^s = f_t^s(u_{\text{AR}}{}_t^s) = a_t^s + b_t^s u_{\text{AR}}{}_t^s, \tag{4}$$

$$\rho_t^s = g_t^s(u_{\text{AR}}{}_t^s) = \Pi_{[0,1]}(c_t^s - d_t^s u_{\text{AR}}{}_t^s) \tag{5}$$

where $u_{\text{AR}}{}_t^s$ denotes the attack resources committed to attack service s at time t, $x_{\text{PD}}{}_t^s$ the potential damage at time t due to an attack to the service s, and P_t^s the probability of realizing this damage. The corresponding *expected damage equation* then becomes:

$$y_{\text{TD}} = \sum_{t=1}^{T} \sum_{s=1}^{S} f_t^s(u_{\text{AR}}{}_t^s) g_t^s(u_{\text{AR}}{}_t^s). \tag{6}$$

3 Optimization

An intelligent attacker would seek to optimally allocate her available resources to maximize the total expected missing damage. We shall consider here several options for this optimization that differ on the information that is available to the attacker.

3.1 Optimization Scheme with Known Mission Damage Data

When all the data $\{a_t^s, b_t^s, c_t^s, d_t^s : \forall s, t\}$ that define the potential damage and uncertainty equations is known a-priori, optimal attack resource allocation can be determined by solving the following optimization.

$$\begin{aligned} \text{maximize} \quad & \sum_{t=1}^{T}\sum_{s=1}^{S} f_t^s(u_{\mathrm{AR}_t^s})g_t^s(u_{\mathrm{AR}_t^s}) \\ \text{subject to} \quad & \sum_{t=1}^{T}\sum_{s=1}^{S} u_{\mathrm{AR}_t^s} \leqslant U_{\mathrm{TR}} \\ \text{w.r.t.} \quad & u_{\mathrm{AR}_t^s} \in [0,\infty),\ \forall t,\ \forall s, \end{aligned} \tag{7}$$

where U_{TR} denotes the total budget of attack resources available to the attacker. As stated in the following proposition, this optimization can be converted into the following concave maximization.

Proposition 1. *When the functions f_t^s, g_t^s are of the form (4)–(5) with a_t^s, b_t^s, c_t^s, $d_t^s \geqslant 0$, $\forall t, s$. The value and optimum of (7) can be obtained through the following concave maximization problem:*

$$\begin{aligned} \textit{maximize} \quad & \sum_{t=1}^{T}\sum_{s=1}^{S} (a_t^s + b_t^s u_{\mathrm{AR}_t^s})(c_t^s - d_t^s u_{\mathrm{AR}_t^s} - \sigma_t^s) \\ \textit{subject to} \quad & \sum_{t=1}^{T}\sum_{s=1}^{S} u_{\mathrm{AR}_t^s} \leqslant U_{\mathrm{TR}}, \quad c_t^s - d_t^s u_{\mathrm{AR}_t^s} - \sigma_t^s \leqslant 1,\ \forall t,\ \forall s \\ \textit{w.r.t.} \quad & u_{\mathrm{AR}_t^s} \in \left[0, \frac{c_t^s}{d_t^s}\right],\ \sigma_t^s \geqslant 0,\ \forall t,\ \forall s. \end{aligned} \tag{8}$$

When $c_t^s \in [0,1]$, one can set the corresponding optimization variable $\sigma_t^s = 0$ in (8). Moreover, when $c_t^s \in [0,1]$, $\forall t, s$ and all the constraints on the $u_{\mathrm{AR}_t^s}$ are inactive, the solution to this optimization can be found in closed form and is equal to

$$u_{\mathrm{AR}_t^s} = \bar{u}_t^s - \bar{\mu}_t^s \max\left\{0, \sum_{\bar{t}}\sum_{s=1}^{S} \bar{u}_{\bar{t}}^s - U_{\mathrm{TR}}\right\}, \quad \bar{u}_t^s := \frac{b_t^s c_t^s - a_t^s d_t^s}{2 b_t^s d_t^s}, \quad \bar{\mu}_t^s := \frac{\frac{1}{2 b_t^s d_t^s}}{\sum_{\bar{t}}\sum_{s=1}^{S} \frac{1}{2 b_{\bar{t}}^s d_{\bar{t}}^s}}.$$

Note that, if any of the constraints on the attack resources are active, a closed-form solution may not be easy and one has to solve the optimization problem (8) instead.

Proof. 1 To prove that (7) and (8) are equivalent, we start by noting that

$$g_t^s(u_{\mathrm{AR}_t^s}) = \begin{cases} 0 & c_t^s - d_t^s u_{\mathrm{AR}_t^s} < 0 \quad \Leftrightarrow \quad u_{\mathrm{AR}_t^s} > \frac{c_t^s}{d_t^s} \\ 1 & c_t^s - d_t^s u_{\mathrm{AR}_t^s} > 1 \quad \Leftrightarrow \quad u_{\mathrm{AR}_t^s} < \frac{c_t^s - 1}{d_t^s} \\ c_t^s - d_t^s u_{\mathrm{AR}_t^s} & u_{\mathrm{AR}_t^s} \in \left[\frac{c_t^s - 1}{d_t^s}, \frac{c_t^s}{d_t^s}\right]. \end{cases}$$

Suppose, by contradiction, that (8) could lead to a larger maximum than (7). The condition $u_{\mathrm{AR}_t^s} \in \left[0, \frac{c_t^s}{d_t^s}\right]$ guarantees that the same set of $u_{\mathrm{AR}_t^s}$ satisfy the constraints of (7) and that

$$g_t^s(u_{\mathrm{AR}_t^s}) = \begin{cases} 1 & c_t^s - d_t^s u_{\mathrm{AR}_t^s} > 1 \quad \Leftrightarrow \quad u_{\mathrm{AR}_t^s} < \frac{c_t^s - 1}{d_t^s} \\ c_t^s - d_t^s u_{\mathrm{AR}_t^s} & c_t^s - d_t^s u_{\mathrm{AR}_t^s} \leqslant 1 \quad \Leftrightarrow \quad u_{\mathrm{AR}_t^s} \geqslant \frac{c_t^s - 1}{d_t^s} \end{cases}$$

and the condition $c_t^s - d_t^s u_{\mathrm{AR}_t^s} - \sigma_t^s \leqslant 1$ guarantees that

$$\begin{cases} c_t^s - d_t^s u_{\mathrm{AR}_t^s} - \sigma_t^s \leqslant 1 = g_t^s(u_{\mathrm{AR}_t^s}) & u_{\mathrm{AR}_t^s} < \frac{c_t^s - 1}{d_t^s} \\ c_t^s - d_t^s u_{\mathrm{AR}_t^s} - \sigma_t^s = g_t^s(u_{\mathrm{AR}_t^s}) - \sigma_t^s \leqslant g_t^s(u_{\mathrm{AR}_t^s}) & u_{\mathrm{AR}_t^s} \geqslant \frac{c_t^s - 1}{d_t^s}, \end{cases}$$

which shows that $c_t^s - d_t^s u_{\mathrm{AR}_t^s} - \sigma_t^s \leqslant g_t^s(u_{\mathrm{AR}_t^s})$ and therefore (8) cannot lead to a larger maximum than (7).

Suppose now, also by contradiction, that (7) could lead to a larger maximum than (8). First note that if a few of the $u_{\mathrm{AR}_t^s}$ that maximize (7) were larger than $\frac{c_t^s}{d_t^s}$, for those $u_{\mathrm{AR}_t^s}$ we would have $g_t^s(u_{\mathrm{AR}_t^s}) = 0$ and the same exact cost could be obtained for (7) by replacing each of these $u_{\mathrm{AR}_t^s}$ with $\frac{c_t^s}{d_t^s}$. So we may assume, without loss of generality, that all the $u_{\mathrm{AR}_t^s}$ are smaller than or equal to $\frac{c_t^s}{d_t^s}$.

In this case, we could use the same $u_{\mathrm{AR}_t^s}$ in (8) and set

$$\sigma_t^s = \begin{cases} 0 & c_t^s - d_t^s u_{\mathrm{AR}_t^s} \leqslant 1 \\ c_t^s - d_t^s u_{\mathrm{AR}_t^s} - 1 & c_t^s - d_t^s u_{\mathrm{AR}_t^s} > 1. \end{cases}$$

This selection of σ_t^s would satisfy the constraints of (8) and guarantee that

$$g_t^s(u_{\mathrm{AR}_t^s}) = c_t^s - d_t^s u_{\mathrm{AR}_t^s} - \sigma_t^s,$$

and therefore (7) and (8) would lead to the same maximum. This completes the proof that (7) and (8) are equivalent.

The optimization scheme just defined is a concave maximization problem (convex minimization) with linear constraints. The dual problem is given by,

$$
\begin{aligned}
J^{\perp} :=& \max_{\lambda_1 \geqslant 0, \eta_t^s \geqslant 0, \zeta_t^s \geqslant 0} \max_{u_{\mathrm{AR}_t^s} \in \mathbb{R}} \sum_{t=1}^{T} \sum_{s=1}^{S} (a_t^s + b_t^s u_{\mathrm{AR}_t^s})(c_t^s - d_t^s u_{\mathrm{AR}_t^s}) \\
&- \lambda_1 \Big(\sum_{t=1}^{T} \sum_{s=1}^{S} u_{\mathrm{AR}_t^s} - U_{\mathrm{TR}} \Big) - \sum_{t=1}^{T} \sum_{s=1}^{S} \eta_t^s \Big(u_{\mathrm{AR}_t^s} - \frac{c_t^s}{d_t^s} \Big) \\
&+ \sum_{t=1}^{T} \sum_{s=1}^{S} \zeta_t^s u_{\mathrm{AR}_t^s} \\
=& \max_{\lambda_1 \geqslant 0, \eta_t^s \geqslant 0, \zeta_t^s \geqslant 0} \max_{u_{\mathrm{AR}_t^s} \in \mathbb{R}} \sum_{t=1}^{T} \sum_{s=1}^{S} \Big(a_t^s c_t^s - a_t^s d_t^s u_{\mathrm{AR}_t^s} + b_t^s c_t^s u_{\mathrm{AR}_t^s} - b_t^s d_t^s u_{\mathrm{AR}_t^s}^{2} \\
&- \lambda_1 u_{\mathrm{AR}_t^s} - \eta_t^s u_{\mathrm{AR}_t^s} + \zeta_t^s u_{\mathrm{AR}_t^s} \Big) + \lambda_1 U_{\mathrm{TR}} + \sum_{t=1}^{T} \sum_{s=1}^{S} \eta_t^s \frac{c_t^s}{d_t^s} \\
=& \max_{\lambda_1 \geqslant 0, \eta_t^s \geqslant 0, \zeta_t^s \geqslant 0} \max_{u_{\mathrm{AR}_t^s} \in \mathbb{R}} \sum_{t=1}^{T} \sum_{s=1}^{S} \Big(a_t^s c_t^s - b_t^s d_t^s u_{\mathrm{AR}_t^s}^{2} + (b_t^s c_t^s - a_t^s d_t^s + \zeta_t^s - \eta_t^s - \lambda_1) u_{\mathrm{AR}_t^s} \Big) \\
&+ \lambda_1 U_{\mathrm{TR}} + \sum_{t=1}^{T} \sum_{s=1}^{S} \eta_t^s \frac{c_t^s}{d_t^s}.
\end{aligned}
$$

The inner maximization can be solved using standard calculus and is achieved for

$$
u_{\mathrm{AR}_t^s} = \frac{b_t^s c_t^s - a_t^s d_t^s + \zeta_t^s - \eta_t^s - \lambda_1}{2 b_t^s d_t^s},
$$

yielding

$$
J^{\perp} := \max_{\lambda_1 \geqslant 0, \eta_t^s \geqslant 0, \zeta_t^s \geqslant 0} \sum_{t=1}^{T} \sum_{s=1}^{S} \Big(\frac{(b_t^s c_t^s - a_t^s d_t^s + \zeta_t^s - \eta_t^s - \lambda_1)^2}{4 b_t^s d_t^s} + \eta_t^s \frac{c_t^s}{d_t^s} \Big) + \lambda_1 U_{\mathrm{TR}}.
$$

For this problem the Karush-Kuhn-Tucker (KKT) conditions [1] lead to

$$
\begin{aligned}
\frac{\partial J^{\perp}}{\partial \lambda_1} = 0 \quad &\Leftrightarrow \quad \lambda_1 = \frac{\sum_{t=1}^{T} \sum_{s=1}^{S} \frac{b_t^s c_t^s - a_t^s d_t^s + \zeta_t^s - \eta_t^s}{2 b_t^s d_t^s} - U_{\mathrm{TR}}}{\sum_{t=1}^{T} \sum_{s=1}^{S} \frac{1}{2 b_t^s d_t^s}} \qquad \text{or} \qquad \lambda_1 = 0 \\
\frac{\partial J^{\perp}}{\partial \eta_t^s} = 0 \quad &\Leftrightarrow \quad \eta_t^s = \zeta_t^s - a_t^s d_t^s - b_t^s c_t^s - \lambda_1 \qquad \text{or} \qquad \eta_t^s = 0 \\
\frac{\partial J^{\perp}}{\partial \zeta_t^s} = 0 \quad &\Leftrightarrow \quad \zeta_t^s = -b_t^s c_t^s + a_t^s d_t^s + \eta_t^s + \lambda_1 \qquad \text{or} \qquad \zeta_t^s = 0.
\end{aligned}
$$

Let us assume that $u_{\mathrm{AR}_t^s}$ is inside the interval $\left[0, \frac{c_t^s}{d_t^s}\right]$, which would lead to all the η_t^s and ζ_t^s equal to zero (inactive constraints) and therefore we would need

$$
\sum_{t=1}^{T} \sum_{s=1}^{S} \frac{b_t^s c_t^s - a_t^s d_t^s - \lambda_1}{2 b_t^s d_t^s} = U_{\mathrm{TR}} \quad \Leftrightarrow \quad \lambda_1 = \frac{\sum_{t=1}^{T} \sum_{s=1}^{S} \frac{b_t^s c_t^s - a_t^s d_t^s}{2 b_t^s d_t^s} - U_{\mathrm{TR}}}{\sum_{t=1}^{T} \sum_{s=1}^{S} \frac{1}{2 b_t^s d_t^s}} \geqslant 0 \tag{9}
$$

$$
\text{or} \qquad \lambda_1 = 0
$$

and

$$u_{\mathrm{AR}_t^s} = \bar{u}_t^s - \bar{\mu}_t^s \max\left\{0, \sum_{\bar{t}} \sum_{s=1}^{S} \bar{u}_{\bar{t}}^s - U_{\mathrm{TR}}\right\}, \quad \bar{u}_t^s := \frac{b_t^s c_t^s - a_t^s d_t^s}{2b_t^s d_t^s}, \quad \bar{\mu}_t^s := \frac{\frac{1}{2b_t^s d_t^s}}{\sum_{\bar{t}} \sum_{s=1}^{S} \frac{1}{2b_{\bar{t}}^s d_{\bar{t}}^s}}.$$

We can view the term being subtracted from $u_{\mathrm{AR}_t^s}$ as a normalizing term that makes sure that the $u_{\mathrm{AR}_t^s}$ add up to the constraint U_{TR}.

Note that if the closed-form formula shown above for $u_{\mathrm{AR}_t^s}$ ever becomes negative, then the corresponding ζ_t^s will become active and we must have

$$\frac{\partial J^{\perp}}{\partial \zeta_t^s} = 0 \quad \Rightarrow \quad \zeta_t^s = \lambda_1 + a_t^s d_t^s - b_t^s c_t^s \quad \Rightarrow \quad u_{\mathrm{AR}_t^s} = 0.$$

Similarly if the formula for $u_{\mathrm{AR}_t^s}$ ever becomes larger than $\frac{c_t^s}{d_t^s}$, then the corresponding η_t^s will become active and we must have

$$\frac{\partial J^{\perp}}{\partial \eta_t^s} = 0 \quad \Rightarrow \quad \eta_t^s = -b_t^s c_t^s - \lambda_1 - a_t^s d_t^s \quad \Rightarrow \quad u_{\mathrm{AR}_t^s} = \frac{c_t^s}{d_t^s}. \qquad \blacksquare$$

Remark 1. Note that, if any of the constraints on the attack resources are active, a closed-form solution is not possible and one has to solve the optimization problem instead. □

3.2 Unknown Mission Damage Data

Often the mission-specific parameters $\{a_t^s, b_t^s, c_t^s, d_t^s : \forall s, t\}$ that define the potential damage and uncertainty equations are not known a-priori and, instead, need to be estimated online. The estimation problem will be posed using a grey-box model that looks like a k-step ahead predictor. The problem is solved at each time instant to incorporate the new information that is obtained in the form of a new measurement of the process variables.

One approach that can be used to address this scenario is to assume that these parameters are generated by linear dynamics of the form

$$x_{at+1}^s = A_a^s x_{at}^s + B_a^s w_t^s, \qquad a_t^s = C_a'^s x_{at}^s, \tag{10}$$
$$x_{bt+1}^s = A_b^s x_{bt}^s + B_b^s w_t^s, \qquad b_t^s = C_b'^s x_{bt}^s, \tag{11}$$
$$x_{ct+1}^s = A_c^s x_{ct}^s + B_c^s w_t^s, \qquad c_t^s = C_c'^s x_{ct}^s, \tag{12}$$
$$x_{dt+1}^s = A_d^s x_{dt}^s + B_d^s w_t^s, \qquad d_t^s = C_d'^s x_{dt}^s, \tag{13}$$

where the $\{w_t^s, \forall s, t\}$ are sequences of zero-mean random processes with variances σ_w^s. One can then use historical data to estimate these dynamics using black-box identification techniques. Once estimates for the dynamics are available, one can use online data to predict future values for the mission-specific parameters $\{a_t^s, b_t^s, c_t^s, d_t^s : \forall s, t\}$, based on past observations.

It is crucial that the data collected from the system conveys continual information on the parameters to be estimated and the identification algorithm can rely on fresh information in forming reliable current estimates (persistence of excitation).

Suppose that at some time $k < T$ the attacker has observed the values of the past mission-specific parameters $\{a_t^s, b_t^s, c_t^s, d_t^s : \forall s, t \leqslant k\}$ and needs to make decisions on the future attack resources $u_{\mathrm{AR}_t^s}$, $t \geqslant k$. One can use (10)–(13) to construct estimates $\{\hat{a}_t^s, \hat{b}_t^s, \hat{c}_t^s, \hat{d}_t^s : \forall s, t > k\}$ for the future mission-specific parameters and obtain the future $u_{\mathrm{AR}_t^s}$, $t \geqslant k$ using the following optimization:

$$\text{maximize} \quad \sum_{t=1}^{k}\sum_{s=1}^{S} f_t^s(u_{\mathrm{AR}_t^s})g_t^s(u_{\mathrm{AR}_t^s}) + \sum_{t=k+1}^{T}\sum_{s=1}^{S} \hat{f}_t^s(u_{\mathrm{AR}_t^s})\hat{g}_t^s(u_{\mathrm{AR}_t^s}) \tag{14}$$

$$\text{subject to} \quad \sum_{t=1}^{T}\sum_{s=1}^{S} u_{\mathrm{AR}_t^s} \leqslant U_{\mathrm{TR}} \tag{15}$$

$$\text{w.r.t.} \quad u_{\mathrm{AR}_t^s} \in [0, \infty),\ \forall t \in \{k, ..., T\},\ \forall s, \tag{16}$$

where f_t^s and g_t^s denote the functions defined in (4) and (5), respectively, whereas $\hat{f}_t^s$ and $\hat{g}_t^s$ are estimates of these functions computed using the estimated mission-specific parameters $\{\hat{a}_t^s, \hat{b}_t^s, \hat{c}_t^s, \hat{d}_t^s : \forall s, t > k\}$.

The optimization (14) can be solved at each time step $k \in \{1, 2, \ldots, T-1\}$, allowing the attacker to improve her allocation of attack resources as new information about the missing parameters becomes available. Note that one could remove from the (double) summations in (14) any terms that do not depend on the optimization variables.

4 iCTF Competition

The international Capture The Flag (iCTF) is a distributed wide-area security exercise to test the security skills of the participants. This contest is organized by the Security Lab of the Department of Computer Science at UCSB and it has been held yearly since 2003. In traditional editions of the iCTF (2003-2007), the goal of each team was to maintain a set of services such that they remain available and uncompromised throughout the contest. Each team also had to attempt to compromise the other teams' services. Since all the teams received an identical copy of the virtual host containing the vulnerable services, each team had to find the vulnerabilities in their copy of the hosts and possibly fix the vulnerabilities without disrupting the services. At the same time, the teams had to leverage their knowledge about the vulnerabilities they found to compromise the servers run by other teams. Compromising a service allowed a team to bypass the service's security mechanisms and to "capture the flag" associated with the service. During the 2008-2010 iCTFs, new competition designs have been introduced. More precisely, in 2008 a separate virtual network was created for each team. The goal was to attack a terrorist network and defuse a bomb after compromising a number of hosts. In 2009, the participants had to compromise the browsers of a large group of simulated users, steal their money, and create a botnet. In 2010, the participants had to attack the rogue nation Litya, ruled by the evil Lisvoy Bironulesk. The teams' goal was to attack the services supporting Litya's infrastructure only at specific times, when certain activities were in progress. In addition, an intrusion detection system would temporarily firewall out the teams whose attacks were detected. The 2011 iCTF competition is briefly summarized below from the perspective of one team playing against the rest of the world. The 2010 [2] and 2011 [5] iCTF competitions were designed closely match practical cyber-security mission scenarios.

4.1 2011 iCTF

The 2011 iCTF was centered around the theme of illegal money laundering. This activity is modeled after cyber-criminal money laundering operations and provided a perfect setting for risk-reward analysis, as the trade-offs are very intuitively understood.

The general idea behind the competition was the conversion ("laundering") of money into points. The money was obtained by the teams by solving security-related challenges (e.g., decrypting an encrypted message, find hidden information in a document, etc.) The conversion of money into points was performed by utilizing data captured from an exploited service. Therefore, first a team had to obtain money by solving challenges, and then the money had to be translated into points by exploiting the vulnerability in a service of another team. Successful conversion of money to points depended on a number of factors, calculated together as the "risk function", which is described in detail below. Note that, at the end of the game, the money had no contribution to the final stand of a team: only points mattered.

One challenge with the formulation "one-against-world" is that in the 2011 iCTF game, winning was not just about maximizing points. Winning was about getting more points than each of the opponents (individually).

The game was played in rounds 255 (each takes about 2min), but we only have data for 248 rounds since the logging server was temporarily down. Each team hosts a server that runs 10 services each with its own (unknown) vulnerabilities. Each service $s \in \{1, 2, \ldots, 10\}$ of each hosting team is characterized by three time-varying quantities $\forall t \in \{1, 2, \ldots, 248\}$:

- the *cut* C_t^s, which is the percentage of money that goes to the team when money is laundered through service s (same values for every team),
- the *payoff* P_t^s, which is the percentage of money that will be transformed into points for the team that launders the money (same value for every team);

$$P_t^s = 0.9e^{-\frac{TicksActive}{10}}$$

- the *risk* R_t^s, which is the probability of losing all the money (instead of getting a conversion to points).

The generation of the time series for the cuts, payoffs, and risks for the different services was based on an underlying set of cyber missions that were running while the game was played. Essentially, when the states of the cyber missions required a particular service, the cut, payoff, and risk would make that service attractive for attackers from the perspective of converting money to points. However, the players were not informed about the state of the cyber-missions and, instead, at the beginning of each round t, the team is informed of the values of C_t^s, P_t^s, R_t^s for every s, and t.

4.2 Actions Available to Every Team

A team (we) has the following key actions in the actual competition:

1. *Defensive actions*: Activate/deactivate one of its own services. In the iCTF competition a team could also correct any vulnerability that it discovered in its services. We assumed here that all known vulnerabilities had been corrected.
2. *Money laundering*: Select
   (a) team to attack (mute decision within the "one-against-world" formulation);
   (b) service s to compromise, which implicitly determines the payoff P_t^s, the risk R_t^s, and the cut C_t^s;
   (c) amount of money to launder $u_{\mathrm{AR}_t^s}$ at time t through the service s.

   This action results in a number of points given by

$$X_t^s = \begin{cases} P_t^s(1 - C_t^s) D_t u_{\mathrm{AR}_t^s} & \text{w.p. } 1 - \min\{\rho_t^s, 1\} \\ 0 & \text{w.p. } \min\{\rho_t^s, 1\} \end{cases} \tag{17}$$

   where D_t is the team's defense level and ρ_t^s is the probability that the conversion of money to points will succeed, as given by the formula

$$\rho_t^s := \frac{R_t^s u_{\mathrm{AR}_t^s}}{30} + \frac{1}{6}\Big(\frac{N_t^j - 700}{300 + |N_t^j - 700|} + 1\Big) + \frac{1}{6}\Big(\frac{Q_t^s - 1500}{300 + |Q_t^s - 1500|} + 1\Big)$$

   where N_t^j is the overall amount of money that has been laundered by the team j through the particular team being exploited and Q_t^s is the overall amount of money that has been laundered by the team through the particular service being exploited. Because we do not model each team individually we will consider the "worst" case scenario for the following quantities, $N = 492$, $Q = 2257$ (according to data from the competition), and defense level of the team as $D = 1$.

To map this game with the general framework described in Section 2, we associate the money to launder $u_{\mathrm{AR}_t^s}$ at time t through service s with the resources $u_{\mathrm{AR}_t^s}$ devoted to attack service s at time t, and associate the points X_t^s in (17) with damage to the mission.

The total attack resources U_{TR} available to each team in the general framework described in Section 2, now corresponds to the money available to each team. While we could model more accurately the process by which teams get money, for simplicity we assumed that each team had available a fixed amount of money ($5060) that could be spend throughout the duration of the game which is given by the average money of all the teams during the competition. The results regarding which services where attacked and when proved to be relatively insensitive to this parameter.

4.3 Optimization Schemes and iCTF

In this section we apply the optimization schemes defined in Sections 3.1 and 3.2 to the iCTF game. We are seeking to optimally allocate our available resources in the competition such that the total number of points is maximized while meeting the specified constraints. The maximization of the expected reward by a team can be formulated as follows

$$\begin{aligned}
\text{maximize} \quad & \sum_{t=1}^{248}\sum_{s=1}^{10} \rho_t^s P_t^s (1 - C_t^s) D_t^s u_{\text{AR}_t^s} \\
\text{subject to} \quad & \sum_{t=1}^{248}\sum_{s=1}^{10} u_{\text{AR}_t^s} \leqslant U_{\text{TR}} := 5060 \\
\text{w.r.t.} \quad & u_{\text{AR}_t^s} \in [0, \infty),\ \forall s \in \{1, 2, \ldots, 10\},\ t \in \{1, 2, \ldots, 248\},
\end{aligned}$$

where,

$$\begin{aligned}
\rho_t^s &= \min\{(\beta_t^s + \frac{R_t^s}{30} u_{\text{AR}_t^s}), 1\}, \\
\beta_t^s &:= \frac{1}{6}\Big(\frac{N_t - 700}{300 + |N_t - 700|} + 1\Big) + \frac{1}{6}\Big(\frac{Q_t - 1500}{300 + |Q_t - 1500|} + 1\Big) = 0.4
\end{aligned}$$

and the parameters $P_t^s, C_t^s, D_t^s, \beta_t^s$ can either be considered known or unknown.

By using Proposition 1, and setting the constraint $\sigma_t^s = 0$ in (8) (since $(1 - \beta_t^s) \in [0, 1]$), we can write the equivalent optimization problem as,

$$\begin{aligned}
\text{maximize} \quad & \sum_{t=1}^{248}\sum_{s=1}^{10} (1 - \beta_t^s - \frac{R_t^s}{30} u_{\text{AR}_t^s}) P_t^s (1 - C_t^s) u_{\text{AR}_t^s} \\
\text{subject to} \quad & \sum_{t=1}^{248}\sum_{s=1}^{10} u_{\text{AR}_t^s} \leqslant U_{\text{TR}} \\
\text{w.r.t.} \quad & u_{\text{AR}_t^s} \in \Big[0, \frac{1 - \beta_t^s}{\frac{R_t^s}{30}}\Big],\ \forall s \in \{1, 2, \ldots, 10\},\ t \in \{1, 2, \ldots, 248\},
\end{aligned}$$

which is a concave maximization problem with linear constraints that is easy to solve numerically as described in Section 3.1.

The above optimization depends on the following assignments, $a_t^s := 0$, $b_t^s := P_t^s(1 - C_t^s)$, $c_t^s := 1 - \beta_t^s$, $d_t^s := \frac{R_t^s}{30}$. When these are not known, one can estimate

$b_t^s := P_t^s(1-C_t^s)$, $c_t^s := 1-\beta_t^s$, $d_t^s := \frac{R_t^s}{30}$ using a low order state space models given by (11)-(13). By then applying the optimization scheme described in Section 3.2, with a horizon of $N = 5$, one can still make accurate predictions of when and how to distribute the available attack resources. The optimization model just described, results in an optimization to obtain the future $u_{\mathrm{AR}_t^s}, \forall t \geqslant k$ and performed under a moving horizon of 5 ticks,

$$\begin{aligned}
\text{maximize} \quad & \sum_{t=1}^{k}\sum_{s=1}^{10} b_t^s(c_t^s - d_t^s u_{\mathrm{AR}_t^s})u_{\mathrm{AR}_t^s} + \sum_{t=k+1}^{248}\sum_{s=1}^{10} \hat{b}_t^s(\hat{c}_t^s - \hat{d}_t^s u_{\mathrm{AR}_t^s})u_{\mathrm{AR}_t^s} \\
\text{subject to} \quad & \sum_{t=1}^{248}\sum_{s=1}^{10} u_{\mathrm{AR}_t^s} \leqslant U_{\mathrm{TR}k} \\
& x_{bt+1}^s = A_b^s x_{at}^s + B_b^s w_t^s \\
& \hat{b}_t^s = C_b^s x_{bt}^s \\
& x_{ct+1}^s = A_c^s x_{ct}^s + B_c^s w_t^s \\
& \hat{c}_t^s = C_c^s x_{ct}^s \\
& x_{dt+1}^s = A_d^s x_{dt}^s + B_d^s w_t^s \\
& \hat{d}_t^s = C_d^s x_{dt}^s \\
\text{w.r.t.} \quad & u_{\mathrm{AR}_t^s} \in \left[0, \frac{\hat{c}_t^s}{\hat{d}_t^s}\right], \ \forall t \in \{k, \dots, 248\}, \forall s \in \{1, 2, \dots, 10\}.
\end{aligned}$$

5 iCTF Results

This section presents numerical results obtained from the optimizations described above to data from the attack logs of the 2011 iCTF competition. All the optimizations have been implemented through a Matlab-based convex optimization solver such as CVX [1]. The optimization scheme described in Section 3.2 yielded very close results to the scheme described in Section 3.1 for a predicting horizon of $N = 5$.

Initially we will assume that a "sophisticated" attacker would be able to compromise any one of the 10 services. Figure 1 show the points and the money collected by such an optimal attacker, whereas Figure 2 shows the same (aggregate) data for the teams that participated in the competition.

One can also consider attackers with different level of sophistication, e.g., attackers that are only able to find vulnerabilities in a subset of the 10 services that the "sophisticated" was able to attack. By observing the data of the top 20 teams in the competition we were able to partition the sophistication in two levels. For comparison, we show the behavior of an attacker A that was only able to attack the services $1, 2, 4, 5, 6, 9$ (similar to the first 10 teams in the competition); and another attacker B that was only able to attack services $1, 2, 5, 6, 7, 8$ (similar to the teams from place 11 to 20 in the competition). The "sophisticated" attacker was able to gather with 1987 points, whereas the two other attackers were able to get 1821 and 1721 points, respectively.

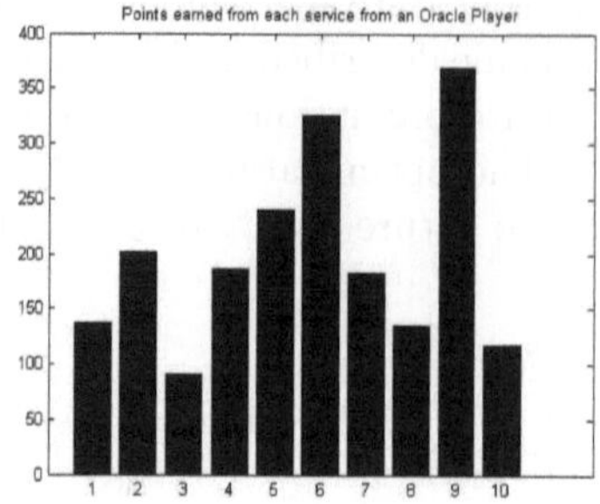

(a) points earned

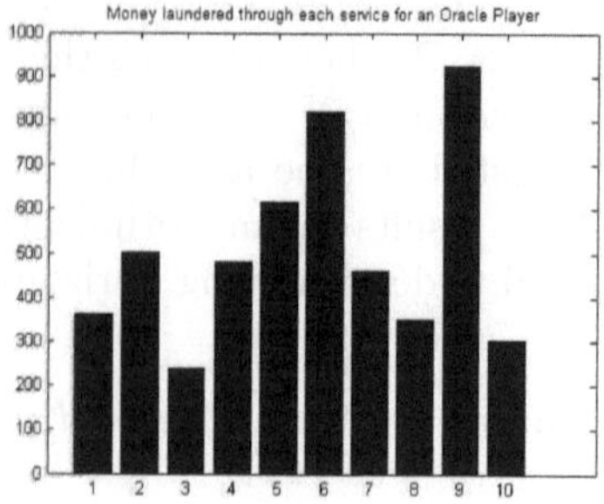

(b) money laundered

Fig. 1. Behavior of an optimal "sophisticated" attacker able to attack all 10 services

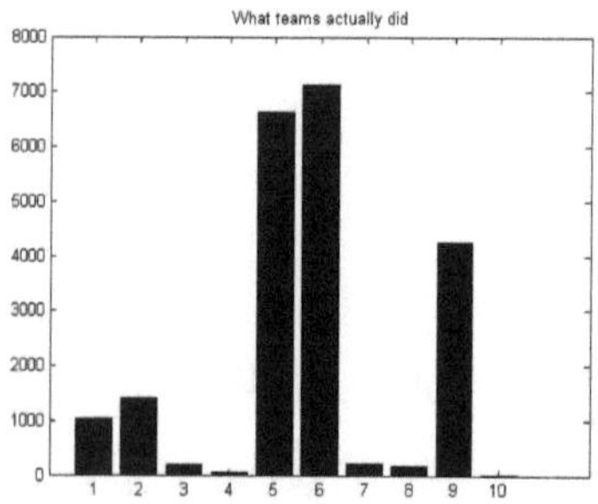

(a) points earned

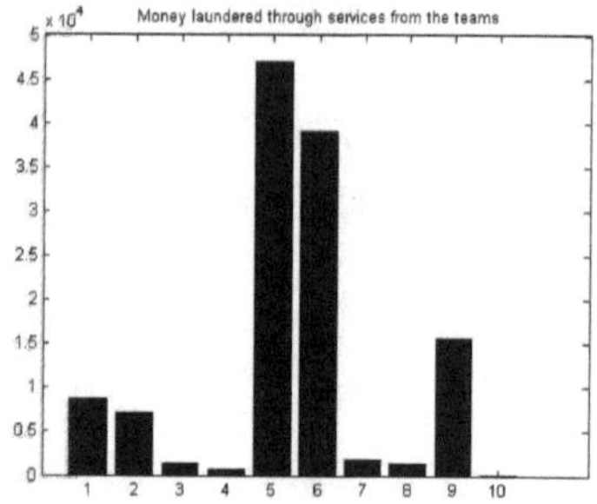

(b) money laundered

Fig. 2. Aggregate behavior of all teams that participated in the competition

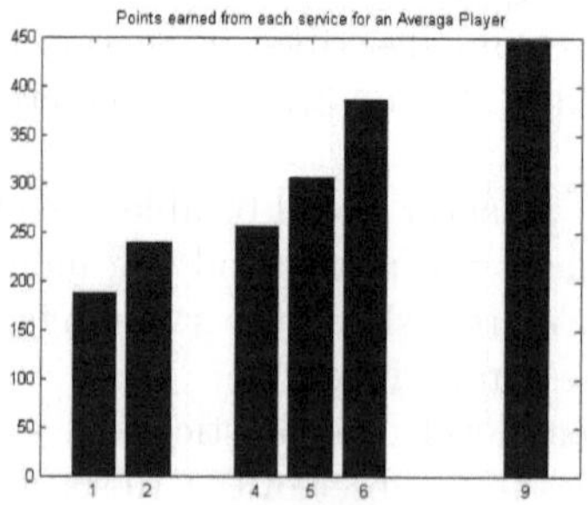

(a) points earned

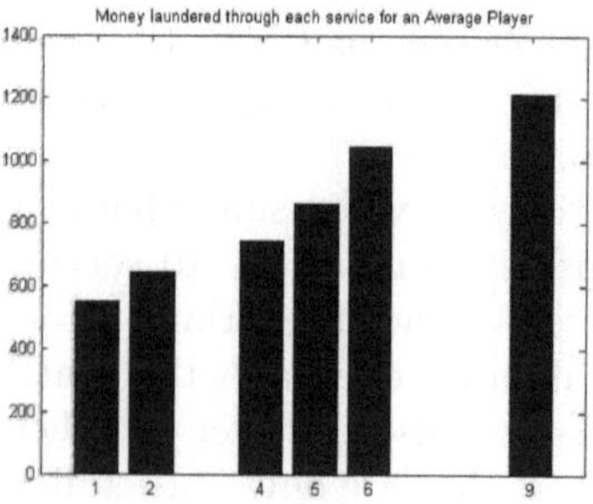

(b) money laundered

Fig. 3. Behavior of an optimal attacker A able to attack services 1,2,4,5,6,9

The results in Figure 1(a) show that the most profitable services to attack were 5, 6 and 9. The top 10 teams in the competition attacked mostly 5 and 6 because 9 was a hard service to get into. Only the top 3 teams discovered

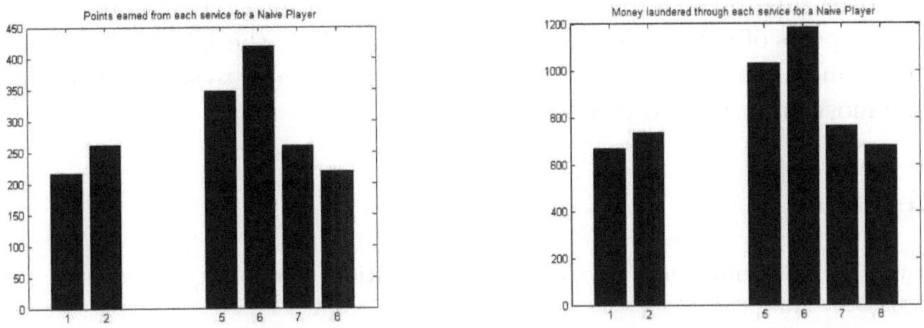

(a) points earned (b) money laundered

Fig. 4. Behavior of an optimal attacker B able to attack services 1,2,5,6,7,8

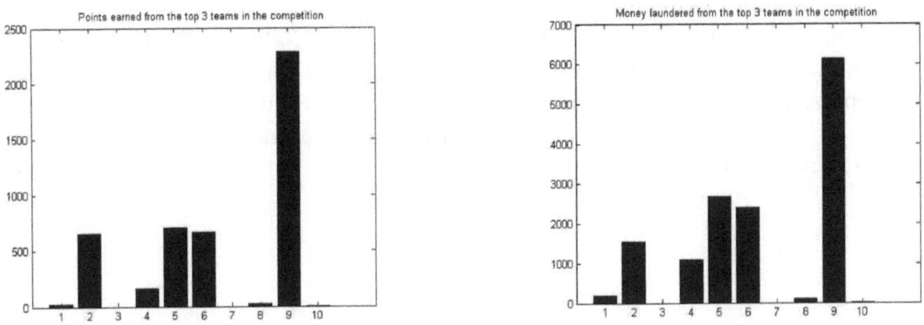

(a) points earned (b) money laundered

Fig. 5. Behavior of the top 3 teams during the competition

how to attack service 9 and only at the end of the game so they had relatively little time to explore that vulnerability. Aside from this, the prediction based on the optimization framework developed here qualitatively reflect the actions of the good teams. In fact, the top two teams in the competition followed attack strategies qualitatively close to that of attacker A in Figure 3 as seen in Figure 5.

6 Future Work

Our future work in this area is focused on developing analysis tools to explore what-if scenarios based on past data and the structure of the cyber-mission. To this end, we are developing optimization schemes for the defender's possible actions, such as taking a service off-line when the service is not needed or extending the duration of a state that would be unable to progress if a certain service is compromised. We are also developing human-computer interfaces to demonstrate the useful of this type of analysis for security analysts.

Moreover this work can be extended to provide a method to analyze cybersecurity aspects of power system state estimators where the attacker has limited resources and an index is introduced to enable the operator to see what resources are the most important to protect.

References

1. Boyd, S., Vandenberghe, L.: Convex Optimization. Cambridge University Press (2004)
2. Doupe, A., Egele, M., Caillat, B., Stringhini, G., Yakin, G., Zand, A., Cavedon, L., Vigna, G.: Hit'em where it hurts: A live security exercise on cyber situational awareness. In: Proceedings of the Annual Computer Security Applications Conference (ACSAC 2011), Orlando, FL (December 2011)
3. Endsley, M.: Theoretical Underpinnings of Situation Awareness: A Critical Review, ch. 1, pp. 3–32. L. Erlbaum Assoc. (2000)
4. Stockman, N., Vamvoudakis, K.G., Devendorf, L., Höllerer, T., Kemmerer, R., Hespanha, J.P.: A Mission-Centric Visualization Tool for Cybersecurity Situation Awareness. Technical Report, University of California, Santa Barbara (August 2012)
5. Vigna, G.: The 2011 UCSB iCTF: Description of the game (2011), `http://ictf.cs.ucsb.edu/`

LQ Control under Security Constraints*

Waseem A. Malik[1,2], Nuno C. Martins[1], and Ananthram Swami[2]

[1] Dept of Electrical and Computer Engineering
University of Maryland, College Park, MD, 20742, USA
[2] Army Research Laboratory
2800 Powder Mill Road, Adelphi, MD, 20783, USA
{wmalik,nmartins}@umd.edu, ananthram.swami.civ@mail.mil

Abstract. In this paper, a LQ optimal control problem in considered under several security constraints. The security constraints depend on the information gathered by an adversary making partial noisy measurements of the state. Control sequences are designed to drive the system to a specified terminal state while minimizing a quadratic cost and satisfying bounds on the information available to the adversary. The resulting optimization problems are shown to be convex. Duality techniques are incorporated to compute the optimal solutions. For a terminal state with a continuous distribution, the optimal solution is shown to be affine.

Keywords: Hypothesis Testing, Lagrangian Duality, Local Theory of Constrained Optimization.

1 Introduction

Control systems are widely used in modern technology and find applications in power systems, nuclear and chemical plants, the aerospace industry, robotics, communication devices, and embedded systems. All these systems typically rely on an underlying computing and networking infrastructure which has considerable security vulnerabilities. Classical control systems were designed without considering such security constraints, network attacks, and other system failures. Therefore one cannot rely on classical control techniques when designing controllers for an application employing *cyber-physical* systems. The term cyber-physical is generically used to describe all such physical systems which rely on a communication network and have computational capabilities. Any successful cyber attack which can undermine normal operation of the control system can create significant limitations in system operation and may cause potential damage in the aforementioned applications. Therefore, incorporation of relevant security constraints is extremely important in modern control system design. This paper is an effort in this research direction.

Control under cyber-physical security constraints (see [1], [2], [3], [4], [5], [6], [7], and [8]) has recently generated a lot of interest in the control, networking, and cyber-security research communities. The common approach has been to study the effect of

* This work has been funded by ONR AppEl center at UMD (Task E12), NSF CPS grant CNS 0931878, AFOSR FA95501110182, and ARL.

D.C. Tarraf (ed.), *Control of Cyber-Physical Systems*,
Lecture Notes in Control and Information Sciences 449,
DOI: 10.1007/978-3-319-01159-2_6,

adversarial attacks against particular control systems. The control systems have themselves been modeled as both deterministic and probabilistic whereas different formulations to model the adversary, attacking the system, have been presented.

A security problem is considered in [4] where a jammer can block the communication link between the controller and the plant. The problem is formulated as a dynamic zero sum game and a saddle-point equilibrium, between the controller and the jammer, is shown to exist. The optimal policy implemented by the jammer is proven to be of a threshold type. The effects of false data injection attacks in control systems are analyzed in [5]. Necessary and sufficient conditions are provided under which the attacker can destabilize the system while evading detection. In [6], system stability and resilience under feedback schemes is characterized in the presence of attacks on sensors and actuators. A game theoretic analysis is provided in [7] where an attacker can physically capture sensors nodes in a wireless network, can replicate captured nodes, and eventually take over the network. Nash equilibrium solutions are provided for both cases when the node capture rate is time invariant and when it is time varying. A geometric control approach is taken in [8] to tackle the problem of cyber-physical security. A mathematical framework for attacks and monitors is presented and fundamental monitoring limitations are characterized from a system theoretic and a graph theoretic perspective.

In this paper, we consider the problem of designing an optimal control system subject to security constraints. A deterministic, linear, and time-invariant system is analyzed for which the terminal state can take a finite number of values. It is assumed that the controller knows the desired terminal state and applies a control sequence accordingly. An adversary makes partial noisy measurements of the state trajectory and wants to estimate the actual value of the terminal state. The adversary has knowledge of the set of possible terminal states. The task of the controller is to develop a strategy to reach the terminal state while providing minimum information to the adversary thereby hindering its ability in estimating the terminal state. Different security metrics like probability of error of the adversary in computing an estimate of the terminal state and conditional mean based security metrics are considered and analyzed. These security metrics are introduced as constraints. Control sequences which minimize a quadratic cost function while satisfying these constraints are computed. The resulting optimization problems are shown to be convex and techniques from Lagrangian duality are employed to compute the optimal solution. The cases of a Gaussian noise distribution and a more general finite mean noise distribution for the measurements made by an adversary are analyzed. This problem has important civilian and military applications. The formulation and the security constraints are general enough to be applied to most systems where security is a concern.

We motivate this problem by presenting some interesting real life applications.

1. **Police Drug Bust:** Consider the situation where a convoy of police vehicles is assigned to inspect certain suspicious neighborhoods in a city. Once the police convoy reaches a neighborhood they completely seal all escape routes and do a complete house to house search for drugs, weapons, and other criminal activity. Also as soon as a neighborhood is sealed the criminals in other neighborhoods get tipped off and escape. The police convoy has several routes to reach each neighborhood.

The criminals have spies posted near the police station and can make some initial measurements regarding the route taken by the convoy. The task of the convoy is to optimize the route such that the criminals get minimum information regarding the actual neighborhood that the police will inspect on any given day.

2. **Soccer Penalty kick:** Consider the situation where a striker takes a penalty kick. He can either hit the ball straight, left, or towards the right side of the net. The goalkeeper can dive on either side of the net or stand in the middle in order to stop the penalty. The goalkeeper can make some measurements by observing how the striker runs in to strike the ball. The task of the striker is to choose an option and execute the kick such that the goalkeeper cannot predict the actual location and score a goal.

The Following Notation Is Adopted:

- Random variables are represented using bold face capital letters, for example $\mathbf{X_T}$ is used to represent the terminal state. Realizations are represented using small letters, like x_T.
- The probability of error is denoted by P_e, the prior distribution by π, the probability density of a random variable $\mathbf{X}$ if it exists by $p(x)$, the joint probability density of $\mathbf{X}$ and $\mathbf{Y}$ by $p(x,y)$, and the conditional probability density of $\mathbf{X}$ given $\mathbf{Y}$ by $p(x|y)$.
- Capital letter H denotes a hypothesis and M is the number of hypotheses.
- $I_{m\times m}$ represents the $m\times m$ identity matrix and $0_{m\times n}$ the $m\times n$ matrix of zeros. Logarithm to the base 2 and natural logarithm are denoted by log and ln respectively.
- The set of real numbers is denoted by $\Re$ and the null set is denoted by ϕ. The standard ℓ_p norm is denoted by $||.||_p$.

The paper is organized in seven sections. In Section 2, we provide a precise mathematical formulation of the problem. Section 3 is focused on designing secure control laws for the case when the terminal state can take two different values with specified probabilities. Under the assumption of a Gaussian distribution for measurement noise, we utilize the probability of error as a security constraint. The optimization problem is shown to be convex. We follow this by considering the case of a more general finite mean noise distribution. A security constraint based on the conditional mean is analyzed. In Section 4, we consider the case when the terminal state can take M different values. A generalization of the conditional mean based security constraint is analyzed and the resulting optimization problem is shown to be convex. Section 5 considers the case when the terminal state has a continuous distribution. A conditional mean based security constraint is analyzed and the optimal solution is shown to be affine in the terminal state. Simulations and conclusions are provided in Section 5 and Section 6, respectively.

2 Problem Formulation

Consider the following linear time-invariant system:

$$x_{k+1} = Ax_k + Bu_k, \quad k = 0, ..., T-1 \tag{1}$$

where $x_k \in \Re^n$ is the state of the system, $u_k \in \Re^m$ is the control input, A is an $n \times n$ matrix, and B is an $n \times m$ matrix. Without loss of generality, the initial state x_0 is assumed to be zero. The cases of both finite and continuous distributions of the terminal state are analyzed. In Section 3, we assume that $x_T \in \{x^0, x^1\}$ and in Section 4 it is assumed that $x_T \in \{x^0, ..., x^{M-1}\}$. The framework for the continuous terminal state distribution will be presented in Section V. In the finite distribution case, the desired terminal state is drawn from $\{x^0, ..., x^{M-1}\}$ with prior distribution $\pi_0, ..., \pi_{M-1}$. The controller knows the desired terminal state and accordingly applies the appropriate control sequence, $\{u_0, ..., u_{T-1}\}$, to reach it.

The adversary does not know the actual value of the terminal state, which is random, but knows its distribution. The adversary is restricted to make only the first $(k+1)$ measurements of the state with $k+1 < T$. These measurements are noisy and are given as follows:

$$\mathbf{Y_i} = Cx_i + \mathbf{V_i}, \quad i = 0, ..., k \tag{2}$$

where C is a $p \times n$ matrix and $\mathbf{V_0}, ..., \mathbf{V_k}$ are $p \times 1$ independent and identically distributed random vectors. Using (1) we can write the measurement model in compact form as follows:

$$\mathbf{Y_{0,k}} = \bar{C}U_{0,k-1} + \mathbf{V_{0,k}} \tag{3}$$

where $\mathbf{Y_{0,k}}$, $\bar{C}$, $U_{0,k-1}$ and $\mathbf{V_{0,k}}$ are given as follows:

$$\mathbf{Y_{0,k}} = \begin{bmatrix} \mathbf{Y_0} \\ \vdots \\ \mathbf{Y_k} \end{bmatrix}, \mathbf{V_{0,k}} = \begin{bmatrix} \mathbf{V_0} \\ \vdots \\ \mathbf{V_k} \end{bmatrix}, U_{0,k-1} = \begin{bmatrix} U_0 \\ \vdots \\ U_{k-1} \end{bmatrix}, \bar{C} = \begin{bmatrix} 0_{p\times m} & 0_{p\times m} & \dots & 0_{p\times m} \\ CB & 0_{p\times m} & \dots & 0_{p\times m} \\ \vdots & \vdots & \ddots & \vdots \\ CA^{k-1}B & CA^{k-2}B & \dots & CB \end{bmatrix} \tag{4}$$

We consider the case when $\mathbf{V_0}, ..., \mathbf{V_k}$ have a Gaussian distribution and the case when they have a general finite mean distribution. We assume that $\pi_i > 0$, where $i = 0, ..., M-1$. The dynamical system is assumed to be controllable and hence there are many control sequences which drive the system to the specified terminal state. We denote any control sequence which drives the system to the terminal state, x^i, by $U^i_{0,T-1}$. The task of the controller is to design the control sequences such that the adversary cannot estimate the actual value of the terminal state. Using the measurement model, the adversary solves a hypothesis testing problem. Under hypothesis H_i, when the terminal state is x^i, the information available to the adversary is as follows:

$$\mathbf{Y_{0,k}} = \bar{C}U^i_{0,k-1} + \mathbf{V_{0,k}} \tag{5}$$

It should be noted that the security constraints are dependent on the information, $\bar{C}U^i_{0,k-1}$, which is provided by the controller to the adversary. The quadratic cost to be minimized is provided below:

$$\mathbb{U}'Q\mathbb{U} = \sum_{i=0}^{M-1} \pi_i {U^i}_{0,T-1}{}' \tilde{Q} U^i_{0,T-1} \tag{6}$$

where $\tilde{Q}$ is a $Tm \times Tm$ symmetric positive definite matrix. The symmetric positive definite matrix Q and the control vector $\mathbb{U}$ are given as follows:

$$\mathbb{U} = \begin{bmatrix} U_{0,T-1}^{0} \\ \vdots \\ U_{0,T-1}^{M-1} \end{bmatrix}, \; Q = \begin{bmatrix} \pi_0 \tilde{Q} & 0_{Tm \times Tm} & \dots & 0_{Tm \times Tm} \\ 0_{Tm \times Tm} & \pi_1 \tilde{Q} & \dots & 0_{Tm \times Tm} \\ \vdots & \vdots & \ddots & \vdots \\ 0_{Tm \times Tm} & 0_{Tm \times Tm} & \dots & \pi_{M-1} \tilde{Q} \end{bmatrix} \tag{7}$$

Since the control sequences $U_{0,T-1}^{0}, \dots, U_{0,T-1}^{M-1}$ must drive the system to the terminal states $x^0, \dots, x^{M-1}$ we need to introduce the following equality constraints:

$$B_T U_{0,T-1}^{i} = x^i, \; i = 0, \dots, M-1 \tag{8}$$

which can be written in compact form as follows:

$$F\mathbb{U} = b \tag{9}$$

$$F = \begin{bmatrix} B_T & 0_{n \times Tm} & \dots & 0_{n \times Tm} \\ 0_{n \times Tm} & B_T & \dots & 0_{n \times Tm} \\ \vdots & \vdots & \ddots & \vdots \\ 0_{n \times Tm} & 0_{n \times Tm} & \dots & B_T \end{bmatrix}, b = \begin{bmatrix} x^0 \\ \vdots \\ x^{M-1} \end{bmatrix}, \; B_T = \left[A^{T-1}B, A^{T-2}B, \dots, B\right] \tag{10}$$

The optimization problem that we will solve will involve minimization of the cost function (6), subject to the equality constraint (9), and the security constraints that we provide in the next two sections. The optimization will be performed with respect to the control variable $\mathbb{U}$. The equality constraints can be removed from this optimization problem by utilizing an appropriate re-parametrization. Let U_b be a suitable control vector which is provided to us and satisfies $FU_b = b$. Let $\tilde{F}$ be the matrix whose columns form a basis for the Null space of F. Then we can write:

$$\mathbb{U} = U_b + \tilde{F}\eta, \quad \eta \in \Re^{dim(Null(F))} \tag{11}$$

It should be noted from the definition of $\tilde{F}$ that for any η, $F(\tilde{F}\eta) = 0$. Now η becomes our optimization variable and we do not need to incorporate the equality constraint (9). The cost function can be written as:

$$\mathbb{U}'Q\mathbb{U} = U_b'QU_b + 2U_b'Q\tilde{F}\eta + \eta'\tilde{F}'Q\tilde{F}\eta \tag{12}$$

3 Secure Control: A Binary Framework

In this section, we consider the framework where the terminal state can take two different values x^0 and x^1 with probabilities π_0 and π_1 respectively. We consider two different cases of measurement noise distribution.

3.1 Gaussian Noise Distribution

Consider the case when $\mathbf{V_0},...,\mathbf{V_k}$ are iid and have a Gaussian $\mathcal{N}(0,\tilde{\Sigma})$ distribution. Then $\mathbf{V_{0,k}}$ is a $\mathcal{N}(0,\Sigma)$ random vector where the covariance matrix Σ is block diagonal and has matrices $\tilde{\Sigma}$ on its diagonal. The minimum probability of error, in the estimate of the actual value of the terminal state using measurements made by the adversary, is introduced as a security constraint. Under H_0 and H_1 the measurements have the following distributions:

$$H_0 : \mathbf{Y_{0,k}} = \bar{C}U^0_{0,k-1} + \mathbf{V_{0,k}} \sim \mathcal{N}(\bar{C}U^0_{0,k-1},\Sigma) \tag{13}$$

$$H_1 : \mathbf{Y_{0,k}} = \bar{C}U^1_{0,k-1} + \mathbf{V_{0,k}} \sim \mathcal{N}(\bar{C}U^1_{0,k-1},\Sigma) \tag{14}$$

We use a Bayesian formulation with a uniform cost and knowledge of the priors to compute the minimum probability of error. The optimal Bayes test is a likelihood ratio test [9], which is given by:

$$L(y_{0,k}) = \frac{p(y_{0,k}|H_1)}{p(y_{0,k}|H_0)}$$

$$= \exp\left\{(U^1_{0,k-1} - U^0_{0,k-1})'\bar{C}'\Sigma^{-1}y_{0,k} - \frac{1}{2}(U^1_{0,k-1} - U^0_{0,k-1})'\bar{C}'\Sigma^{-1}\bar{C}(U^1_{0,k-1} + U^0_{0,k-1})\right\} \tag{15}$$

The optimal Bayes test γ_B is given as follows:

$$\gamma_B(y_{0,k}) = \begin{Bmatrix} H_1 \text{ if } L(y_{0,k}) \geq \frac{\pi_0}{\pi_1} \\ H_0 \text{ if } L(y_{0,k}) < \frac{\pi_0}{\pi_1} \end{Bmatrix} \tag{16}$$

By taking ln on both sides we can write the inequality $L(y_{0,k}) \geq \frac{\pi_0}{\pi_1}$ as:

$$(U^1_{0,k-1} - U^0_{0,k-1})'\bar{C}'\Sigma^{-1}y_{0,k} \geq \tau$$

$$\tau = \frac{1}{2}(U^1_{0,k-1} - U^0_{0,k-1})'\bar{C}'\Sigma^{-1}\bar{C}(U^1_{0,k-1} + U^0_{0,k-1}) + \ln(\frac{\pi_0}{\pi_1})$$

Now $Y_{0,k}$ is a Gaussian random vector whose linear transformation is also Gaussian. By computing the mean and variance of $(U^1_{0,k-1} - U^0_{0,k-1})'\bar{C}'\Sigma^{-1}\mathbf{Y_{0,k}}$ under H_0 and H_1 we get:

$$H_0 : \; (U^1_{0,k-1} - U^0_{0,k-1})'\bar{C}'\Sigma^{-1}\mathbf{Y_{0,k}} \sim \mathcal{N}\left((U^1_{0,k-1} - U^0_{0,k-1})'\bar{C}'\Sigma^{-1}\bar{C}U^0_{0,k-1}, d^2\right) \tag{17}$$

$$H_1 : \; (U^1_{0,k-1} - U^0_{0,k-1})'\bar{C}'\Sigma^{-1}\mathbf{Y_{0,k}} \sim \mathcal{N}\left((U^1_{0,k-1} - U^0_{0,k-1})'\bar{C}'\Sigma^{-1}\bar{C}U^1_{0,k-1}, d^2\right) \tag{18}$$

$$d^2 = (U^1_{0,k-1} - U^0_{0,k-1})'\bar{C}'\Sigma^{-1}\bar{C}(U^1_{0,k-1} - U^0_{0,k-1})$$

The probability of error in estimating the true value of the terminal state is given by:

$$P_e = \pi_0 P\left(L(\mathbf{Y_{0,k}} \geq \frac{\pi_0}{\pi_1}) \middle| H_0\right) + \pi_1 P\left(L(\mathbf{Y_{0,k}} < \frac{\pi_0}{\pi_1}) \middle| H_1\right)$$

$$= \pi_0 - \pi_0 \Phi\left(\frac{1}{d}\ln\frac{\pi_0}{\pi_1} + \frac{d}{2}\right) + \pi_1 \Phi\left(\frac{1}{d}\ln\frac{\pi_0}{\pi_1} - \frac{d}{2}\right) \tag{19}$$

where Φ is the pdf of a standard normal $\mathcal{N}(0,1)$ distribution. For the special case of equi-probable priors, $\pi_0 = \pi_1 = \frac{1}{2}$, the above expression simplifies to give:

$$P_e = \rho\left(\frac{d}{2}\right) \tag{20}$$

Here ρ denotes the Q-function, the tail probability of the standard normal distribution. Consider the following optimization problem which introduces the probability of error as a security constraint:

Problem 3.1:
Minimize the cost function

$$U_b' Q U_b + 2U_b' Q \tilde{F} \eta + \eta' \tilde{F}' Q \tilde{F} \eta$$

subject to the constraint:

$$P_e \geq \alpha$$

where $\alpha \geq 0$ is a constraint parameter that we choose. This constraint basically tells how inaccurate the estimate of the adversary is in determining the actual value of the terminal state. The value of α provides a measure on the security level of the control sequences. The following result shows that this security constraint is convex in η.

Proposition 3.1.1:
The probability of error constraint is convex in η and Problem 3.1 is a convex program.

Proof: From (19), we note that P_e is decreasing in d which is nonnegative. Therefore:

$$P_e \geq \alpha \;\Leftrightarrow\; d^2 \leq \alpha_1$$

where α_1 can be determined from the values of α, π_0, and π_1. For the special case where $\pi_0 = \pi_1$ we have $\alpha_1 = 4(\rho^{-1}(\alpha))^2$. Consider the following notation which enables us to write d^2 in terms of η.

$$U_{0,k-1}^i = GS_i(U_b + \tilde{F}\eta),\; G = [I_{km\times km} \;\; 0_{km\times(T-k)m}]$$

$$S_0 = [I_{Tm\times Tm} \;\; 0_{Tm\times Tm}],\; S_1 = [0_{Tm\times Tm} \;\; I_{Tm\times Tm}] \tag{21}$$

$$d^2 = (U_b + \tilde{F}\eta)'(S_1 - S_0)' G' \bar{C}' \Sigma^{-1} \bar{C} G (S_1 - S_0)(U_b + \tilde{F}\eta)$$

Clearly d^2 is convex in η, $d^2 \leq \alpha_1$ forms a convex set, and the cost is strictly convex in η. □

Using Proposition 3.1.1 and (21) we solve the following convex program:

$$\min_{\eta} \; U_b' Q U_b + 2U_b' Q \tilde{F} \eta + \eta' \tilde{F}' Q \tilde{F} \eta$$

subject to the constraint:

$$(U_b + \tilde{F}\eta)'(S_1 - S_0)' G' \bar{C}' \Sigma^{-1} \bar{C} G (S_1 - S_0)(U_b + \tilde{F}\eta) \leq \alpha_1 \tag{22}$$

We can practically solve this convex program by using standard convex optimization software like cvx [10]. By making some constraint qualification assumptions we characterize the optimal solution using Lagrangian duality.

Assumption I: We assume that α is selected such that: $\Big\{ \eta \in \Re^{dim(Null(F))} \mid (U_b + \tilde{F}\eta)'(S_1 - S_0)' G' \bar{C}' \Sigma^{-1} \bar{C} G (S_1 - S_0)(U_b + \tilde{F}\eta) < \alpha_1 \Big\} \neq \phi$.

This assumption ensures that there exists a η such that the constraint in (22) is satisfied with strict inequality. This is precisely Slater's condition (see [11] for details) for this problem. Therefore, Assumption I implies that the duality gap is zero.

Proposition 3.1.2: The optimal solution to Problem 3.1, under Assumption I, is nonlinear in b and is given by:

$$\eta^* = -\left(\tilde{F}' Q \tilde{F} + \lambda^* \tilde{F}' \mathscr{L}' \mathscr{L} \tilde{F} \right)^{-1} \left(\tilde{F}' Q + \lambda^* \tilde{F}' \mathscr{L}' \mathscr{L} \right) U_b$$

where $\lambda^* \geq 0$ is the solution to the following equation:

$$\left\| \mathscr{L} U_b - \mathscr{L} \tilde{F} \left(\tilde{F}' Q \tilde{F} + \lambda \tilde{F}' \mathscr{L}' \mathscr{L} \tilde{F} \right)^{-1} \left(\tilde{F}' Q + \lambda \tilde{F}' \mathscr{L}' \mathscr{L} \right) U_b \right\|_2 = \sqrt{\alpha_1}$$

where $\mathscr{L} = W' \bar{C} G (S_1 - S_0)$, $\Sigma^{-1} = WW'$, and b is the vector of terminal states.

Proof: Problem 3.1.2 is a Quadratically Constrained Quadratic Program (QCQP). Such problems have previously been solved in the literature (See Ch 4 in [11]). We provide a proof here for the sake of completeness.

Let $\Sigma^{-1} = WW'$ be the Cholesky decomposition of the inverse of the covariance matrix. The Lagrangian can be written as follows:

$$L(\eta, \lambda) = U_b' Q U_b + 2U_b' Q \tilde{F} \eta + \eta' \tilde{F}' Q \tilde{F} \eta + \lambda \Big(U_b' \mathscr{L}' \mathscr{L} U_b + 2\eta' \tilde{F}' \mathscr{L}' \mathscr{L} U_b + \eta' \tilde{F}' \mathscr{L}' \mathscr{L} \tilde{F} \eta - \alpha_1 \Big) \tag{23}$$

It should be noted that the Lagrangian, $L(\eta, \lambda)$, is strictly convex in η. This is due to the columns of $\tilde{F}$ being linearly independent. Strong duality holds from Slater's conditions and we can solve the dual problem to get the optimal cost. Strict convexity, strong duality, and the fact that the optimal cost is finite ensure that we can get a solution of the primal problem through the dual problem [11] (See Ch. 5). The Lagrange dual function is given by:

$$g(\lambda) = \min_{\eta} \left\{ U_b' Q U_b + 2U_b' Q \tilde{F} \eta + \eta' \tilde{F}' Q \tilde{F} \eta + \lambda \left(U_b' \mathscr{L}' \mathscr{L} U_b + 2\eta' \tilde{F}' \mathscr{L}' \mathscr{L} U_b + \eta' \tilde{F}' \mathscr{L}' \mathscr{L} \tilde{F} \eta - \alpha_1 \right) \right\} \tag{24}$$

Computing the gradient of the Lagrangian with respect to η and setting it equal to zero we get:

$$\eta = -\left(\tilde{F}' Q \tilde{F} + \lambda \tilde{F}' \mathscr{L}' \mathscr{L} \tilde{F} \right)^{-1} \left(\tilde{F}' Q + \lambda \tilde{F}' \mathscr{L}' \mathscr{L} \right) U_b \tag{25}$$

Plugging (25) into (24) we get:

$$\begin{aligned} g(\lambda) = {} & U_b' Q U_b - 2U_b' Q \tilde{F} \left(\tilde{F}' Q \tilde{F} + \lambda \tilde{F}' \mathscr{L}' \mathscr{L} \tilde{F} \right)^{-1} \left(\tilde{F}' Q + \lambda \tilde{F}' \mathscr{L}' \mathscr{L} \right) U_b + \\ & U_b' (\tilde{F}' Q + \lambda \tilde{F}' \mathscr{L}' \mathscr{L})' (\tilde{F}' Q \tilde{F} + \lambda \tilde{F}' \mathscr{L}' \mathscr{L} \tilde{F})^{-1} (\tilde{F}' Q + \lambda \tilde{F}' \mathscr{L}' \mathscr{L}) U_b + \lambda U_b' \mathscr{L}' \mathscr{L} U_b \\ & -2\lambda U_b' (\tilde{F}' Q + \lambda \tilde{F}' \mathscr{L}' \mathscr{L})' (\tilde{F}' Q \tilde{F} + \lambda \tilde{F}' \mathscr{L}' \mathscr{L} \tilde{F})^{-1} \tilde{F}' \mathscr{L}' \mathscr{L} U_b - \lambda \alpha_1 \end{aligned}$$

Now the optimal cost can be obtained by maximizing the Lagrange dual function with respect to λ which is assumed to be non-negative. Differentiating g with respect to λ and setting the derivative equal to zero we get:

$$\begin{aligned} & \Big(U_b' (\tilde{F}' Q + \lambda \tilde{F}' \mathscr{L}' \mathscr{L})' (\tilde{F}' Q \tilde{F} + \lambda \tilde{F}' \mathscr{L}' \mathscr{L} \tilde{F})^{-1} \tilde{F}' \mathscr{L}' \mathscr{L} \tilde{F} (\tilde{F}' Q \tilde{F} + \lambda \tilde{F}' \mathscr{L}' \mathscr{L} \tilde{F})^{-1} \times \\ & (\tilde{F}' Q + \lambda \tilde{F}' \mathscr{L}' \mathscr{L}) U_b \Big) + U_b' \mathscr{L}' \mathscr{L} U_b - \Big(2U_b' (\tilde{F}' Q + \lambda \tilde{F}' \mathscr{L}' \mathscr{L})' (\tilde{F}' Q \tilde{F} + \lambda \tilde{F}' \mathscr{L}' \mathscr{L} \tilde{F})^{-1} \\ & \times \tilde{F}' \mathscr{L}' \mathscr{L} U_b \Big) = \alpha_1 \end{aligned}$$

Simplifying this equation we get:

$$\left\| \mathscr{L} U_b - \mathscr{L} \tilde{F} \left(\tilde{F}' Q \tilde{F} + \lambda \tilde{F}' \mathscr{L}' \mathscr{L} \tilde{F} \right)^{-1} \left(\tilde{F}' Q + \lambda \tilde{F}' \mathscr{L}' \mathscr{L} \right) U_b \right\|_2 = \sqrt{\alpha_1} \tag{26}$$

Plugging the value of $\lambda \geq 0$ which solves the above equation into (25) provides an optimal solution to the problem, proving the claim in the proposition. It should be noted from (26) that λ^* is non-linear in U_b. Therefore, the optimal solution of the problem is non-linear in b. □

Proposition 3.1.2 gives us a form of the optimal control sequences which will satisfy the probability of error security constraint. The convexity of this problem makes it very easy to practically implement these results using commercial optimization solvers.

3.2 Finite Mean Noise Distribution

In this section, we consider the case where $\mathbf{V_0},...,\mathbf{V_k}$ are iid and have a general distribution with finite mean. Computing a closed form expression for the probability of error is a very difficult problem even for the specific case of the exponential family of distributions [12].

We consider a new security framework based on the conditional mean which leads to a constraint very similar in structure to (22). Let μ be the mean of the noise vector $\mathbf{V_{0,k}}$, defined in (3). Under the hypothesis H_0 and H_1 the measurement model is given as follows:

$$H_0 : \mathbf{Y_{0,k}} = \bar{C}GS_0(U_b + \tilde{F}\eta) + \mathbf{V_{0,k}},\ H_1 : \mathbf{Y_{0,k}} = \bar{C}GS_1(U_b + \tilde{F}\eta) + \mathbf{V_{0,k}} \quad (27)$$

Problem 3.2:
Minimize the cost function:

$$U_b'QU_b + 2U_b'Q\tilde{F}\eta + \eta'\tilde{F}'Q\tilde{F}\eta$$

subject to the constraint:

$$\Big(E(\mathbf{Y_{0,k}}\,|H_1) - E(\mathbf{Y_{0,k}}\,|H_0)\Big)'\Big(E(\mathbf{Y_{0,k}}\,|H_1) - E(\mathbf{Y_{0,k}}\,|H_0)\Big) \le \alpha_2\,,\quad \alpha_2 \ge 0 \quad (28)$$

The security constraint in (28) has a nice operational interpretation. It basically measures how far apart the means of the observations are under the two different hypothesis. The further apart the means the easier it will be for the adversary to determine the true value of the terminal state. It has the same intuitive interpretation as the probability of error constraint utilized in the previous section. It should be noted that, $\alpha_2 = 0$, implies that the adversary does not get any useful information from the partial state observations.

Proposition 3.2.1: Problem 3.2 is a convex program.

Proof:
Now the difference of the conditional means is given as follows:

$$E(\mathbf{Y_{0,k}}\,|H_1) - E(\mathbf{Y_{0,k}}\,|H_0) = \bar{C}G(S_1 - S_0)(U_b + \tilde{F}\eta) \quad (29)$$

Using (29) the constraint in (28) can be written as:

$$U_b'\mathscr{X}'\mathscr{X}U_b + 2\eta'\tilde{F}'\mathscr{X}'\mathscr{X}U_b + \eta'\tilde{F}'\mathscr{X}'\mathscr{X}\tilde{F}\eta \le \alpha_2 \quad (30)$$

where $\mathscr{X} = \bar{C}G(S_1 - S_0)$. Clearly this security constraint is convex and Problem 3.2 is a convex program. □

The security constraint (30) is very similar to (22), as $\mathscr{L}'\mathscr{L} = \mathscr{X}'\Sigma^{-1}\mathscr{X}$, and hence this section generalizes the work in section 3.1. In order to use Lagrangian duality techniques we make the following assumption.

Assumption II: We assume that α_2 is selected such that: $\Big\{\eta \in \Re^{dim(Null(F))} \mid U_b'\mathscr{L}'\mathscr{L}U_b + 2\eta'\tilde{F}'\mathscr{L}'\mathscr{L}U_b + \eta'\tilde{F}'\mathscr{L}'\mathscr{L}\tilde{F}\eta < \alpha_2\Big\} \neq \phi$.

Proposition 3.2.2: The optimal solution to Problem 3.2, under Assumption II, is non-linear in b and is given by:

$$\eta^* = -\left(\tilde{F}'Q\tilde{F} + \lambda^*\tilde{F}'\mathscr{L}'\mathscr{L}\tilde{F}\right)^{-1}\left(\tilde{F}'Q + \lambda^*\tilde{F}'\mathscr{L}'\mathscr{L}\right)U_b$$

where $\lambda^* \geq 0$ is the solution to the following equation:

$$\left\|\mathscr{L}U_b - \mathscr{L}\tilde{F}\left(\tilde{F}'Q\tilde{F} + \lambda\tilde{F}\mathscr{L}'\mathscr{L}\tilde{F}\right)^{-1}\left(\tilde{F}'Q + \lambda\tilde{F}'\mathscr{L}'\mathscr{L}\right)U_b\right\|_2 = \sqrt{\alpha_2}$$

Proof:
The proof follows exactly along the same lines as the proof of Proposition 3.1.2. □

Using these results we can design secure controls for any system for which the adversary makes observations where the additive noise is drawn from a general finite mean distribution. The aforementioned security constraint can also be extended to the M-ary framework. We explore such an extension in the next section.

4 Secure Control: A M-ary Framework

In this section, we consider the framework where the terminal state can take M different values $x^0, x^1, ..., x^{M-1}$ with probabilities $\pi_0, \pi_1, ..., \pi_{M-1}$ respectively. We assume that the measurement noise vector $\mathbf{V_{0,k}}$ has a distribution where the mean vector, μ, has finite components.

We consider a constraint defined using the conditional means. Now under the hypothesis H_i, $i = 0, ..., M-1$, the measurement model is given by:

$$H_i : \mathbf{Y_{0,k}} = \bar{C}GS_i(U_b + \tilde{F}\eta) + \mathbf{V_{0,k}}, \;\; i = 0, ..., M-1 \tag{31}$$

Consider the following security metric:

$$\sum_{i=0}^{M-1} \pi_i \left(E(\mathbf{Y_{0,k}}|H_i) - \sum_{j=0}^{M-1} \pi_j E(\mathbf{Y_{0,k}}|H_j)\right)' \left(E(\mathbf{Y_{0,k}}|H_i) - \sum_{j=0}^{M-1} \pi_j E(\mathbf{Y_{0,k}}|H_j)\right) \tag{32}$$

This security metric can be considered to be a generalization of the difference of conditional means which was employed for the binary framework. The distribution π_i is used to assign weights to each quadratic quantity in (32), which provides a difference of the conditional mean given one hypothesis with the weighted sum of the conditional means given other hypotheses. It should be noted that $Y_{0,k}$ is independent of terminal state X_T if and only if this security metric is zero. A lower level of dependence between the observations and the terminal state is indicated by a small value of this metric. A smaller value makes it more difficult for the adversary to predict the terminal state and indicates a higher level of security of the control sequence. Utilizing this security metric we state the following optimization problem:

Problem 4.1:
Minimize the cost function

$$U_b'QU_b + 2U_b'Q\tilde{F}\eta + \eta'\tilde{F}'Q\tilde{F}\eta$$

subject to the constraint:

$$\sum_{i=0}^{M-1} \pi_i \left(E(\mathbf{Y_{0,k}}|H_i) - \sum_{j=0}^{M-1} \pi_j E(\mathbf{Y_{0,k}}|H_j) \right)' \left(E(\mathbf{Y_{0,k}}|H_i) - \sum_{j=0}^{M-1} \pi_j E(\mathbf{Y_{0,k}}|H_j) \right) \leq \alpha_3 \tag{33}$$

The parameter α_3 is selected to be non-negative. A small α_3 indicates that a lower level of useful information can be transmitted to the adversary. For the case, $\alpha_3 = 0$, the controller will be restricted to those control sequences which generate the same state trajectory for the first k+1 steps. Using the measurement model we get:

$$E(\mathbf{Y_{0,k}}|H_i) - \sum_{j=0}^{M-1} \pi_j E(\mathbf{Y_{0,k}}|H_j) = \bar{C}G(S_i - \sum_{j=0}^{M-1} \pi_j S_j)(U_b + \tilde{F}\eta) \tag{34}$$

Using (34), and $\mathscr{Z}_i = \bar{C}G(S_i - \sum_{j=0}^{M-1} \pi_j S_j)$, we can write (33) as:

$$\sum_{i=0}^{M-1} \pi_i \left(U_b' \mathscr{Z}_i' \mathscr{Z}_i U_b + 2\eta'\tilde{F}' \mathscr{Z}_i' \mathscr{Z}_i U_b + \eta'\tilde{F}' \mathscr{Z}_i' \mathscr{Z}_i \tilde{F}\eta \right) \leq \alpha_3 \tag{35}$$

Clearly (35) is convex in η and hence we can conclude that Problem 4.1 is a convex program. By making similar assumptions we can extend the results stated in section 3.2 to the M-ary framework.

Assumption III: We assume that α_3 is selected such that: $\Big\{ \eta \in \Re^{dim(Null(F))} \mid$
$\sum_{i=0}^{M-1} \pi_i \left(U_b' \mathscr{Z}_i' \mathscr{Z}_i U_b + 2\eta'\tilde{F}' \mathscr{Z}_i' \mathscr{Z}_i U_b + \eta'\tilde{F}' \mathscr{Z}_i' \mathscr{Z}_i \tilde{F}\eta \right) < \alpha_3 \Big\} \neq \phi$.

Proposition 4.1: The optimal solution of Problem 4.1, under Assumption III, is non-linear in b and is given by:

$$\eta^* = -\left(\tilde{F}'Q\tilde{F} + \lambda^* \sum_{i=0}^{M-1} \pi_i \tilde{F}' \mathscr{Z}_i' \mathscr{Z}_i \tilde{F} \right)^{-1} \left(\tilde{F}'Q + \lambda^* \sum_{i=0}^{M-1} \pi_i \tilde{F}' \mathscr{Z}_i' \mathscr{Z}_i \right) U_b$$

where $\lambda^* \geq 0$ is the solution to the following equation:

$$\sum_{k=0}^{M-1} \pi_k \left\| \mathscr{Z}_k U_b - \mathscr{Z}_k \tilde{F} \left(\tilde{F}'Q\tilde{F} + \lambda \sum_{i=0}^{M-1} \pi_i \tilde{F}' \mathscr{Z}_i' \mathscr{Z}_i \tilde{F} \right)^{-1} \times \right.$$
$$\left. \times \left(\tilde{F}'Q + \lambda \sum_{i=0}^{M-1} \pi_i \tilde{F}' \mathscr{Z}_i' \mathscr{Z}_i \tilde{F} \right) U_b \right\|_2 = \sqrt{\alpha_3}$$

Proof: The proof follows exactly like the proof of Proposition 3.2 to which the reader is referred. □

The results stated in Proposition 4.1 are very similar to the results presented in Propositions 3.1 and 3.2. These results imply that the optimal solution is non-linear in the vector of terminal states. In order to find the optimal solution in these problems we need to solve for the equality of a norm to a design parameter. In Proposition 4.1, we have the weighted sum of such norms equaling $\sqrt{\alpha_3}$.

5 Secure Control: Terminal State with a Continuous Distribution

In this section, we consider the situation where the terminal state has a continuous finite mean distribution with given density function $p(x_T)$. In addition, the components of the terminal state are assumed to have a finite variance. It is assumed that the adversary knows the distribution of the terminal state. We first provide a definition of the Gateaux differential which will be used repeatedly in the sequel.

Definition 5.1: **(Gateaux Differential [13])**
Let X be a vector space and let T be a transformation defined on a domain $D \subset X$. Let $x \in D$, $\gamma \in \Re$, and let h be arbitrary in X. If the limit

$$\delta T(x;h) = \lim_{\gamma \to 0} \frac{1}{\gamma}[T(x+\gamma h) - T(x)]$$

exists, it is called the Gateaux differential of T at x with increment h. If the limit exists for each $h \in X$, then T is said to be Gateaux differentiable at x.

Let $U(x_T)$ be the control sequence which drives the terminal state to x_T, $B_T U(x_T) = x_T$. Such a control sequence exists because of the controllability assumption. It should be noted that $U(.)$ is a function of the terminal state. Consider the following cost functional:

$$\mathbb{J}(U(.)) = \int_{\Re^n} U(x_T)'\tilde{Q}U(x_T)p(x_T)\,dx_T \tag{36}$$

The functions $U(.)$ are $\mathbb{L}^2$ integrable and belong to the space $\mathbb{L}^2(\Re^n, \mathscr{B}(\Re^n), \mu_p)$. Note that $\mathscr{B}(\Re^n)$ is the Borel σ-algebra on $\Re^n$ and μ_p is the probability measure corresponding to the distribution of the terminal state. We use the same technique which we employed in the previous sections to remove the equality constraint, $B_T U(x_T) = x_T$. Let $\tilde{U}(x_T)$ be a given control function such that $B_T\tilde{U}(x_T) = x_T, \forall x_T \in \Re^n$. We assume that $\tilde{U}(x_T)$ is linear in x_T. This is possible due to the controllability assumption and one possible choice for this control function is $\tilde{U}(x_T) = B_T^{\dagger}x_T$. $B_T^{\dagger}$ is the Moore-Penrose pseudoinverse of B_T. Let $\tilde{B}$ be a basis for the null space of B_T. Then we can write:

$$U(x_T) = \tilde{U}(x_T) + \tilde{B}\eta(x_T),\ \eta(x_T) \in \Re^q,\ q = dim(Null(B_T)) \tag{37}$$

Using (37), $\eta(.)$ becomes our new optimization variable and the cost functional can be written as:

$$\mathbb{J}(\eta(.)) = \int_{\Re^n} \Big(\tilde{U}(x_T) + \tilde{B}\eta(x_T)\Big)'\tilde{Q}\Big(\tilde{U}(x_T) + \tilde{B}\eta(x_T)\Big)p(x_T)\,dx_T \tag{38}$$

We assume that the observation noise has a general finite mean noise distribution. Under the hypothesis H_{x_T}, the measurements made by the adversary are given by:

$$H_{x_T}: \quad \mathbf{Y}_{0,k} = \bar{G}U(x_T) + \mathbf{V}_{0,k}, \quad \bar{G} = \bar{C}G \tag{39}$$

We now introduce the following security metric based on the difference of conditional means:

$$\int_{\Re^n}\int_{\Re^n} p(x_T)p(y_T)\Big(E\big(Y_{0,k}\big|H_{x_T}\big) - E\big(Y_{0,k}\big|H_{y_T}\big)\Big)' \times \Big(E\big(Y_{0,k}\big|H_{x_T}\big) - E\big(Y_{0,k}\big|H_{y_T}\big)\Big) dy_T dx_T \tag{40}$$

where y_T is another realization of the terminal state. This security metric basically provides a measure on the difference of the conditional means. The higher the value of this metric the easier it will be for the adversary to estimate the terminal state. Using the measurement model this metric can be simplified as follows:

$$\int_{\Re^n}\int_{\Re^n} p(x_T)p(y_T)\Big(U(x_T) - U(y_T)\Big)'\bar{G}'\bar{G}\Big(U(x_T) - U(y_T)\Big) dy_T dx_T \tag{41}$$

Using the cost functional (38), and the security constraint (41), we state the following optimization problem:

Problem 5.1:

$$\min_{\eta(.)} \int_{\Re^n}\Big(\tilde{U}(x_T) + \tilde{B}\eta(x_T)\Big)'\tilde{Q}\Big(\tilde{U}(x_T) + \tilde{B}\eta(x_T)\Big) p(x_T)\, dx_T$$

subject to the constraint:

$$\int_{\Re^n}\int_{\Re^n} p(x_T)p(y_T)\Big(\tilde{U}(x_T) - \tilde{U}(y_T) + \tilde{B}\big(\eta(x_T) - \eta(y_T)\big)\Big)'\bar{G}'\bar{G} \times \Big(\tilde{U}(x_T) - \tilde{U}(y_T) + \tilde{B}\big(\eta(x_T) - \eta(y_T)\big)\Big)\, dy_T dx_T \le \alpha_4$$

where α_4 is assumed to be nonnegative. A small value of α_4 indicates that a lower level of useful information is provided to the adversary. If $\alpha_4 = 0$, then the first k control inputs of all admissible control sequences will be the same. Problem 5.1 is a convex infinite dimensional optimization problem. We provide a solution to this problem by utilizing the Generalized Kuhn Tucker Conditions [13] (See Ch. 9).

Assumption IV: We consider only those values of α_4 for which the appropriate Lagrange multiplier, λ^*, is small enough such that $\Big[I_{q\times q} - 2\lambda^*\tilde{B}'\bar{G}'\bar{G}\tilde{B}\Big(\tilde{B}'(\tilde{Q} + 2\lambda^*\bar{G}'\bar{G})\tilde{B}\Big)^{-1}\Big]$ is nonsingular.

Assumption V: We assume that the constraint parameter, α_4, and the system dynamics are selected such that the optimal solution $\eta^*(.)$ to Problem 5.1 satisfies the security constraint and that there exists a $h(.) \in \mathbb{L}^2(\Re^n, \mathscr{B}(\Re^n), \mu_p)$ such that:

$$\int_{\Re^n}\int_{\Re^n} p(x_T)p(y_T)\Big(\tilde{U}(x_T) - \tilde{U}(y_T) + \tilde{B}(\eta^*(x_T) - \eta^*(y_T))\Big)' \bar{G}'\bar{G}\Big(\tilde{U}(x_T) - \tilde{U}(y_T) +$$
$$\tilde{B}(\eta^*(x_T) - \eta^*(y_T))\Big)\, dy_T dx_T + 4\int_{\Re^n}\Big(\tilde{U}(x_T) + \tilde{B}\eta^*(x_T)\Big)' \bar{G}'\bar{G}\tilde{B}h(x_T)p(x_T)dx_T$$
$$-4\int_{\Re^n}\int_{\Re^n}\Big(\tilde{U}(x_T) + \tilde{B}\eta^*(x_T)\Big)' \bar{G}'\bar{G}\tilde{B}h(y_T)p(y_T)p(x_T)dy_T dx_T < \alpha_4$$

Using Gateaux differentials, Assumption IV, Assumption V, and the generalized Kuhn Tucker conditions we obtain the following result.

Proposition 5.1: The optimal solution to Problem 5.1, under Assumptions IV and V, is affine in x_T and is given by:

$$\eta^*(x_T) = -\Big(\tilde{B}'\tilde{Q}\tilde{B} + 2\lambda^*\tilde{B}'\bar{G}'\bar{G}\tilde{B}\Big)^{-1}\Big(\tilde{B}'\tilde{Q} + 2\lambda^*\tilde{B}'\bar{G}'\bar{G}\Big)\tilde{U}(x_T) \ +$$
$$2\lambda^*\Big(\tilde{B}'\tilde{Q}\tilde{B} + 2\lambda^*\tilde{B}'\bar{G}'\bar{G}\tilde{B}\Big)^{-1}\Big(\tilde{B}'\bar{G}'\bar{G} - (I_{q\times q} - 2\lambda^*\Gamma_{\lambda^*})^{-1}\Gamma_{\lambda^*}\tilde{B}'\tilde{Q}\Big)\int p(y_T)\tilde{U}(y_T)dy_T$$

where $\Gamma_{\lambda^*} = \tilde{B}'\bar{G}'\bar{G}\tilde{B}\Big(\tilde{B}'(\tilde{Q} + 2\lambda^*\bar{G}'\bar{G})\tilde{B}\Big)^{-1}$ and $\lambda^* \geq 0$ is the solution to the following equation:

$$\lambda\Big[-2\int_{\Re^n}\int_{\Re^n} p(x_T)p(y_T)\Big(\tilde{U}(x_T) + \tilde{B}\eta^*(x_T)\Big)' \bar{G}'\bar{G}\Big(\tilde{U}(y_T) + \tilde{B}\eta^*(y_T)\Big)dy_T dx_T +$$
$$2\int_{\Re^n} p(x_T)\Big(\tilde{U}(x_T) + \tilde{B}\eta^*(x_T)\Big)' \bar{G}'\bar{G}\Big(\tilde{U}(x_T) + \tilde{B}\eta^*(x_T)\Big)dx_T - \alpha_4\Big] = 0$$

Proof:
We will first compute the Gateaux differential of the Lagrangian and then use the Kuhn Tucker Conditions to compute the optimal solution. Now the Lagrangian can be written as follows:

$$J_\lambda(\eta(.)) = \int_{\Re^n} p(x_T)\Big(\tilde{U}(x_T) + \tilde{B}\eta(x_T)\Big)'\Big(\tilde{Q} + 2\lambda\bar{G}'\bar{G}\Big)\Big(\tilde{U}(x_T) + \tilde{B}\eta(x_T)\Big)dx_T$$
$$-2\lambda\int_{\Re^n}\int_{\Re^n} p(x_T)p(y_T)\Big(\tilde{U}(x_T) + \tilde{B}\eta(x_T)\Big)' \bar{G}'\bar{G}\Big(\tilde{U}(y_T) + \tilde{B}\eta(y_T)\Big)dy_T dx_T \quad (42)$$

For any admissible variation $h(.) \in \mathbb{L}^2(\Re^n, \mathscr{F}, \mu_p)$ the Gateaux differential of the Lagrangian is given by:

$$\delta J_\lambda(\eta(.),h(.)) = 2\int_{\Re^n} p(x_T)\tilde{U}(x_T)'\left(\tilde{Q}+2\lambda\bar{G}'\bar{G}\right)\tilde{B}h(x_T)dx_T + 2\int_{\Re^n} p(x_T)\times$$
$$\eta(x_T)'\tilde{B}'\left(\tilde{Q}+2\lambda\bar{G}'\bar{G}\right)\tilde{B}h(x_T)dx_T - 4\lambda\int_{\Re^n}\int_{\Re^n} p(x_T)p(y_T)\tilde{U}(x_T)'\bar{G}'\bar{G}\tilde{B}h(y_T)dy_T dx_T$$
$$-4\lambda\int_{\Re^n}\int_{\Re^n} p(x_T)p(y_T)\eta(x_T)'\tilde{B}'\bar{G}'\bar{G}h(y_T)\,dy_T dx_T \quad (43)$$

Setting the Gateaux differential equal to zero we get:

$$\int_{\Re^n} p(x_T)h(x_T)'\Big(\tilde{B}'\tilde{Q}\tilde{U}(x_T)+2\lambda\tilde{B}'\bar{G}'\bar{G}\tilde{U}(x_T)+\tilde{B}'\tilde{Q}\tilde{B}\eta(x_T)+2\lambda\tilde{B}'\bar{G}'\bar{G}\tilde{B}\eta(x_T)$$
$$-2\lambda\int_{\Re^n} p(y_T)\tilde{B}'\bar{G}'\bar{G}\tilde{U}(y_T)dy_T - 2\lambda\int_{\Re^n} p(y_T)\tilde{B}'\bar{G}'\bar{G}\tilde{B}\eta(y_T)dy_T\Big)dx_T = 0,\ \forall\, h(.) \quad (44)$$

Now (44) holds if and only if:

$$\tilde{B}'\left(\tilde{Q}+2\lambda\bar{G}'\bar{G}\right)\tilde{U}(x_T)+\tilde{B}'\left(\tilde{Q}+2\lambda\bar{G}'\bar{G}\right)\tilde{B}\eta(x_T) - 2\lambda\int_{\Re^n} p(y_T)\tilde{B}'\bar{G}'\bar{G}\tilde{U}(y_T)dy_T$$
$$-2\lambda\int_{\Re^n} p(y_T)\tilde{B}'\bar{G}'\bar{G}\tilde{B}\eta(y_T)dy_T = 0, \quad \forall x_T \in \Re^n \quad (45)$$

Now multiplying (45) throughout by $p(x_T)\Gamma_\lambda$ and integrating over x_T we get:

$$\left(I_{q\times q}-2\lambda\Gamma_\lambda\right)\int_{\Re^n} p(x_T)\tilde{B}'\bar{G}'\bar{G}\tilde{B}\eta(x_T)dx_T = -\int_{\Re^n} p(x_T)\Gamma_\lambda\tilde{B}'\tilde{Q}\tilde{U}(x_T)dx_T$$

Using Assumption IV we get:

$$\int_{\Re^n} p(x_T)\tilde{B}'\bar{G}'\bar{G}\tilde{B}\eta(x_T)dx_T = -\left(I_{q\times q}-2\lambda\Gamma_\lambda\right)^{-1}\int_{\Re^n} p(x_T)\Gamma_\lambda\tilde{B}'\tilde{Q}\tilde{U}(x_T)dx_T \quad (46)$$

Now plugging (46) into (45) we get:

$$\eta^*(x_T) = -\left(\tilde{B}'\tilde{Q}\tilde{B}+2\lambda^*\tilde{B}'\bar{G}'\bar{G}\tilde{B}\right)^{-1}\left(\tilde{B}'\tilde{Q}+2\lambda^*\tilde{B}'\bar{G}'\bar{G}\right)\tilde{U}(x_T)+$$
$$2\lambda^*\left(\tilde{B}'\tilde{Q}\tilde{B}+2\lambda^*\tilde{B}'\bar{G}'\bar{G}\tilde{B}\right)^{-1}\left(\tilde{B}'\bar{G}'\bar{G}-(I_{q\times q}-2\lambda^*\Gamma_{\lambda^*})^{-1}\Gamma_{\lambda^*}\tilde{B}'\tilde{Q}\right)\int_{\Re^n} p(y_T)\tilde{U}(y_T)dy_T \quad (47)$$

which from the Kuhn Tucker conditions, used in conjunction with Assumption V, is the optimal solution. Also from Kuhn Tucker conditions λ^* is given by the solution of the following equation:

$$\lambda\Big[-2\int_{\Re^n}\int_{\Re^n} p(x_T)p(y_T)\Big(\tilde{U}(x_T)+\tilde{B}\eta^*(x_T)\Big)'\bar{G}'\bar{G}\Big(\tilde{U}(y_T)+\tilde{B}\eta^*(y_T)\Big)dy_T dx_T+$$
$$2\int_{\Re^n} p(x_T)\Big(\tilde{U}(x_T)+\tilde{B}\eta^*(x_T)\Big)'\bar{G}'\bar{G}\Big(\tilde{U}(x_T)+\tilde{B}\eta^*(x_T)\Big)dx_T-\alpha_4\Big]=0 \quad (48)$$

It should be noted that λ^* depends upon $\tilde{U}(.), p(.)$, and α_4 but not on x_T. Also $\tilde{U}(x_T)$ is selected to be linear in x_T. Therefore, we conclude from (47) that the optimal solution to Problem 5.1 is affine in x_T. □

Proposition 5.1 is an important result and shows that the optimal solution to the problem is affine in the terminal state. Contrary to the results in the previous sections we obtain an affine solution in the continuous terminal distribution case. Assumption IV is somewhat restrictive as it allows us to operate only over particular regimes of the parameter α_4. Such an assumption can be removed if we consider the following simpler security constraint which also has an affine solution in the terminal state.

$$\int_{\Re^n} p(x_T)U(x_T)'\bar{G}'\bar{G}U(x_T)\,dx_T \leq \alpha_5,\ \ \alpha_5 \geq 0 \quad (49)$$

The main difference between these constraints is that in (49) we compare the first k components of a control sequence, which drive the state to a particular terminal state, to the zero control sequence. For the security constraint employed in Problem 5.1, we compare the first k components of two control sequences which drive the system to different terminal states. Both constraints are then constructed by multiplying with relevant probability densities and integrating over $\Re^n$.

6 Simulations

In this section, we provide an analysis of the behavior of the cost function as the security constraint parameter is varied. In addition, the optimal solution is analyzed through simulations. We consider the framework of Problem 3.1 and make the additional assumption that the priors are equi-probable. Consider the system dynamics:

$$A=\begin{bmatrix}1 & 23 & 4\\ 5 & 7 & 12\\ 1 & 23 & 16\end{bmatrix},\ B=\begin{bmatrix}1 & 4 & 4.3\\ 6 & 2.3 & 8\\ 12 & 7 & 1.8\end{bmatrix},b_1=\begin{bmatrix}53\\ 75\\ 100\\ 32\\ 37\\ -8\end{bmatrix}$$

Let $\Sigma = 2\times I_{12\times 12}$, $Q=5\times I_{30\times 30}$, and $\alpha=0.45$. We assume that $T=5$ and $k=2$. Using these values α_1 is calculated to be $4\big(\rho^{-1}(0.45)\big)^2$. We use the standard optimization software cvx to compute the optimal control sequences. The optimal solution, $\mathbb{U}$, is given in (50).

$$\mathbb{U}(1:15) = \begin{bmatrix} -0.0118 \\ 0.0128 \\ 0.0076 \\ -0.0048 \\ -0.0023 \\ 0.0099 \\ \hdashline 0.0032 \\ 0.0052 \\ -0.0221 \\ 0.5366 \\ 0.0016 \\ -0.3726 \\ 0.1369 \\ 0.0453 \\ -0.1924 \end{bmatrix}, \quad \mathbb{U}(16:30) = \begin{bmatrix} -0.0118 \\ 0.0127 \\ 0.0076 \\ -0.0051 \\ -0.0020 \\ 0.0093 \\ \hdashline -0.0063 \\ -0.0058 \\ 0.0251 \\ -0.5374 \\ -0.0018 \\ 0.3736 \\ -0.1371 \\ -0.0454 \\ 0.1927 \end{bmatrix} \tag{50}$$

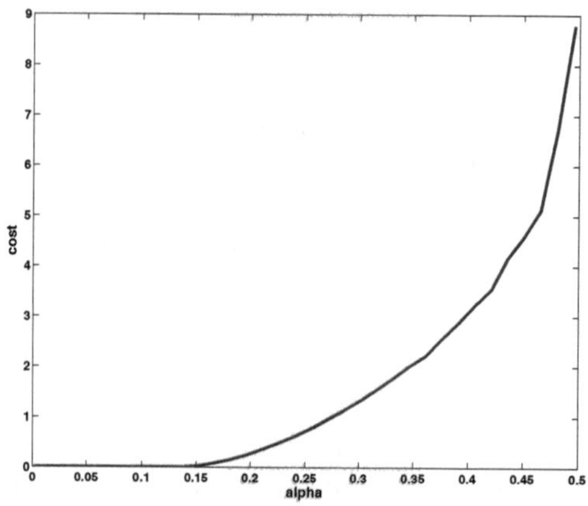

Fig. 1. Optimal Cost vs α for the system with terminal state vector b_1

It should be noted that $\mathbb{U}(1:15)$ correspond to the first 15 elements of the control vector $\mathbb{U}$ which drive the system to the terminal state $\begin{bmatrix} 53 & 75 & 100 \end{bmatrix}'$. Similarly $\mathbb{U}(16:30)$ correspond to the control inputs which drive the system to the terminal state $\begin{bmatrix} 32 & 37 & -8 \end{bmatrix}'$, respectively. It should be noted that $k = 2$, $m = 3$, and hence the adversary can only make noisy measurements of the first 2 control inputs, which are colored red, of these control vectors. Note that the control inputs which the adversary measures in both cases are very similar to one another thereby providing minimum useful information to the adversary. This is what we expected and this further signifies the importance of incorporating security constraints in control problems.

The simulation in Fig. 1 utilizes the system described above. We plot the optimal cost against the constraint parameter α, where $P_e \geq \alpha$. As discussed in Section III, if $\alpha > 0.5$ then the security constraint becomes infeasible. Clearly as α is increased the problem becomes more constrained and cost increases respectively. An exponential increase in the cost is observed by an increase in the value of α, for $\alpha \geq .14$. In Fig. 1, $\begin{bmatrix}53 & 75 & 100 & 32 & 37 & -8\end{bmatrix}'$ was utilized as the vector of terminal states. In order to avoid any confusion we clarify that the value of the optimal cost, for $0 \leq \alpha \leq 0.15$, is approximately 0.01 and not zero.

We now consider a different dynamical system and a different vector of terminal states to analyze the increase in the optimal cost with the constraint parameter. Let:

$$A = \begin{bmatrix} 1 & 3 & 4 \\ 3.1 & 7 & 2 \\ 5 & 3 & 6 \end{bmatrix}, \ B = \begin{bmatrix} 1 & 4 & 0.3 \\ 6 & 3 & 2 \\ 2 & 7 & 9 \end{bmatrix}, b_2 = \begin{bmatrix} 10 \\ 12 \\ -8 \\ 4 \\ 7 \\ 5 \end{bmatrix}$$

We take $\Sigma = 0.5 \times I_{12\times 12}$, $Q = 10 \times I_{30\times 30}$, $T = 5$ and $k = 3$. Fig. 2 shows the increase in the optimal cost with the value of α for this system. Clearly in this case the optimal cost increases less rapidly than in the case of Fig. 1. Unlike the previous system, the optimal cost does not stay constant over a large range of values of α. Therefore, we can conclude that for this system the probability of error constraint is more tightly enforced as compared to the previous system. Also the rate of change in the optimal cost is approximately constant. These simulations further show the role of constraint parameters in the optimal solution. Similar results can be obtained when we employ constraints based on the conditional mean.

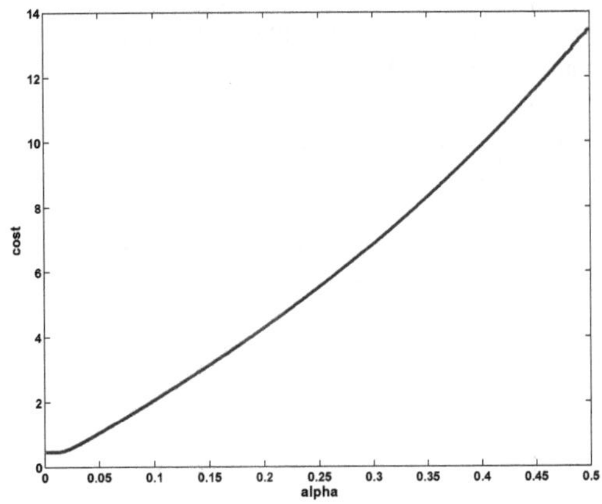

Fig. 2. Optimal Cost vs α for the system with terminal state vector b_2

7 Conclusions

Several security constraints have been incorporated in the control synthesis of a linear quadratic optimal control problem. The resulting optimization problems are shown to be convex. Lagrangian duality techniques have been used to compute and characterize the optimal solutions and their properties. The optimal solution is shown to be affine for the case when the terminal state has a continuous distribution. Utilizing the standard optimization software cvx, we have computed the optimal control sequences and have also validated the results via numerical simulations.

References

1. Cardenas, A., Amin, S., Sastry, S.: Research challenges for the security of control systems. In: Proceedings of the 3rd conference on Hot Topics in Security. USENIX Association, San Jose (2008)
2. Cardenas, A., Amin, S., Sastry, S.: Secure Control: Towards survivable cyber-physical systems. In: 28th International Conference Distributed Computing Systems Workshops, Beijing, China, pp. 495–500 (June 2008)
3. Mo, Y., Kim, T., Brancik, K., Dickinson, D., Lee, H., Perrig, A., Sinopoli, B.: Cyber-Physical Security of a Smart Grid Infrastructure. Proceedings of the IEEE: Special Issue on Cyber-Physical Systems 100, 195–209 (2012)
4. Gupta, A., Langbort, C., Basar, T.: Optimal Control in the presence of an intelligent jammer with limited actions. In: 49th IEEE Conference on Decision and Control, pp. 1096–1101 (2010)
5. Mo, Y., Sinopli, B.: False Data Injection Attacks in Control Systems. In: 1st Workshop on Secure Control Systems (2010)
6. Fwazi, H., Tabuada, P., Diggavi, S.: Security of control systems under sensors and actuator attacks. In: 51st IEEE Conference on Decision and Control, pp. 3412–3417 (2012)
7. Zhu, Q., Bushnell, L., Basar, T.: Game Theoretic Analysis of Node Capture and Cloning Attack with Multiple Attackers in Wireless Sensor Networks. In: 51st IEEE Conference on Decision and Control, pp. 3404–3411 (2012)
8. Pasqualetti, F., Dorfler, F., Bullo, F.: Cyber-Physical Security via Geomtric Control: Distributed Monitoring and Mailicious Attacks. In: 51st IEEE Conference on Decision and Control, pp. 3418–3425 (2012)
9. Poor, V.: An Introduction to Signal Detection and Estimation, 2nd edn. Springer (1994)
10. Grant, M., Boyd, S.: CVX: Matlab software for disciplined convex programming (2000), `http://stanford.edu/~boyd/cvx`
11. Boyd, S., Vandenberghe, L.: Convex Optimization. Cambridge University Press (2004)
12. Bickel, P., Doksum, K.: Mathematical Statistics: Basic Ideas and Selected Topics, 2nd edn., vol. 1. Pearson Prentice Hall (2007)
13. Luenberger, D.: Optimization by Vector Space Methods. Wiley (1969)

Part III
Cyber-Security of Networked Systems

Quantifying Cyber-Security for Networked Control Systems

André Teixeira[1], Kin Cheong Sou[2], Henrik Sandberg[1], and Karl H. Johansson[1]

[1] ACCESS Linnaeus Centre and Automatic Control Lab,
KTH Royal Institute of Technology, Stockholm, Sweden
{andretei,hsan,kallej}@kth.se
[2] Mathematical Sciences,
Chalmers University of Technology and University of Gothenburg,
Gothenburg, Sweden
kincheong.sou@chalmers.se

Abstract. In this paper we consider a typical architecture for a networked control system under false-data injection attacks. Under a previously proposed adversary modeling framework, various formulations for quantifying cyber-security of control systems are proposed and formulated as constrained optimization problems. These formulations capture trade-offs in terms of attack impact on the control performance, attack detectability, and adversarial resources. The formulations are then discussed and related to system theoretic concepts, followed by numerical examples illustrating the various trade-offs for a quadruple-tank process.

Keywords: Security, Networked Control Systems, Impact Analysis.

1 Introduction

Critical infrastructure security is of utmost importance in modern society and has been a major concern in recent years. The increasing complexity of these systems and the desire to improve their efficiency and flexibility has led to the use of heterogeneous IT systems, which support the timely exchange of data among and across different system layers, from the corporate level to the local control level. Furthermore, IT infrastructures are composed of heterogeneous components from several vendors and often use non-proprietary communication networks. Therefore the amount of cyber threats to these IT infrastructures has greatly increased over the past years, given their large number of possible attack points across the system layers. There are several examples of cyber threats being exploited by attackers to disrupt the behavior of physical processes, including a staged attack on a power generator [10] and the recent Stuxnet virus attack on centrifuges' control system [16,12]. Hence monitoring and mitigating cyber attacks to these systems is crucial, since they may bring disastrous consequences to society. This is well illustrated by recalling the consequences of the US-Canada 2003 blackout [19], partially due to lack of awareness in the control center.

A particular type of complex cyber attack is that of false-data injection, where the attacker introduces corrupted data in the communication network. Several

D.C. Tarraf (ed.), *Control of Cyber-Physical Systems*,
Lecture Notes in Control and Information Sciences 449,
DOI: 10.1007/978-3-319-01159-2_7,

instances of this scenario have been considered in the context of control systems, see [2,4,15] and references therein. In this paper we address stealthy false-data injection attacks that are constructed so that they are not detected based on the control input and measurement data available to anomaly detectors. A sub-class of these attacks have been recently addressed from a system theoretic perspective. In [14] the author characterizes the set of attack policies for covert (stealthy) false-data injection attacks with detailed model knowledge and full access to all sensor and actuator channels, while [11] described the set of stealthy false-data injection attacks for omniscient attackers with full-state information, but possibly compromising only a subset of the existing sensors and actuators. Similarly, the work in [5] considers a finite time-interval and characterizes the number of corrupted channels that cannot be detected during that time-interval. In the previous approaches the control input and measurement data available to the anomaly detector with and without the attack were the same, thus rendering the attack undetectable. Instead in this paper we allow more freedom to the adversary and consider attacks that may be theoretically detectable, but are still stealthy since they do not trigger any alarm by the anomaly detector.

Contributions and Outline

In this paper we consider the typical architecture for a networked control system under false-data injection attacks and adversary models presented in [17]. Under this framework, various formulations for quantifying cyber-security of control systems are proposed and formulated as constrained optimization problems. These formulations capture trade-offs in terms of impact on the control system, attack detectability, and adversarial resources. In particular, one of the formulations considers the minimum number of data channels that need to be corrupted so that the adversary remains stealthy, similarly to the security index for static systems proposed in [13]. The formulations are related to system theoretic concepts.

The outline of the paper is as follows. The control system architecture and model are described in Section 2, followed by the adversary model in Section 3. Different formulations quantifying cyber-security of control systems are introduced in Section 4 for a given time-horizon and in Section 5 for steady-state. A particular formulation is posed as a mixed integer linear program and illustrated through numerical examples in Section 6, followed by conclusions in Section 7.

Notation and Preliminaries

Let $\mathbf{x}_{[k_0,\,k_f]} = \{x_{k_0},\; x_{k_0+1},\; \ldots,\; x_{k_f}\}$ be a discrete-time signal in the time-interval $[k_0,\; k_f] = \{k_0,\; \ldots,\; k_f\}$ with $x_k \in \mathbb{R}^n$ for $k \in [k_0,\; k_f]$. For simplicity, we also denote the time-domain signal of x_k in vector form as $\mathbf{x}_{[k_0,\,k_f]} \in \mathbb{R}^{n(k_f-k_0+1)}$, with $\mathbf{x}_{[k_0,\,k_f]} = [x_{k_0}^\top,\; \ldots,\; x_{k_f}^\top]^\top$. When the time-interval at consideration is clear, the short-form notation $\mathbf{x}$ will be used in place of $\mathbf{x}_{[k_0,\,k_f]}$.

For $y \in \mathbb{C}^n$, denote the p-norm of y as $\|y\|_p \triangleq \left(\sum_{i=1}^n |y_{(i)}|^p\right)^{1/p}$ for $1 \le p < \infty$, where $y_{(i)}$ is the i-th entry of the vector y, and let $\|y\|_\infty \triangleq \max_i\; |y_{(i)}|$.

Additionally, we denote $\|y\|_0$ as the number of non-zero elements of y and define $\mathbb{S} = \{z \in \mathbb{C} : |z| = 1\}$ as the unit circle in the complex plane.

As for the discrete-time signal $\mathbf{x}$, denote its ℓ_p-norm in the time-interval $[k_0,\ k_f]$ as $\|\mathbf{x}\|_{\ell_p[k_0,\ k_f]} \triangleq \|\mathbf{x}_{[k_0,\ k_f]}\|_p = \left(\sum_{k=k_0}^{k_f} \|x_k\|_p^p\right)^{1/p}$ for $1 \leq p < \infty$, and let $\|\mathbf{x}\|_{\ell_\infty[k_0,\ k_f]} \triangleq \sup_{k\in[k_0,\ k_f]} \|x_k\|_\infty$.

For a given matrix $G \in \mathbb{C}^{n\times m}$, denote its Hermitian conjugate as G^H and, supposing G is full-column rank, let $G^\dagger = (G^H G)^{-1}G^H$ be its pseudo-inverse.

2 Networked Control System

In this section we describe the networked control system structure, where we consider three main components as illustrated in Fig. 1: the physical plant and communication network, the feedback controller, and the anomaly detector.

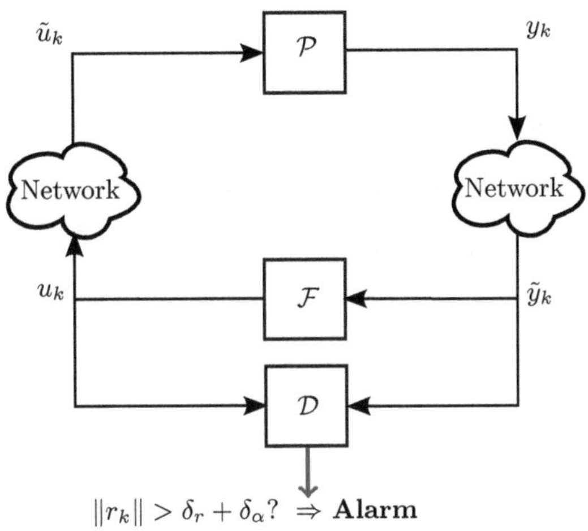

Fig. 1. Schematic of networked control system

2.1 Physical Plant and Communication Network

The physical plant is modeled in a discrete-time state-space form

$$\mathcal{P} : \begin{cases} x_{k+1} = Ax_k + B\tilde{u}_k + Gw_k + Ff_k \\ y_k = Cx_k + v_k \end{cases}, \tag{1}$$

where $x_k \in \mathbb{R}^n$ is the state variable, $\tilde{u}_k \in \mathbb{R}^{n_u}$ the control actions applied to the process, $y_k \in \mathbb{R}^{n_y}$ the measurements from the sensors at the sampling instant $k \in \mathbb{Z}$, and $f_k \in \mathbb{R}^d$ is the unknown signal representing the effects of anomalies, usually denoted as fault signal in the fault diagnosis literature [3]. The process and measurement Gaussian noise, $w_k \in \mathbb{R}^n$ and $v_k \in \mathbb{R}^{n_y}$, represent the discrepancies between the model and the real process, due to unmodeled dynamics or disturbances, for instance, and we assume their means are respectively bounded by δ_w and δ_v, i.e. $\bar{w} = \|\mathbb{E}\{w_k\}\| \leq \delta_w$ and $\bar{v} = \|\mathbb{E}\{v_k\}\| \leq \delta_v$.

The physical plant operation is supported by a communication network through which the sensor measurements and actuator data are transmitted, which at the plant side correspond to y_k and $\tilde{u}_k$, respectively. At the controller side we denote the sensor and actuator data by $\tilde{y}_k \in \mathbb{R}^{n_y}$ and $u_k \in \mathbb{R}^{n_u}$, respectively. Since the communication network may be unreliable, the data exchanged between the plant and the controller may be altered, resulting in discrepancies in the data at the plant and controller ends. In this paper we do not consider the usual communication network effects such as packet losses and delays. Instead we focus on data corruption due to malicious cyber attacks, as described in Section 3. Therefore the communication network *per se* is supposed to be reliable, not affecting the data flowing through it.

Given the physical plant model (1) and assuming an ideal communication network, the networked control system is said to have a *nominal behavior* if $f_k = 0$, $\tilde{u}_k = u_k$, and $\tilde{y}_k = y_k$. The absence of either one of these condition results in an abnormal behavior of the system.

2.2 Feedback Controller

In order to comply with performance requirements in the presence of the unknown process and measurement noises, we consider that the physical plant is controlled by an appropriate linear time-invariant feedback controller [20]. The output feedback controller can be written in a state-space form as

$$\mathcal{F} : \begin{cases} z_{k+1} = A_c z_k + B_c \tilde{y}_k \\ u_k = C_c z_k + D_c \tilde{y}_k \end{cases} \tag{2}$$

where the states of the controller, $z_k \in \mathbb{R}^{n_z}$, may include the process state and tracking error estimates. Given the plant and communication network models, the controller is supposed to be designed so that acceptable performance is achieved under nominal behavior.

2.3 Anomaly Detector

In this section we consider the anomaly detector that monitors the system to detect possible anomalies, i.e. deviations from the nominal behavior. The anomaly detector is supposed to be collocated with the controller, therefore it only has access to $\tilde{y}_k$ and u_k to evaluate the behavior of the plant.

Several approaches to detecting malfunctions in control systems are available in the fault diagnosis literature [3,6]. Here we consider the following observer-based fault detection filter

$$\mathcal{D} : \begin{cases} \hat{x}_{k+1|k} = A\hat{x}_{k|k} + Bu_k \\ \hat{x}_{k|k} = \hat{x}_{k|k-1} + K(\tilde{y}_k - C\hat{x}_{k|k-1}) \, , \\ r_k = V(\tilde{y}_k - \hat{y}_{k|k}) \end{cases} \tag{3}$$

where $\hat{x}_{k|k} \in \mathbb{R}^n$ and $\hat{y}_{k|k} = C\hat{x}_{k|k} \in \mathbb{R}^{n_y}$ are the state and output estimates given measurements up until time k, respectively, and $r_k \in \mathbb{R}^{n_r}$ the residue evaluated to detect and locate existing anomalies. The previous filter dynamics can be rewritten as

$$\mathcal{D}: \begin{cases} \hat{x}_{k+1|k} = A(I - KC)\hat{x}_{k|k-1} + Bu_k + AK\tilde{y}_k \\ \quad r_k = V[(I - CK)\tilde{y}_k - (I - CK)C\hat{x}_{k|k-1}]. \end{cases} \tag{4}$$

The anomaly detector is designed by choosing K and V such that

1. under nominal behavior of the system (i.e., $f_k = 0$, $u_k = \tilde{u}_k$, $y_k = \tilde{y}_k$), the expected value of r_k converges asymptotically to a neighborhood of zero, i.e., $\lim_{k\to\infty} \mathbb{E}\{r_k\} \in \mathcal{B}_{\delta_r}$, with $\delta_r \in \mathbb{R}^+$ and $\mathcal{B}_{\delta_r} \triangleq \{r \in \mathbb{R}^{n_r} : \|r\|_p \leq \delta_r\}$;
2. the residue is sensitive to the anomalies ($f_k \not\equiv 0$).

The characterization of $\mathcal{B}_{\delta_r}$ depends on the noise terms and can be found in [3] for particular values of p. Given the residue signal over the time-interval $[d_0,\ d_f]$, $\mathbf{r}_{[d_0,\ d_f]}$, an alarm is triggered if

$$\mathbf{r}_{[d_0,\ d_f]} \notin \mathcal{U}_{[d_0,\ d_f]}, \tag{5}$$

where the set $\mathcal{U}_{[d_0,\ d_f]}$ is chosen so that the false-alarm rate does not exceed a given threshold $\alpha \in [0,\ 1]$. This necessarily requires no alarm to be triggered in the noiseless nominal behavior i.e., $\mathbf{r}_{[d_0,\ d_f]} \in \mathcal{U}_{[d_0,\ d_f]}$ if for all $k \in [d_0,\ d_f]$ it holds that $r_k \in \mathcal{B}_{\delta_r}$. For instance, one can take $\mathcal{U}_{[d_0,\ d_f]}$ to be a bound on the energy of the residue signal over the time-interval $[d_0,\ d_f]$, resulting in $\mathcal{U}_{[d_0,\ d_f]} = \{\mathbf{r} : \|\mathbf{r}\|_{\ell_2[d_0,\ d_f]} \leq \delta\}$.

3 Adversary Model

In this section we discuss the adversary models composed of adversarial goals and the system dynamics under attack. In particular, we consider attack scenarios where the adversary's goal is to drive the system to an unsafe state while remaining stealthy. Below we describe the networked control system under attack with respect to the attack vector $a_k \in \mathbb{R}^{q_a}$.

3.1 Networked Control System under Attack

The system components under attack are now characterized for the attack vector a_k. Considering the plant and controller states to be augmented as $\eta_k = [x_k^\top \ \ z_k^\top]^\top$, the dynamics of the closed-loop system composed by $\mathcal{P}$ and $\mathcal{F}$ under the effect of a_k can be written as

$$\begin{aligned} \eta_{k+1} &= \mathbf{A}\eta_k + \mathbf{B}a_k + \mathbf{G}\begin{bmatrix} w_k \\ v_k \end{bmatrix} \\ \tilde{y}_k &= \mathbf{C}\eta_k + \mathbf{D}a_k + \mathbf{H}\begin{bmatrix} w_k \\ v_k \end{bmatrix}, \end{aligned} \tag{6}$$

where the system matrices are

$$\mathbf{A} = \begin{bmatrix} A + BD_cC & BC_c \\ B_cC & A_c \end{bmatrix}, \mathbf{G} = \begin{bmatrix} G & BD_c \\ 0 & B_c \end{bmatrix},$$

$$\mathbf{C} = \begin{bmatrix} C & 0 \end{bmatrix}, \qquad \mathbf{H} = \begin{bmatrix} 0 & I \end{bmatrix},$$

and $\mathbf{B}$ and $\mathbf{D}$ capture how the attack vector a_k affects the plant and controller.

Similarly, using $\mathcal{P}$ and $\mathcal{D}$ as in (1) and (4), respectively, the anomaly detector error dynamics under attack are described by

$$\begin{aligned} \xi_{k+1|k} &= \mathbf{A}_e \xi_{k|k-1} + \mathbf{B}_e a_k + \mathbf{G}_e \begin{bmatrix} w_k \\ v_k \end{bmatrix} \\ r_k &= \mathbf{C}_e \xi_{k|k-1} + \mathbf{D}_e a_k + \mathbf{H}_e \begin{bmatrix} w_k \\ v_k \end{bmatrix}, \end{aligned} \tag{7}$$

where $\xi_{k|k-1} \in \mathbb{R}^n$ is the estimation error and

$$\begin{aligned} \mathbf{A}_e &= A(I - KC), & \mathbf{G}_e &= \begin{bmatrix} G & -AK \end{bmatrix}, \\ \mathbf{C}_e &= VC(I - KC), & \mathbf{H}_e &= \begin{bmatrix} 0 & V(I - CK) \end{bmatrix}. \end{aligned}$$

The matrices $\mathbf{B}_e$ and $\mathbf{D}_e$ are specific to the available disruptive resources and are characterized below for data deception attacks.

Data Deception Resources. The deception attacks modify the control actions u_k and sensor measurements y_k from their calculated or real values to the corrupted signals $\tilde{u}_k$ and $\tilde{y}_k$, respectively. Denoting $\mathcal{R}_I^u \subseteq \{1, \dots, n_u\}$ and $\mathcal{R}_I^y \subseteq \{1, \dots, n_y\}$ as the deception resources, i.e. set of actuator and sensor channels that can be affected, and $|\mathcal{R}_I^u|$ and $|\mathcal{R}_I^y|$ as the respective cardinality of the sets, the deception attacks are modeled as

$$\tilde{u}_k \triangleq u_k + \Gamma^u b_k^u, \quad \tilde{y}_k \triangleq y_k + \Gamma^y b_k^y, \tag{8}$$

where the signals $b_k^u \in \mathbb{R}^{|\mathcal{R}_I^u|}$ and $b_k^y \in \mathbb{R}^{|\mathcal{R}_I^y|}$ represent the data corruption and $\Gamma^u \in \mathbb{B}^{n_u \times |\mathcal{R}_I^u|}$ and $\Gamma^y \in \mathbb{B}^{n_y \times |\mathcal{R}_I^y|}$ ($\mathbb{B} \triangleq \{0, 1\}$) are the binary incidence matrices mapping the data corruption to the respective data channels. The matrices Γ^u and Γ^y indicate which data channels can be accessed by the adversary and are therefore directly related to the adversary resources in deception attacks. Recalling that $a \in \mathbb{R}^{q_a}$, the number of data channels that may be compromised by the adversary are given by $q_a = |\mathcal{R}_I^u| + |\mathcal{R}_I^y|$. Defining $a_k = [b_k^{u\top} \; b_k^{y\top}]^\top$, the system dynamics are given by (6) and (7) with

$$\mathbf{B} = \begin{bmatrix} B\Gamma^u & BD_c\Gamma^y \\ 0 & B_c\Gamma^y \end{bmatrix}, \quad \mathbf{D} = \begin{bmatrix} 0 & \Gamma^y \end{bmatrix},$$

$$\mathbf{B}_e = \begin{bmatrix} B\Gamma^u & -AK\Gamma^y \end{bmatrix}, \quad \mathbf{D}_e = \begin{bmatrix} 0 & V(I - CK)\Gamma^y \end{bmatrix}.$$

3.2 Attack Goals and Constraints

In addition to the attack resources, the attack scenarios need to include the adversary's intent, namely the attack goals and constraints shaping the attack policy. The attack goals can be stated in terms of the attack impact on the system operation, while the constraints may be related to the attack detectability.

Several physical systems have tight operating constraints which if not satisfied might result in physical damage to the system. In this work we use the concept of safe sets to characterize the safety constraints.

Definition 1. *For a given time-interval* $[k_0,\ k_f]$*, the system is said to be safe if* $\mathbf{x}_{[k_0,\ k_f]} \in \mathcal{S}_{[k_0,\ k_f]}$*, where* $\mathcal{S}_{[k_0,\ k_f]}$ *is a compact set with non-empty interior.*

The above definition of safe set $\mathcal{S}_{[k_0,\ k_f]}$ allows one to consider both time-interval and time-instant characterizations of safe regions, for instance signal energy and safe regions of the state space, respectively.

Assumption 1. *The system is in a safe state at the beginning of the attack, i.e.* $\mathbf{x}_{(-\infty,\ k_0-1]} \in \mathcal{S}_{(-\infty,\ k_0-1]}$.

The physical impact of an attack can be evaluated by assessing whether or not the state of the system remained in the safe set during and after the attack. The attack is considered successful if the state is driven out of the safe set. For simplicity of notation, the safe set $\mathcal{S}_{[k_0,\ k_f]}$ will be simply denoted as $\mathcal{S}$ whenever the time-interval is not ambiguous. Moreover, the safe sets considered in the remaining of this paper are of the form $\mathcal{S}^p_{[k_0,\ k_f]} = \{\mathbf{x} :\ \|\mathbf{x}\|_{\ell_p[k_0,\ k_f]} \leq 1\}$.

Regarding the attack constraints, we consider that attacks are constrained to remain stealthy. Furthermore, we consider the disruptive attack component consists of only physical and data deception attacks, and thus we have the attack vector $a_k = [b_k^{u\top}\ \ b_k^{y\top}]^\top$. Given the anomaly detector described in Section 2, denoting $\mathbf{a}_{[k_0,\ k_f]} = \{a_{k_0}, \ldots, a_{k_f}\}$ as the attack signal, and recalling that the residue signal $\mathbf{r}_{[k_0,\ +\infty)}$ is a function of the attack signal, the set of stealthy attacks are defined as follows.

Definition 2. *The attack signal* $\mathbf{a}_{[k_0,\ k_f]}$ *is stealthy over the time-interval* $[k_0,\ d_f]$ *with* $d_f \geq k_f$ *if* $\mathbf{r}_{[k_0,\ d_f]} \in \mathcal{U}_{[k_0,\ d_f]}$.

Note that the above definition is dependent on the initial state of the system at k_0, as well as the noise terms w_k and v_k. Furthermore, it also requires the attack to be stealthy even after it has been performed, as $d_f \geq k_f$.

Since the closed-loop system (6) and the anomaly detector (7) under linear attack policies are linear systems, each of these systems can be separated into two components, the nominal component with $a_k = 0\ \forall k$ and the following systems with zero initial conditions $\eta_0^a = \xi_{0|0}^a = 0$

$$\begin{aligned} \eta_{k+1}^a &= \mathbf{A}\eta_k^a + \mathbf{B}a_k \\ \tilde{y}_k^a &= \mathbf{C}\eta_k^a + \mathbf{D}a_k, \end{aligned} \tag{9}$$

$$\begin{aligned}\xi^a_{k|k} &= \mathbf{A}_e\xi^a_{k-1|k-1} + \mathbf{B}_e a_{k-1}\\ r^a_k &= \mathbf{C}_e\xi^a_{k-1|k-1} + \mathbf{D}_e a_{k-1}.\end{aligned} \tag{10}$$

Assuming the system is behaving nominally before the attack and that, given the linearity of (7), there exists a set $\mathcal{U}^a_{[k_0,\, d_f]} \triangleq \{\mathbf{r} :\ \|\mathbf{r}\|_{\ell_p[k_0,\, d_f]} \leq \delta_\alpha\}$ such that $\mathbf{r}^a_{[k_0,\, d_f]} \in \mathcal{U}^a_{[k_0,\, d_f]} \Rightarrow \mathbf{r}_{[k_0,\, d_f]} \in \mathcal{U}_{[k_0,\, d_f]}$, we have the following definition:

Definition 3. *The attack signal* $\mathbf{a}_{[k_0,\, k_f]}$ *is stealthy over the time-interval* $[k_0,\ d_f]$ *if* $\mathbf{r}^a_{[k_0,\, d_f]} \in \mathcal{U}^a_{[k_0,\, d_f]}$.

Albeit more conservative than Definition 2, this definition only depends on the attack signals $\mathbf{a}_{[k_0,\, k_f]}$. Similarly, the impact of attacks on the closed-loop system can also be analyzed by looking at the linear system (9).

4 Quantifying Cyber-Security: Transient Analysis

As mentioned in Section 3.2, the adversary aims at driving the system to an unsafe state while remaining stealthy. Additionally we consider that the adversary also has resource constraints, in the sense that only a small number of attack points to the system are available. In the following, several formulations for quantifying cyber-security of networked control systems are discussed.

Consider the dynamical system in (9) and the time-interval $[0,\ N]$ with $d_0 = k_0 = 0$ and $k_f = d_f = N$. Defining $\mathbf{n} = [\eta_0^\top\ \dots\ \eta_N^\top]^\top$, $\mathbf{a} = [a_0^\top\ \dots\ a_N^\top]^\top$, and $\mathbf{y} = [y_0^\top\ \dots\ y_N^\top]^\top$, the state and output trajectories can be described by the following mappings

$$\begin{aligned}\mathbf{n} &= \mathcal{O}_\eta\eta_0 + \mathcal{T}_\eta\mathbf{a}\\ \mathbf{y} &= \mathcal{C}_\eta\mathbf{n} + \mathcal{D}_\eta\mathbf{a},\end{aligned} \tag{11}$$

where

$$\mathcal{O}_\eta = \begin{bmatrix} I\\ \mathbf{A}\\ \mathbf{A}^2\\ \vdots\\ \mathbf{A}^N\end{bmatrix},\quad \mathcal{T}_\eta = \begin{bmatrix}\mathbf{D} & 0 & \dots & 0\\ \mathbf{B} & 0 & \dots & 0\\ \mathbf{AB} & \mathbf{B} & \dots & 0\\ \vdots & \vdots & \ddots & 0\\ \mathbf{A}^{N-1}\mathbf{B} & \mathbf{A}^{N-2}\mathbf{B} & \dots & \mathbf{B}\end{bmatrix}, \tag{12}$$
$$\mathcal{C}_\eta = I_{N+1}\otimes\mathbf{C},\quad \mathcal{D}_\eta = I_{N+1}\otimes\mathbf{D}$$

Similarly for (10), defining $\mathbf{e} = [\xi^\top_{-1|-1}\ \cdots\ \xi^\top_{N-1|N-1}]^\top$, $\mathbf{r} = [r_0^\top\ \dots\ r_N^\top]^\top$ yields

$$\begin{aligned}\mathbf{e} &= \mathcal{O}_\xi\xi_{-1|-1} + \mathcal{T}_\xi\mathbf{a}\\ \mathbf{r} &= \mathcal{C}_\xi\mathbf{e} + \mathcal{D}_\xi\mathbf{a}.\end{aligned} \tag{13}$$

Recall that the system is operating safely during the time-interval $[k_0,\ k_f]$ if $\mathbf{x} \in \mathcal{S}_{[k_0,\, k_f]}$. Supposing $\mathcal{S}^p_{[k_0,\, k_f]} = \{\mathbf{x} :\ \|\mathbf{x}\|_{\ell_p[k_0,\, k_f]} \leq 1\}$ for $p \geq 1$, the system is safe during the time-interval $\{0, 1, \dots, N\}$ if

$$\mathbf{x} \triangleq \mathcal{C}_x\mathbf{n} \in \mathcal{S}^p_{[0,\, N]}, \tag{14}$$

where $\mathcal{C}_x = I_{N+1} \otimes [I_n \ 0]$. In particular, for $p = \infty$ we have that the system is safe if $\|\mathbf{x}\|_\infty = \|\mathcal{C}_x \mathbf{n}\|_\infty \leq 1$.

4.1 Maximum-Impact Attacks

One possible way to quantify cyber-security is by analyzing the impact of attacks on the control system, given some pre-defined resources available to the adversary. Recalling the safe set introduced earlier, $\mathcal{S}^p_{[0,\,N]} = \{\mathbf{x} : \ \|\mathbf{x}\|_{\ell_p[0,\,N]} \leq 1\}$, the attack impact during the time-interval $[0,\ N]$ is characterized by

$$g_p(\mathbf{n}) = \begin{cases} \|\mathcal{C}_x \mathbf{n}\|_p \ , \text{ if } \ \mathcal{C}_x \mathbf{n} \in \mathcal{S}^p_{[0,\,N]} \\ \ +\infty \qquad \ , \text{ otherwise,} \end{cases} \tag{15}$$

since the adversary aims at driving the system to an unsafe state. Similarly, recall the set of stealthy attacks $\mathbf{a}$ such that $\mathbf{r} \in \mathcal{U}^a_{[k_0,\,d_f]} \triangleq \{\mathbf{r} : \ \|\mathbf{r}\|_{\ell_p[k_0,\,d_f]} \leq \delta_\alpha\}$.

The attack yielding the maximum impact can be computed by solving

$$\begin{aligned} \max_{\mathbf{a}} \quad & g_p(\mathbf{n}) \\ \text{s.t.} \quad & \|\mathcal{C}_\xi \mathbf{e} + \mathcal{D}_\xi \mathbf{a}\|_q \leq \delta_\alpha, \\ & \mathbf{e} = \mathcal{O}_\xi \xi_{-1|-1} + \mathcal{T}_\xi \mathbf{a}, \\ & \mathbf{n} = \mathcal{O}_\eta \eta_0 + \mathcal{T}_\eta \mathbf{a}, \end{aligned} \tag{16}$$

with p and q possibly different. Given the objective function $g_p(\mathbf{n})$, the adversary's optimal policy is to drive the system to an unsafe state while keeping the residue below the threshold. When the unsafe state is not reachable while remaining stealthy, the optimal attack drives the system as close to the unsafe set as possible by maximizing $\|\mathbf{x}\|_{\ell_p[0,\,N]} = \|\mathcal{C}_x \mathbf{n}\|_p$.

Letting $\xi_{-1|-1} = 0$ and $\eta_0 = 0$, the optimal values of (16) can be characterized by analyzing the following modified problem

$$\begin{aligned} \max_{\mathbf{a}} \quad & \|\mathcal{T}_x \mathbf{a}\|_p \\ \text{s.t.} \quad & \|\mathcal{T}_r \mathbf{a}\|_q \leq \delta_\alpha, \end{aligned} \tag{17}$$

where $\mathcal{T}_x = \mathcal{C}_x \mathcal{T}_\eta$ and $\mathcal{T}_r = \mathcal{C}_\xi \mathcal{T}_\xi + \mathcal{D}_\xi$. The conditions under which (17) admits bounded optimal values are characterized in the following result.

Lemma 1. *The problem* (17) *is bounded if and only if* $\ker(\mathcal{T}_r) \subseteq \ker(\mathcal{T}_x)$.

Proof. Suppose that $\ker(\mathcal{T}_r) \neq \emptyset$ and consider the subset of solutions where $\mathbf{a} \in \ker(\mathcal{T}_r)$. For this subset of solutions, the optimization problem then becomes $\max_{\mathbf{a} \in \ker(\mathcal{T}_r)} \|\mathcal{T}_x \mathbf{a}\|_p$. Since the latter corresponds to a maximization of a convex function, its solution is unbounded unless $\mathcal{T}_x \mathbf{a} = 0$ for all $\mathbf{a} \in \ker(\mathcal{T}_r)$ i.e., $\ker(\mathcal{T}_r) \subseteq \ker(\mathcal{T}_x)$. For $\mathbf{a} \notin \ker(\mathcal{T}_r)$ the feasible set is compact and thus the objective function over the feasible set is bounded, which concludes the proof.

Supposing that the optimization problem (17) is bounded and $p = q = 2$, (17) can be rewritten as a generalized eigenvalue problem and solved analytically.

Theorem 1. *Let $p = q = 2$ and suppose that $\ker(\mathcal{T}_r) \subseteq \ker(\mathcal{T}_x)$. The optimal attack policy for (17) is given by*

$$\mathbf{a}^\star = \frac{\delta_\alpha}{\|\mathcal{T}_r \mathbf{v}^\star\|_2} \mathbf{v}^\star, \tag{18}$$

where $\mathbf{v}^\star$ is the eigenvector associated with λ^*, the largest generalized eigenvalue of the matrix pencil $(\mathcal{T}_x^\top \mathcal{T}_x,\ \mathcal{T}_r^\top \mathcal{T}_r)$. Moreover, the corresponding optimal value is given by $\|\mathcal{T}_x \mathbf{a}^\star\|_2 = \sqrt{\lambda^*}\delta_\alpha$.

Proof. The proof is similar to that of [17, Thm. 12].

Given the solution to (17) characterized by the previous result, the maximum impact with respect to (16) is given by

$$g_p(\mathcal{T}_x \mathbf{a}^\star) = \begin{cases} \sqrt{\lambda^*}\delta_\alpha & \text{, if } \sqrt{\lambda^*}\delta_\alpha \leq 1 \\ +\infty & \text{, otherwise.} \end{cases}$$

4.2 Minimum-Resource Attacks

Cyber-security of control systems can also be quantified by assessing the number of resources needed by the adversary to perform a given set of attacks, without necessarily taking into account the attack impact, as formulated below.

Consider the set of attacks $\mathcal{G}$ such that $\mathbf{a} \in \mathcal{G}$ satisfies the goals of a given attack scenario. Recall that $a_k \in \mathbb{R}^{q_a}$ for all $k \in [k_0,\ k_f]$ and denote $\mathbf{a}_{(i),\,[k_0,\,k_f]} = \{a_{(i),k_0},\ \ldots,\ a_{(i),k_f}\}$ as the signal corresponding to the $i-$th attack resource. Consider the function

$$h_p(\mathbf{a}) = [\|\mathbf{a}_{(1)}\|_{\ell_p}\ \cdots\ \|\mathbf{a}_{(q_a)}\|_{\ell_p}]^\top \tag{19}$$

with $1 \leq p \leq +\infty$. The number of resources employed in a given attack are $\|h_p(\mathbf{a})\|_0$. For the set of attacks $\mathcal{G}$, the minimum-resource attacks are computed by solving the following optimization problem

$$\begin{aligned} \min_{\mathbf{a}} \quad & \|h_p(\mathbf{a})\|_0 \\ \text{s.t.} \quad & \|\mathcal{C}_\xi \mathbf{e} + \mathcal{D}_\xi \mathbf{a}\|_q \leq \delta_\alpha, \\ & \mathbf{e} = \mathcal{O}_\xi \xi_{-1|-1} + \mathcal{T}_\xi \mathbf{a}, \\ & \mathbf{a} \in \mathcal{G}. \end{aligned} \tag{20}$$

Although the set $\mathcal{G}$ may be chosen depending on the attack impact $g_p(\mathbf{n})$, i.e., $\mathcal{G} = \{\mathbf{a} :\ g_p(\mathbf{n}) = \|\mathcal{T}_x \mathbf{a}\|_{\ell_p} > \gamma\}$, this generally results in non-convex constraints that increase the computational complexity of the problem. As an example, the set $\mathcal{G} = \{\mathbf{a} :\ \|\mathcal{T}_x \mathbf{a}\|_{\ell_\infty} > \gamma\}$ is formulated as a set of linear constraints with binary variables in (35). However, $\mathcal{G}$ might not be directly related to the impact of the attack in terms of $g_p(\mathbf{n})$. For instance, the formulation (20) captures the security-index proposed for static systems in [13], where the adversary aims at corrupting a given measurement i without being detected. The security-index formulation is retrieved by having $\xi_{-1|-1} = 0$, $N = 0$, $\delta_\alpha = 0$, and $\mathcal{G} = \{\mathbf{a} \in \mathbb{R}^{q_a} : \mathbf{a}_{(i)} = 1\}$. However, for dynamic systems when $N > 0$, the specification of the attack scenario and corresponding set of attacks $\mathcal{G}$ is more involved. The same scenario where the adversary aims at corrupting a given channel i can be formulated by having $\delta_\alpha = 0$ and $\mathcal{G} = \{\mathbf{a} : \|\mathbf{a}_{(i)}\|_{\ell_p} = \epsilon\}$. However, for positive values of δ_α the feasibility of the problem depends on both δ_α and ϵ, which need to be carefully chosen.

The optimization problem (20) is also related to the estimation of sparse unknown-input signals. For instance, the ℓ_1/ℓ_q decoder proposed in [5] relaxes (20) using a ℓ_1/ℓ_q-norm regularization [8] and can be obtained by having $p \geq 1$ and solving the following modified problem for given $\mathbf{r}$

$$\begin{aligned} \min_{\mathbf{a},\, \xi_{-1|-1}} \quad & \|h_p(\mathbf{a})\|_1 \\ \text{s.t.} \quad & \mathbf{r} = \mathcal{C}_\xi \mathbf{e} + \mathcal{D}_\xi \mathbf{a}, \\ & \mathbf{e} = \mathcal{O}_\xi \xi_{-1|-1} + \mathcal{T}_\xi \mathbf{a} \end{aligned}$$

The optimal solution $\mathbf{a}^\star$ and $\xi^\star_{-1|-1}$ can then be used reconstruct the state trajectory according to (9). However, note that using $\|h_p(\mathbf{a})\|_1$ as the objective function instead of $\|h_p(\mathbf{a})\|_0$ may lead to substantially different solutions, since $\|h_p(\mathbf{a})\|_1$ mixes the time and physical dimensions of the attack signal. In fact, letting $p = 1$ so that $\|h_p(\mathbf{a})\|_1 = \|\mathbf{a}\|_1$ and supposing the number of available channels is given by $q_a = 2$ and $N = 1$, having $\mathbf{a} = [\mathbf{a}_{(1)}^\top\ \mathbf{a}_{(2)}^\top]^\top = [1\ 0\ 0\ 1]$ leads to $\|h_p(\mathbf{a})\|_1 = \|h_p(\mathbf{a})\|_0 = 2$ and corrupts two channels, while $\mathbf{a} = [1\ 1\ 0\ 0]^\top$ yields $\|h_p(\mathbf{a})\|_1 = 2$ and $\|h_p(\mathbf{a})\|_0 = 1$, thus corrupting only one channel. These attacks are significantly different in terms of adversarial resources, as corrupting two channels requires much larger effort than corrupting only one.

4.3 Maximum-Impact Minimum-Resource Attacks

The previous formulations considered impact and resources independently when quantifying cyber-security. Here the impact and resources and addressed simultaneously by considering the multi-objective optimization problem

$$\begin{aligned} \max_{\mathbf{a}} \quad & [g_p(\mathbf{n}),\ -\|h_p(\mathbf{a})\|_0]^\top \\ \text{s.t.} \quad & \left\|\mathcal{C}_\xi \mathbf{e} + \mathcal{D}_\xi \mathbf{a}\right\|_q \leq \delta_\alpha, \\ & \mathbf{e} = \mathcal{O}_\xi \xi_{-1|-1} + \mathcal{T}_\xi \mathbf{a}, \\ & \mathbf{n} = \mathcal{O}_\eta \eta_0 + \mathcal{T}_\eta \mathbf{a}. \end{aligned} \tag{21}$$

The vector-valued objective function indicates that the adversary desires to simultaneously maximize and minimize $g_p(\mathbf{n})$ and $\|h_p(\mathbf{a})\|_0$, respectively. Solutions to multi-objective problems are related to the concept of Pareto optimality [9] and correspond to the optimal trade-off manifold between the several objectives. These solutions can be obtained through several techniques, for instance the bounded objective function method in which all but one of the objectives are posed as constraints, thus obtaining a scalar-valued objective function. Applying this method to (21) and constraining $\|h_p(\mathbf{a})\|_0$ yields

$$\begin{aligned}
\max_{\mathbf{a}} \quad & g_p(\mathbf{n}) \\
\text{s.t.} \quad & \left\|\mathcal{C}_\xi \mathbf{e} + \mathcal{D}_\xi \mathbf{a}\right\|_q \leq \delta_\alpha, \\
& \mathbf{e} = \mathcal{O}_\xi \xi_{-1|-1} + \mathcal{T}_\xi \mathbf{a}, \\
& \mathbf{n} = \mathcal{O}_\eta \eta_0 + \mathcal{T}_\eta \mathbf{a}, \\
& \|h_p(\mathbf{a})\|_0 < \epsilon,
\end{aligned} \tag{22}$$

which can be interpreted as a maximum-impact resource-constrained attack policy. The Pareto frontier that characterizes the optimal trade-off manifold can be obtained by iteratively solving (22) for $\epsilon \in \{1, \ldots, q_a\}$. This approach is illustrated in Section 6 for the quadruple-tank process.

5 Quantifying Cyber-Security: Steady-State Analysis

Here we consider the steady-state of the system under attack. Let $z \in \mathbb{C}$ and define

$$\begin{aligned}
G_{xa}(z) &= [I_n \; 0](zI - \mathbf{A})^{-1}\mathbf{B} + \mathbf{D}, \\
G_{ra}(z) &= C_e(zI - A_e)^{-1}B_e + D_e,
\end{aligned} \tag{23}$$

which correspond to the transfer functions from a_k to x_k and r_k respectively. Considering exponential attack signals of the form $a_k = gz^k$ for fixed z, denote $a(z) = g \in \mathbb{C}^{q_a}$, $x(z) = G_{xa}(z)a(z)$, and $r(z) = G_{ra}(z)a(z)$ as the phasor notation of a_k, x_k, and r_k, respectively. Since the analysis in this section is restricted to steady-state, we consider z to be on the unit circle, $z \in \mathbb{S}$, and thus $a(z)$ corresponds to sinusoidal signals of constant magnitude. Defining the frequency-domain safe set as $\mathcal{S}_\infty^p = \{x \in \mathbb{C}^n : \|x\|_p \leq 1\}$, the system under attack is said to be safe at steady-state if $x(z) = G_{xa}(z)a(z) \in \mathcal{S}_\infty^p$.

5.1 Maximum-Impact Attacks

For a given $z \in \mathbb{S}$, the steady-state attack impact is characterized by

$$g_p(x(z)) = \begin{cases} \|x(z)\|_p \text{ , if } x(z) \in \mathcal{S}_\infty^p \\ +\infty \quad \text{, otherwise.} \end{cases} \tag{24}$$

Similarly, recall the set of steady-state stealthy attacks $a(z)$ such that $r(z) \in \mathcal{U}^a \triangleq \{r \in \mathbb{C}^{p_d} : \|r\|_p \leq \delta_\alpha\}$, where $r(z) = G_{ra}(z)a(z)$.

The attack yielding the maximum impact can be computed by solving

$$\begin{aligned}
\sup_{z \in \mathbb{S}} \max_{a(z)} \quad & g_p(G_{xa}(z)a(z)) \\
\text{s.t.} \quad & \left\|G_{ra}(z)a(z)\right\|_p \leq \delta_\alpha.
\end{aligned} \tag{25}$$

The maximum impact over all stealthy attacks can be computed by replacing the objective function $g_p(G_{xa}(z)A(z))$ with $\|G_{xa}(z)a(z)\|_p$, solving

$$\begin{aligned} \sup_{z\in\mathbb{S}}\ \max_{a(z)} \quad & \|G_{xa}(z)a(z)\|_p \\ \text{s.t.} \quad & \|G_{ra}(z)a(z)\|_q \leq \delta_\alpha, \end{aligned} \tag{26}$$

and evaluating $g_p(G_{xa}(z)a(z))$ for the obtained solution. The conditions under which (26) admits bounded optimal values are characterized as follows.

Lemma 2. *The optimization problem* (26) *is bounded if and only if* $\ker(G_{ra}(z)) \subseteq \ker(G_{xa}(z))$ *for all* $z \in \mathbb{S}$.

Proof. The proof follows the same reasoning as that of Lemma 1.

The previous statement is related to the concept of invariant-zeros of dynamical systems [18] as discussed below.

Definition 4. *Consider a linear time-invariant system in discrete-time with the state-space realization* (A, B, C, D) *and the equation*

$$\begin{bmatrix} zI - A & -B \\ C & D \end{bmatrix} \begin{bmatrix} x_0 \\ u_z \end{bmatrix} = \begin{bmatrix} 0 \\ 0 \end{bmatrix}, \tag{27}$$

with $z \in \mathbb{C}$ *and* $x_0 \neq 0$. *For a given solution to the previous equation* (z, u_z, x_0), *denote* z *as the invariant-zero,* u_z *as the input-zero direction, and* x_0 *as the state-zero direction.*

Lemma 3. *The optimization problem* (26) *is bounded if and only if either of the following hold:*

1. *the transfer function* $G_{ra}(z)$ *does not contain invariant-zeros on the unit circle;*
2. *all the invariant-zeros of the transfer function* $G_{ra}(z)$ *on the unit circle are also invariant-zeros of* $G_{xa}(z)$, *with the same input-zero direction.*

Proof. For the first statement, note that if $G_{ra}(z)$ does not contain invariant-zeros on the unit circle, then $\ker(G_{ra}(z)) = \emptyset$ for $z \in \mathbb{S}$ and thus (26) is bounded. As for the second statement, suppose that $G_{ra}(z)$ contains an invariant-zero $\bar{z} \in \mathbb{S}$ and recall that (A_e, B_e, C_e, D_e) is the state-space realization of $G_{ra}(z)$. Applying the Schur complement to (27) we see that, for a non-zero state-zero direction x_0, (27) can be rewritten as

$$\begin{aligned} (\bar{z}I - A_e)x_0 - B_e u_z &= 0, \\ C_e x_0 + D_e u_z &= 0. \end{aligned} \tag{28}$$

Since A_e is stable and $|\bar{z}| = 1$ we have that $\bar{z}I - A_e$ is invertible and thus (28) can be rewritten as $\left(C_e(\bar{z}I - A_e)^{-1}B_e + D_e\right)u_z = G_{ra}(\bar{z})u_z = 0$. Hence we conclude that the input-zero direction u_z lies in the null-space of $G_{ra}(\bar{z})$. In this case, applying Lemma 2 shows that the problem is bounded if and only if u_z also lies in the null-space of $G_{xa}(\bar{z})$, which concludes the proof.

Supposing that the optimization problem (26) is bounded and $p = 2$, (26) can be rewritten as a generalized eigenvalue problem and solved analytically.

Theorem 2. *Let $p = q = 2$ and suppose that $\ker(G_{ra}(z)) \subseteq \ker(G_{xa}(z))$ for all $z \in \mathbb{S}$. The optimal maximum-impact attack policy is given by*

$$a^\star(z^\star) = \frac{\delta_\alpha}{\|G_{ra}(z^\star)\mathbf{v}^\star\|_2}\mathbf{v}^\star, \tag{29}$$

where $\mathbf{v}^\star$ is the eigenvector associated with λ^*, the largest generalized eigenvalue of the matrix pencil $\left(G_{xa}^H(z)G_{xa}(z),\ G_{ra}^H(z)G_{ra}(z)\right)$ maximized over $z \in \mathbb{S}$. Moreover, the corresponding impact is given by $\|G_{xa}(z^\star)a^\star(z^\star)\|_2 = \sqrt{\lambda^*}\delta_\alpha$.

Proof. The proof is similar to that of [17, Thm. 12].

Given the solution to (26) characterized by the previous result, the maximum impact with respect to (25) is given by

$$g_p(G_{xa}(z^\star)a^\star(z^\star)) = \begin{cases} \sqrt{\lambda^*}\delta_\alpha\ , & \text{if } \sqrt{\lambda^*}\delta_\alpha \le 1 \\ +\infty & , \text{otherwise.} \end{cases}$$

Theorem 3. *Supposing $G_{ra}(z)$ is left-invertible for all $z \in \mathbb{S}$, the largest generalized eigenvalue of the matrix pencil $\left(G_{xa}^H(z)G_{xa}(z),\ G_{ra}^H(z)G_{ra}(z)\right)$, $\lambda^*(z^*)$, maximized over $z^\star \in \mathbb{S}$ corresponds to the $\mathcal{H}_\infty$-norm of $G_{xa}(z)G_{ra}^\dagger(z)$ with $G_{ra}^\dagger(z) = \left(G_{ra}^H(z)G_{ra}(z)\right)^{-1} G_{ra}^H(z)$.*

Proof. First observe that $\ker\left(G_{ra}(z)\right) = \emptyset$, since $G_{ra}(z)$ is left-invertible for all $z \in \mathbb{S}$. Letting $\delta_\alpha = 1$, from Theorem 2 we then have that

$$\lambda^\star(z^\star) = \sup_{z\in\mathbb{S}} \max_{a(z):\, \|G_{ra}(z)a(z)\|_2=1} \|G_{xa}(z)a(z)\|_2.$$

The proof concludes by noting that, since $G_{ra}(z)$ is left-invertible and $G_{xa}(z)$ and $G_{ra}(z)$ are stable, we have $a(z) = G_{ra}^\dagger(z)b(z)$ for some $b(z) \in \mathbb{C}^{n_r}$ and so $\lambda^\star(z^\star)$ can be rewritten as

$$\lambda^\star(z^\star) = \sup_{z\in\mathbb{S}} \max_{b(z):\, \|b(z)\|_2=1} \|G_{xa}(z)G_{ra}^\dagger(z)b(z)\|_2^2 \triangleq \|G_{xa}(z)G_{ra}^\dagger(z)\|_\infty.$$

5.2 Minimum-Resource Attacks

Consider the set of attacks $\mathcal{G}$ such that $a(z) \in \mathcal{G}$ satisfies the goals of a given attack scenario. For the set of attacks $\mathcal{G}$, the minimum-resource steady-state attacks are computed by solving the following optimization problem

$$\begin{aligned} \inf_{z\in\mathbb{S}}\ \min_{a(z)}\quad & \|a(z)\|_0 \\ \text{s.t.}\quad & \left\|G_{ra}(z)a(z)\right\|_q \le \delta_\alpha, \\ & a(z) \in \mathcal{G}. \end{aligned} \tag{30}$$

As in the security-index formulation for a given channel i [13], one can define $\mathcal{G} \triangleq \{a(z) \in \mathbb{C}^{q_a} :\ a_{(i)}(z) = 1\}$.

5.3 Maximum-Impact Minimum-Resource Attacks

Similarly as for the transient analysis, the impact and adversarial resources can be treated simultaneously in the multi-objective optimization problem

$$\begin{aligned} \sup_{z\in\mathbb{S}} \max_{a(z)} \quad & [g_p(G_{xa}(z)a(z)),\ -\|a(z)\|_0]^\top \\ \text{s.t.} \quad & \|G_{ra}(z)a(z)\|_q \leq \delta_\alpha. \end{aligned} \tag{31}$$

Using the bounded objective function method [9], the Pareto frontier can be obtained by iteratively solving the following problem for $\epsilon \in \{1,\ \ldots,\ q_a\}$

$$\begin{aligned} \sup_{z\in\mathbb{S}} \max_{a(z)} \quad & g_p(G_{xa}(z)a(z)) \\ \text{s.t.} \quad & \|G_{ra}(z)a(z)\|_q \leq \delta_\alpha, \\ & \|a(z)\|_0 < \epsilon. \end{aligned} \tag{32}$$

6 Computational Algorithms and Examples

In this section the maximum-impact resource-constrained formulation proposed in the transient analysis with $p = \infty$ is formulated as a mixed integer linear programming problem. Numerical examples are also presented to illustrate some of the proposed formulations for quantifying cyber-security of control systems.

6.1 Mixed Integer Linear Programming

Consider the maximum-impact resource-constrained formulation from the transient analysis (22) reproduced below

$$\begin{aligned} \max_{\mathbf{a}} \quad & g_p(\mathbf{n}) \\ \text{s.t.} \quad & \|\mathcal{C}_\xi \mathbf{e} + \mathcal{D}_\xi \mathbf{a}\|_p \leq \delta_\alpha, \\ & \|h_p(\mathbf{a})\|_0 \leq \epsilon, \\ & \mathbf{e} = \mathcal{O}_\xi \xi_{-1|-1} + \mathcal{T}_\xi \mathbf{a}, \\ & \mathbf{n} = \mathcal{O}_\eta \eta_0 + \mathcal{T}_\eta \mathbf{a}. \end{aligned}$$

For $1 \leq p \leq \infty$, the constraint $\|h_p(\mathbf{a})\|_0 \leq \epsilon$ models the fact that the number of channels the adversary can compromise is upper bounded by epsilon. By introducing the binary decision variables $\mathbf{z}_i$, one for each channel, the constraint can be modeled as follows:

$$\begin{aligned} \mathbf{a}_{(i)} &\leq M_h \mathbf{z}_i \mathbf{1} && \forall\, i = 1,\ \ldots,\ q_a \\ -\mathbf{a}_{(i)} &\leq M_h \mathbf{z}_i \mathbf{1} && \forall\, i = 1,\ \ldots,\ q_a \\ \sum_{i=1}^{q_a} \mathbf{z}_i &\leq \epsilon \\ \mathbf{z}_i &\in \{0, 1\} && \forall\, i = 1,\ \ldots,\ q_a. \end{aligned} \tag{33}$$

In (33), $\mathbf{1}$ is a vector of ones of appropriate dimension. M_h is a given large number used to model "infinity". Its value is typically chosen according to the physical limitation of the system. The binary decision variables $\mathbf{z}_i$ serve to count the number of channels the adversary can compromise. That is, $\mathbf{z}_i = 1$ if and only if channel i can be compromised. Once a channel is compromised, the adversary is expected to be able to modify the time signal in that channel in any way he desires. This is modeled by the first two sets of constraints in (33).

In the constraint $\|C_\xi \mathbf{e} + D_\xi \mathbf{a}\|_p \leq \delta_\alpha$, the ℓ_p norm is chosen to be the ℓ_∞ norm modeling a constraint on the worst case output violation. This constraint can be modeled as

$$\begin{aligned} C_\xi \mathbf{e} + D_\xi \mathbf{a} &\leq \delta_\alpha \mathbf{1} \\ -C_\xi \mathbf{e} - D_\xi \mathbf{a} &\leq \delta_\alpha \mathbf{1}. \end{aligned} \tag{34}$$

In the objective function $g_p(n)$, the safety set $\mathcal{S}^p$ is chosen to be a ℓ_∞ norm ball. That is, $\mathcal{C}_x \mathbf{n} \in \mathcal{S}^p$ if and only if $\|\mathcal{C}_x \mathbf{n}\|_\infty \leq M_\mathcal{S}$ for some given safety tolerance $M_\mathcal{S}$. This is to model the fact that if any component of $\mathcal{C}_x \mathbf{n}$ is too large, then the system is considered to be unsafe. Consequently, the adversary's goal is to maximize $g_p(\mathbf{n})$ so that at least one component of $\mathcal{C}_x \mathbf{n}$ is larger than the safety tolerance $M_\mathcal{S}$. In hypograph form [1], maximizing $g_p(\mathbf{n})$ amounts to maximizing a slack variable γ with the additional constraint that $g_p(\mathbf{n}) \geq \gamma$. The later constraint can be modeled as

$$\begin{aligned} \mathcal{C}_x \mathbf{n} &\geq +\gamma \mathbf{1} - M_{\mathcal{C}_x} (\mathbf{1} - \mathbf{z}^+) \\ \mathcal{C}_x \mathbf{n} &\leq -\gamma \mathbf{1} + M_{\mathcal{C}_x} (\mathbf{1} - \mathbf{z}^-) \\ \mathbf{z}_i^+ + \mathbf{z}_i^- &\leq 1 \qquad \forall\, i \\ \sum_i (\mathbf{z}_i^+ + \mathbf{z}_i^-) &\geq 1 \\ \mathbf{z}_i^+ &\in \{0,1\} \quad \forall\, i \\ \mathbf{z}_i^- &\in \{0,1\} \quad \forall\, i. \end{aligned} \tag{35}$$

In (35), $M_{\mathcal{C}_x}$ is another given large number used to represent "infinity". For each i, when the binary decision variable $\mathbf{z}_i^+ = 1$, the ith constraint of $\mathcal{C}_x \mathbf{n} \geq \gamma \mathbf{1} - M_{\mathcal{C}_x} (\mathbf{1} - \mathbf{z}^+)$ implies that the ith component of $\mathcal{C}_x \mathbf{n}$ is greater than or equal to γ. On the other hand, if $\mathbf{z}_i^+ = 0$ then this constraint component can be ignored. A similar interpretation holds for the combination of $\mathbf{z}^-$ and $\mathcal{C}_x \mathbf{n} \leq -\gamma \mathbf{1} + M_{\mathcal{C}_x} (\mathbf{1} - \mathbf{z}^-)$. Furthermore, the constraint $\mathbf{z}_i^+ + \mathbf{z}_i^- \leq 1$ models the fact that the ith component of $\mathcal{C}_x \mathbf{n}$ cannot be both greater than γ and less than $-\gamma$ when $\gamma > 0$. Together with the above discussion, the constraint $\sum_i (\mathbf{z}_i^+ + \mathbf{z}_i^-) \geq 1$ indicates that at least one component of $\mathcal{C}_x \mathbf{n}$ must be greater than or equal to γ in absolute value. Since the objective is to maximize γ, it holds that $\gamma = \|\mathcal{C}_x \mathbf{n}\|_\infty$ at optimality. Finally, to model the fact that once the goal $\|\mathcal{C} \mathbf{n}\|_\infty > M_\mathcal{S}$ is achieved the adversary no longer needs to maximize γ. An additional constraint

$$\gamma \leq M_\mathcal{S} \tag{36}$$

can be imposed.

In conclusion, the maximum-impact resource-constrained attack can be modeled by the following mixed integer linear program:

$$\begin{aligned} \max_{\mathbf{a},\gamma,\mathbf{z},\mathbf{z}^+,\mathbf{z}^-} \quad & \gamma \\ \text{s.t.} \quad & \mathbf{e} = \mathcal{O}_\xi \xi_{-1|-1} + \mathcal{T}_\xi \mathbf{a}, \\ & \mathbf{n} = \mathcal{O}_{\eta_0} \eta_0 + \mathcal{T}_\eta \mathbf{a}, \\ & (33), (34), (35), (36). \end{aligned} \tag{37}$$

6.2 Numerical Example

Next we illustrate some of the proposed formulations for the Quadruple-Tank Process (QTP) illustrated in Fig. 2. The plant model can be found in [7]

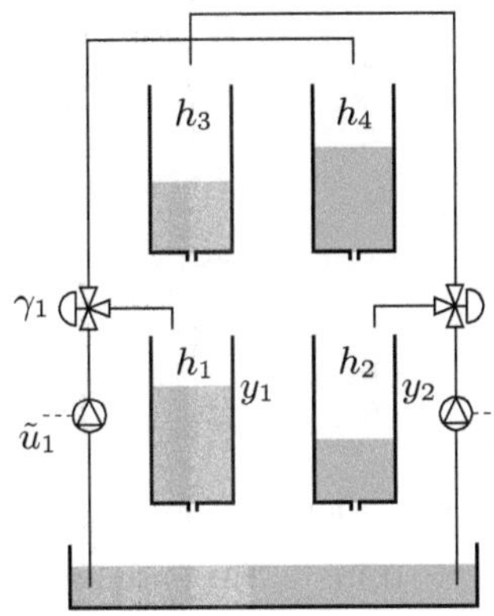

Fig. 2. Schematic of the Quadruple-Tank Process

$$\begin{aligned} \dot{h}_1 &= -\frac{a_1}{A_1}\sqrt{2gh_1} + \frac{a_3}{A_1}\sqrt{2gh_3} + \frac{\gamma_1 k_1}{A_1} u_1, \\ \dot{h}_2 &= -\frac{a_2}{A_2}\sqrt{2gh_2} + \frac{a_4}{A_2}\sqrt{2gh_4} + \frac{\gamma_2 k_2}{A_2} u_2, \\ \dot{h}_3 &= -\frac{a_3}{A_3}\sqrt{2gh_3} + \frac{(1-\gamma_2)k_2}{A_3} u_2, \\ \dot{h}_4 &= -\frac{a_4}{A_4}\sqrt{2gh_4} + \frac{(1-\gamma_1)k_1}{A_4} u_1, \end{aligned} \tag{38}$$

where h_i are the heights of water in each tank, A_i the cross-section area of the tanks, a_i the cross-section area of the outlet hole, k_i the pump constants, γ_i the flow ratios and g the gravity acceleration. The nonlinear plant model is linearized for a given operating point and sampled with a sampling period $T_s = 2\,s$. The QTP is controlled using a centralized LQG controller with integral action and

Table 1. Values of $\|\mathbf{x}\|_p$ for the maximum-impact formulation with $p = q = 2$ and $\delta_\alpha = 0.15$.

	$\|h_p(\mathbf{a})\|_0$			
	1	**2**	**3**	**4**
Minimum phase	1.15	140.39	∞	∞
Non-minimum phase	2.80	689.43	∞	∞

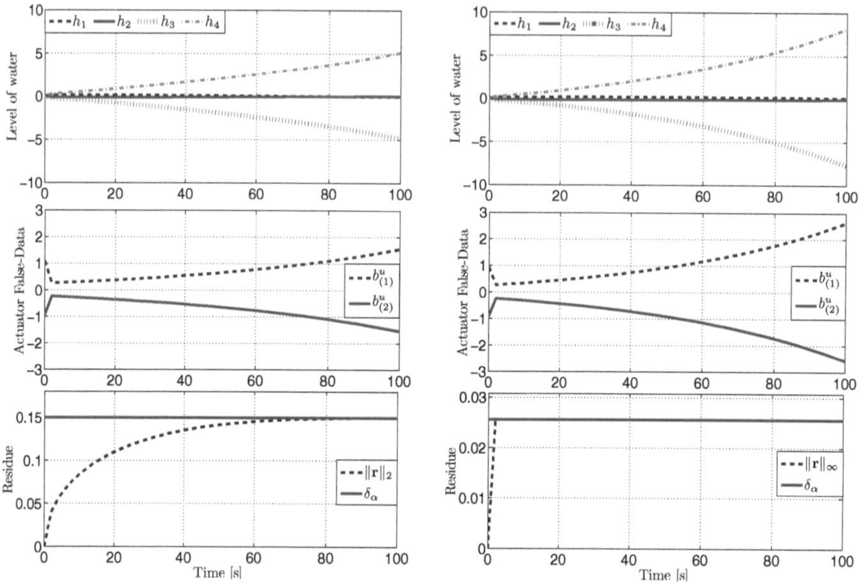

(a) Parameters: $p = q = 2$, $\delta_\alpha = 0.15$. (b) Parameters: $p = q = \infty$, $\delta_\alpha = 0.025$.

Fig. 3. Simulation results of the multi-objective problem (22) with $\epsilon = 2$ for the non-minimum phase system

a Kalman-filter-based anomaly detector is used so that alarms are triggered according to (5), for which we chose $\delta_\alpha = 0.25$ for illustration purposes.

For the time-interval $[0,\ 50]$, the maximum-impact minimum-resource attacks were computed for the process in minimum and non-minimum phase settings by iteratively solving (22) with $p = q = 2$. The respective impacts are presented in Table 1. As expected, the non-minimum phase system is less resilient than the minimum-phase one. In both settings the attack impact can be made arbitrarily large by corrupting 3 or more channels and thus the adversary can drive the state out of the safe set while remaining stealthy.

The maximum-impact attack signal for the non-minimum phase system with $\epsilon = 2$, $\delta_\alpha = 0.15$, and $p = q = 2$ is presented in Fig. 3(a). For the parameters $\epsilon = 2$, $\delta_\alpha = 0.025$, and $p = q = \infty$, the maximum-impact attack signal was computed using the mixed-integer linear programming problem (37) and is shown in Fig. 3(b). In both cases the optimal attack corrupts both actuator channels and ensures $\|\mathbf{r}\|_{\ell_p} \leq \delta_\alpha$.

7 Conclusions

Several formulations for quantifying cyber-security of networked control systems were proposed and formulated as constrained optimization problems, capturing trade-offs among adversary goals and constraints such as attack impact on the control system, attack detectability, and adversarial resources. Although the formulations are non-convex, some can be related to system theoretic concepts such as invariant-zeros and weighted $\mathcal{H}_\infty$ norm of the closed-loop system and thus may be solved efficiently. The maximum-impact resource-constrained attack policy was also formulated as a mixed-integer linear program for a particular choice of parameters. The results were illustrated for the quadruple-tank process.

Acknowledgments. This work was supported in part by the European Commission through the HYCON2 project, the Swedish Research Council under Grants 2007-6350 and 2009-4565, the Swedish Foundation for Strategic Research, and the Knut and Alice Wallenberg Foundation.

References

1. Boyd, S., Vandenberghe, L.: Convex Optimization. Cambridge University Press (2004)
2. Cárdenas, A., Amin, S., Lin, Z., Huang, Y., Huang, C., Sastry, S.: Attacks against process control systems: risk assessment, detection, and response. In: Proceedings of the 6th ACM Symposium on Information, Computer and Communications Security, ASIACCS 2011, pp. 355–366. ACM, New York (2011)
3. Ding, S.X.: Model-based Fault Diagnosis Techniques: Design Schemes. Springer (2008)
4. Esfahani, P., Vrakopoulou, M., Margellos, K., Lygeros, J., Andersson, G.: Cyber attack in a two-area power system: Impact identification using reachability. In: American Control Conference, pp. 962–967 (July 2010)
5. Fawzi, H., Tabuada, P., Diggavi, S.: Security for control systems under sensor and actuator attacks. In: Proceedings of the 51st IEEE Conference on Decision and Control, Maui, Hawaii, USA (December 2012)
6. Hwang, I., Kim, S., Kim, Y., Seah, C.E.: A survey of fault detection, isolation, and reconfiguration methods. IEEE Transactions on Control Systems Technology 18(3), 636–653 (2010)
7. Johansson, K.: The quadruple-tank process: a multivariable laboratory process with an adjustable zero. IEEE Transactions on Control Systems Technology 8(3), 456–465 (2000)
8. Liu, J., Ye, J.: Efficient L1/Lq Norm Regularization. ArXiv e-prints (September 2010)
9. Marler, R.T., Arora, J.S.: Survey of multi-objective optimization methods for engineering. Structural and Multidisciplinary Optimization 26(6), 369–395 (2004)

10. Meserve, J.: Sources: Staged cyber attack reveals vulnerability in power grid. CNN (2007), http://edition.cnn.com/2007/US/09/26/power.at.risk/index.html
11. Pasqualetti, F., Dorfler, F., Bullo, F.: Cyber-physical attacks in power networks: Models, fundamental limitations and monitor design. In: Proc. of the 50th IEEE Conf. on Decision and Control and European Control Conference, Orlando, FL, USA (December 2011)
12. Rid, T.: Cyber war will not take place. Journal of Strategic Studies 35(1), 5–32 (2011)
13. Sandberg, H., Teixeira, A., Johansson, K.H.: On security indices for state estimators in power networks. In: Preprints of the First Workshop on Secure Control Systems, CPSWEEK 2010, Stockholm, Sweden (April 2010)
14. Smith, R.: A decoupled feedback structure for covertly appropriating networked control systems. In: Proc. of the 18th IFAC World Congress, Milano, Italy (August-September 2011)
15. Sundaram, S., Hadjicostis, C.: Distributed function calculation via linear iterative strategies in the presence of malicious agents. IEEE Transactions on Automatic Control 56(7), 1495–1508 (2011)
16. Symantec: Stuxnet introduces the first known rootkit for industrial control systems. Symantec (August 6, 2010), http://www.symantec.com/connect/blogs/stuxnet-introduces-first-known-rootkit-scada-devices
17. Teixeira, A., Shames, I., Sandberg, H., Johansson, K.H.: A Secure Control Framework for Resource-Limited Adversaries. Submitted to Automatica (December 2012) ArXiv e-prints
18. Tokarzewski, J.: Finite zeros in discrete time control systems. LNCIS, vol. 338. Springer, Heidelberg (2006)
19. U.S.-Canada PSOTF: Final report on the August 14th blackout in the United States and Canada. Tech. rep., U.S.-Canada Power System Outage Task Force (April 2004)
20. Zhou, K., Doyle, J.C., Glover, K.: Robust and Optimal Control. Prentice-Hall, Inc., Upper Saddle River (1996)

Security and Infiltration of Networks: A Structural Controllability and Observability Perspective

Airlie Chapman and Mehran Mesbahi*

William E. Boeing Department of Aeronautics and Astronautics,
University of Washington, Seattle USA
{airliec,mesbahi}@uw.edu

Abstract. This paper examines the role of structural controllability (s-controllability) in the design of secure linear-time-invariant networked systems. We reason about secure network design in the face of two attack vectors: a "Disrupt" attack where the infiltrator's objective is to perturb the network to render it unusable, and a "Highjack and eavesdrop" attack to actively control and probe the network. For the former attack, strong s-controllable input sets are chosen to control the network to provide robustness to these attacks. Weak s-controllable input sets are selected by infiltrators for the "Highjack and eavesdrop" attack so as to generically guarantee a successful attack. We provide necessary and sufficient conditions for weak and strong s-controllability involving matchings over a bipartite graph representation of the network. We also provide a brief analysis of s-controllability over special families of networks.

Keywords: Strong structural controllability, Weak structural controllability, Network controllability, Network observability, Constrained matching.

1 Introduction

Network security and its complement, network infiltration, are of paramount importance in design and analysis of complex dynamic networks. Dynamic networks are core to the natural world, such as biological, chemical and social networks, and our technological world with networks such as the internet, power grids, and robotic networks. In recent years, there has been a myriad of research in the area of network systems [1,2,3]. Of increasing importance is securing networks from outside manipulation and monitoring. The complementary problem is identifying effective means to infiltrate, influence or identify networks. At the cornerstone of these problems is the interplay between system performance and network structure.

Most modern day technological networks rely on access security to protect the network. An alternative method to generate a secure network is intrusion

* The research of the authors was supported by AFOSR grant FA9550-12-1-0203 DEF.

D.C. Tarraf (ed.), *Control of Cyber-Physical Systems*,
Lecture Notes in Control and Information Sciences 449,
DOI: 10.1007/978-3-319-01159-2_8, © Springer International Publishing Switzerland 2013

detection[1] coupled with either inter-agent security through each agent's dynamics or intra-agent security via the network topology. The former includes implementation of disturbance rejection or agent disabling techniques, e.g., noise canceling systems and power grid "brown outs". The latter involves global or local network design to improve security, e.g., TCP network re-routing. This topology-centric approach for security is the main focus of the present work. Specifically, the following two attack vectors are the focus of this paper:

- Disrupt: An infiltrator perturbs the coupling strength between nodes in the network to disrupt the user's ability to control and/or monitor the network.
- Highjack and eavesdrop: An infiltrator with knowledge of the inter-agent dependencies, i.e., who effects whom, but not the coupling strength between nodes, attaches to a subset of "infiltrated" nodes in the graph. The infiltrator then applies signals and/or monitors the state of these nodes to control and/or identify the network.

The current work is part of a more general effort that aims to identify fundamental bounds on the security of coordination algorithms for dynamic systems when infiltrated by an adversary. As such, our work is related to a number of other research works such those in computer network security [8], disease control [9,10], and predator/prey swarming [11]. The paper presents the problem of network security from both the infiltrator's perspective complementing work on infiltration detection such as [4].

Fruitful investigation into the effectiveness of controlling a network has occurred in the area of controllability. In controllability, the importance of network topology has appeared through investigation of the roles of symmetries in the network [12,13]. There has been a strong focus on consensus-based networks with controllability established for many special families of graphs including circulants, grids, distance regular and Cartesian products [13,14,15].

Research in weak and strong structural controllability (s-controllability) has a rich history since its inception [16,17], and by its very nature exposes the role of structure in network control. Structural controllability establishes generic (weak) and complete (strong) controllability of a network based solely on the direct coupling between nodes appearing as a distinct pattern of zeros and nonzeros in the network dynamics. This is irrespective of the magnitude of these couplings. Weak and strong s-controllability also provide lower and upper bounds, respectively, on the cardinality of a minimum input set for controllability. Figure 1 illustrates conceptually this approach to controllability.

Recently, a result by Liu *et al.* has linked weak s-controllability to matchings in a bipartite graph representation of the network [18]. This provides an attractive and efficient way to check generically controllable input sets. We recently presented a similar matching approach to establish strong s-controllability of the network [19]. Further, unlike controllability, methods to establish s-controllability are typically numerically robust.

[1] Techniques for intrusion or fault detection on consensus-type networks include those based on reachability analysis [4], and the more popular unknown-input observers [5,6,7].

As s-controllability only requires knowledge of interconnections and not the associated coupling strengths, establishing weak and strong s-controllability removes the accurate modeling requirement of controllability. A such, s-controllability is an attractive tool for infiltrators to establish low cardinality s-controllable input sets to highjack and eavesdrop on the network without accurate knowledge of the network parameters. Typically, infiltrating more nodes in the network is difficult and exposes an infiltrator to a higher risk of detection. Thus, weak s-controllability provides a more cost effective measure than strong s-controllability and for this reason will be used as the main tool in this paper for analysis of the "Highjack and eavesdrop" attack scenario.

One attraction of s-controllability is that, independent of small variations in the coupling strengths, controllability can be guaranteed. This provides network controllability robust to parameter uncertainty and floating point errors. Although it is atypical that a weakly s-controllable network has a coupling strength realization rendering it uncontrollable, there are systems that are inadvertently designed or maliciously altered with such a coupling. Examples with such a coupling is unweighted undirected graphs such as the aforementioned consensus-based networks and the "Disrupt" attack scenario. Such homogeneity generates symmetry in the network which typically renders a system uncontrollable. For these cases, strong s-controllability is a useful alternative to weak s-controllability, and will be used in this paper as a method to secure networks against "Disrupt" attacks.

The organization of the paper is as follows. We commence with the introduction of notation, graphs, bipartite matching, and pattern matrices, and describe the linear-time-invariant model that is examined. We present an equivalent representation of the network through a bipartite graph representation and relate matching conditions on this graph to requirements for weak and strong s-controllability. These conditions are exploited to reason about a secure design against "Disrupt" and "Highjack and eavesdrop" attacks. Finally, we analyze minimum cardinality input sets for some common families of networks, reasoning about the different levels of security that certain network structures exhibit.

2 Notation, Preliminaries and Model

We provide a brief background on constructs that will be used in this paper.

For the column vector $v \in \mathbb{R}^p$, v_i or $[v]_i$ denotes its ith entry, e_i denotes the column vector which contains all zero entries except $[e_i]_i = 1$. For matrix $M \in \mathbb{R}^{p \times q}$, $[M]_{ij}$ denotes the entry in its ith row and jth column. We form the submatrix $A(\alpha|\beta)$ from $A \in \mathbb{R}^{m \times n}$, where $\alpha \subseteq \{1, \dots, m\}$ and $\beta \subseteq \{1, \dots, n\}$, by removing rows with indices α and columns with indices β.

2.1 Graphs

A succinct way to represent the interactions of agents in a network is through a graph. A graph $\mathcal{G} = (V, E, W)$ is defined by a node set $V = \{1, \dots, n\}$ where n is

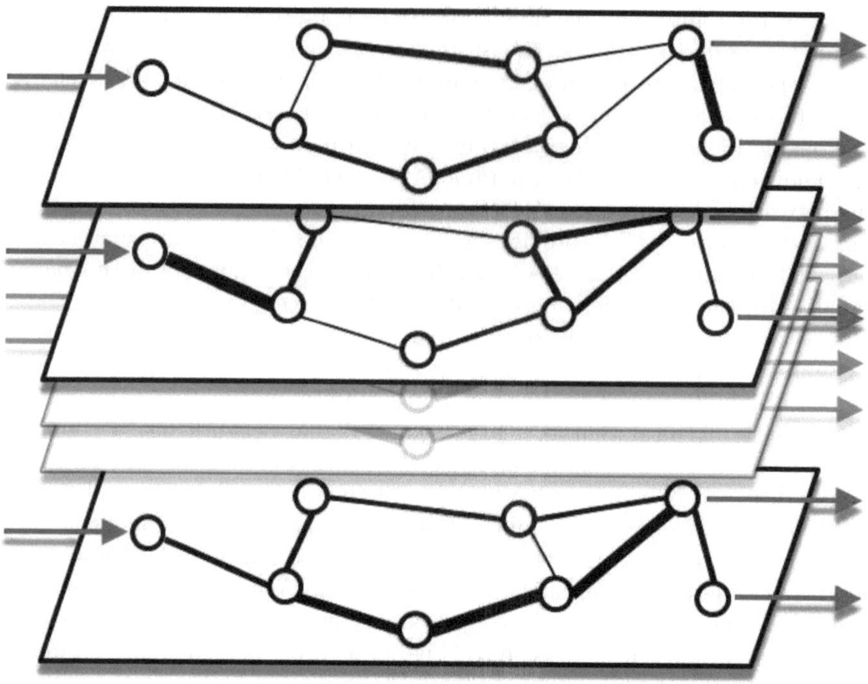

Fig. 1. An illustration of the s-controllability concept. Weak and strong s-controllability establish generic and absolute controllability and observability over all realizations of a network.

the number of nodes n in the graph, representing the agents in the network, an edge set E comprising of pairs of nodes which represents the agents interactions, i.e., agent i affects agent j's dynamics if there an edge from i to j, i.e., $(i,j) \in E$, and a weight set W containing weights w_{ji} for each edge $(i,j) \in E$. A graph is undirected if $(i,j) \in E$ implies $(j,i) \in E$ and $w_{ji} = w_{ij}$. Special undirected graphs are the complete graph with edges between every pair of nodes and the path graph where $(i,j) \in E$ if and only if $|i-j| = 1$. A self-loop is an edge $(i,i) \in E$ and the subset of nodes in V with self-loops will be denoted as V_s. The adjacency matrix $A(\mathcal{G})$ is a matrix representation of $\mathcal{G}$ with $[A(\mathcal{G})]_{ji} = w_{ji}$ for $(i,j) \in E$ and $[A(\mathcal{G})]_{ji} = 0$, otherwise. The node set $V_1 \subseteq V$ is a rooted node set for $V_2 \subseteq V$ if every node in V_2 can be reached from at least one node in V_1 along a directed path. Further, if V_1 is (not) a rooted node set for V then $\mathcal{G}$ is *input (in)accessible* from V_1.

2.2 Model

The system model considered is a linear-time-invariant networked system of single-integrator agents. The graph $\mathcal{G} = (V, E, W)$, represents this network with agents corresponding to the node set V, and interactions between agents by edges E and the strength of such interactions by the edge weights W. The state matrix can be formed from the graph through its adjacency matrix, namely $A(\mathcal{G})$. Control of these networks involves applying a unique signal to a subset of agents, represented by the *input node set* $S \subseteq V$, i.e., the nodes in S are controlled directly. The input node set $S = \{i_1, i_2, \dots, i_m\}$, where $m \leq n$, defines the input matrix $B(S) = [e_{i_1}, e_{i_2}, \dots, e_{i_m}] \in \mathbb{R}^{n\times m}$. The monitoring of the network is accomplished by directly sensing a subset of agents, represented by the *output node set* $R \subseteq V$. The output node set $R = \{j_1, j_2, \dots, j_p\}$, where $p \leq n$, defines the output matrix $C(R) = \left[e_{j_1}, e_{j_2}, \dots, e_{j_p}\right]^T \in \mathbb{R}^{p\times n}$. The states of the agents and the inputs and outputs at time t, are denoted as $x(t) \in \mathbb{R}^n$, $u(t) \in \mathbb{R}^m$ and $y(t) \in \mathbb{R}^p$, respectively.

The abbreviated form of the dynamics is consequently,

$$\begin{aligned}\dot{x}(t) &= A(\mathcal{G})x(t) + B(S)u(t)\\ y(t) &= C(R)x(t).\end{aligned}$$

This problem setup is popular in networked system and has been surveyed by Mesbahi and Egerstedt [20].

We consider a family of these dynamics, with members defined by $(A(\mathcal{G}), B(S), C(R))$, by generalizing the weight set W, considering E the set nonzero weight assigned edges, and the variable scaling of the input and output nodes signal. This has the effect of preserving the zero structure of $A(\mathcal{G})$, $B(S)$ and $C(R)$ across all network realizations; we represent this family as all realizations of a trio of pattern matrices.

A *pattern matrix* $\boldsymbol{A}$ (or Boolean structure matrix) is a matrix with each element set as either a zero or a star, denoted throughout by an $\times$. A numerical matrix A is called a *realization* of the pattern $\boldsymbol{A}$ if A can be obtained by assigning nonzero numerical values to the star entries of $\boldsymbol{A}$, in short $A \in \boldsymbol{A}$. The pattern matrix $\boldsymbol{A}(\mathcal{G})$ of a graph $\mathcal{G}$ can be formed by replacing the nonzero entries of $A(\mathcal{G})$ with stars. Applying the same process to matrices $B(S)$ and $C(R)$ form the pattern matrices $\boldsymbol{B}(S)$ and $\boldsymbol{C}(R)$, respectively.

The family of dynamics we are considering is consequently all systems of the form

$$\begin{aligned}\dot{x}(t) &= Ax(t) + Bu(t)\\ y(t) &= Cx(t),\end{aligned} \tag{1}$$

where $A \in \boldsymbol{A}(\mathcal{G})$, $B \in \boldsymbol{B}(S)$ and $C \in \boldsymbol{C}(R)$. For succinctness, if $\mathcal{G}$ and S are clear from the context, then $\boldsymbol{A}$, $\boldsymbol{B}$ and $\boldsymbol{C}$ will replace $\boldsymbol{A}(\mathcal{G})$, $\boldsymbol{B}(S)$ and $\boldsymbol{C}(R)$, respectively.

Example 1. Consider the graph $\mathcal{G}$ in Figure 2a) . The corresponding pattern matrix is

$$\boldsymbol{A}(\mathcal{G}) = \begin{bmatrix} \times & 0 & \times \\ \times & 0 & \times \\ 0 & \times & 0 \end{bmatrix}.$$

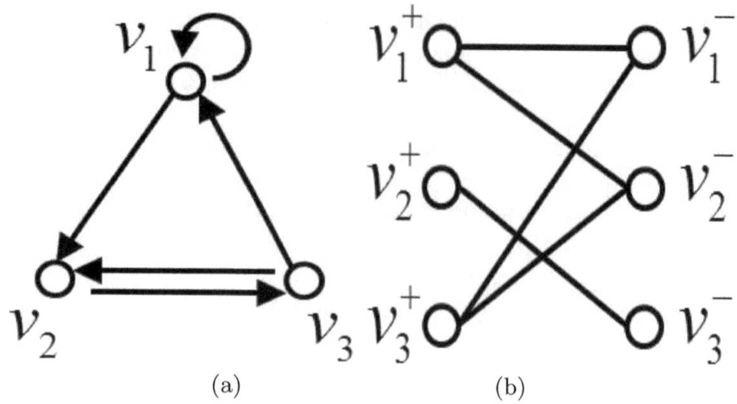

Fig. 2. (a) Graph $\mathcal{G}$ and (b) an equivalent bipartite graph $\mathcal{H}$

3 Structural Controllability and Observability

Controllability, and its dual system property observability, is often a necessary condition for the application of control system tools, such as stabilization and optimal control. There are many techniques for establishing controllability of linear systems. One such approach for establishing controllability of a pair (A, B) in (1) is through the full column rank of the matrix $\left[A_\lambda \; B \right]$ for every eigenvalue λ of A, where $A_\lambda = A - \lambda I$ [21]. Due to the duality between controllability and observability, the pair (A, B) is controllable if and only if the pair (A, C) is observable, where $B = C^T$, and so a similar approach for establishing observability can be performed.

Unfortunately, the above approach, like other controllability criteria, have computational drawbacks, falling victim to rounding errors, and requiring an accurate knowledge of the system dynamics [22]. A more numerically stable approach, which relaxes the modeling precision requirements on the system, is the establishment of s-controllability and observability.

Lin [16] defined a pair $(\boldsymbol{A}, \boldsymbol{B})$ as weak s-controllable if it admits some controllable numerical realization (A, B). From a generic analysis of this setup [23], it follows that weak s-controllability implies that *almost all* numerical realizations of $(\boldsymbol{A}, \boldsymbol{B})$ are controllable. Mayeda and Yamada [17] adapted this by defining the pair $(\boldsymbol{A}, \boldsymbol{B})$ as strongly s-controllable system if *all* numerical realizations (A, B) are controllable. We state that for a given $\boldsymbol{A}$, the input set S is weakly (strongly) s-controllable if the pair $(\boldsymbol{A}, \boldsymbol{B}(S))$ is weakly (strongly) s-controllable.

It is clear that in general a smaller input set S is required to meet the conditions for weak s-controllability as compared to strong s-controllability. This benefit comes with the loss of guaranteed controllability that strong s-controllability provides over all realizations. It is this reduced network attachment cost (small $|S|$) for a high chance of controllability, compared to a conservative certainty of controllability that makes weak and strong s-controllability ideal for reasoning about infiltration and security. More specifically, an infiltrator attempting to control an unknown weighted network will try to establish a cheap weak input set while, for security, a larger strong input set investment is worthwhile for the guarantee, under fluctuations in edge weights that controllability will be maintained.

Extending the aforementioned rank condition for controllability, the pattern matrix pair $(\boldsymbol{A}, \boldsymbol{B})$ in Model (1) is weakly (strongly) s-controllable if and only if $\left[\boldsymbol{A}_\lambda \ \boldsymbol{B}\right]$ is full rank for some (all) realizations of $\boldsymbol{A}$ and $\boldsymbol{B}$, and $\lambda \in \mathbb{C}$. In other words, s-controllability can be established through a pattern matrix rank condition. In turn pattern matrix rank conditions can be extracted by examining matchings on an equivalent bipartite graph representation of $(\boldsymbol{A}, \boldsymbol{B})$ - a numerically robust combinatorial property. This connection was recently explored by Liu *et al.* [18] for weak s-controllability. A similar set of results can be obtained for the case of strong s-controllability of $(\boldsymbol{A}(\mathcal{G}), \boldsymbol{B}(S))$ [19]. It is with this in mind that we now discuss relevant background on bipartite graphs and matching.

3.1 Bipartite Graphs and Matching

A different graph representation of the network, and in fact any $n \times m$ matrix, can be formed using a bipartite graph representation of the interactions. A bipartite graph $\mathcal{H} = (V^+, V^-, E)$ is an undirected graph on independent node sets V^+ and V^-, where the edge set E connects nodes in V^+ to V^-. A bipartite graph representation of a matrix $M \in \mathbb{R}^{p \times q}$ can be formed by setting $V^+ = \{1, \ldots, q\}$ and $V^- = \{1, \ldots, p\}$ and having an edge $\{i, j\} \in E$ if and only if $[M]_{ij} \neq 0$. The bipartite graph of a pattern matrix of an equivalent $A(\mathcal{G})$ can be formed in this way.

Let $\mathcal{H} = (V^+, V^-, E)$ be a bipartite graph. A set of r edges $\left\{\left(i_1^+, i_1^-\right), \ldots, \left(i_t^+, i_r^-\right)\right\}$ in $\mathcal{H}$ is said to be a *t-matching* (between $I^+ = \left\{i_1^+, \ldots, i_r^+\right\} \subseteq V^+$ and $I^- = \left\{i_1^-, \ldots, i_r^-\right\} \subseteq V^-$) if $i_1^+, \ldots, i_r^+$ are distinct and $i_1^-, \ldots, i_r^-$ are distinct. Such a t-matching is said to be a *constrained t-matching* (or uniquely restricted matching) if it is the only t-matching in $\mathcal{H}$ between I^+ and I^-. Those nodes in $\mathcal{H}$ that are in I^- are referred to as *matched*, and those that are not in I^- are called *unmatched*. A matching is *T-less* if it contains no edges of the form (j^+, j^-), where j is in the set T. A T-less matching where $T = \{1, \ldots, \min\{|V^+|, |V^-|\}\}$, is referred to as a *self-less* matching. A (constrained) (T-less) t_1-matching in $\mathcal{H}$ is *maximum* if there is no (constrained) (T-less) t_2-matching in $\mathcal{H}$ with $t_2 > t_1$.

We say that a pattern matrix $\boldsymbol{A}$ has a (constrained) t-matching if the associated bipartite graph $\mathcal{H}$ has a (constrained) t-matching. An example of a pattern matrix of a graph and associated maximum t-matchings is presented in the following example.

Example 2. Consider again the graph $\mathcal{G}$ in Figure 2a) with the associated bipartite graph $\mathcal{H}$ in Figure 2b). The pattern matrix $\boldsymbol{A}(\mathcal{G})$ appearing in Example 1 has two maximum 3-matchings, $\left\{\left(v_1^+, v_1^-\right), \left(v_2^+, v_3^-\right), \left(v_3^+, v_2^-\right)\right\}$

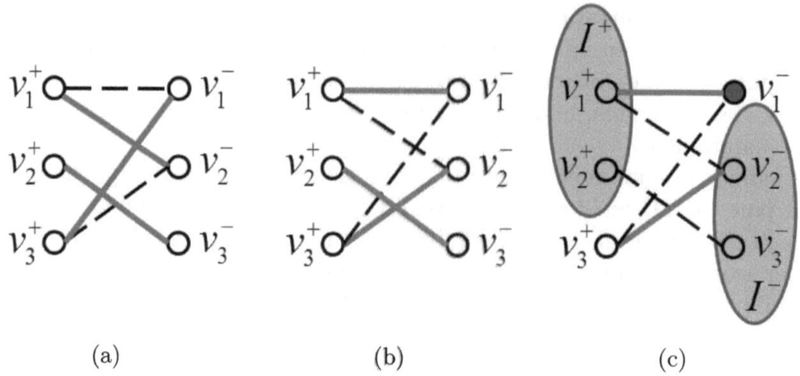

Fig. 3. (a-b) Unconstrained 3-matchings and (c) constrained 2-matching with corresponding sets I^+, I^- and unmatched node v_1 (solid red) on the bipartite graph $\mathcal{H}$ appearing in Example 2

and $\{(v_1^+,v_2^-),(v_2^+,v_3^-),(v_3^+,v_1^-)\}$ and four maximum constrained 2-matchings $\{(v_1^+,v_1^-),(v_2^+,v_3^-)\}$, $\{(v_1^+,v_2^-),(v_2^+,v_3^-)\}$, $\{(v_2^+,v_3^-),(v_3^+,v_1^-)\}$ and $\{(v_2^+,v_3^-),(v_3^+,v_2^-)\}$. A selection of these are shown in Figure 3.

There is an intimate relationship between the matrix rank of realizations of $\boldsymbol{A}$ and the (constrained) t-matchings of $\boldsymbol{A}$. The following theorem summarizes some of these links and will be subsequently used in this paper.

Theorem 1. *[24] Let $\boldsymbol{A}$ be an $m \times n$ pattern matrix, and let t be a nonnegative integer, $1 \leq t \leq \min\{m,n\}$.*

(1) If $\boldsymbol{A}$ has a t-matching, then there exists a matrix $A \in \boldsymbol{A}$ with rank$(A) \geq t$.

(2) If $\boldsymbol{A}$ has a constrained t-matching, then every matrix $A \in \boldsymbol{A}$ has rank$(A) \geq t$.

Specifically, conditions 1) and 2) provide a mechanism to establish weak s-controllability and strong s-controllability, respectively. For an in depth study of the pattern matrix rank and matching relationship we refer the reader to Hershkowitz and Schneider [25] and Golumbic *et al.* [26].

3.2 Weak Structural Controllability – A cautious Lower Bound

As previously mentioned, weak s-controllability is a generic condition on controllability. If the family of systems in Model (1) is weak s-controllable, a given realization will be controllable with high probability. For non-homogeneous networks with independent edge weights, if an input set satisfies weak s-controllability it will, in general, also satisfy general controllability. For homogenous networks, such as unweighted networks, symmetries inherent in the network tend to break the controllability of weak inputs. For example, from a single input node the undirected complete graph is weak s-controllable but its unweighted realization requires $n-1$ nodes to control completely.

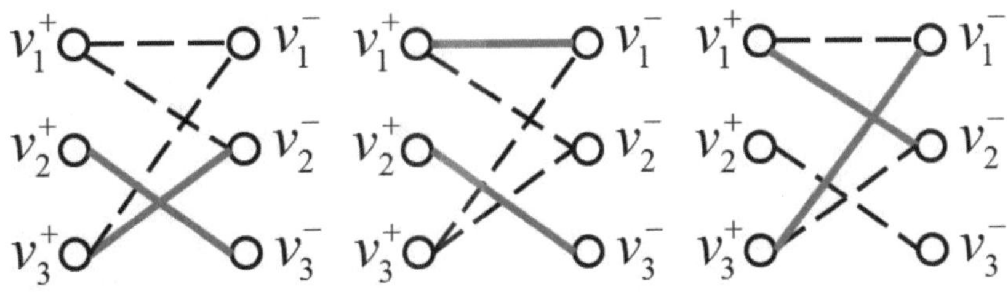

Fig. 4. Three 2-matchings relevant to Example 3 satisfying the matching condition in Theorem 2 on the bipartite graph $\mathcal{H}$ appearing in Example 2

From an infiltrator's perspective, if the network has unknown weights, weak s-controllability presents a useful method to isolate, with high probability, controllable or observable network attachment points. The necessity of controllability and observability for effective network identification and control make weak s-controllability ideal for infiltrator "Highjack and Eavesdrop" scenarios. From a different perspective, a cautious lower bound suitable for network security is additional protection of weak input attachment nodes to prevent such an attack. To this end, the following theorem provides a matching method to find and check for weak input sets [18].

Theorem 2. *Given a pattern matrix* $\boldsymbol{A}(\mathcal{G})$*:*

(1) (check) The pair $(\boldsymbol{A}, \boldsymbol{B}(S))$ *is weakly s-controllable from a nonempty m-input set* S *if and only if* $\boldsymbol{A}(S|\cdot)$ *has an* $(n-m)$*-matching and* S *is input accessible.*

(2) (find) Given a maximum matching with unmatched nodes S'*: If* S' *is empty* $S = \{i\}$ *for any* $i = 1, \dots, n$*, otherwise* $S = S'$*. If* T *is the rooted set for those nodes in* $\mathcal{G}$ *(input) inaccessible to* S*, then* $(\boldsymbol{A}, \boldsymbol{B}(S \bigcup T))$ *is weakly s-controllable.*

The strength of this result is that the structure of the graph can be directly linked to a controllable input set. Further, this graph feature described in Theorem 2(1) can be efficiently checked via the method of maximum matching in bipartite graphs which can be calculated deterministically in $\mathcal{O}(\sqrt{n}\,|E|)$ time [27] and probabilistically in $\mathcal{O}(n^{2.376})$ time [28], coupled with a depth first search to determine input accessibility in $\mathcal{O}(|E|)$ time [29]. Theorem 2 is demonstrated in the following example.

Example 3. Referring to Example 2, for $\boldsymbol{A}(S|\cdot)$ the set $S = \{i\}$ is input accessible and has a 2-matching for $i = 1, 2$ and 3, namely $\{(v_2^+, v_3^-), (v_3^+, v_2^-)\}$, $\{(v_1^+, v_1^-), (v_2^+, v_3^-)\}$ and $\{(v_1^+, v_2^-), (v_3^+, v_1^-)\}$, respectively. These matchings appear in Figure 4. As each of these input sets satisfy the condition in Theorem 2, the pairs $(\boldsymbol{A}, \boldsymbol{B}(\{1\}))$, $(\boldsymbol{A}, \boldsymbol{B}(\{2\}))$ and $(\boldsymbol{A}, \boldsymbol{B}(\{3\}))$ are weakly s-controllable.

The following example examines a "Highjack and Eavesdrop" scenario for an unknown realization of Model (1). Theorem 2 is applied to acquire the most vulnerable weakly controllable attack vectors and reason about the relative node security in the network.

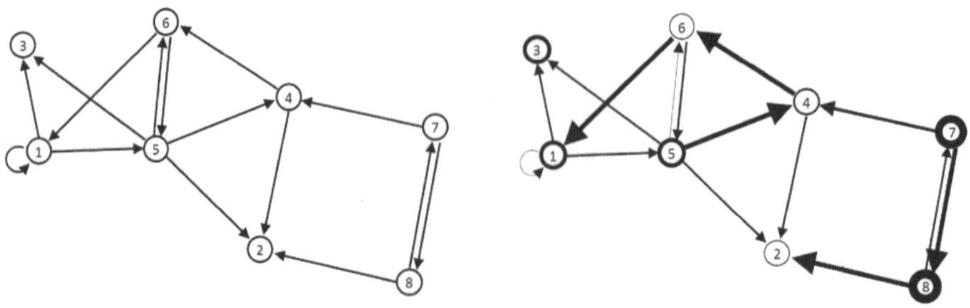

(a) (b)

Fig. 5. (a) Graph $\mathcal{G}$ and (b) the resultant node and edge weights appearing in Examples 4 and 6

Example 4. (Highjack and Eavesdrop) Consider the graph $\mathcal{G}$ in Figure 5(a) and network dynamics corresponding to a realization of $(\boldsymbol{A}(\mathcal{G}), \boldsymbol{B}(S), \boldsymbol{C}(S))$ in Model (1). A weakly controllable inputs set S presents a particularly attractive attack vector for an infiltrator. The nodes set S almost always provide a controllable input and observable output sets to effectively control (highjack) the network and/or monitor (eavesdrop on) the network through system identification.

Applying Theorem 2 to all possible input sets the smallest cardinality attack vectors involve the successful infiltration of two nodes. Specifically, the nine smallest attack vectors are of the form $\{j, 7\}$ and $\{i, 8\}$, where $j \in \{1, 3, 4, 5\}$ and $i \in \{1, 2, 3, 5, 6, 7\}$. All larger attack vectors contain one of these pairs. In aggregate there are 821 attack vectors, the 345 attacks involving at most 4 nodes are distributed as $\mathcal{A} = \{38, 36, 38, 37, 38, 36, 60, 62\}$ where the number of attacks involving node i is $\mathcal{A}(i)$.

From a network design perspective the most to least vulnerable node sets are $\{8\}, \{7\}, \{1, 3, 5\}, \{4\}, \{2, 6\}$, providing a priority ordering for security. Figure 5(b) indicates this ordering. Further, if nodes 7 and 8 are completely secured against attachment then there will be no input set that will render the system controllable or observable.

3.3 Strong Structural Controllability – Guaranteed Security

An attractive feature of s-controllability is the provided controllability guarantees in the face of perturbations. Specifically, as long as the interconnections in the graph remain intact and no new ones are added controllability will be maintained. One can consider this a type of *controllability robustness* - an often elusive feature of system dynamics, with progress made in the area of controllability of interval matrices [30]. For arbitrary edge weight perturbations the benefits of weak s-controllability are similar to that of strong s-controllability. In a malicious infiltrator scenario where edge weights may be perturbed to actively disrupt the network then edge weights rendering a weak input set uncontrollable may be intentionally selected. For this "Disrupt" scenario, an input set that is strong is preferred as, up to the removal of edges, the input set remains controllable, guaranteeing security.

Noting the similarities in the definitions for weak and strong s-controllability, it is not surprising that structural features that are equivalent to weak s-controllability share similarities to structural features that are equivalent to strong s-controllability. It is with this in mind that we explore the role of t-matchings in strong s-controllability.

Validating that a given set of inputs is controllable is computationally distinct from searching for a minimum strongly s-controllable input set. This section focuses on this search problem. The following theorem provides a bipartite matching approach to find strongly s-controllable inputs.

Theorem 3. *Given a pattern matrix* $\boldsymbol{A}(\mathcal{G})$.

(1) (check) The pair $(\boldsymbol{A}, \boldsymbol{B}(S))$ *is strongly s-controllable from a nonempty* m*-input set* S *if and only if* $\boldsymbol{A}(S|\cdot)$ *has a constrained* $(n-m)$*-matching and* $\boldsymbol{A}_\times(S|\cdot)$ *has a constrained* V_s*-less* $(n-m)$*-matching*

(2) (find) Given a constrained $(n-m_1)-$*matching on* $\boldsymbol{A}$ *with unmatched nodes* S_1 *and a constrained* V_s*-less* $(n-m_2)-$*matching on* $\boldsymbol{A}_\times$ *with unmatched nodes* S_2*, then* $(\boldsymbol{A}, \boldsymbol{B}(S_1 \bigcup S_2))$ *is strongly s-controllable.*

Fundamentally, the "check" condition in Theorem 3 require a validation that there exists a maximum n-matching in $\boldsymbol{A}(S|\cdot)$ and in $\boldsymbol{A}_\times(S|\cdot)$ that is constrained. A maximum bipartite matching can be found deterministically in $\mathcal{O}(\sqrt{n}\,|E|)$ time [27], and testing whether a given bipartite matching is constrained can be checked in $\mathcal{O}(n+|E|)$ time [26]. Reinschke *et al.* [31] provided a $\mathcal{O}(n^3)$ algorithm to check the conditions in Theorem 3. We have presented a similar approach performing the check in $\mathcal{O}(n^2)$ [19].

Consequently, a check that an input set S is strongly s-controllable reduces to a problem of finding constrained $(n-m)$-matchings. This is demonstrated in the following example.

Example 5. Returning to Example 2, for $i=1$ and 2 (but not 3), $\boldsymbol{A}(S|\cdot)$ with input set $S=\{i\}$ has a constrained 2-matching, satisfying the first condition in Theorem 3(1). Examining the matrix

$$\boldsymbol{A}_\times = \begin{bmatrix} \times & 0 & \times \\ \times & \times & \times \\ 0 & \times & \times \end{bmatrix},$$

and setting $V_s = \{1\}$, the pattern matrix $\boldsymbol{A}_\times$ has three constrained V_s-less 2-matchings, $\{(v_1^+, v_2^-), (v_2^+, v_3^-)\}$, $\{(v_2^+, v_3^-), (v_3^+, v_1^-)\}$ and $\{(v_2^+, v_3^-), (v_3^+, v_2^-)\}$. Thus for $i=1$ and 2 (but not 3), the input set $S=\{i\}$ with $\boldsymbol{A}_\times(S|\cdot)$ has a constrained V_s-less 2-matchings satisfying the second condition in Theorem 3(1). Therefore, the pairs $(\boldsymbol{A}, \boldsymbol{B}(\{1\}))$ and $(\boldsymbol{A}, \boldsymbol{B}(\{2\}))$ are strongly s-controllable. The matchings associated with input set $\{2\}$ appear in Figure 6.

We now apply Theorem 3 to a "Disrupt" scenario where an infiltrator perturbs edge weights, potentially removing them so as to reduce the effectiveness of

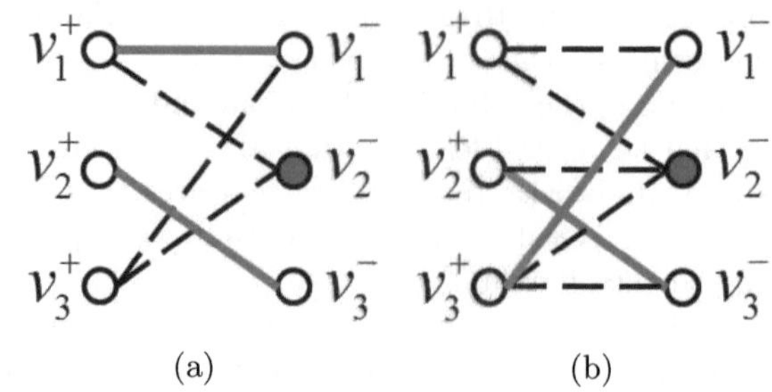

Fig. 6. (a) Constrained 2-matching on $\boldsymbol{A}(\{2\}|\cdot)$ and (b) V_s-less constrained 2-matching on $\boldsymbol{A}_\times(\{2\}|\cdot)$ relevant to Example 5 satisfying the conditions of Theorem 3 with input set $S = \{2\}$ on the bipartite graph $\mathcal{H}$ appearing in Example 2

the control input into network running a realization of Model (1). Examining the effect of edge failures on strong s-controllability of the input set one can identify the critical edges in the network and identify the most significant security vulnerabilities.

Example 6. (Disrupt) Consider the graph $\mathcal{G}$ in Figure 5(a) and network dynamics corresponding to a realization of $(\boldsymbol{A}(\mathcal{G}), \boldsymbol{B}(S), \boldsymbol{C}(S))$ in Model (1), where $S = \{3, 5, 7\}$. Applying Theorem 3 reveals that the system is strongly s-controllable and observable. If an infiltrator's objective is to disrupt the effective control and monitoring of the network, a viable strategy is to perturb the network's interconnection strengths, i.e., its edge weights. As long all interconnections are maintained then the input set S will maintain the integrity of the network. But, in the event that edges can fail, it is important to establish the most critical.

Applying Theorem 3 to every combination of edge failures there are 241216 successful attack combinations that cause the loss of strong s-controllability. The most critical edges, whose individual failure removes strong s-controllability, are edges in $E_1 = \{4 \to 6, 5 \to 4, 6 \to 1, 7 \to 8, 8 \to 2\}$ and $E_2 = \{7 \to 4\}$. All other successful combinations involve edges in $E_1 \bigcup E_2$. It is important to note that unlike Example 4, where if S_1 is weak and $S_1 \subseteq S_2$ then S_2 is weak, the same condition does not hold for edge failure attack vectors. For example the removal of E_2 renders the system no longer strong s-controllable but also removing $1 \to 1$ returns strong s-controllability.

Examining all 1208 successful attacks involving 3 edges, there were 106, 89, 67 and 60 successful attacks involving each of the edges in E_1, E_2, E_3 and E_4, respectively where $E_3 = E \setminus \left\{\bigcup_{i=\{1,2,4\}} E_i\right\}$ and $E_4 = \{1 \to 1, 5 \to 6\}$. Consequently, the network edge sets $E_1, \ldots, E_4$ present a preferential ordering of the distribution of security resources. Figure 5(b) depicts this ordering.

As previously mentioned, "checking" for a strong input is computationally tractable. With the objective to establish the complexity of "finding" a minimum cardinality strong input we examine the following corollary.

Corollary 1. *Given a state matrix $\boldsymbol{A}$ with diagonal elements all nonzero or all zero, consider the maximum constrained self-less $(n-m)$-matching of $\boldsymbol{A}_\times$, with unmatched nodes S^-. Then the m inputs associated with S^- is a minimum cardinality strongly s-controllable input set.*

Proof. For the case where $\boldsymbol{A}$ has all diagonal elements nonzero. As pattern matrices $\boldsymbol{A}$ and $\boldsymbol{A}_\times$ are equal, they share a maximum constrained self-less $(n-m)$-matching.

For the case where $\boldsymbol{A}$ has all diagonal elements zero. A constrained t-matching is invariant to the removal of edges from a bipartite graph that are not members of the matching. Thus, a constrained self-less t-matching is invariant to the removal of edges in the bipartite graph corresponding to the diagonal elements of $\boldsymbol{A}_\times$. Consequently, $\boldsymbol{A}$ can be formed from $\boldsymbol{A}_\times$ by removing the edges associated its diagonal elements. Therefore, it follows that $\boldsymbol{A}$ and $\boldsymbol{A}_\times$ share the same constrained self-less $(n-m)$-matching. From Theorem 3, the corollary follows.

This corollary states that for the special case where $\boldsymbol{A}$ has all diagonal elements nonzero, i.e., self-damped, and the case where $\boldsymbol{A}$ has all diagonal elements zero, i.e., no node is self-damped, only a single constrained t-matching needs to be found for strong s-controllability. From Theorem 3 and Corollary 1, the problem of finding the minimum cardinality strongly s-controllable input set is equivalent to finding the maximum constrained matching on a bipartite graph. This is computationally hard; Golumbic *et al.* have shown that finding a maximum constrained matching in bipartite graphs is NP-complete [26]. Further, Mishra has recently shown that even approximating a maximum constrained matching is hard demonstrating that such a matching can not be approximated with a factor of $\frac{1}{2\sqrt[3]{9}}n^{\frac{1}{3}-\epsilon}$ for any $\epsilon > 0$, in polynomial times unless NP=ZPP [32].

Corollary 1 also serves to motivate the following section which examines families of networked dynamics, including members of the family described in this corollary. Specifically, the section explores features of the smallest strong and weak input sets.

4 Special Networks

In this section we explore the smallest strong and weak s-controllable input set for self-damped undirected networks and Erdős-Rényi random networks. These networks include popularly implemented and naturally occurring networks such as consensus and randomly modeled biological and social networks [2,20]. From a network security design perspective, this analysis provides bounds on the smallest number of attachment points required for a successful attack in the "Highjack and eavesdrop" scenario. In the "Disrupt" scenario, it also provides alternative cheap strong input sets to maintain controllability in the event edges critical to one set of strong inputs are lost.

4.1 Self-damped Undirected Networks

We commence with the study of the smallest strongly s-controllable input set for a special class of connected networks, dubbed self-damped undirected networks. First, a general result pertaining to the frequency of single strongly s-controllable inputs.

Proposition 1. *There exists no network that is strongly s-controllable from all single inputs sets $S = \{i\}$, for $i = 1, \ldots, n$.*

Proof. Assume otherwise and consider the pattern matrix $\boldsymbol{A}_\times$. By Theorem 3, $\boldsymbol{A}_\times(\{i\}\,|\cdot)$ has a constrained V_s-less $(n-1)$-matching for every $i \in \{1, \ldots, n\}$. Consequently by Theorem 1, every realization $A \in \boldsymbol{A}_\times$ has a rank $n-1$ submatrix when an arbitrary row is removed. Therefore, every such A has rank n. From Theorem 1 then $\boldsymbol{A}_\times$ has a constrained n-matching. By Theorem 3.1 [26], there exists permutation matrices P_1 and P_2 such that $P_1 \boldsymbol{A}_\times P_2$ is lower triangular with nonzero diagonal entries, i.e., the elements corresponding to V_s must lie below the diagonal. As there must be exactly one term of the form a_{ii} in every row and column of $P_1 \boldsymbol{A}_\times P_2$, the diagonal of $P_1 \boldsymbol{A}_\times P_2$ is composed of elements $a_{11}, \ldots, a_{nn}$ of $\boldsymbol{A}_\times$. Therefore V_s is empty otherwise one of its elements would lie on the diagonal. Similar to the argument for $\boldsymbol{A}_\times$, the pattern matrix $\boldsymbol{A}$ must also be triangulizable with nonzero diagonal elements corresponding to $a_{11}, \ldots, a_{nn}$, but if V_s is empty, then $a_{ii} = 0$ for all i, contradicting that $\boldsymbol{A}$ has has nonzero diagonal elements.

Proposition 1 is in stark contrast to weak s-controllability where there are many networks which exhibit single input controllability from an arbitrary node. One such class of networks that falls in this category for weak s-controllability is the family of self-damped networks. These are networks where every node's state directly damps itself, i.e., for every node i, $\dot{x}_i = \alpha_i x_i + \sum_{j \neq i} \alpha_j x_j$, where $\alpha_i \neq 0$. The popular consensus (Laplacian) dynamics fall into this class for connected undirected graphs. The following formally states this single input property for weak s-controllability.

Proposition 2. *A self-damped undirected network is weakly s-controllable for any single inputs set $S = \{i\}$, where $i = 1, \ldots, n$ if and only if it is connected.*

Proof. The network is input accessibility for some $S = \{i\}$ if and only if the network is connected. Further, the bipartite realization of the network contains the maximum matching $E = \{v_i^+, v_i^- | i = 1, \ldots, n\}$ and thus by Theorem 2 the result follows.

The following two propositions pertain to strong s-controllability of self-damped undirected networks. Proposition 3 illustrates the rarity of single strongly s-controllable inputs.

Proposition 3. *The only connected self-damped undirected network strongly s-controllable from a single input is the path graph, controllable from either end node.*

Proof. A consequence of Theorem 1, described in [33] (Theorem 2.4), is that if $\boldsymbol{A}$ is self-damped and $S = \{i\}$ then $(\boldsymbol{A}, \boldsymbol{B}(S))$ is strongly s-controllable if and only if there exists a permutation matrix P such that $P\boldsymbol{A}P^T$ is in unreduced upper-Hessenberg[2] form and $P\boldsymbol{B}(S) = \boldsymbol{B}(S)$. The first condition is equivalent to the graph bandwidth[3] of realizations of $\boldsymbol{A}$ being 2. The only undirected graph with bandwidth 2 is the path graph. In bandwidth form (i.e., with a bandwidth labeling) $P\boldsymbol{B}(S) = \boldsymbol{B}(S)$ if and only if i is either end node.

On the other extreme, the following proposition indicates that there is only one graph strongly s-controllable from all but one node.

Proposition 4. *The only connected self-damped undirected network strongly s-controllable requiring $n-1$ inputs to be strongly s-controllable is the complete graph.*

Proof. If $n-1$ inputs are required, by Theorem 3, the largest constrained matching in $\boldsymbol{A}$ is a 1-matching. If the network is not a complete graph then there exists some edge $\{i, j\} \notin E$, where because it is self-damped, $i \neq j$. As the network is connected, there exists some edge $\{i, p\} \in E$ with $i \neq p$, similarly there exists some edge $\{j, q\} \in E$ with $j \neq q$. Consequently, there is a constrained V_s-less 2-matching $\{\{v_i^+, v_p^-\}, \{v_q^+, v_j^-\}\}$. This satisfies Theorem 3, for $m = 2$, i.e., the network is strongly s-controllable from $n-2$ inputs. Further, it can be shown that the largest constrained matching in $\boldsymbol{A}$ for a complete graph is a V_s-less 1-matching. From Theorem 3, the proposition now follows.

4.2 Bounds on General Controllability

An attraction of weak and strong s-controllability is that they provide lower and upper bounds, respectively, on the minimum number of inputs n_D required for general controllability. Consequently, it is a fruitful exercise to compare n_D for weak and strong s-controllability on the two families of graphs, namely self-damped undirected graphs and Erdős-Rényi random networks.

As there always exists an n-matching involving the self-damped edges of the network, by Theorem 2, the network is weakly s-controllable from any arbitrary single node. Figure 7 compares the average n_D for weak and strong s-controllability. Though the weak s-controllability result implies that almost all graphs are controllable from a single node, we note that from the strong s-controllability bound that on average the worst case on some graphs can require far more inputs.

Erdős-Rényi networks are randomly generated graphs on n nodes with $(i, j) \in E$ with probability p. The mean degree is defined as $\langle k \rangle = 2np$. Liu *et al.* [18] examined directed Erdős-Rényi random networks, finding that as n tends to infinity, $n_D \approx e^{-\langle k \rangle/2}$. Figure 8 compares this result with the sampled average

[2] A matrix is unreduced upper-Hessenberg if all entries on the first superdiagonal nonzero and all entries above this diagonal are zero.

[3] The bandwidth of a graph is the minimum max $\{|i - j| \mid \{i, j\} \in E\}$ over all labeling of the nodes.

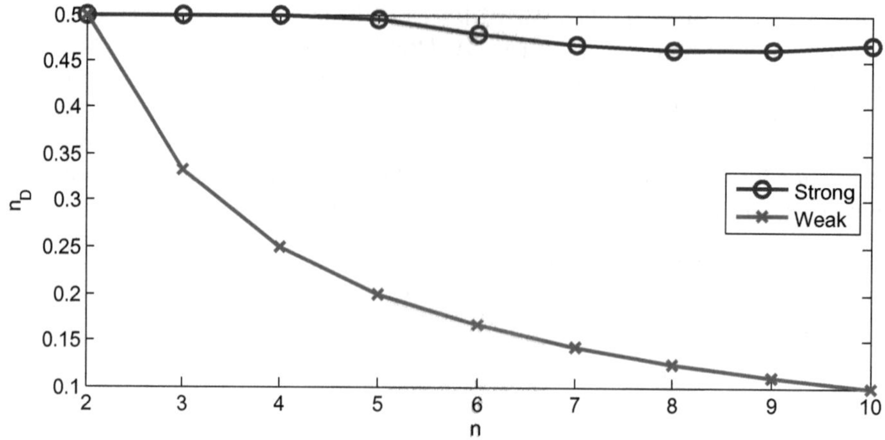

Fig. 7. The average n_D for weak and strong s-controllability for all self-damped undirected graphs for 2 to 10 nodes

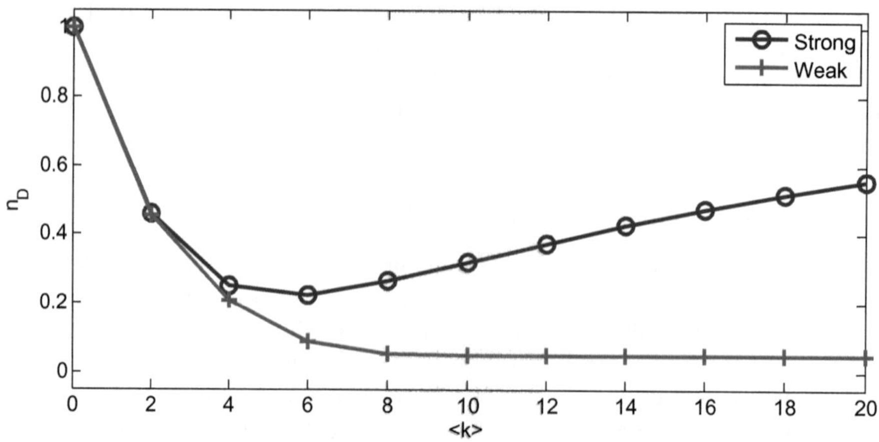

Fig. 8. The average n_D for weak and strong s-controllability for a sampling of directed Erdős-Rényi random networks on 20 nodes. Each $\langle k \rangle$ value is sampled 1200 times.

n_D for strong s-controllability. We observe that for small values of $\langle k \rangle$, the weak and strong bounds are close but as $\langle k \rangle$ increase, n_D for weak s-controllability tends to 0 and n_D for strong s-controllability tends to 1. This is not surprising and fundamental to the difference between weak and strong s-controllability, in that weak s-controllability requires the existence of t-matchings while strong s-controllability requires constrained t-matchings. The addition of edges in a bipartite graph, for example as $\langle k \rangle$ increases, promotes t-matchings while discouraging constrained t-matchings, separating the corresponding bounds.

5 Conclusion

This paper presents an analysis of the security of networked system topologies using weak and strong s-controllability. Focusing on "Disrupt" and "Highjack and eavesdrop" attack scenarios, we propose controllability metrics to identify vulnerable nodes and critical edges of a network. This was accomplished through a computationally efficient matching condition on weak and strong s-controllability. We accompanied this work with a brief controllability examination of self-damped undirected networks and Erdős-Rényi random networks. Future work of particular interest involves establishing conditions for output weak and strong s-controllability and their implications for network security.

References

1. Boccaletti, S., Latora, V., Moreno, Y., Chavez, M., Hwang, D.: Complex networks: Structure and dynamics. Physics Reports 424(4-5), 175–308 (2006)
2. Ganguly, N., Deutsch, A., Mukherjee, A.: Dynamics On and Of Complex Networks: Applications to Biology, Computer Science, and the Social Sciences. Birkhauser, Boston (2009)
3. Kocarev, L., Vattay, G.: Complex Dynamics in Communication Networks. Springer, Berlin (2005)
4. Fagiolini, A., Valenti, G., Pallottino, L., Dini, G., Bicchi, A.: Decentralized intrusion detection for secure cooperative multi-agent systems. In: Proc. 46th IEEE Conference on Decision and Control, pp. 1553–1558 (2007)
5. Sundaram, S., Hadjicostis, C.N.: Distributed function calculation via linear iterations in the presence of malicious agents - part I: attacking the network. In: Proc. American Control Conference, pp. 1350–1355 (2008)
6. Pasqualetti, F., Bicchi, A., Bullo, F.: Distributed intrusion detection for secure consensus computations. In: Proc. 46th IEEE Conference on Decision and Control, pp. 5594–5599 (2007)
7. Pasqualetti, F., Bicchi, A., Bullo, F.: On the security of linear consensus networks. In: Proc. 48th IEEE Conference on Decision and Control, pp. 4894–4901 (2009)
8. Gueye, A., Walrand, J.C.: Security in networks: A game-theoretic approach. In: Proc. 47th IEEE Conference on Decision and Control, pp. 829–834 (2008)
9. Bloem, M., Alpcan, T., Basar, T.: Optimal and robust epidemic response for multiple networks. In: Proc. 46th IEEE Conference on Decision and Control, pp. 5074–5079 (2007)
10. Wang, Y., Chakrabarti, D., Wang, C., Faloutsos, C.: Epidemic spreading in real networks: an eigenvalue viewpoint. In: Proc. 22nd International Symposium on Reliable Distributed Systems, pp. 25–34 (2003)
11. Olfati-Saber, R.: Flocking for multi-agent dynamic systems: algorithms and theory. IEEE Transactions on Automatic Control 51(3), 401–420 (2006)
12. Rahmani, A., Mesbahi, M.: Pulling the strings on agreement: anchoring, controllability, and graph automorphisms. In: Proc. American Control Conference, pp. 2738–2743 (2007)
13. Parlangeli, G., Notarstefano, G.: Observability and reachability of grid graphs via reduction and symmetries. In: 50th IEEE Conference on Decision and Control, pp. 5923–5928 (2011)

14. Nabi-Abdolyousefi, M., Mesbahi, M.: On the controllability properties of circulant Networks. IEEE Transactions on Automatic Control (to appear)
15. Zhang, S., Camlibel, M., Cao, M.: Controllability of diffusively-coupled multi-agent systems with general and distance regular coupling topologies. In: 50th IEEE Conference on Decision and Control, pp. 759–764 (2011)
16. Lin, C.-T.: Structural Controllability. IEEE Transactions on Automatic Control AC-19(3), 201–208 (1974)
17. Mayeda, H., Yamada, T.: Strong structural controllability. SIAM Journal on Control and Optimization 17(1), 123–138 (1979)
18. Liu, Y.-Y., Slotine, J.-J., Barabási, A.-L.: Controllability of complex networks. Nature 473(7346), 167–173 (2011)
19. Chapman, A., Mesbahi, M.: Strong structural controllability of networked dynamics. In: Proc. American Control Conference (to appear, 2013)
20. Mesbahi, M., Egerstedt, M.: Graph Theoretic Methods in Multiagent Networks. Princeton University Press, NJ (2010)
21. Chen, C.-T.: Linear Systems Theory and Design. Oxford University Press, New York (1999)
22. Paige, C.: Properties of numerical algorithms related to computing controllability. IEEE Transactions on Automatic Control 26(1), 130–138 (1981)
23. Shields, R., Pearson, J.: Structural controllability of multiinput linear systems. IEEE Transactions on Automatic Control 21(2), 203–212 (1976)
24. Cantó, R., Boix, M., Ricarte, B.: On the Minimal Rank Completion Problem for Pattern Matrices. WSEAS Transactions on Mathematics 3(3), 711–716 (2004)
25. Hershkowitz, D., Schneider, H.: Ranks of zero patterns and sign patterns. Linear and Multilinear Algebra 34(90), 3–19 (1993)
26. Golumbic, M., Hirst, T., Lewenstein, M.: Uniquely restricted matchings. Algorithmica 31(2), 139–154 (2001)
27. Micali, S., Vazirani, V.V.: An $O(\sqrt{|V|}|E|)$ algoithm for finding maximum matching in general graphs. In: 21st Annual Symposium on Foundations of Computer Science, pp. 17–27 (October 1980)
28. Mucha, M., Sankowski, P.: Maximum Matching via Gaussian Elimination. In: 45th IEEE Symp. Foundations of Computer Science, pp. 248–255 (2004)
29. Even, S.: Graph Algorithms. Cambridge University Press, New York (2011)
30. Wang, K., Michel, A.: Necessary and sufficient conditions for the controllability and observability of a class of linear, time-invariant systems with interval plants. IEEE Transactions on Automatic Control 39(7), 1443–1447 (1994)
31. Reinschke, K.J., Svaricek, F., Wend, H.-D.: On strong structural controllability of linear systems. In: Proceedings of the 31st IEEE Conference on Decision and Control, vol. (1), pp. 203–206 (1992)
32. Mishra, S.: On the Maximum Uniquely Restricted Matching for Bipartite Graphs. Electronic Notes in Discrete Mathematics 37, 345–350 (2011)
33. Olesky, D.D., Tsatsomeros, M., van den Driessche, P.: Qualitative controllability and uncontrollability by a single entry. Linear Algebra and its Applications 187, 183–194 (1993)

Modeling and Designing Network Defense against Control Channel Jamming Attacks: A Passivity-Based Approach

Phillip Lee, Andrew Clark, Linda Bushnell, and Radha Poovendran

Dept. of Electrical Engineering, University of Washington, Seattle, WA, 98195, USA
{leep3,awclark,lb2,rp3}@uw.edu

Abstract. Cyber-physical systems rely on distributed embedded wireless nodes for sensing, computation, and control, often leaving them vulnerable to intelligent attacks. Detecting and mitigating such attacks in a resource-efficient manner requires the network to identify the components under attack and adapt its response based on the observed network performance and attack impact. In this paper, we present our ongoing work on a passivity-based, control-theoretic framework for designing and evaluating adaptive network defenses against the control channel jamming attack. In this attack, an adversary makes use of information gathered from the compromised nodes to identify the location and disrupt the control messages. We present a dynamical model of the attack impact over time, and represent the network response as a feedback control action. Using passivity theory, we identify a class of linear controllers that correspond to network responses that guarantee convergence to a desired operating point. Through a numerical study, we analyze network characteristics such as robustness to uncertain observations of the network state, stability in the presence of delayed observations, and rate of convergence to the desired state. We find that, in general, higher-order controllers provide increased robustness and stability, at the cost of slower overall convergence.

Keywords: Cyber-physical system, Node capture attacks, Security, Control-channel jamming, Passivity.

1 Introduction

Cyber-physical systems in a variety of settings are controlled by wireless networks that observe states of physical processes and transmit control signals in a distributed fashion [1]. When deployed in critical applications over wide geographic areas, these networks are susceptible to attacks due to their unattended nature. Moreover, the networks consist of resource-constrained low-cost embedded devices that can be physically captured and tampered by the adversary. These vulnerabilities can lead to a compromise of nodes [2] or external attacks that exploit the open wireless medium [3].

D.C. Tarraf (ed.), *Control of Cyber-Physical Systems*,
Lecture Notes in Control and Information Sciences 449,
DOI: 10.1007/978-3-319-01159-2_9,

One such attack is the control-channel jamming attack where the adversary prevents nodes from receiving control packets by broadcasting an interfering signal during control packet transmission [4]. Control packets contain system parameters such as routing table update, or the access control messages and hence jamming these messages severely degrades network performance. Furthermore, since control messages are transmitted infrequently, and contained in short packets, control channel jamming can be performed by resource constrained mobile adversaries. The adversary can enhance the effectiveness of the control channel jamming attack by compromising valid nodes and extracting the location of control channels.

In order to maintain resilient network performance and functionality, defense mechanisms are needed to detect and remove compromised network nodes and update the control channels of valid nodes. Achieving high detection rate, however, can be costly in terms of network bandwidth, communications/storage overhead and energy consumption for resource constrained devices [5], creating a trade-off between resilience and cost of defense. Defense costs can be minimized by adaptively performing the detection and removal based on the observed system states, in order to drive the system to an operating point where the desired trade-off is achieved.

An adaptive defense against control channel jamming will be affected by several network properties. The effectiveness of defense mechanisms is impacted by delayed observations due to dropped packets. These observations may also be subjected to uncertainties in adversarial parameters such as the location and jamming power. Furthermore, different defense methods achieve varying speeds at which the network recovers to its desired state. An analytical approach is therefore required to compare and evaluate possible defense methods based on these criteria and determine which is most suitable for given network characteristics.

In this paper, we develop a control-theoretic framework for modeling and analyzing control channel jamming attacks and network defenses in cyber-physical systems. Under this framework, the attack and its impact are modeled as a dynamical system while the network response is represented by a feedback control action. We make the following contributions:

- We propose a passivity-based approach to design the network response based on the observed impact of attacks. We prove that the class of network responses corresponding to Strictly Positive Real (SPR) systems guarantee the convergence to a desired operating point, defined as the fraction of control messages that are correctly received.
- We show how to design SPR controllers to achieve and maintain a given operating point.
- We numerically evaluate a set of SPR controllers based on the criteria of (i) delay robustness, representing the robustness of the system to delayed observations due to packet drops, (ii) rejection of disturbances due to the uncertainties in the observations caused by unpredictable adversaries, and

(iii) convergence rate, which corresponds to the speed of recovery of the system.
- We numerically show that the proportional controller achieves the higher convergence rate while the higher order controllers provide robustness to uncertainties. We also show that the delay robustness of the controller is determined by the performance at the steady state.

The paper is organized as follows. We review the related work in Section 2. The system model and background on passivity theory are given in Section 3. We present our control-theoretic framework in Section 4. Numerical results are presented in Section 5. Section 6 concludes the paper.

2 Related Work

Control channel jamming, in which the adversary broadcasts interference signal in order to disrupt control messages, has been identified as a threat to the functionality of wireless network. In [6], a channel assignment scheme was developed that enables the base station to determine the compromised nodes, up to a fixed threshold, based on the reported jammed channels. The channel assignment problem was mapped to key assignment problem in [4], providing non-zero detection probability even when a fraction of compromised nodes exceeds a given threshold. The channel assignment problem has also been studied using a coding theoretic framework in [7]. These mechanisms [4,6,7], however, do not model the progress of the attack over time. By providing such a temporal model, our approach enables the defense to change adaptively based on the observed impact of the attack, such as the reduced throughput of the control message or the fraction of jammed control channels.

A dynamical system model of the node capture/cloning attack in sensor networks was presented in [8]. The network response was derived as the solution of LQR/LQG problems. The derived network response, however, only utilizes the current observed impact of the attack and hence neglects previously observed information of the attack impact. In [9], a passivity framework was proposed which models the adversarial actions and the network response jointly. Sufficient conditions were derived that stabilize the network performance under different adversarial models. Design of specific control laws that satisfy the derived conditions, however, was not considered in [9].

3 Model and Preliminaries

In this section, we describe our models for the networked system, adversary, and system defense. In particular, we describe how the adversary can increase the impact of control jamming attack by jointly performing node capture attack. We also give relevant background on passive systems.

3.1 Networked System Model

We consider a system of n nodes, which communicate via a wireless network using a time-slotted communication protocol such as orthogonal frequency-division multiple access (OFDMA). The network is facilitated by a base station, which sends necessary information for network functionality, such as scheduling data and public encryption keys, in control messages. Control messages are sent via dedicated control channels. In order to provide such channels, we assume that, periodically, the initial portion of a time slot is dedicated to control messages, while the remainder of the slot is used for data transmission. The portion of the slot dedicated to control messages is further divided into S sub-slots. During each sub-slot, control messages can be transmitted on one or more frequency bands, out of M available frequencies. A (sub-slot, band) pairing defines a control channel (Figure 1).

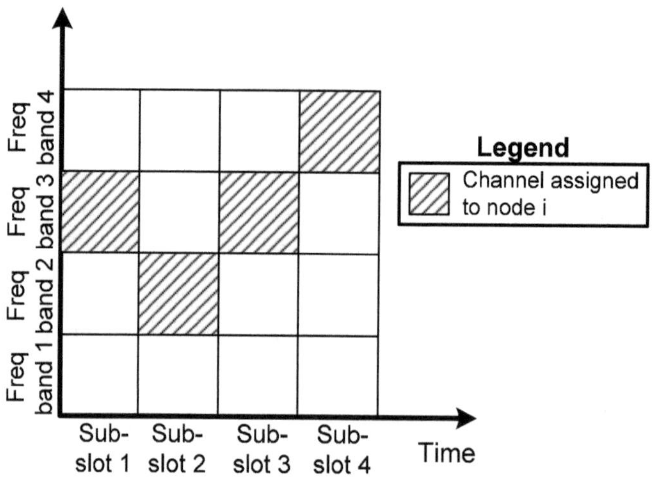

Fig. 1. Illustration of control channel assignment. Each (sub-slot, band) pairing defines a control channel. Node i is assigned the channels (sub-slot 1, band 3), (sub-slot 2, band 2), (sub-slot 3, 3), (sub-slot 4, band 4).

In order to prevent jamming of control messages, the control channel used by each node varies between time slots. Letting c_{ij} denote the control channel of node i during time slot j, the channel hopping sequence of node i is given by $\mathcal{C}_i = \{c_{i1}, c_{i2}, \ldots, \}$. We assume that $\mathcal{C}_i$ repeats itself with period p, so that the sequence i is uniquely defined by the period $\overline{\mathcal{S}}_i = \{c_{i1}, \ldots, c_{ip}\}$. In order to determine c_{ij}, each node is assumed to have access to a secret seed K_i, known only to node i and the base station. Node i computes c_{ij} as $h(K_i, j)$, where $h(\cdot) : \mathcal{K} \times \mathbb{Z} \rightarrow \mathcal{M}$ is a pseudorandom function, $j = t \bmod p$, $\mathcal{M}$ is the set of available frequencies, and $\mathcal{K}$ is the set of possible seeds [4]. To reduce the overhead at the base station, the seeds $\{K_i : i = 1, \ldots, n\}$ may be chosen so that the hopping sequences overlap, i.e., $\mathcal{C}_i \cap \mathcal{C}_j \neq \emptyset$, reducing the number of times that the base station must retransmit each control message.

3.2 Adversary Model

The network operates in the presence of one or more mobile, intelligent adversaries, whose goal is to prevent a subset of victim nodes from receiving control messages. In order to obtain the hopping sequences used by the base station, the adversary compromises a set of nodes, denoted $\mathcal{A}$, and thereby gains access to the seeds $\mathcal{K}_{\mathcal{A}} = \{K_i : i \in \mathcal{A}\}$. After capturing a seed K_i, the adversary computes the publicly available function $h(K_i, j)$ at time slot j, and can then jam all control channels in the sequence $\mathcal{C}_i$. Due to the overlap between channel hopping sequences of different nodes, a jamming attack on the channels $\mathcal{C}_i$ may prevent other, valid nodes from receiving control messages.

The base station and nodes detect and respond to the attack as follows. Each valid node is assumed to have a secure feedback channel to the base station for reporting errors in received control messages. Based on this feedback received from valid nodes, the base station identifies a set of control channels $\hat{\mathcal{C}}$ that may be under jamming attack. Since a control channel can be jammed when one of the corresponding seeds is compromised, the set of compromised nodes is then estimated based on the set $\hat{\mathcal{C}}$. In particular, the base station identifies a set of nodes $\hat{\mathcal{A}}$ such that $\hat{\mathcal{C}} \subseteq \mathcal{K}_{\hat{\mathcal{A}}}$. Nodes that are suspected of being compromised are removed from the network, while any node i satisfying $\mathcal{C}_i \cap \mathcal{C}_j \neq \emptyset$ for some compromised node j is issued a new seed K_i'.

3.3 Background on Passivity

In what follows, we consider a state-space model (Σ), with state $x(t)$, input $u(t)$, and output $y(t)$, defined by

$$(\Sigma) \begin{cases} \dot{x}(t) = f(x(t), u(t)) \\ y(t) = g(x(t), u(t)) \end{cases}$$

The definitions and results in this subsection can be found in [10]. A passive system is defined as follows.

Definition 1. *The system (Σ) is passive iff there exists a nonnegative C^1 function $V : \mathbb{R} \rightarrow \mathbb{R}_{\geq 0}$ satisfying $V(0) = 0$ and*

$$\dot{V}(t) \leq -S(x(t)) + u(t)^T y(t) \tag{1}$$

where $S(\cdot)$ is a nonnegative continuous function. If $S(x) > 0$ for all $x \neq 0$, then the system is strictly passive. *A function V satisfying (1) for a system (Σ) is a* storage function *for (Σ).*

A passive system response should be efficiently computable by the nodes and base station. One such class of passive systems consists of linear systems that satisfy the *strict positive realness (SPR)* property, defined below.

Definition 2. *Consider the linear system* (Σ_L)*, defined by*

$$(\Sigma_L) \begin{cases} \dot{x}(t) = Ax(t) + Bu(t) \\ y(t) = Cx(t) + Du(t) \end{cases}$$

Then (Σ_L) *is positive real if its transfer function* $h(s)$ *is real for positive real* s *and satisfies* $\mathbf{Re}[h(s)] \geq 0$ *for all* $\mathbf{Re}[s] > 0$. Σ_L *is strictly positive real if* $h(s-\epsilon)$ *is positive real for some* $\epsilon > 0$.

A linear system is passive if and only if it is positive real, and is strictly passive if and only if it is strictly positive real. The Kalman-Yakubovich-Popov lemma gives equivalent conditions for a linear system to be SPR.

Lemma 1 (KYP). *The transfer function of* (Σ_L) *is strictly positive real if and only if there exist matrices* $P = P^T > 0$, L, W, *and a positive constant* $\epsilon > 0$ *such that*

$$\begin{aligned} PA + A^T P &= -LL^T - \epsilon P \\ PB - C^T &= -LW \\ D + D^T &= W^T W \end{aligned}$$

Passivity leads to a variety of techniques for guaranteeing stability of dynamical systems, such as the following proposition.

Proposition 1. *A negative feedback interconnection between two strictly passive systems is globally asymptotically stable.*

In particular, a negative feedback interconnection between a strictly passive system and an SPR system is globally asymptotically stable. For a negative feedback interaction between two strictly passive systems with storage functions V_1 and V_2, the function $V = V_1 + V_2$ is a Lyapunov function for the combined system.

4 Control-Theoretic Framework for Mitigating Control Channel Jamming

In this section, we present a control-theoretic framework for modeling and mitigating control-channel jamming attacks, in which the attack is represented as a nonlinear dynamical system and the network response is the feedback control. We analyze the steady-state behavior and prove that the class of linear SPR controllers leads to network responses that converge to a desired steady-state value, which can be tuned by choosing the parameters of the controller.

4.1 Dynamical Model of Attack and Network Response

Our dynamical modeling approach consists of two components, namely, a model of the attack dynamics and a model of the system's defense mechanisms. For the attack dynamics, we consider an affine model first introduced in [9]. Let $x(t)$ denote the fraction of nodes that have been compromised at time t. Define $f : [0,1] \rightarrow \mathbb{R}_{\geq 0}$ to be a nonnegative, continuous, and increasing function, such that $f(x)$ quantifies the impact of the attack on the system performance when the fraction of compromised nodes is x. The dynamical model of the attack is given by

$$(\Sigma_{\text{sys}}) \begin{cases} \dot{x}(t) = \lambda(1 - x(t)) + u_1(t) \\ y_1(t) = f(x(t)) \end{cases} \tag{2}$$

where λ represents the rate at which nodes are captured. The model (2) reflects the fact that, as the adversary compromises nodes, the incremental effort to locate and compromise an additional node increases, thus reducing the capture rate.

The network response is defined by the rate at which compromised nodes are detected and revoked and removed from the system, denoted $y_2(t)$, so that $\dot{x}(t) = \lambda(1 - x(t)) - y_2(t)$. We consider a revocation rate $y_2(t)$ that is given as the output of a linear system of the form

$$(\Sigma_{\text{control}}) \begin{cases} \dot{\xi}(t) = A\xi(t) + Bf(x) \\ y_2(t) = C\xi(t) + Df(x) \end{cases} \tag{3}$$

The $Df(x)$ term represents the network response to the currently-observed impact of the attack, $f(x)$. The state variable $\xi(t)$ is based on the attack dynamics from $[0,t]$, and hence represents the past observations. Tuning the parameters (A, B, C, D) therefore determines how rapidly the system will respond to changes in network performance caused by the attack.

Based on (2) and (3), the combined system consisting of the control channel jamming attack and network response is interpreted as a negative feedback interconnection, where the network response acts as a feedback control, as illustrated in Figure 2. In what follows, we first derive the steady-state operating point (x^*, ξ^*) of the system in Figure 2. The steady-state x^* represents the fraction of nodes that are compromised, and hence the fraction of channels that are jammed $f(x^*)$, after the system has been deployed for a sufficiently long time period. We show that an arbitrary value of $x^* \in (0,1)$ can be achieved by selecting the

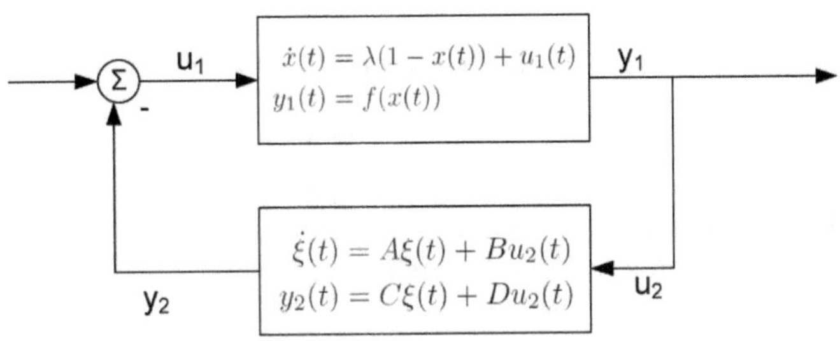

Fig. 2. Attack and response dynamics

matrices (A, B, C, D). We then prove that, for the class of strictly positive real systems Σ_{control}, the combined system is globally asymptotically stable.

4.2 Steady-State Value

The desired steady-state of the system can be determined by trading off the impact of the attack on the network performance with the resource cost of mitigating the attack [5]. Once a desired steady-state x^* has been chosen, the goal is to design a control law defined by Σ_{control} that achieves this value.

The steady-state is derived of (2) and (3) is derived as follows. $\dot{\xi}(t)$ is equal to zero when $\xi = \xi^* = -A^{-1}Bf(x^*)$. Substituting into (2) and setting $\dot{x}(t) = 0$ yields

$$\lambda(1 - f^{-1}(y^*)) + (CA^{-1}B - D)y^* = 0,$$

where we use the fact that f is continuous and increasing, and hence invertible. Set $r = \lambda(1 - f^{-1}(y^*))$ and $\psi = (CA^{-1}B - D)$. Then y^* is the steady-state if $\psi = -r/y^*$, i.e., if $CA^{-1}B - D = -r/y^*$. This can be accomplished in different ways, such as selecting D in order to ensure that the condition is satisfied, or by scaling the matrices (A, B, C, D) by the same value in order to achieve the desired ratio.

4.3 Global Asymptotic Stability

In order to prove global asymptotic stability of (2) and (3), we first introduce new variables representing the deviation of the state from the steady-state value, denoted as $\hat{x} \triangleq x - x^*$, $\hat{\xi} \triangleq \xi - \xi^*$. Rewriting the dynamics in terms of $\hat{x}$ and $\hat{\xi}$ yields

$$\begin{aligned}
\dot{\hat{x}}(t) = \dot{x}(t) &= \lambda(1 - \hat{x}(t) - x^*) - C(\hat{\xi}(t) + \xi^*) \\
&\quad - D(f(\hat{x}(t) + x^*) - f(x^*) + f(x^*)) \\
&= -\lambda\hat{x}(t) - C\hat{\xi}(t) - D(f(\hat{x}(t) + x^*) - f(x^*)) \qquad (4) \\
\dot{\hat{\xi}}(t) = \dot{\xi}(t) &= A(\hat{\xi}(t) + \xi^*) + B(f(\hat{x}(t) + x^*) - f(x^*) + f(x^*)) \\
&= A\hat{\xi}(t) + B(f(\hat{x}(t) + x^*) - f(x^*)). \qquad (5)
\end{aligned}$$

Here, (4) follows from the fact that, since (x^*, ξ^*) is a steady-state value for (2) and (3), $\lambda(1 - x^*) - C\xi^* - Df(x^*) = 0$. Similarly, (5) follows from the fact that $A\xi^* + Bf(x^*) = 0$.

The equivalent dynamics (4) and (5) are described by the block diagram in Figure 3. In essence, the dynamics are equivalent to a negative feedback interconnection between a system H_1, defined by

$$(H_1) \begin{cases} \dot{\hat{x}}(t) = -\lambda\hat{x}(t) + u_1(t) \\ y_1(t) = f(\hat{x} + x^*) - f(x^*) \end{cases}$$

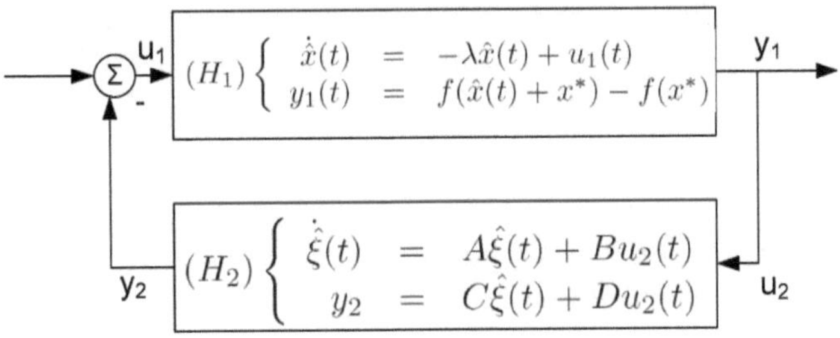

Fig. 3. Equivalent representation as negative feedback interconnection between two passive systems.

and a system H_2 defined by

$$(H_2)\left\{\begin{array}{l}\dot{\hat{\xi}}(t)=A\hat{\xi}(t)+Bu_2(t)\\ y_2(t)=C\hat{\xi}(t)+Du_2(t)\end{array}\right.$$

The following lemma concerns the system H_1.

Lemma 2. *The system H_1 is passive from the input u_1 to the output y_1.*

Proof. In order to prove passivity, the goal is to find a function $V(\hat{x})$ such that $V(0)=0$, $V(\hat{x})\geq 0$ for all $\hat{x}\in\mathbb{R}$, and $\dot{V}(\hat{x}(t))\leq u_1(t)y_1(t)$. The candidate function is given by

$$V(\hat{x})=\int_{x^*}^{\hat{x}+x^*} f(a)\,da-\hat{x}f(x^*).$$

First, note that $V(0)=0$. To see that $V(\hat{x})\geq 0$, first consider the case that $\hat{x}>0$. Since $V(0)=0$, it suffices to show that $\frac{\partial V}{\partial\hat{x}}\geq 0$ for $\hat{x}>0$. In this case, $\frac{\partial V}{\partial\hat{x}}=f(\hat{x}+x^*)-f(x^*)\geq 0$, where the inequality holds because $\hat{x}>0$ and f is increasing.

Now, when $\hat{x}<0$, we write $\hat{x}=-s$ for some $s>0$. Thus $V(\hat{x})$ is given by

$$\begin{aligned}V(\hat{x})&=\int_{x^*}^{x^*-s} f(a)\,da+sf(x^*)=-\int_{x^*-s}^{x^*} f(a)\,da+sf(x^*)\\&=-\int_{x^*-s}^{x^*} f(a)\,da+\int_{x^*-s}^{x^*} f(x^*)\,da=\int_{x^*-s}^{x^*} f(x^*)-f(a)\,da.\end{aligned}$$

Since f is increasing, $f(x^*)\geq f(a)$ for all $a\leq x^*$. Hence $V(\hat{x})$ is the integral of a nonnegative function, and is therefore nonnegative.

Finally, we want to show that $\dot{V}(\hat{x}(t))\leq u_1(t)y_1(t)$ for all $\hat{x}(t)\in\mathbb{R}$. We have

$$\begin{aligned}\dot{V}(\hat{x}(t))&=(f(\hat{x}(t)+x^*)-f(x^*))(-\lambda\hat{x}(t)+u_1(t))\\&=-\lambda(f(\hat{x}(t)+x^*)-f(x^*))\hat{x}(t)+u_1(t)y_1(t)\end{aligned}\tag{6}$$

If $\hat{x}(t) > 0$, then $f(\hat{x}(t)+x^*) > f(x^*)$, and hence the first term of (6) is negative. If $\hat{x}(t) < 0$, then $f(\hat{x}(t) + x^*) < f(x^*)$, and the first term of (6) is therefore negative. Thus $\dot{V}(\hat{x}(t)) \leq u_1(t)y_1(t)$ for all $\hat{x}(t) \in \mathbb{R}$, which establishes the passivity of H_1.

Theorem 1. *Suppose that the matrices (A, B, C, D) in H_2 define a strictly positive real transfer function. Then the dynamics of (2) and (3) are globally asymptotically stable with steady state (x^*, ξ^*).*

Proof. If (A, B, C, D) define a strictly positive real system, then H_2 is passive, and hence the joint dynamics (2) and (3) are equivalent to a negative feedback interconnection between two strictly passive systems, and is therefore passive.

4.4 Ensuring Correct Values of $x(t)$

The last remaining task is to ensure that the dynamics of $x(t)$ does not leave the feasible region $x(t) \in [0, 1]$. This is accomplished by using the following modified control law

$$(\Sigma_{mod}) \begin{cases} \dot{x}(t) = \lambda(1 - x(t)) - C\xi(t)\mathbb{1}(\mathcal{R}) - Df(x) \\ \dot{\xi}(t) = A\xi(t) + Bf(x) \end{cases}$$

where $\mathbb{1}(\cdot)$ denotes the indicator function and

$$\mathcal{R} \triangleq \{f(x) \in (0, f(1))\} \cup \{f(x) \geq f(1), C\xi > 0\} \cup \{f(x) \leq 0, C\hat{<}0\},$$

noting that the control action described can be implemented with knowledge only of $f(x)$. We now show that the modified dynamics of Σ_{mod} satisfy $x(t) \in [0, 1]$ $\forall t \geq 0$ and global asymptotic stability of the same equilibrium point (x^*, ξ^*).

Proposition 2. *The dynamics of Σ_{mod} satisfy $x(t) \in [0, 1]$, provided that (A, B, C, D) is strictly positive real, f is increasing, and $x(0) \in [0, 1]$.*

Proof. Since the dynamics is continuous, it suffices to show that $\dot{x}(t) > 0$ when $x(t) = 0$ and $\dot{x}(t) < 0$ when $x(t) = 1$. If $x(t) = 0$, we have

$$\dot{x}(t) = \lambda - C\xi(t)\mathbb{1}(\mathcal{R})$$

since $f(0) = 0$. Now, $C\xi(t) > 0$, then $\mathbb{1}(\mathcal{R}) = 0$. Hence $\dot{x}(t) \geq \lambda > 0$, as desired. Similarly, if $x(t) = 1$, then

$$\dot{x}(t) = -C\xi(t)\mathbb{1}(\mathcal{R}) - Df(1).$$

If $C\xi(t) < 0$, then $\mathbb{1}(\mathcal{R}) = 0$, and hence $\dot{x}(t) \leq -Df(1)$. Since (A, B, C, D) is strictly positive real, the KYP lemma implies that $D > 0$, and so $Df(1) > 0$. Thus $\dot{x}(t) \leq -Df(1) < 0$ when $x(t) = 1$.

It remains to show that the global asymptotic stability still holds for the modified dynamics Σ_{mod}. This result is provided by the following proposition.

Proposition 3. *The dynamics of Σ_{mod} are globally asymptotically stable with equilibrium point (x^*, ξ^*).*

Proof. As in Lemma 2 and Theorem 1, we consider the stability of the system defined by $\hat{x}(t) = x(t) - x^*$ and $\hat{\xi}(t) = \xi(t) - \xi^*$. In order to prove global asymptotic stability, it suffices to find a Lyapunov function $V(x(t), \xi(t))$ such that $\frac{dV}{dt} < 0$. Since the H_1 and H_2 of Figure 3 are passive, the function $V(\hat{x}(t), \hat{\xi}(t)) = V_1(\hat{x}(t)) + V_2(\hat{\xi}(t))$, where V_1 and V_2 are storage functions for H_1 and H_2, respectively, is a Lyapunov function for the system in Figure 3. We prove that $V(\hat{x}(t), \hat{\xi}(t))$ is also a Lyapunov function for the system defined by Σ_{mod}.

For the system defined by Σ_{mod}, we have

$$\begin{aligned}
\dot{V}(\hat{x}(t), \hat{\xi}(t)) &= \frac{\partial V}{\partial \hat{x}}\dot{\hat{x}}(t) + \frac{\partial V}{\partial \xi}\dot{\xi}(t) \\
&= (f(x(t)) - f(x^*))(\lambda(1 - x(t)) - [C\xi(t)]\mathbb{1}(\mathcal{R}) - Df(x)) + \frac{\partial V}{\partial \xi}\dot{\xi}(t) \\
&\leq (f(x(t)) - f(x^*))(\lambda(1 - x(t)) - C\xi(t) - Df(x)) + \frac{\partial V}{\partial \xi}\dot{\xi}(t) < 0
\end{aligned}$$

where the final inequality follows from the global asymptotic stability of the system in Figure 3. Hence $V(\hat{x}(t), \hat{\xi}(t))$ is a Lyapunov function for Σ_{mod}, implying that the system is globally asymptotically stable.

5 Numerical Study

We conduct a numerical study using Matlab. The goal of this study is to evaluate different SPR controllers using the metrics 1) convergence rate, 2) disturbance rejection in the observation errors, and 3) robustness to delay in observation.

We consider the capture rate $\lambda = 0.1$. The impact function $f(x)$ has been linearized around the x^*, and assume that $f(x) \approx x$. The initial condition $x(0) = 0$. Our results for each of the metrics are as follows.

5.1 Robustness to Imperfect Observation

Given different network responses, we first compare the ability of the network to reach the desired operating state when observations $f(x)$ are imperfect. Imperfect observations may be due to uncertainty in the adversarial model or noisy communication channels. Robustness to imperfect observations is simulated by comparing the stability performance between the proportional controller and the higher order SPR controller in the presence of additive noise $\omega(t)$ introduced in the observation. The simulated noise is white zero mean Gaussian with variance of 4. Proportional controller has parameter values $A = B = C = 0$, and $D = 0.4$.

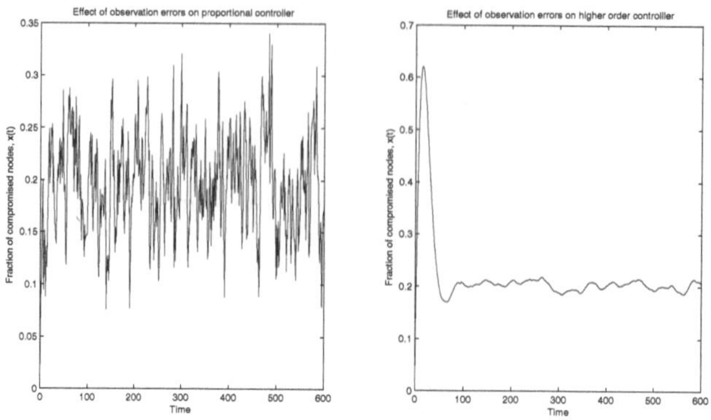

Fig. 4. Simulation results comparing the stability performance given additive white Gaussian noise in the observation. Gaussian noise had statistical property of zero mean and variance 4. $x^* = 0.2$, which resulted in $D = 0.4$ for the proportional controller (left plot). For the higher order controller, $A = -\frac{1}{64}, B = \frac{1}{16}, C = \frac{1}{10}$ values were found heuristically in order to achieve the same $x^* = 0.2$. The higher order controller provides superior noise rejection at the steady state at the cost of higher attack impact prior to convergence.

The higher order SPR controller has parameter values $A = -\frac{1}{64}, B = \frac{1}{16}, C = \frac{1}{10}$ and $D = 0$. The steady state value is $x^* = 0.2$ in both cases. The average stability performance is estimated via Monte-Carlo methods. We performed 300 trials for both controllers with independent noise. The results are illustrated in Figure 4. The higher order controller is more robust to observation errors than the proportional controller. In the proportional case, D is determined uniquely by x^*. Since the impact of observation error introduced in the state dynamics is additive error of $D\omega(t)$, the impact of additive error cannot be decreased without changing the steady state value x^*. On the other hand, in the higher order SPR controller, $x^* = \frac{1}{1-\frac{1}{\lambda}\frac{BC}{A}}$. This enables higher degree of freedom in choosing A, B, C values to achieve x^* while minimizing the impact of observation error.

5.2 Robustness to Delay in Observation

Given different network responses, we next compare the ability of the network to reach the desired operating state when observations are delayed due to packet drops. Robustness to delay is simulated by comparing the stability performance between the proportional controller and the higher order SPR controller in the presence of delay in observation. A delay of 5 time units was introduced in the observation data. The proportional controller has parameter values $A = B = C = 0$, and $D = 0.4$. The higher order SPR controller has parameter values $A = -\frac{1}{64}, B = \frac{1}{16}, C = \frac{1}{10}$ and $D = 0$. The steady state value is $x^* = 0.2$ in both cases. The results are illustrated in Figure 5. Given fixed delay and steady state

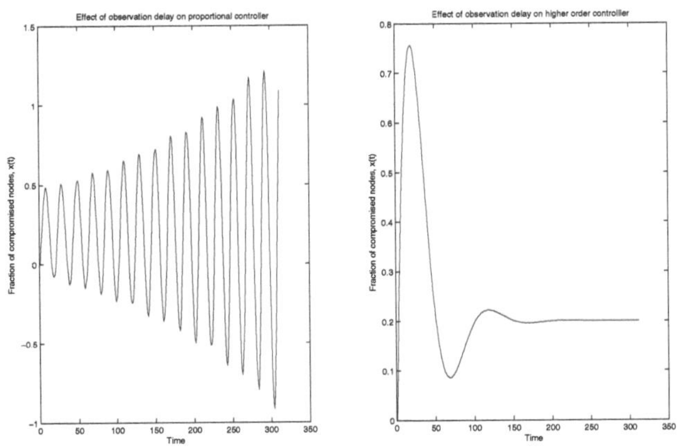

Fig. 5. Simulation results comparing the robustness to delay in observation. A delay of 5 time units was introduced in the observation data. The steady-state value is $x^* = 0.2$, which resulted in $D = 0.4$ for the proportional controller (left plot). For the higher order controller, $A = -\frac{1}{64}, B = \frac{1}{16}, C = \frac{1}{10}$ values were found heuristically in order to achieve the same $x^* = 0.2$. The higher order controller provides superior performance in satisfying convergence to x^* in the presence of delay.

value x^*, the higher order SPR controller enables convergence to the desired steady state value, while convergence to the steady state value can no longer be guaranteed in the proportional controller case. Since D is uniquely determined by x^* in the proportional controller case, the simulation result proves that there exists a set of higher order SPR controllers which provide robustness to delay given a fixed x^*. It should be noted the higher order SPR controller with the parameters given above is quite robust to delay. Simulation results, which are omitted here, suggest that the higher order controller guarantee convergence to the steady state of 0.2 up to approximately 20 units of time delay in observation.

5.3 Convergence Rate

We compare how fast the network reaches the desired fraction of compromised nodes for given network responses. In the case of no noise in the observation or delay, we compare the convergence rate of the higher order controller with same parameter values given above ($A = -\frac{1}{64}, B = \frac{1}{16}, C = \frac{1}{10}, D = 0$) to the proportional controller with $D = 0.4$. The results are illustrated in Figure 6. The same higher order controller that provided robustness to delay and noisy observation has a much slower convergence rate than the proportional controller. The higher order SPR controller utilizes the past observations and hence is slower to respond to current observation. This suggests that there is a trade-off between the convergence rate and the robustness to delayed and noisy observations.

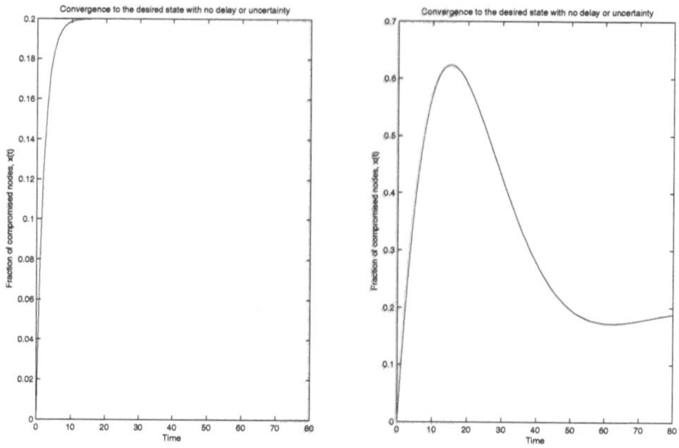

Fig. 6. Simulation results comparing the convergence rate. $x^* = 0.2$, which resulted in $D = 0.4$ for the proportional controller (left plot). For the higher order controller, $A = -\frac{1}{64}, B = \frac{1}{16}, C = \frac{1}{10}$ values were found heuristically in order to achieve the same $x^* = 0.2$. The proportional controller achieves steady state within 15 time units while the higher order SPR controller does not achieve the steady state within 80 time units.

6 Conclusion

In this paper, we presented a control-theoretic framework for modeling and analyzing defenses against attacks on cyber-physical systems. We considered the control channel jamming attack, in which an adversary compromises a set of nodes and uses information extracted from the nodes to jam control messages sent from a base station. The attack dynamics and impact were modeled as a dynamical system, while the network response was represented as a feedback control action. Using a passivity-based approach, we identified a class of response dynamics, corresponding to the set of strictly positive real controllers, that guarantee convergence to a steady-state value, and proved that a controller can be chosen to achieve any steady state. We further showed that different control laws within this space result in different network performance characteristics, including robustness to delays and observation errors, as well as rate of network recovery from the attack. In particular, through a numerical study we found that choosing a detection and removal rate that is proportional to the fraction of jammed channels leads to faster recovery, but is less robust to delays and uncertainties than a higher-order control law. Analytically verifying these results by finding the maximum observation delay that can be tolerated while maintaining stability, as well as deriving bounds on the convergence rate and noise robustness, will be the subject of future work. Moreover, we will study a general set of attack dynamics such as replicator-type dynamics while incorporating the multi-slot/multi-band control channel allocation model.

References

1. Pajic, M., Sundaram, S., Pappas, G., Mangharam, R.: The wireless control network: A new approach for control over networks. IEEE Transactions on Automatic Control 56(10), 2305–2318 (2011)
2. Mo, Y., Kim, T., Brancik, K., Dickinson, D., Lee, H., Perrig, A., Sinopoli, B.: Cyber–physical security of a smart grid infrastructure. Proceedings of the IEEE 100(1), 195–209 (2012)
3. Amin, S., Cárdenas, A.A., Sastry, S.S.: Safe and secure networked control systems under denial-of-service attacks. In: Majumdar, R., Tabuada, P. (eds.) HSCC 2009. LNCS, vol. 5469, pp. 31–45. Springer, Heidelberg (2009)
4. Tague, P., Li, M., Poovendran, R.: Mitigation of control channel jamming under node capture attacks. IEEE Transactions on Mobile Computing 8(9), 1221–1234 (2009)
5. Bonaci, T., Lee, P., Bushnell, L., Poovendran, R.: A convex optimization approach for clone detection in wireless sensor networks. In: Pervasive and Mobile Computing (2012)
6. Chan, A., Liu, X., Noubir, G., Thapa, B.: Broadcast control channel jamming: Resilience and identification of traitors. In: IEEE International Symposium on Information Theory, pp. 2496–2500 (2007)
7. Lazos, L., Liu, S., Krunz, M.: Mitigating control-channel jamming attacks in multi-channel ad hoc networks. In: Second ACM Conference on Wireless Network Security, pp. 169–180 (2009)
8. Bonaci, T., Bushnell, L., Poovendran, R.: Node capture attacks in wireless sensor networks: A system theoretic approach. In: 49th IEEE Conference on Decision and Control (CDC), pp. 6765–6772 (2010)
9. Clark, A., Bushnell, L., Poovendran, R.: A passivity-based framework for composing attacks on networked control systems. In: 50th Allerton Conference on Communication, Control, and Computing (2012)
10. Brogliato, B., Egeland, O., Lozano, R., Maschke, B.: Dissipative Systems Analysis and Control: Theory and Applications. Springer (2007)

[illegible]

Part IV

Fault Diagnosis and Sensor Fusion in CPS

Parameter-Invariant Actuator Fault Diagnostics in Cyber-Physical Systems with Application to Building Automation

James Weimer[1], José Araujo[2], Mani Amoozadeh[2], Seyed Alireza Ahmadi[2], Henrik Sandberg[2], and Karl Henrik Johansson[2]

[1] Department of Computer and Information Sciences
School of Engineering and Applied Sciences
University of Pennsylvania
Philadelphia, PA 19104, USA
weimerj@seas.upenn.edu

[2] ACCESS Linnaeus Centre, School of Electrical Engineering
KTH Royal Institute of Technology
Stockholm, Sweden 10044
{araujo,maniam,saahmadi,hsan,kalle}@kth.se

Abstract. This paper introduces a robust method for performing active actuator fault detection and diagnostics (FDD) in heating ventilation and air conditioning (HVAC) systems. The proposed actuator FDD strategy, for testing whether an actuator is stuck in a given position, is designed on using an invariant hypothesis testing approach and is an improvement of a previous strategy that employed an adaptive detection strategy. The parameter-invariant detector is formulated to provide a constant detection performance, invariant to unknown building parameters, and it is described how this approach can replace the adaptive detector in the previous work. A closed-loop experimental HVAC testbed at the KTH Royal Institute of Technology campus in Stockholm, Sweden is introduced and employed to evaluate the parameter-invariant detector.

Keywords: building automation, fault detection and diagnostics (FDD), invariant hypothesis testing, heating ventilation and air conditioning (HVAC) systems.

1 Introduction

Heating, ventilation and cooling (HVAC) are known to be the largest consumer of energy in buildings, accounting for 43% of U.S. residential energy consumption. The design of energy-efficient HVAC systems has therefore become a worldwide research priority. In the U.S. and U.K., buildings consume nearly 40% and 47% of the national energy, respectively [28,27]. Due to this high usage, there exists a high potential for energy consumption improvement, which has thrust HVAC

D.C. Tarraf (ed.), *Control of Cyber-Physical Systems*,
Lecture Notes in Control and Information Sciences 449,
DOI: 10.1007/978-3-319-01159-2_10,

system operations to the forefront of world-wide research agendas. Recently, several researchers have studied how to improve the control of HVAC systems by deploying more embedded sensors to monitor temperature, humidity, and CO_2 levels [17], using information about occupant behavior [18,11,4], and improving the modeling and control approaches [21,19,25,20,24,23,5].

To achieve an increase in building efficiency requires an increase in the number of sensors and actuators deployed. While the inclusion of these smart devices enables low cost and environmentally friendly building energy management systems, undetected sensor and actuator failures can result in poor temperature and air quality management. Moreover, HVAC Fault Detection and Diagnostic (FDD) schemes which result in unpredictable or erratic performance can deter building managers from investigating potential failures. For these reasons, technological development of FDD schemes tailored for HVAC systems is paramount and has received much research interest in the recent years [16,12,10,14].

The study of HVAC FDD systems has only been investigated since the late 1980s, with a particular interest in identifying low-cost, timely, and accurate methods for detecting actuator faults. A thorough review of approaches to HVAC actuator fault detection, diagnostics, and prognostics prior to 2006 is provided in [16,15]. In general, approaches to HVAC actuator fault detection can be classified as either hardware-based or software-based solutions [16]. The hardware-based solutions introduce additional smart components strictly for the purposes of actuator fault detection and provide accurate detection capabilities; however, hardware solutions are far more expensive to both deploy and maintain than software-based approaches, and are much more difficult to reconfigure with the introduction of additional smart-actuator devices [15]. Moreover, the inclusion of additional hardware has the added drawback of further increasing the complexity of the HVAC system itself. Software-based actuator FDD approaches are attractive in theory, but suffer from either a reliance on unknown (and difficult to learn) physical models or system-specific detector design specifications [12,16,15,14].

Modern building energy management systems require accurate HVAC control to minimize energy usage while maintaining an acceptable level of comfort for the building occupants. Thus, actuator fault detection is necessary to ensure proper building operation as HVAC systems are subject to various aging and operation errors which can lead to hardware malfunction. A common failure in HVAC systems occurs when the actuator "sticks" and no longer changes its set point, despite controller requests. This type of actuator failure can occur in any position. For example, a valve can be stuck fully open, fully shut, or at any intermediate setting. Additionally, being able to isolate specific actuator failures is paramount to performing timely maintenance.

In [1], an HVAC FDD architecture is introduced that utilizes a fast-deciding steady-state detector and an adaptive model-based detector that are applied to a cooling vent fault detection. The primary difference between the previous approaches, and other model-based software approaches described in [16,15] is the use of a two-tiered detection approach containing a distributed quantitative model-based approach and a distributed qualitative model-based approach to provide quick inference when an actuator is working and provide accurate detection when an actuator has failed. Logic indicates that in the event that an actuator is working, applying a significant change in the actuation input results in a measurable change in the temperature. Under this reasoning, the steady-state detector quickly identifies operational actuators, but it tends to result in a high probability of false alarm when the actuator is operational, but its effect on the temperature is dampened (i.e. a window is open such that the temperature is not significantly affected through actuation). To reduce the probability of false alarm in the HVAC FDD strategy, an adaptive model-based detector is employed based on a first-order building thermal dynamic model. The model-based detector utilizes a history of measurements to estimate the unknown model parameters, then employs the estimated model to detect faults. While the adaptive model-based detector was shown to have a significantly lower probability of false alarm in comparison to the steady-state detector, the resulting performance was highly variant depending on the environmental variables (windows open vs. closed, outside air temperature, etc.) which directly affected the underlying parameter estimation. This sporadic false alarm behavior is undesirable in HVAC FDD schemes as it leads to mistrust by building managers.

In this work, the primary contribution is a parameter-invariant detector (to replace the previously developed model-based detector [1]) that maintains a constant probability of miss across all operating conditions of the HVAC FDD strategy. Additionally, the parameter-invariant detector does not require full model knowledge as it only relies on the structure of a simplified thermodynamic model. The fact that the detector not only detects, but also isolates individual actuator failures is an added contribution. A second contribution of this work is the introduction and description of a closed-loop HVAC monitoring and control system that interfaces directly with the KTH Royal Institute of Technology HVAC system, located in Stockholm, Sweden. An evaluation of the system is provided using the experimental testbed that illustrates the performance of the parameter-invariant detector.

In terms of notation, we use plain lower case italic fonts to indicate scalars or functions with scalar range, bold lower case italic fonts to indicate vectors or functions with vector range, and plain upper case italic fonts to indicate matrices. We also use $\otimes$ to denote Kronecker products, and $\boldsymbol{e}_{i,j}$ to denote the elementary vector of dimension i consisting of all zeros with a single unit entry in the j-th position.

In the following section, we motivate and formulate the actuator FDD problem for building automation. Section 3 introduces the parameter-invariant detector and a novel diagnostic input design. The KTH Royal Institute of Technology HVAC test bed is described in section 4 and an evaluation of the parameter-invariant detector is provided in section 5. The concluding section provides discussion and insight into future work.

2 Problem Formulation

In this section, we formulate a distributed actuator fault detection problem for HVAC systems. While precise thermal modeling of buildings is an ongoing science [6,8], it has repeatedly been experimentally demonstrated that a first principle of the thermal dynamics model is accurate for zone-level temperature evolution in buildings [26,13,9,3]. We consider a building with M interconnected temperature zones for which there exists an underlying interconnection graph, $\mathcal{G}(\mathcal{V},\mathcal{E})$, between the M zones, where $\mathcal{V} := \{1,\ldots,M\}$ is the vertex set, with $i \in \mathcal{V}$ corresponding to zone i, and $\mathcal{E} \subseteq \mathcal{V} \times \mathcal{V}$ is the edge set of the graph. The undirected edge $\{i,j\}$ is incident on vertices i and j if zones i and j interact. The neighborhood of zone i, $\mathcal{N}_i$, is defined as

$$\mathcal{N}_i := \{j \in \mathcal{V} \mid \{i,j\} \in \mathcal{E}\} \tag{1}$$

A generic thermodynamic model of the zone interactions is provided through a spatial and temporal discretization of the first-order heat equation as

$$\begin{aligned} x_j(k+1) &= x_j(k) + m_j \sum_{i\in\mathcal{N}_j} a_{ji}\Big(x_i(k) - x_j(k)\Big) + b_j d_j(k) + w_j(k) \\ y_j(k) &= x_j(k) + v_j(k) \end{aligned} \tag{2}$$

where:

- $k = 0,\ldots,T$ is the time index (T even for notational simplicity[1]);
- $j = 1,\ldots,M$ is the zone index;
- the temperature $x_j(k)$'s, measurements $y_j(k)$'s and actuator inputs $d_j(k)$'s are scalar;
- m_j is the volume of air contained in zone j;
- $a_{ji} = a_{ij} \in \mathbb{R}$ and $b_j \in \mathbb{R}$ denote respectively the gains between $x_i(k)$ and $x_j(k+1)$, and between $d_j(k)$ and $x_j(k+1)$;
- $w_j(k), v_j(k) \in \mathbb{R}$ are uncorrelated i.i.d. Gaussian process noise and measurement noise with moments

$$\mathbb{E}\left[w_j(k)\right] = \chi_{j,w} \quad \mathbb{E}\left[v_j(k)\right] = \chi_{j,v} = 0,$$

$$\mathbb{E}\left[\left(w_j(k) - \chi_{j,w}\right)^2\right] = \sigma^2_{j,w} \quad \mathbb{E}\left[\left(v_j(k) - \chi_{j,v}\right)^2\right] = \sigma^2_{j,v}.$$

We note that in the event of a zone containing multiple actuators and a single sensor, actuator FDD can be performed by allowing only a single actuator to vary. In doing this, the stationary actuators are effectively lumped into the process noise mean since it is a constant input into the thermal model. To compact the notation we let, for $j = 1,\ldots,M$,

[1] For ease of notation and without loss of generality we assume that the available measurements are over a given period whose length is fixed *ex ante*.

$$A := \left[\, \alpha_{ij} \,\right]$$
$$\alpha_{ij} := \begin{cases} 1 - m_j \sum\limits_{n \in \mathcal{N}_j} a_{nj} & \text{if } i = j \\ m_j a_{ij} & \text{if } i \in \mathcal{N}_j, \quad i \neq j \\ 0 & \text{otherwise} \end{cases}$$
$$B := \operatorname{diag}\left[b_1, \ldots, b_M\right]$$
$$\boldsymbol{y}_j := \left[y_j(0), \ldots, y_j(T)\right]^\top$$
$$\boldsymbol{d}_j := \left[d_j(0), \ldots, d_j(T)\right]^\top .$$

Additionally, we consider the following quantities, assuming $\mathcal{N}_j = \{i_1, \ldots, i_J\}$ is the sorted list of neighboring zones of zone j. Then

$$\vec{\alpha}_j := \left[\alpha_{i_1 j}, \ldots, \alpha_{i_J j}\right]^\top ,$$
$$\vec{y}_j(k) := \left[y_{i_1}(k), \ldots, y_{i_J}(k)\right]^\top ,$$
$$\vec{\boldsymbol{y}}_j := \left[\boldsymbol{y}_{i_1}^T, \ldots, \boldsymbol{y}_{i_J}^T\right]^\top ,$$

where, $\vec{y}_j(k)$ is the set of the measurements of agent j *and* its neighbors (sorted lexicographically) at time k, while $\vec{\boldsymbol{y}}_j$ is the set of *all* the measurements of agent j *and* its neighbors (again sorted lexicographically).

Consider then a *specific* zone $\ell \in \{1, \ldots, M\}$, containing an actuator. The structure of the actual actuator input $\boldsymbol{d}_\ell$ is assumed to be as follows:

- $\boldsymbol{u}_\ell := \left[u_\ell(0), \ldots, u_\ell(T)\right]^\top$ is a *desired* and *known* actuation signal;
- $\theta_\ell \in \mathbb{R}$ is an unknown (but constant) input.

Then

$$\boldsymbol{d}_\ell = \theta_\ell \mathbf{1} + \mu_\ell \boldsymbol{u}_\ell \tag{3}$$

where the binary scalar $\mu_\ell \in \{0, 1\}$ is an unknown test parameter indicating whether the actuation signal is present ($\mu_\ell = 1$) or absent ($\mu_\ell = 0$).

We summarize the available information for detecting an actuator failure in zone ℓ as follows:

Assumption 1. Available information:

- the time-series measurements $\vec{\boldsymbol{y}}_\ell$
- the local desired actuation signal $\boldsymbol{u}_\ell$;
- the local zone air-volume weight m_ℓ;
- when an actuator fails, its input to the system remains constant;
- the fact that the state dynamics are LTI-Gaussian, constant in time, and with $b_\ell \neq 0$.

For completeness, we summarize the unavailable information as:

Assumption 2. Unavailable information:

- all the time-series measurements except $\vec{y}_\ell$
- all the local desired input signals except $\boldsymbol{u}_\ell$;
- all the local weights except m_ℓ;
- the weights A and B;
- the moments of the process and measurement noises $\chi_{j,w}$, $\sigma^2_{j,w}$, $\sigma^2_{j,v}$, for all $j = 1, \ldots, M$;
- the actuation parameters θ_j and μ_j for all $j = 1, \ldots, M$;
- the initial conditions $x_1(0), \ldots, x_M(0)$;
- the input signals $\boldsymbol{d}_1, \ldots, \boldsymbol{d}_M$.

We then assume the unknown μ_ℓ to be either 0 (actuator ℓ is at fault) or 1 (actuator ℓ not at fault) and pose the following binary hypothesis testing problem:

Assumption 3. Structure of the actuator fault μ_ℓ satisfies either one of the two following hypotheses:

H_0 **(null hypothesis):**	$\mu_\ell = 1$	(no fault)
H_1 **(alternative hypothesis):**	$\mu_\ell = 0$	(fault)

In words, both hypotheses assume the actual $\boldsymbol{d}_\ell$ to be unknown, since θ_ℓ is unknown, but with a fixed and known functional structure. H_0 additionally assumes the presence of the known actuation input $\boldsymbol{u}_\ell$. Our aim is the following: develop a distributed test that considers a **specific** zone $\ell \in \{1, \ldots, M\}$, and decides among the hypotheses H_0 vs. H_1 in Assumption 3 using only the information in Assumption 1 and, at the same time, being invariant to the unavailable information in Assumption 2. Thus, we state the following problem:

Problem 4

Find a test that detects whether zone ℓ has an actuator fault independently of whether a fault exists at any other zone $j \neq \ell$ (fault isolation) *and* minimizes the probability of false alarm while maintaining a constant probability of detection.

The following section presents a solution to the problem introduced in this section.

3 Parameter-Invariant Actuator FDD

In this section, we introduce a distributed HVAC actuator FDD strategy, tailored for detecting and isolating whether actuators are stuck in an unknown position such that a constant level of detection is maintained. To achieve this goal, the following subsections introduces a parameter-invariant detector and an actuator diagnostic input, respectively.

3.1 Parameter-Invariant Detector

In this section, we recall the test developed for distributed detection of inputs in networked systems in [2]. This test, in the context of the HVAC detection problem, is designed to minimize the probability of false alarm, subject to a constraint on the probability of detection. We state the primary result of [2], augmented for the actuation detection problem, in the following lemma:

Lemma 5. A maximally invariant statistic for Problem 4 is

$$T[\boldsymbol{z}_\ell] = \frac{\boldsymbol{z}_\ell^\top P_\ell \boldsymbol{z}_\ell}{\frac{1}{N_\ell - 1}\boldsymbol{z}_\ell^\top (I_{N_\ell} - P_\ell)\boldsymbol{z}_\ell} \tag{4}$$

with

$$\begin{aligned} \boldsymbol{z}_\ell &:= F_\ell Q \boldsymbol{y}_\ell \\ P_\ell &:= \frac{F_\ell Q \boldsymbol{u}_\ell \boldsymbol{u}_\ell^\top Q^\top F_\ell^\top}{\boldsymbol{u}_\ell^\top Q^\top F_\ell^\top F_\ell Q \boldsymbol{u}_\ell} \\ N_\ell &:= \frac{T}{2} - \|\mathcal{N}_\ell\|_0 - 1 \end{aligned} \tag{5}$$

and where the exploited quantities satisfy

$$\begin{aligned} F_\ell^\top F_\ell &= I_{\frac{T}{2}} - \vec{Y}_\ell(\vec{Y}_\ell^\top \vec{Y}_\ell)^{-1}\vec{Y}_\ell^\top \\ Q &= I_{\frac{T}{2}} \otimes [\,0 \;\; 1\,] \\ \vec{Y}_\ell &= \begin{bmatrix} \vec{y}_\ell^\top(0) & 1 \\ \vec{y}_\ell^\top(2) & 1 \\ \vec{y}_\ell^\top(4) & 1 \\ \vdots & \vdots \\ \vec{y}_\ell^\top(T) & 1 \end{bmatrix} \end{aligned} \tag{6}$$

Applying the maximally invariant statistic, and following the same reasoning as in [2], we write the test for detecting actuator failures as:

Corollary 6. A distributed test minimizing the probability of false alarm and providing a constant probability of missed detection of α for Problem 4 is

$$\phi_\ell(\boldsymbol{z}_\ell) = \begin{cases} H_0 \text{ if } T_\ell[\boldsymbol{z}_\ell] > \mathcal{F}^{-1}_{1,N_\ell-1}(\alpha) \\ H_1 \text{ otherwise} \end{cases} \tag{7}$$

where $\mathcal{F}^{-1}_{n,m}(\alpha)$ is the inverse central cumulative F-distribution of dimensions n and m.

We remark that test (7) can be performed in simultaneously across multiple zones and it is invariant to the non-local measurements. This comes with a price, namely, the test exploits half the measurements for testing (the other half are used to establish invariance). To maximize the performance across this reduced data set, the following subsection introduces an adaptive actuator diagnostic signal.

3.2 Diagnostic Input Design

The performance of the detector is significantly affected by the actuation input driving the test. In this subsection, and motivated by the performance of the adaptive model-based detector in [1], we design a diagnostic input that attempts to maximize the divergence of test.

To design the diagnostic actuation input, we observe that the discrete-time dynamics for measurement of the j-th zone can be written as

$$\begin{aligned} \boldsymbol{z}_j(k) &= \boldsymbol{G}_k \boldsymbol{z}_j(k) + \boldsymbol{n}_j(k) \\ y_j(k) &= \boldsymbol{C}\boldsymbol{z}_j(k) + v_j(k) \end{aligned} \tag{8}$$

where

$$\begin{aligned} \boldsymbol{z}_j(k) &= \left[x_j(k),\, m_j\boldsymbol{\alpha}_j^T,\, b_j,\, \chi_{j,w} + b_j\theta_j\right]^T \\ \boldsymbol{G}_k &= \begin{bmatrix} 1 & \vec{\boldsymbol{y}}_j^\top(k) - \boldsymbol{1}^\top y_j(k) & \mu_j u_j(k) & 1 \\ 0 & \boldsymbol{I} & 0 & 0 \\ 0 & \boldsymbol{0}^\top & 1 & 0 \\ 0 & \boldsymbol{0}^\top & 0 & 1 \end{bmatrix} \\ \boldsymbol{n}_j(k) &= \left[w_j(k) + \textstyle\sum_{i\in \mathrm{N}_j} \alpha_{ij} v_i(k),\, \boldsymbol{0}^\top, 0, 0\right]^T \\ \boldsymbol{C} &= \left[1\; \boldsymbol{0}^\top\; 0\; 0\right] \end{aligned} \tag{9}$$

As a heuristic, we assume (strictly for the purpose of designing an actuation input) that the true values of $\boldsymbol{\alpha}_{ij}$ are equal to the corresponding values provided through $z_j(k)$. Under this assumption, the measurements have a Gaussian distribution, parameterized by μ_j, written as

$$f_j(y_j(k)) = \frac{1}{\sqrt{2\pi\left(\boldsymbol{C}\boldsymbol{\Sigma}_{k,j}\boldsymbol{C}^\top + \sigma^2_{j,v}\right)}} \exp\left\{-\frac{1}{2}\frac{\left(y_j(k) - \boldsymbol{C}\boldsymbol{m}_{k,j}\right)^2}{\left(\boldsymbol{C}\boldsymbol{\Sigma}_{k,j}\boldsymbol{C}^T + \sigma^2_{j,v}\right)}\right\} \tag{10}$$

where, assuming $\boldsymbol{\Sigma}_{j,n} = \mathbb{E}\left[n_{k,j} n_{k,j}^{\top}\right]$

$$\begin{aligned} \boldsymbol{m}_{k+1,j} &= \left(\boldsymbol{G}(k) - \boldsymbol{K}_{k,j}\boldsymbol{C}\right)\boldsymbol{m}_{k,j} + \boldsymbol{K}_{k,j} y_j(k) \\ \boldsymbol{\Sigma}_{k+1,j} &= \left(\boldsymbol{G}(k) - \boldsymbol{K}_{k,j}\boldsymbol{C}\right)\boldsymbol{\Sigma}_{k,j}\boldsymbol{G}^{\top}(k) + \boldsymbol{\Sigma}_{j,n} \\ \boldsymbol{K}_{k,j} &= \boldsymbol{G}(k)\boldsymbol{\Sigma}_{k,j}\boldsymbol{C}^{\top}\left(\boldsymbol{C}\boldsymbol{\Sigma}_{k,j}\boldsymbol{C}^{\top} + \sigma_{j,v}^{2}\right)^{-1} \end{aligned} \tag{11}$$

are the mean and covariance of $\boldsymbol{z}_j(k)$ and the observer gain, respectively.

To identify the actuator input for evaluating the detection problem, we utilize an information-theoretic approach and choose the actuator input to maximize the next step Kulbach-Liebner [7] divergence according to

$$u_k = \arg\max_{0 \le u \le 1} -\mathbb{E}\left[l_j(y_k)\right] \tag{12}$$

where $l_j(y_k)$ is the log-likelihood ratio,

$$l_j(k) = l_j(k-1) + \ln\frac{f_j(y_j(k)|\mu_j = 0)}{f_j(y_j(k)|\mu_j = 1)}. \tag{13}$$

This approach is common in information theory as it results in the control sequence that maximizes the next step log-likelihood ratio. Since the log-likelihood is a convex function of the control sequence, it is maximized at the extreme points of the range of the control sequence as denoted as follows

$$u_k = \begin{cases} 1 \text{ if } \; \mathbb{E}\left[l_j(k+1|u_j(k)=1)\right] > \mathbb{E}\left[l_j(k+1|u_j(k)=0)\right] \\ 0 \text{ if } \; \mathbb{E}\left[l_j(k+1|u_j(k)=1)\right] \le \mathbb{E}\left[l_j(k+1|u_j(k)=0)\right] \end{cases} \tag{14}$$

In an HVAC system this equates to either turning the HVAC actuator completely on or completely off. While this control input is advantageous for fault detection and diagnostics, it comes at a trade-off with the performance of the HVAC system since the control input does not correspond to the optimal building operation set-point.

It will be shown in the experimental evaluation section that the parameter-invariant detector requires significant monitoring periods to accurately determine whether an actuator has failed. Moreover, fault detection schemes that require long monitoring periods may not be necessary to identify a working actuator if the actuator has a significant effect on the temperature. For this reason, the parameter-invariant detector is best suited to replace the model-based adaptive detector in the previous work.

4 Experimental Testbed

The KTH Royal Institute of Technology main campus in Stockholm, Sweden consists of over 45 buildings which house roughly 559 laboratories, 2569 office rooms and 87 lecture halls. The campus has an HVAC system managed by a centralized SCADA system. The SCADA map of the KTH campus is depicted

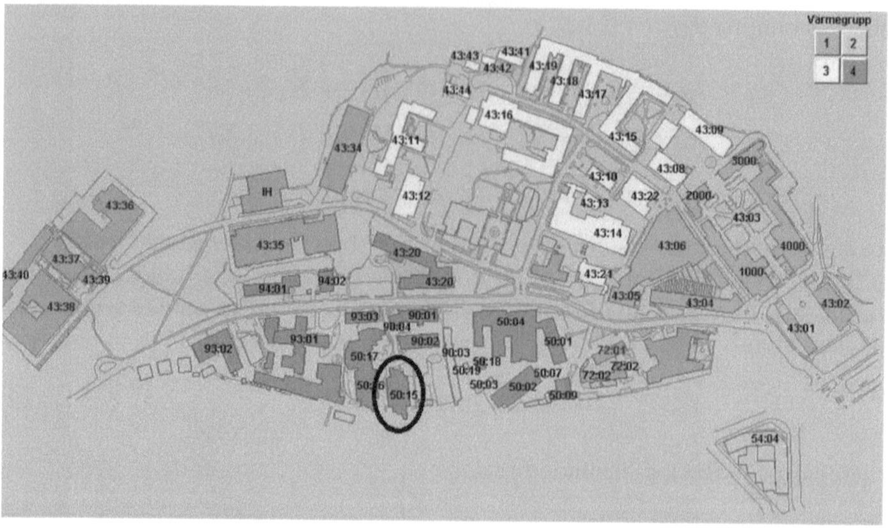

Fig. 1. KTH Royal Institute of Technology campus view from the SCADA system map

in Fig. 1. Each building is contains at least one Programming Logic Controller (PLC) units which measure and control the local HVAC system components. Each of the PLC units in the campus communicates to the SCADA system through an OPC client/server interface. The KTH HVAC test bed is currently deployed in the Q-building (denoted by the black circle in Fig. 1). The Q-building is a multi-story building housing the School of Electrical Engineering with multiple academic departments, lecture rooms, and laboratories. This building is equipped with three separate ventilation units for fresh air supply and relies on a district-managed water supply for induction-based heating and cooling. The fresh air supply takes place from 7 : 00 AM to 4 : 00 PM, and can be set manually by demand at other times of the day and weekends.

The HVAC testbed is currently comprised of the second floor of the Q-building and is depicted in Fig. 2. This floor houses four laboratories (rooms A225, A213, B230 and the PCB Lab), an office room (A235), the Q2 lecture hall, one storage room and a boiler room.

Each room in the testbed is considered to be a thermal zone and has a set of sensors and actuators that can be individually controlled. In the figure, the red circles depict sensor locations, the green circles illustrate the actuator locations and the blue circle denotes the external temperature sensor. The available sensors are temperature (GQ101) and CO2 sensors (GT101)[2]. The actuators are the flow valve of the heating radiator (SV201), the flow valve for the air conditioning system (SV401), the air vent for fresh air flow at constant temperature of 21 °C

[2] In this section we note, in parentheses, the PLC tags corresponding to each sensor and actuator such that they can be referenced in the downloaded KTH HVAC testbed data.

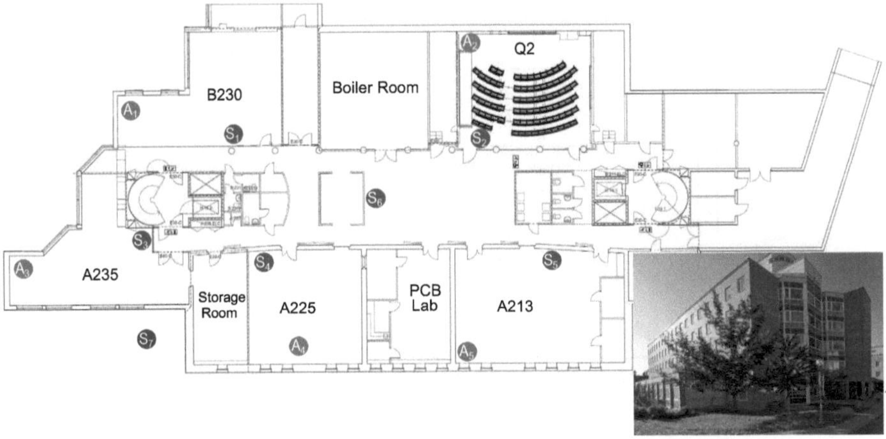

Fig. 2. KTH HVAC testbed at the second floor of the Q-building at KTH. Each of the five rooms considered contain sensors and actuators used for HVAC control. Additional sensors are located in the corridor and outside of the building.

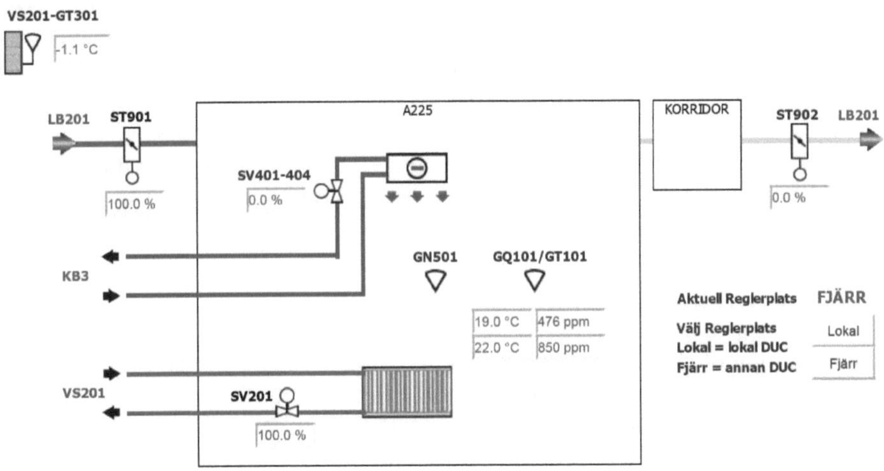

Fig. 3. The HVAC system components in room A225, the Automatic Control experimental lab. Various sensors and actuators are available allowing for the control of ventilation and heating.

(ST901) and the air vent for air exhaust to the corridor (ST902). Each actuator can be set between 0% and 100%. As an illustration, the HVAC system in room A225 is depicted in Fig. 3.

The HVAC testbed is developed in LabVIEW and is comprised of two separate components: the experimental application and a database/web server system. The database is responsible for logging the data from all HVAC

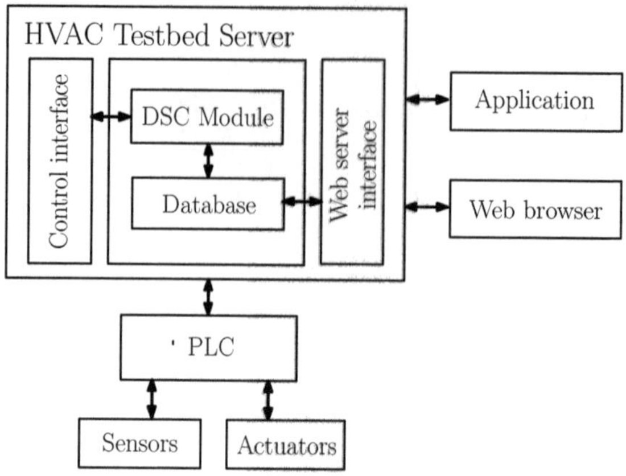

Fig. 4. The HVAC system architecture. Users are able to design experiments through a Labview application and remotely connect to the HVAC testbed. Additionally, through a web browser any user can download experimental data from the testbed database.

components in real-time, which is publicly available through a web server (`http://hvac.ee.kth.se`). Additionally, the experimental application is developed by each user and interacts with the Data-logging and Supervisory Control (DSC) module in the HVAC Testbed Server, which connects to the PLC through and OPC client/server. This component allows for real-time sensing, computation and actuation. Even though the application is developed in LabVIEW, MATLAB code is integrated in the application through a Mathscript zone. An overview of the testbed architecture is shown in Fig. 4.

5 Experimental Results

The evaluation of the actuator fault detector was performed in the KTH HVAC testbed. To evaluate the parameter-invariant detector performance, multiple experiments were performed utilizing room A225 in Figs. 2 and 3 as the test room. The air mass of room A225 interacts with the outdoor and corridor air masses as well as the adjacent rooms, the PCB lab and the storage room (each representing a unique thermal zone). Since the PCB Lab and Storage room do not have sensors, we neglect their effect on room A225's temperature. The effect of ignoring the potential thermal contribution of these unobservable air masses is minimized by including room A235 and room A213 as *adjacent* rooms (or zones). We note that room A225 has four exterior windows and one door connecting to the common corridor.

To evaluate the actuator fault detector, we attempt to detect an actuator failure in the air conditioning system, namely whether the fresh air vent, actuator ST901 in Fig. 3, is stuck in a given position. The system is set to measure the temperature at five minute intervals. In the following experiments, we neglect the tests using measurements gathered for less than 30 minutes (6 sampling periods), as these tests yield irrelevant results since it requires at least 6 sets of measurements to calculate the test statistic for the parameter-invariant detector.

To emulate an actuator failure, we simply do not apply the control value given by the diagnostic input mechanism and leave the air vent closed (as opposed to physically breaking the actuator). The cooling actuator SV401, the radiator valve SV201, and the exhaust air vent ST902 were kept at constant values throughout each experiment. Additionally, the actuators in the adjacent rooms were allowed to operate normally, as in a normal operating scenario where we wish to not only detect the actuator failure, but do so in a distributed manner without PLC coordination. Under these testing conditions, the parameter-invariant detector was evaluated for the scenarios which exhibited erratic performance in [1]. Specifically, these scenarios are (1) detecting an actuator fault when the windows are open/closed and (2) detecting there is not an actuator fault when the windows are open. The reason for not evaluating the parameter-invariant detector when the actuator is working and the windows are closed is because the steady-state detector in [1] is very accurate under this scenario, thus the parameter-invariant detector is unlikely to be utilized.

To evaluate the parameter-invariant detector when an actuator is in fault and the windows are closed, 100 unique experiments were performed, each lasting three hours. For the parameter-invariant detector, we specify a probability of miss (probability of deciding there was no fault when there actually was a fault) of 0.10. The statistical results of these tests are shown in Fig. 5, where the upper subplot illustrates the average value of the test statistic (solid black line) versus the maximum and minimum value of the test statistic (dotted black lines) and the test decision threshold (dashed red line) for the tests, while the lower subplot illustrates the actual rate of miss by the black x's and the specified probability of miss of 0.10. When a test statistic is above the the threshold, then the test decides there was no fault (which is incorrect in these experiments), and by specification should happen with a probability of 0.10, regardless of the test time. From Fig. 5, we observe that, the actual rate of missed detection varies between 0.06 to 0.16, which is very near the specified rate of 0.10, for all time. Averaging the miss rate over the three hours yields a rate of 0.101, which is nearly identical to the probability of miss specification. These results illustrate that when the windows are closed, the parameter-invariant detector accurately detects actuator faults as specified by the probability of miss.

Consistent with previous experiments, to evaluate the parameter-invariant detector when an actuator is in fault and the windows are open, 100 unique experiments were performed, each lasting three hours and employed the same specified probability of miss (0.10). The statistical results of these tests are shown in Fig. 6, where the plots follow the same structure as in Fig. 5. In Fig. 6, we observe very similar performance as in Fig. 5. This similarity is expected since by changing the state of the window (opening the window), we have merely changed the interaction between room A225 with the outside. Since the parameter-invariant detector is designed to be invariant to different thermal zone

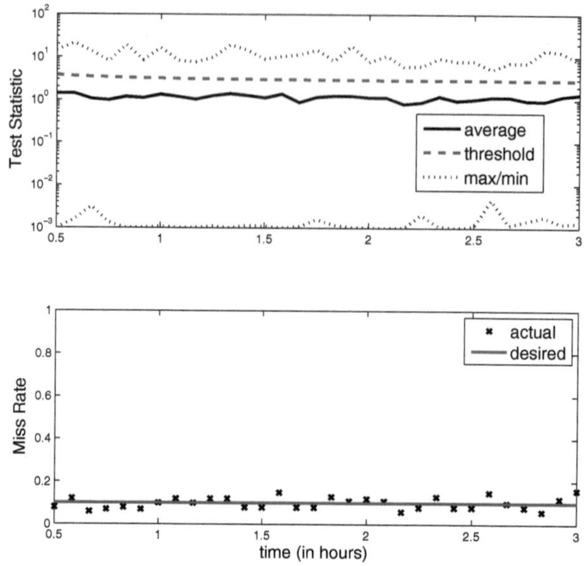

Fig. 5. Experimental parameter-invariant detector results when the actuator is at fault and the window is closed.

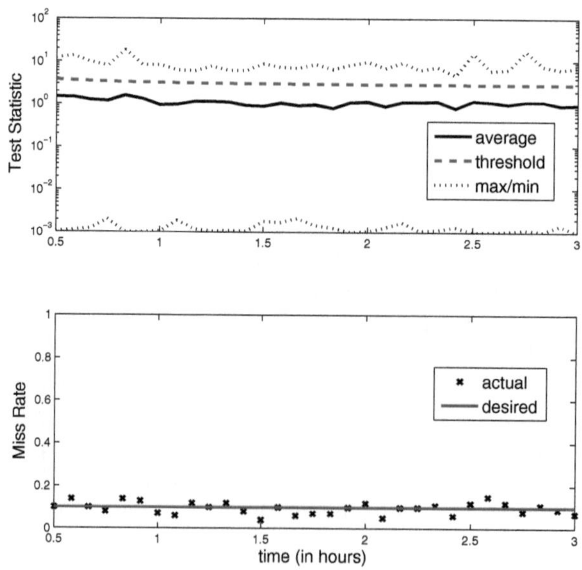

Fig. 6. Experimental parameter-invariant detector results when the actuator is at fault and the window is open.

interactions, the performance (in terms of the probability of miss) should be unaffected. In comparison to the performance of the model-based detector in [1], we observe that the parameter-invariant detector has nearly constant performance (in terms of probability of miss) with both the state of the window and with time, while the model-based detector exhibits varying performance with both the state of the window and with time. Having near-constant performance (which matches the specification) is preferred as it allows a building manager to reliable select the probability of missed detection of actuator fault.

To evaluate the parameter-invariant detector when an actuator is working properly and the windows are open, 25 experiments were performed, each lasting three hours. For the parameter-invariant detector, and consistent with the previous experiments, we specify a probability of miss to be 0.10. The statistical results of these tests are shown in Fig. 7, where the upper subplot follows the same structure as in Figs. 5 and 6. The lower subplot illustrates the rate of false alarm (deciding there is an actuator fault when in reality there is not a fault) versus time. In these experiments (and opposite of the previous experiments) a false alarm occurs when the test statistic is below the threshold. In Fig. 7, we immediately observe that the test statistic is (in general) increasing as the test time increases. This is desirable since the larger the test statistic, the more likely it is to claim no fault (which is true in these experiments). As time increases, we observe from the lower subplot that the rate of miss is (in general) decreasing. The reason for non-monotonic performance is explained by the fact only 25

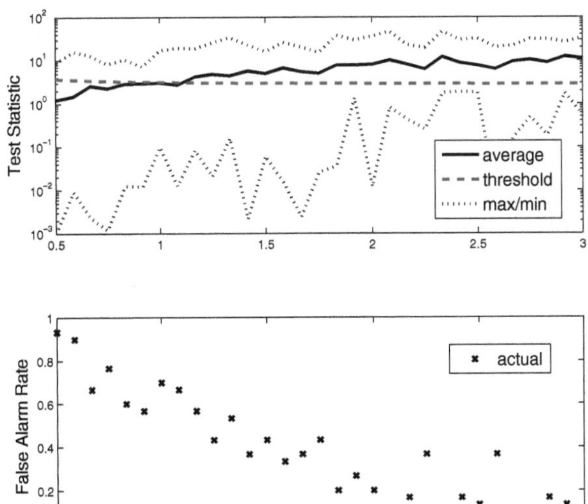

Fig. 7. Experimental parameter-invariant detector results when the actuator is working and the window is open.

experiments were used to evaluate the parameter-invariant detector when the actuator is working and the windows are open. These results illustrate that performance (in terms of false alarm rate) can be improved (decreased) by allowing the parameter-invariant detector to run for longer time periods. For this room and configuration, a false alarm rate of 0.05 can be achieved by allowing the parameter-invariant detector to run for 3 hours. Depending on the zone (room) and its interactions with the adjacent zones, the probability of false alarm will vary. However, this variance is generally acceptable in practice so long as the probability of miss remains constant since it implies that by simply letting the test run longer will yield improved performance.

6 Discussion and Future Work

The parameter-invariant detector introduced in this work for building HVAC systems is based on a previously designed CFAR detector for networked systems, where the parameter-invariant detector with constant performance is designed to replace the model-based detector with unpredictably varying performance in the HVAC FDD scheme previously developed. The parameter-invariant detector is designed to maintain a constant probability of missing a fault, invariant to the unknown and time-varying building parameters. An experimental testbed using a real HVAC system is described that allows automatic sensing and actuation of several HVAC components. Future work includes a full evaluation of the two-tier fault detection strategy on the KTH HVAC testbed and extending the detection theory to handle the detection of faulty sensors and imminent actuator failure (i.e. detecting whether the actuator range of motion has decreased).

References

1. Weimer, J., Ahmadi, A., Araujo, J., Mele, F., Papale, D., Shames, I., Sandberg, H., Johansson, K.H.: Active Actuator Fault Detection and Diagnostics in HVAC Systems. In: 4th ACM Workshop on Embedded Sensing Systems For Energy-Efficiency in Buildings, BuildSys (2012)
2. Weimer, J., Varagnolo, D., Johansson, K.: Distributed Model-Invariant Detection of Unknown Inputs in Networked Systems. In: 2nd ACM International Conference on High Confidence Networked Systems, HiCoNS (2013)
3. Široký, J., Oldewurtel, F., Cigler, J., Prívara, S.: Experimental analysis of model predictive control for an energy efficient building heating system. Applied Energy 88(9), 3079–3087 (2011)
4. Agarwal, Y., Balaji, B., Dutta, S., Gupta, R., Weng, T.: Duty-cycling buildings aggressively: The next frontier in HVAC control. In: 2011 10th International Conference on Information Processing in Sensor Networks (IPSN), pp. 246–257 (April 2011)
5. Aswani, A., Master, N., Taneja, J., Culler, D., Tomlin, C.: Reducing transient and steady state electricity consumption in HVAC using learning-based model-predictive control. Proceedings of the IEEE 100(1), 240–253 (2012)
6. Chow, W.: Application of computational fluid dynamics in building services engineering. Building and Environment 31(5), 425–436 (1996)

7. Cover, T.M., Thomas, J.A.: Elements of Information Theory. Wiley-Interscience, New York (1991)
8. Crawley, D., Lawrie, L., Winkelmann, F., Buhl, W., Huang, Y., Pedersen, C., Strand, R., Liesen, R., Fisher, D., Witte, M., et al.: Energyplus: creating a new-generation building energy simulation program. Energy and Buildings 33(4), 319–331 (2001)
9. Deng, K., Barooah, P., Mehta, P., Meyn, S.: Building thermal model reduction via aggregation of states. In: IEEE Proceedings of the American Control Conference (ACC), pp. 5118–5123 (July 2010)
10. Djuric, N., Novakovic, V.: Review of possibilities and necessities for building lifetime commissioning. Renewable and Sustainable Energy Reviews 13(2), 486–492 (2009)
11. Erickson, V., Carreira-Perpinan, M., Cerpa, A.: Observe: Occupancy-based system for efficient reduction of HVAC energy. In: 10th International Conference on Information Processing in Sensor Networks (IPSN), pp. 258–269 (April 2011)
12. Fernandez, N., Brambley, M., Katipamula, S., Cho, H., Goddard, J., Dinh, L.: Self correcting HVAC controls project final report PNNL-19074. Technical report, Pacific Northwest National Laboratory, Richland, WA (2009)
13. Gouda, M., Danaher, S., Underwood, C.: Building thermal model reduction using nonlinear constrained optimization. Building and Environment 37(12), 1255–1265 (2002)
14. Jagemar, L., Olsson, D.: The EPBD and continuous commisioning. Technical report, CIT Energy Mangement AB, Goteborg, Sweden (October 2007)
15. Katipamula, S., Brambley, M.R.: Methods for fault detection, diagnostics, and prognostics for building systems - a review, part ii. HVAC&R Research 11(2), 169–187 (2005)
16. Katipamula, S., Brambley, M.R.: Methods for fault detection, diagnostics, and prognostics for building systems - a review, part i. HVAC &R Research 11(1), 3–25 (2005)
17. Kim, Y., Schmid, T., Srivastava, M.B., Wang, Y.: Challenges in resource monitoring for residential spaces. In: Proceedings of the First ACM Workshop on Embedded Sensing Systems for Energy-Efficiency in Buildings, BuildSys 2009, New York, NY, USA, pp. 1–6 (2009)
18. Lu, J., Sookoor, T., Srinivasan, V., Gao, G., Holben, B., Stankovic, J., Field, E., Whitehouse, K.: The smart thermostat: using occupancy sensors to save energy in homes. In: Proceedings of the 8th ACM Conference on Embedded Networked Sensor Systems, SenSys 2010, New York, NY, USA, pp. 211–224 (2010)
19. Ma, J., Qin, J., Salsbury, T., Xu, P.: Demand reduction in building energy systems based on economic model predictive control. Chemical Engineering Science 67(1), 92–100 (2012)
20. Ma, Y., Kelman, A., Daly, A., Borrelli, F.: Predictive control for energy efficient buildings with thermal storage: Modeling, stimulation, and experiments. IEEE Control Systems 32(1), 44–64 (2012)
21. Marchiori, A., Han, Q.: Distributed wireless control for building energy management. In: Proceedings of the 2nd ACM Workshop on Embedded Sensing Systems for Energy-Efficiency in Building, BuildSys 2010, New York, NY, USA, pp. 37–42 (2010)
22. Melman, A.: Geometry and convergence of euler's and halley's methods. SIAM Review 39, 728–735 (1997)

23. Nghiem, T.X., Behl, M., Mangharam, R., Pappas, G.J.: Scalable scheduling of building control systems for peak demand reduction. In: American Control Conference (June 2012)
24. Oldewurtel, F., Parisio, A., Jones, C.N., Gyalistras, D., Gwerder, M., Stauch, V., Lehmann, B., Morari, M.: Use of model predictive control and weather forecasts for energy efficient building climate control. Energy and Buildings 45, 15–27 (2012)
25. Goyal, H.A.I.S., Barooah, P.: Zone level control algorithms based on occupancy information for energy efficient buildings. In: American Control Conference (June 2012)
26. Seem, J., Klein, S., Beckman, W., Mitchell, J.: Transfer functions for efficient calculation of multidimensional transient heat transfer. Journal of Heat Transfer 111, 5 (1989)
27. UK Department of Trade and Industry. DTI, Energy Trends 2005. Department of Trade and Industry, London (2011)
28. U.S. Department of Energy. Energy Efficiency Trends in Residential and Commercial Buildings. U.S. Department of Energy (October 2008)
29. Wald, A.: Sequential Analysis. John Wiley & Sons, Inc., New York (1947)

Resilient Parameter-Invariant Control with Application to Vehicle Cruise Control

James Weimer, Nicola Bezzo, Miroslav Pajic,
George J. Pappas, Oleg Sokolsky, and Insup Lee

School of Engineering and Applied Sciences
University of Pennsylvania
Philadelphia, PA 19104, USA
{weimerj,nicbezzo,pajic,pappasg}@seas.upenn.edu,
{sokolsky,lee}@cis.upenn.edu

Abstract. This work addresses the general problem of resilient control of unknown stochastic linear time-invariant (LTI) systems in the presence of sensor attacks. Motivated by a vehicle cruise control application, this work considers a first order system with multiple measurements, of which a bounded subset may be corrupted. A frequency-domain-designed resilient parameter-invariant controller is introduced that simultaneously minimizes the effect of corrupted sensors, while maintaining a desired closed-loop performance, invariant to unknown model parameters. Simulated results illustrate that the resilient parameter-invariant controller is capable of stabilizing unknown state disturbances and can perform state trajectory tracking.

Keywords: Secure Cyber-Physical Systems, Robust Control, Resilient Sensor Fusion.

1 Introduction

Modern large-scale control systems are becoming more and more integrating into our daily lives. As the integration of smart devices in modern control systems increases, so does the need for defenses against potential attacks. Today, our dependence on integrated controllers automates everything from inter-home appliances to nation-wide power distribution, where the effect of unpredicted behaviors can range from a minor inconvenience of resetting a smart device to a regional blackout. Our dependence on these closed-loop automated systems requires that their performance be resilient to malicious behavior and robust to non-malicious behavior.

With respect to the vehicle cruise control system, non-malicious agents include environmental variables (gravity, wind speed, parts fatigue and failure, etc.) while malicious behavior can be introduced through, for example, sensor spoofing. Employing redundant measurements is a well established method of providing better estimates of control variables and model parameters; however, when attacked, a redundant measurement can be used as a means to destabilize a

D.C. Tarraf (ed.), *Control of Cyber-Physical Systems*,
Lecture Notes in Control and Information Sciences 449,
DOI: 10.1007/978-3-319-01159-2_11,

control system. Thus, to ensure safe performance of the vehicle cruise controller requires securing the sensory data, while simultaneously designing controllers robust to unknown environmental parameters.

Literature review: The design of algorithms which are resilient against faults or unknown parameters has been addressed from many points of view, including fault detection [1], robust control [2], adaptive control [3], and more generally from estimation and hypothesis testing [4]. In general, these approaches address the issue of maximizing some performance measure with respect to known or bounded disturbances. In the context of security against malicious attacks, many of these approaches are not applicable because of their assumption that the attack is either known or bounded, with notable exceptions being approaches which ask for invariance to the unknown parameters [5]. The remainder of this literature review focuses on secure estimation/control and control of unknown systems, respectively.

Secure estimation and control system design in the presence of disturbances or attacks has received increasing research interest [6,7,8,9,10,11]. Most closely related to the work presented herein is [12], which addresses the secure estimation and control of linear deterministic systems under malicious sensor attacks. While the approach in [12] is shown to stabilize the systems under consideration, their approach requires full knowledge of the underlying system dynamics in order to secure the closed-loop system. When the underlying dynamics are unknown (and potentially stochastic) more robust detection and control algorithms are needed.

Control of unknown linear systems can be approached through adaptive control techniques [3], typically based on a Lyapunov stability requirement. Adaptive control is generally classified as either indirect (estimation of model parameters) or direct (estimation of the control sequence), where direct adaptive control is more robust than indirect adaptive control, since the parameter estimation may not be accurate over all frequencies [13,14,15]. However, adaptive control techniques ultimately require an estimation of either the unknown parameters or unknown control feedback gain, which dictates the performance of the closed-loop system. In contrast, in this work we wish to directly calculate the controller outputs for systems with unknown parameters, without requiring the intermediate estimation of the unknown parameters or feedback control gain, such that a closed-loop specification is satisfied.

Statement of contributions: Beyond the previous approaches, this work focuses on the specification-based design of resilient parameter-invariant controllers for stochastic linear time-invariant systems, with specific interest in vehicle cruise control systems. The primary technical contributions of this work are: (a) a resilient sensor fusion strategy for unknown attacks on noisy measurements; (b) a finite-horizon mean-stabilizing parameter-invariant controller; (c) a sensor fusion and controller co-design requirement that satisfies a stochastic Lyapunov criteria.

Structure of the paper: Section 2 formulates precisely the problem considered in this work. We introduce the resilient sensor fusion strategy and parameter-invariant controller in Section 3. A co-design requirement for satisfying the performance specification is provided in Section 4. Section 5 presents numerical evaluations of the resilient parameter-invariant controller in the presence of sensor attacks for vehicle cruise control. The concluding section provides discussion and proposes future extensions. For readability, the notation and constant definitions employed in this paper are included as an appendix.

2 Problem Formulation

This section introduces a resilient control problem for a system with unknown LTI-Gaussian dynamics and (potentially) corrupted measurements.

Specifically, we consider an LTI system with a single state that evolves according to

$$\begin{aligned} x(k+1) &= ax(k) + bu(k) + w(k) \\ \boldsymbol{y}(k) &= \boldsymbol{c}x(k) + \boldsymbol{v}(k) + \boldsymbol{d}(k) \end{aligned} \tag{1}$$

where:

- $x, u \in \mathbb{R}$, are the state and control input, respectively;
- $a, b \in \mathbb{R}$ are the state dynamic and control input gains;
- $\boldsymbol{y}, \boldsymbol{c}, \boldsymbol{d} \in \mathbb{R}^{\mathrm{N}}$, are the measurements, state measurement gain, and corruption, respectively;
- $w \in \mathbb{R}$ and $\boldsymbol{v} \in \mathbb{R}^N$ are uncorrelated i.i.d. Gaussian process noise and measurement noise with central moments[1]:

$$\mathbb{E}[w] = \mu \qquad \mathbb{E}[\boldsymbol{v}] = \mathbf{0}$$
$$\mathbb{E}\left[(w-\mu)^2\right] = \sigma \qquad \mathbb{E}\left[(\boldsymbol{v}\boldsymbol{v}^\top)^2\right] = \boldsymbol{I}$$

At time k, the model information available in this work is summarized in the following assumption:

Assumption 1 - Available Information:

- the time-series measurements, $\mathbf{y} = \left[\boldsymbol{y}^\top(0), \ldots, \boldsymbol{y}^\top(k)\right]^\top$;
- the time-series control inputs, $\mathbf{u} = [u(0), \ldots, u(k)]^\top$;
- the variance of the process noise, σ;
- the state measurement gain, $\boldsymbol{c}$;
- the state dynamics, control gain, and process noise mean are constant.

[1] Without loss of generality, we assume the measurement noise is white and normalized to unit variance, where colored noise and non-unit variance white noise can be whitened by applying a normalizing pre-whitening filter.

In words, we assume that the measurements, measurement state gain, inputs, and noise covariances are known; however, the dynamics and process bias governing the evolution of the state are unknown, but constant. For completeness, we summarize the unavailable information in the following assumption[2]:

Assumption 2 - Unavailable Information:

- the state dynamics, a
- the control gain, b;
- the process noise mean, μ;
- the time-series measurement corruption, $\mathbf{d} = \left[\boldsymbol{d}^\top(0), \ldots, \boldsymbol{d}^\top(k)\right]^\top$.

Although we assume the measurement corruption is unknown, we assume a maximum of M measurements are corrupted, as defined in the following assumption:

Assumption 3 - Measurement Corruption Structure: At each time step, at most M measurements are corrupted, $\|\boldsymbol{d}(k)\|_0 \leq M$, where

$$M = \begin{cases} \frac{N}{2} - 1\,, & N \text{ even} \\ \frac{N-1}{2}\,, & N \text{ odd} \end{cases}$$

such that for

$$\left\{\boldsymbol{P}_d^\perp\right\} := \left\{\boldsymbol{Q}^\top\boldsymbol{Q} \middle| \ \boldsymbol{Q} \in \{0,1\}^{T\times N},\ \boldsymbol{Q}\boldsymbol{Q}^\top = \boldsymbol{I},\ T > N - M\right\}$$

the following is true:

$$\boldsymbol{F}\boldsymbol{d}(k) = \boldsymbol{0},\ \exists \boldsymbol{F} \in \left\{\boldsymbol{P}_d^\perp\right\}$$

Assumption 3 is consistent with the assumption in [12] and implies a maximum of M corrupted measurements since elements of $\boldsymbol{d}(k)$ which equal zero, imply no corruption is applied to the corresponding measurement. Under the assumptions introduced in this section, we wish to solve the following problem:

Problem 1 - Stochastic Boundedness: Given (1) and assumptions 1-3, show that

$$\mathbb{E}\left[\|x(k+1)\|^2\right] \leq \alpha^\star \|x(k)\|^2,\ \ \forall\ \ \|x(k)\|^2 \geq \eta \tag{2}$$

where $\eta \in \mathbb{R}$ is a desired state convergence threshold and $\alpha^\star \in [0,1]$ denotes a desired state convergence rate.

[2] The notation $\|\|_0$ denotes the zero-norm, while $\|\|$ denotes the two-norm.

The problem introduced in this section is addressed in the following section by introducing criteria for resilient sensor fusion and finite-horizon parameter-invariant control.

3 Main Contributions

By extensively employing the notation and definitions introduced in the appendix, the main contributions of this work are summarized in the following propositions:

Proposition 1 - Resilient Sensor Fusion: Given (1), then

$$\hat{\boldsymbol{P}}_{\boldsymbol{d}}^{\perp} = \arg \min_{\boldsymbol{Q} \in \{\boldsymbol{P}_{\boldsymbol{d}}^{\perp}\}} \frac{\left(\|\boldsymbol{P}_{\boldsymbol{Q}^c}^{\perp} \boldsymbol{Q} \boldsymbol{y}(k)\|^2 - \|\boldsymbol{Q}\|_0 + 1\right)^2}{2\left(\|\boldsymbol{Q}\|_0 - 1\right)} \implies \mathbb{E}\left[\hat{\boldsymbol{P}}_{\boldsymbol{d}}^{\perp} \boldsymbol{d}(k)\right] = \boldsymbol{0}$$

Proposition 2 - Finite-Horizon Parameter-Invariant Control: Given (1), $\alpha \in [0, 1]$, positive integers $k_0, k, \kappa \in \mathbb{N}$ satisfying

$$0 < k - k_0 \leq \kappa \leq 2(k - k_0) - 4,$$

a $(\kappa - k_0 + 1)$-point DFT matrix, $\boldsymbol{V}$, and

$$\boldsymbol{s}_{n-k_0} = \begin{cases} \mathbb{E}\left[x(n)|\boldsymbol{y}(n)\right], & n \in \{k_0, \ldots, k-1\} \\ \alpha^{n-k}\mathbb{E}\left[x(k)|\boldsymbol{y}(k)\right], & n \in \{k, \ldots, \kappa\} \end{cases}$$

$$\boldsymbol{H} = \left[\boldsymbol{\Lambda V s}, \quad \boldsymbol{V s}, \quad \textstyle\sum_{n=k_0}^{\kappa} \boldsymbol{V}_n, \quad \boldsymbol{V}_0\right]$$

then, assuming $\mathbf{u}_{k_0:k-1} \neq \mathbf{0}$,

$$\boldsymbol{P}_{\boldsymbol{H}}^{\perp} \boldsymbol{V} \mathbf{u}_{k_0:\kappa-1} = \mathbf{0} \implies \mathbb{E}\left[x(t)|\mathbf{y}_{k_0:k}\right] = \alpha^{t-k}\mathbb{E}\left[x(k)|\boldsymbol{y}(k)\right] \quad \forall k \leq t \leq \kappa$$

The following subsections discuss proposition 1 and proposition 2, respectively. The propositions discussed in this section are utilized in the following section to design a resilient model-invariant controller that satisfies the stochastic boundedness constraint in problem 1.

3.1 Resilient Sensor Fusion

Designing a stabilizing controller for problem 1 requires information feedback through sensor measurements. When corrupted by an unmodeled attacker, the sensor contains no information with respect to the system state. Moreover, inclusion of corrupted measurements in state estimation infects the healthy information provided by the uncorrupted sensors. The purpose of resilient sensor fusion is to identify a set of sensors that are expected to be unaffected by the measurement corruption vector, $\boldsymbol{d}(k)$, (i.e. reside in the null space of $\boldsymbol{d}(k)$) and to generate a minimum mean squared estimate of the state, $x(k)$. From assumption 3, and consistent with the standard assumptions in the related work [12], we assume that there are at least $M+1$ sensors which are uncorrupted. However, unlike the previous work, this problem considers uncorrupted measurements that are inherently noisy and that the specific sensors which are corrupted can change at each time-step. Coupled with the fact that the underlying dynamics are unknown (by assumption 2), the resilient sensor fusion strategy at time k is limited to using only sensor measurements provided at time k, and no prior information on the state (as prior information on the state will propagate previously corrupted information). In this subsection, we develop a resilient sensor fusion strategy that estimates the null space of the corruption, denoted as $\hat{\boldsymbol{P}}_{\boldsymbol{d}}^{\perp} \in \{\boldsymbol{P}_{\boldsymbol{d}}^{\perp}\}$, invariant to the value of the state and potential corruption, which is employed to generate a minimum mean squared error (MMSE) state estimate. Specifically, the design of the resilient sensor fusion strategy is organized into three steps:

1. formulate a test statistic;
2. test for the *best* corruption null space;
3. generate a MMSE state estimate.

These steps are respectively addressed in the remainder of this subsection.

Test Statistic: To determine the corrupted sensors requires identifying a test statistic that is invariant to the unknown system state which can be tested. Assuming a candidate corruption null space, $\boldsymbol{Q} \in \{\boldsymbol{P}_{\boldsymbol{d}}^{\perp}\}$, and consistent with optimal signal detection theory [5], we write a maximally invariant statistic for $\boldsymbol{d}(k)$, invariant to the unknown state, $x(k)$, as

$$\boldsymbol{P}_{\boldsymbol{Qc}}^{\perp}\boldsymbol{Q}\boldsymbol{y}(k) = \boldsymbol{P}_{\boldsymbol{Qc}}^{\perp}\boldsymbol{Q}\left(\boldsymbol{d}(k)+\boldsymbol{v}(k)\right). \tag{3}$$

The maximally invariant statistic is a statistic of the measurements which contains all the information with respect to $\boldsymbol{Q}\boldsymbol{d}(k)$ that is invariant to the unknown state, $x(k)$. It is known that there exists no uniformly most powerful (UMP) test for detecting an unknown vectored signal [5], where the UMP test is considered optimal in hypothesis testing theory. Following well established practices employed in bad-data detection [5], we utilize the inner product of the maximally invariant statistic in (3),

$$\|\boldsymbol{P}_{\boldsymbol{Qc}}^{\perp}\boldsymbol{Q}\boldsymbol{y}(k)\|^2 = \|\boldsymbol{P}_{\boldsymbol{Qc}}^{\perp}\boldsymbol{Q}\left(\boldsymbol{d}(k)+\boldsymbol{v}(k)\right)\|^2, \tag{4}$$

as a test statistic, in place of the maximally invariant statistic.

Corruption Null Space Test: To determine the corruption null space, we seek to identify the candidate $\boldsymbol{Q}$ for the test statistic that *best fits* the measurements, assuming $\boldsymbol{Q}\boldsymbol{d}(k) = 0$, as implemented in proposition 1, the implication of which is described in the following proof:

Proof of proposition 1. From assumption 3, we observe that the test statistic in (4) has the property $\|\boldsymbol{P}^{\perp}_{\boldsymbol{Qc}}\boldsymbol{Q}\boldsymbol{d}(k)\|^2 = \boldsymbol{0}, \;\; \exists \boldsymbol{Q}^{\top}\boldsymbol{Q} \in \{\boldsymbol{P}^{\perp}_{\boldsymbol{d}}\}$. It is expected that the above property is best satisfied, by a candidate $\boldsymbol{Q}$, when the goodness-of-fit of the test statistic with respect to its corresponding distribution, assuming $\boldsymbol{Q}\boldsymbol{d}(k) = \boldsymbol{0}$,

$$\frac{\left(\|\boldsymbol{P}^{\perp}_{\boldsymbol{Qc}}\boldsymbol{Q}\boldsymbol{y}(k)\|^2 - \mathbb{E}\left[\|\boldsymbol{P}^{\perp}_{\boldsymbol{Qc}}\boldsymbol{Q}\boldsymbol{y}(k)\|^2 \middle| \boldsymbol{Q}\boldsymbol{d}(k) = \boldsymbol{0}\right]\right)^2}{\mathrm{Cov}\left[\|\boldsymbol{P}^{\perp}_{\boldsymbol{Qc}}\boldsymbol{Q}\boldsymbol{y}(k)\|^2 \middle| \boldsymbol{Q}\boldsymbol{d}(k) = \boldsymbol{0}\right]}$$

is minimized. We conclude the proof by observing that when $\boldsymbol{Q}\boldsymbol{d}(k) = \boldsymbol{0}$, it is true that $\|\boldsymbol{P}^{\perp}_{\boldsymbol{Qc}}\boldsymbol{Q}\boldsymbol{y}(k)\|^2$ has a central chi-square distribution with central moments:

$$\begin{aligned}
\mathbb{E}\left[\boldsymbol{P}^{\perp}_{\boldsymbol{Qc}}\boldsymbol{Q}\boldsymbol{y}(k) | \boldsymbol{Q}\boldsymbol{d}(k) = 0\right] &= \|\boldsymbol{Q}\|_0 - 1 \\
\mathrm{Cov}\left[\boldsymbol{P}^{\perp}_{\boldsymbol{Qc}}\boldsymbol{Q}\boldsymbol{y}(k) | \boldsymbol{Q}\boldsymbol{d}(k) = 0\right] &= 2(\|\boldsymbol{Q}\|_0 - 1).
\end{aligned}$$

□

State Estimation: Given the expected null space of the corruption, $\hat{\boldsymbol{P}}^{\perp}_{\boldsymbol{d}}$, from proposition 1, the minimum mean-squared error estimate of the state is

$$\mathbb{E}\left[x(k) | \boldsymbol{P}^{\perp}_{\boldsymbol{Qc}}\boldsymbol{Q}\boldsymbol{y}(k)\right] = \frac{\boldsymbol{c}^{\top}\boldsymbol{Q}^{\top}\boldsymbol{Q}\boldsymbol{y}(k)}{\boldsymbol{c}^{\top}\boldsymbol{Q}^{\top}\boldsymbol{Q}\boldsymbol{c}}$$

where $\boldsymbol{Q}^{T}\boldsymbol{Q} = \hat{\boldsymbol{P}}^{\perp}_{\boldsymbol{d}}$.

3.2 Finite-Horizon Parameter-Invariant Control

The parameter-invariant controller utilizes a time-history of the state estimates and control inputs to design a finite-horizon controller that stabilizes the mean and is invariant to the unknown system dynamics.

Proof of proposition 2. Defining $\hat{x}(k) = \mathbb{E}\left[x(k) | \boldsymbol{y}(k)\right]$, we consider the following time-series generated by (1),

$$\begin{bmatrix} \hat{x}(k_0) \\ \hat{x}(k_0+1) \\ \vdots \\ \hat{x}(k) \\ \alpha\hat{x}(k) \\ \vdots \\ \alpha^{\kappa-k}\hat{x}(\kappa) \end{bmatrix} = \begin{bmatrix} 1 & & & & & \\ -a & 1 & & & & \\ & \ddots & \ddots & & & \\ & & -a & 1 & & \\ & & & -a & 1 & \\ & & & & \ddots & \ddots \\ & & & & & -a \; 1 \end{bmatrix}^{-1} \begin{bmatrix} \hat{x}(k_0) \\ \mu + bu(k_0) \\ \vdots \\ \mu + bu(k-1) \\ \mu + bu(k) \\ \vdots \\ \mu + bu(\kappa-1) \end{bmatrix}$$

$$\Longleftrightarrow \begin{bmatrix} \hat{x}(k_0) \\ \hat{x}(k_0+1) \\ \vdots \\ \hat{x}(k) \\ \alpha\hat{x}(k) \\ \vdots \\ \alpha^{\kappa-k}\hat{x}(\kappa) \end{bmatrix} = \begin{bmatrix} 1 & & & & & -a \\ -a & 1 & & & & \\ & \ddots & \ddots & & & \\ & & -a & 1 & & \\ & & & -a & 1 & \\ & & & & \ddots & \ddots \\ & & & & -a & 1 \end{bmatrix}^{-1} \begin{bmatrix} \hat{x}(k_0) - a\alpha^{\kappa-k}\hat{x}(k) \\ \mu + bu(k_0) \\ \vdots \\ \mu + bu(k-1) \\ \mu + bu(k) \\ \vdots \\ \mu + bu(\kappa-1) \end{bmatrix}$$

$$\Longleftrightarrow \boldsymbol{s} = \bar{\boldsymbol{V}}^\top (\boldsymbol{I} + a\boldsymbol{\Lambda})^{-1} \boldsymbol{V} \begin{bmatrix} \hat{x}(k_0) - a\alpha^{\kappa-k}\hat{x}(k) \\ b\mathbf{u}_{k_0:\kappa-1} + \mu\mathbf{1} \end{bmatrix}$$

$$\Longleftrightarrow \boldsymbol{V}\boldsymbol{s} = (\boldsymbol{I} + a\boldsymbol{\Lambda})^{-1} \boldsymbol{V} \begin{bmatrix} \hat{x}(k_0) - a\alpha^{\kappa-k}\hat{x}(k) \\ b\mathbf{u}_{k_0:\kappa-1} + \mu\mathbf{1} \end{bmatrix}$$

$$\Longleftrightarrow \mathbf{0} = \boldsymbol{H}\boldsymbol{\theta} + \boldsymbol{V}_{1:\kappa-k_0}\mathbf{u}_{k_0:\kappa-1}$$

$$\Longleftrightarrow \boldsymbol{P}_{\boldsymbol{H}}^{\perp}\boldsymbol{V}_{1:\kappa-k_0}\mathbf{u}_{k_0:\kappa-1} = \mathbf{0} \qquad \text{(by Nyquist-Shannon Theorem)}$$

where

$$\boldsymbol{\theta} = \left[-\tfrac{a}{b},\ -\tfrac{1}{b},\ \tfrac{\mu}{b},\ \tfrac{x(k_0)-ax(\kappa)-\mu}{b} \right]^\top .$$

The first implication transforms the time-series signal into an equivalent $(\kappa - k_0+1)$-step periodic signal by augmenting the initial condition. By the Nyquist-Shannon Theorem, the future control inputs, $\mathbf{u}_{k:\kappa-1}$, can be reconstructed from the previous measurements, a desired state trajectory, and the past control inputs, $\mathbf{u}_{k_0:k-1}$ if $\kappa - k_0 < 2(k-k_0) - \|\Theta\|_0$. Since this criteria is satisfied in the worst-case by assuming all parameters are non-zero, the control sequence that generates the desired future state trajectory is reconstructed by solving

$$\boldsymbol{P}_{\boldsymbol{H}}^{\perp}\boldsymbol{V}_{1:\kappa-k_0}\mathbf{u}_{k_0:\kappa-1} = \mathbf{0}$$

assuming $\kappa - k_0 < 2(k-k_0) - 4$. □

Recalling that the controller is design to stabilize the mean through an additive control input, the predicted covariance of the state is given in the following corollary:

Corollary 1 - Covariance of Predicted State: Assuming $\boldsymbol{Q}^\top \boldsymbol{Q} \in \left\{\hat{\boldsymbol{P}}_{\boldsymbol{d}}^\perp\right\}$, then

$$\mathrm{Cov}\left[x(t)|\boldsymbol{P}_{\boldsymbol{Q}\mathbf{c}}^\perp \boldsymbol{Q}\boldsymbol{y}(k)\right] = \mathrm{Cov}\left[x(k)|\boldsymbol{P}_{\boldsymbol{Q}\mathbf{c}}^\perp \boldsymbol{Q}\boldsymbol{y}(k)\right] + a^{t-k-1}\sigma, \ \ \forall t \geq k$$

Satisfying the equality constraint in proposition 2 restricts the horizon for which the controller can be designed. As a best estimate of the control sequence which satisfies proposition 2, we employ a maximum likelihood estimate of the future control sequence as

$$\mathbf{u}_{k:\kappa-1} = -\boldsymbol{G}\mathbf{u}_{k_0:k-1} \tag{5}$$

where

$$\boldsymbol{G} = \left(\bar{\boldsymbol{V}}_{k-k_0+1:\kappa}^\top \boldsymbol{P}_{\boldsymbol{H}}^\perp \boldsymbol{V}_{k-k_0+1:\kappa}\right)^{-1} \bar{\boldsymbol{V}}_{k-k_0+1:\kappa}^\top \boldsymbol{P}_{\boldsymbol{H}}^\perp \boldsymbol{V}_{1:k-k_0} \tag{6}$$

The resilient sensor fusion strategy presented in this section identifies a maximum likelihood estimate of the corruption null space, consistent with commonly adopted goodness-of-fit approaches. The resulting estimator is employed to design an parameter-invariant control sequence which stabilizes the mean of the estimate at a convergence rate of α.

4 Resilient Parameter-Invariant Controller Design

From the previous section, it is clear that the performance of the resilient state estimator affects the controller. Since it is a primary concern to secure the measurements against malicious attacks, and a secondary concern to maximize the performance with respect to the environmental unknowns, we introduce the following proposition to design the parameter-invariant controller convergence rate, α:

Proposition 3 - Controller Design : Given problem 1, $\hat{\boldsymbol{P}}_{\boldsymbol{d}}^\perp$ from proposition 1, then a control sequence $\mathbf{u}$ in proposition 2 designed assuming α will satisfy (2) if and only if

$$0 \leq \alpha \leq \sqrt{\frac{\alpha^\star - \frac{\sigma + \frac{1}{\mathbf{c}^\top \hat{\boldsymbol{P}}_{\boldsymbol{d}}^\perp \mathbf{c}}}{\eta}}{\frac{1}{\eta \mathbf{c}^\top \hat{\boldsymbol{P}}_{\boldsymbol{d}}^\perp \mathbf{c}} + 1}}$$

Proof . We define $\boldsymbol{l} = \hat{\boldsymbol{P}}_{\boldsymbol{d}}^{\perp}\boldsymbol{c}(\boldsymbol{c}^{\top}\hat{\boldsymbol{P}}_{\boldsymbol{d}}^{\perp}\boldsymbol{c})^{-1}$, and write

$$\alpha \leq \sqrt{\frac{\alpha^{\star} - \frac{\sigma + \boldsymbol{l}^{\top}\boldsymbol{l}}{\eta}}{\frac{1}{\eta}\boldsymbol{l}^{\top}\boldsymbol{l} + 1}} \quad \Longleftrightarrow \quad \alpha^2\left(\frac{\boldsymbol{l}^{\top}\boldsymbol{l}}{\eta} + 1\right) + \frac{\sigma + \boldsymbol{l}^{\top}\boldsymbol{l}}{\eta} \leq \alpha^{\star}$$

$$\Longleftrightarrow \alpha^2\left(\boldsymbol{l}^{\top}\boldsymbol{l} + \|x(k)\|^2\right) + \left(\boldsymbol{l}^{\top}\boldsymbol{l} + \sigma\right) \leq \alpha^{\star}\|x(k)\|^2, \quad \forall \ \|x(k)\|^2 \geq \eta$$

$$\Longleftrightarrow \mathbb{E}\left[\|\alpha\mathbb{E}\left[x(k)|\boldsymbol{y}(k)\right]\|^2 + \mathrm{Cov}\left[x(k+1)|\boldsymbol{y}(k)\right]\right] \leq \alpha^{\star}\|x(k)\|^2, \quad \forall \ \|x(k)\|^2 \geq \eta$$

$$\Longleftrightarrow \mathbb{E}\left[\|\mathbb{E}\left[x(k+1)|\boldsymbol{y}(k)\right]\|^2 + \mathrm{Cov}\left[x(k+1)|\boldsymbol{y}(k)\right]\right] \leq \alpha^{\star}\|x(k)\|^2, \quad \forall \ \|x(k)\|^2 \geq \eta$$

$$\Longleftrightarrow \mathbb{E}\left[\|x(k+1)\|^2\right] \leq \alpha^{\star}\|x(k)\|^2, \quad \forall \ \|x(k)\|^2 \geq \eta$$

□

A direct consequence of proposition (3) is the following corollary identifying when a resilient parameter-invariant controller exists that satisfies the performance criteria in (2).

Corollary 2 - Controller Existence : A resilient parameter-invariant controller satisfying (2) exists if and only if

$$\frac{1}{\boldsymbol{c}^{\top}\hat{\boldsymbol{P}}_{\boldsymbol{d}}^{\perp}\boldsymbol{c}} \leq \alpha^{\star}\eta - \sigma$$

By applying Markov's inequality to proposition 3, a probabilistic bound on the likelihood the state diverges is provided in the following corollary:

Corollary 3 - Probability of Divergence: Assuming proposition 3,

$$\Pr\left[\|x(k+1)\|^2 \geq \|x(k)\|^2\right] \leq \alpha^{\star}, \quad \forall \ \|x(k)\|^2 \geq \eta \quad \text{(Markov's inequality)}$$

The resilient parameter-invariant controller formulated in this section is evaluated through simulation in the following section.

5 Simulation Results

This section provides a qualitative evaluation of the resilient parameter-invariant controller. This evaluation is presented in two subsections. The following subsection presents disturbance rejection results considering first order systems, both unstable and stable, when sensor corruption is both present and absent. The final subsection presents a simulated cruise control scenario for robotic platforms.

5.1 Disturbance Rejection

In this subsection, we evaluate the resilient parameter-invariant controller with respect to disturbance rejection. We consider the following global variables for all simulations in this subsection

- $b = 1$ (input gain);
- $C = [2, 3, 4, 5, 10]^\top$ (measurement gain);
- $x(0) = 5$ (initial condition);
- $\mu = -1$ (process noise mean);
- $\sigma = 0.01$ (process noise covariance);
- $\alpha^\star = 0.1$, $\eta = 1$ (stochastic boundedness parameters);

Additionally, we use a windowed approach for the controller design, where we use the last 20 measurements (19 previous inputs), to design a 10-step finite horizon control sequence at each time step. Assuming this system and controller design strategy, we evaluate the resilient controller when the system is stable ($a = 0.8$) and unstable ($a = 1.8$) when all the sensors are uncorrupted, and when the first and last sensors are corrupted by a random attacker using a zero-mean Gaussian attack with variance 10. The remainder of this subsection addresses these four scenarios: (a) stable, no corruption; (b) stable, with corruption; (c) unstable, no corruption; (d) unstable, with corruption.

All figures in this section (i.e. Fig. 1 - 4), have the same subplot structure, as demonstrated by Fig. 1. The upper most subplot denotes the state trajectory (i.e. the value of $x(k)$ during the entire simulation). The second subplot denotes the two-norm of the state trajectory, $\|x\|$, as a solid line, the crosses at each time step identify the two-norm of the state-estimate, and the dashed line denotes the stochastic boundedness threshold, η. The third subplot identifies the control input, u, at each time step. The fourth subplot identifies whether a sensor (1 to 5) was selected at a specific time step (denoted by a cross) or identified as corrupted at each time step (no cross). The final subplot illustrates the convergence rate, α, chosen by the parameter-invariant controller at each time step such that the desired convergence rate $\alpha^\star$ is achieved.

Fig. 1 presents the results for a stable system with no sensor attacks. In Fig. 1, we simulate a state disturbance injection (beyond the process noise) occurring every 30 time steps. It is clear that the controller is capable of stabilizing the system, and achieves the performance bound on the norm of the state; however, we note that despite no attacks being present in this simulation it is assumed that some of the sensor measurements are corrupted at most time steps. This results from the fact that unlike classical state-estimators, which minimize the mean-squared error of the state estimate, the resilient sensor fusion minimizes a normalized deviation of the measured sensor noise from a weighted average of the sensor measurements. This results in the rejection of measurements which are significantly different from the mean of the sensor measurements. As a

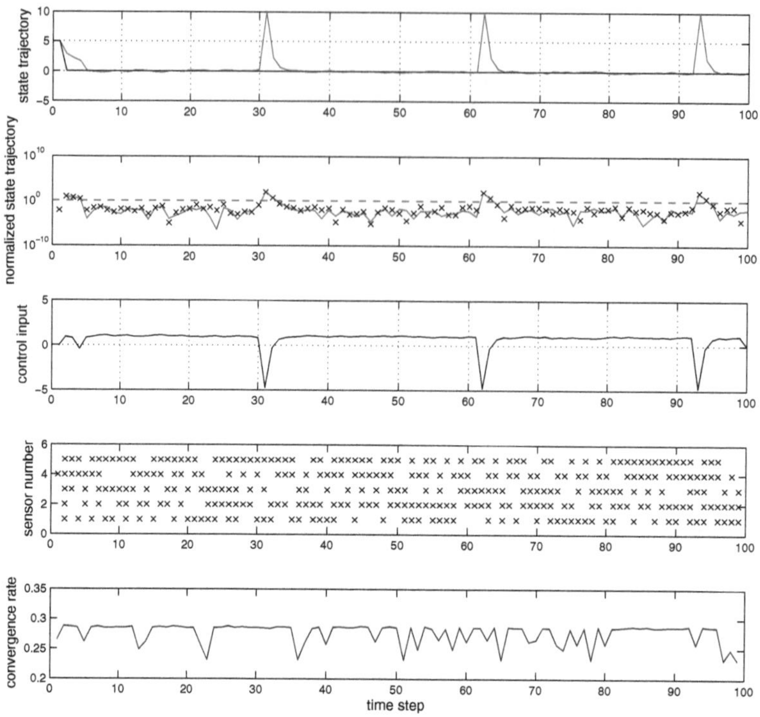

Fig. 1. Stable system ($a = 0.8$) with no sensor attacks

consequence, depending on the specific subset of measurements accepted as secure, the controller convergence rate changes to maintain the performance specification.

Simulated results for a stable system in the presence of sensor attacks is presented in Fig. 2. We consider the same state disturbance injection as in Fig. 1, and observe a very similar performance in terms of disturbance rejection. However, this comes at the cost of an increased variance in the controller sequence (as compared to Fig. 1). The increased variance is undesirable in most physical actuators since this results in increased strain and fatigue on mechanical parts. The increased variance is a direct result of designing the controller sequence to match the performance specification, and can be reduced by upper bounding the desired state convergence rate, α, employed by the parameter-invariant controller. Through a comparison between Fig. 1 and Fig. 2 we notice that sensor one is selected significantly less when under attack and sensor five is selected only marginally less, despite being corrupted by the same attack. This is expected since given the same attack, signal-to-corruption ratio is greater in sensor five than in sensor one. Recalling that the measurement model in (1) is normalized such that all the sensors have the same noise profile, sensor five having a higher

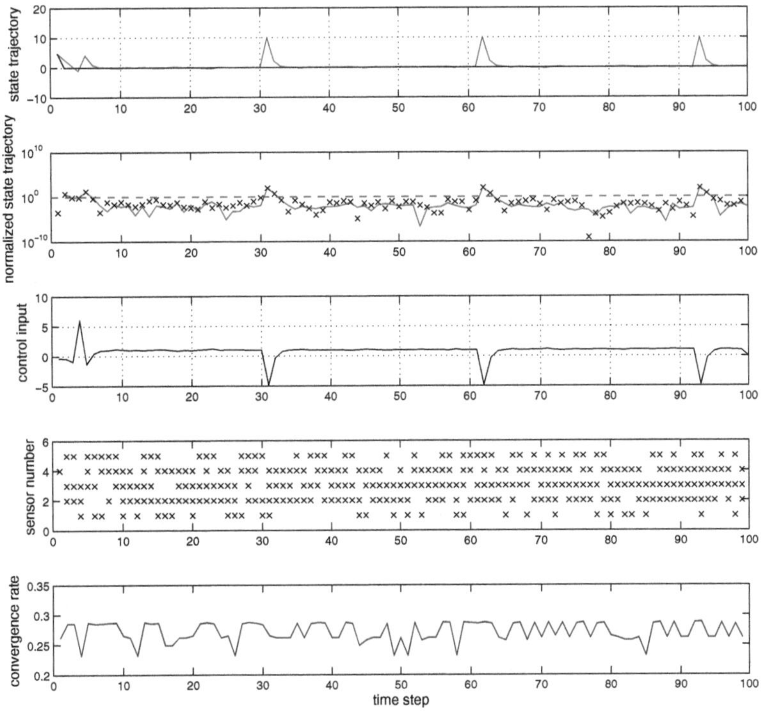

Fig. 2. Stable system ($a = 0.8$) with sensor 1 and 5 under attack

signal-to-corruption ratio (assume the same attack) than sensor one is equivalent to stating that since sensor five is less noisy than sensor one, sensor five is more likely to be trusted.

A unstable system with no sensor attacks is considered in Fig. 3. Similar to the results for the stable system, here we observe that resilient parameter-invariant controller stabilizes the system with respect to unknown disturbances, and has a response very similar to the stable system results in Fig. 1. Although the control input is different, the performance is nearly identical, which is a direct result of the controller design requirement to be invariant to the unknown system parameters, regardless of stability. Similarly, we notice a similar sensor selection and rejection profile as in Fig. 1. This result is consistent with the fact that the sensor fusion strategy is design invariant to the unknown state, which orthogonalizes the issues of stability and resilience for the purposes of identifying corrupted sensors.

The final figure in this subsection, Fig. 4, illustrates the results when considering an unstable system and with sensor attacks. Consistent with the results

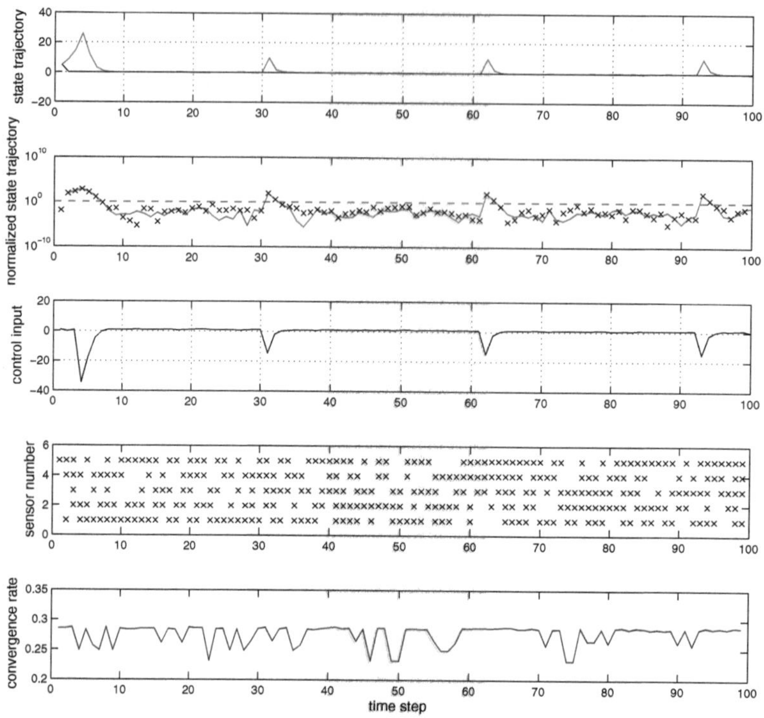

Fig. 3. Unstable system ($a = 1.8$) with no sensor attacks

when no sensor attacks are present, we observe nearly identical results when sensors are attacked, regardless of the underlying system stability. This further illustrates that the specification based-design of the resilient parameter-invariant controller can be achieved regardless of the underlying system parameters. We note, however, a multi-step design specification will vary with the system stability since the multi-step predicted state covariance is a non-linear function of the process noise and state dynamic gain, a. For this reason, it is necessary that the resilient parameter invariant controller be designed at each time step in order to satisfy the performance specification invariant to the unknown state dynamics.

5.2 Resilient Cruise Control

To evaluate the resilient parameter-invariant controller as a potential cruise control mechanism, we consider a simplified first-order model of a mobile robot as

$$\begin{aligned} x(k+1) &= 0.95x(k) + u(k) + w(k) \\ y(k) &= [2, 3, 4]^\top x(k) + v(k) \end{aligned} \tag{7}$$

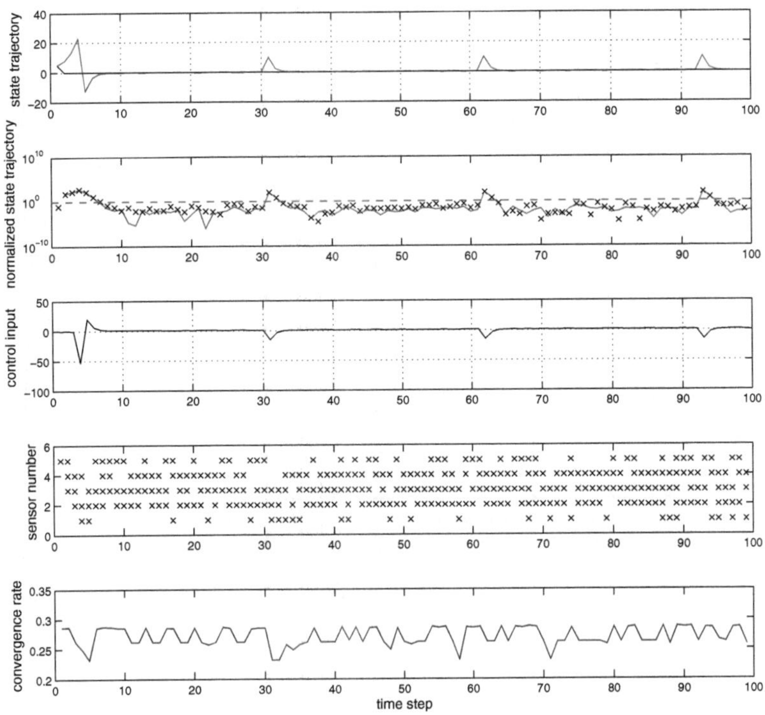

Fig. 4. Unstable system ($a = 1.8$) with sensor 1 and 5 under attack

where $w(k)$ $[-1, .01]$, and we assume $\alpha^{\star} = .1$, and $\eta = 1$ for the performance constraint. The parameter choices in this simulation are motivated by the fact that the velocity of surface robots is accurately modeled as a stable first-order system when set-point control is performed at a rate an order of magnitude greater than the electro-mechanical system time-constant.

In this simulation, we assume the initial vehicle speed is 5 km/h ($x(0) = 5$) and at time step 50 the speed is desired to increase to 15 km/h. Additionally, we assume the control input is bounded on the interval of -1 to 5. The result of employing the resilient parameter-invariant controller for cruise control is provided in Fig. 5 without sensor attacks. We observe in Fig. 5 that the vehicle speed is stabilized with a lag in the state trajectory. The lag is a result of the fact that

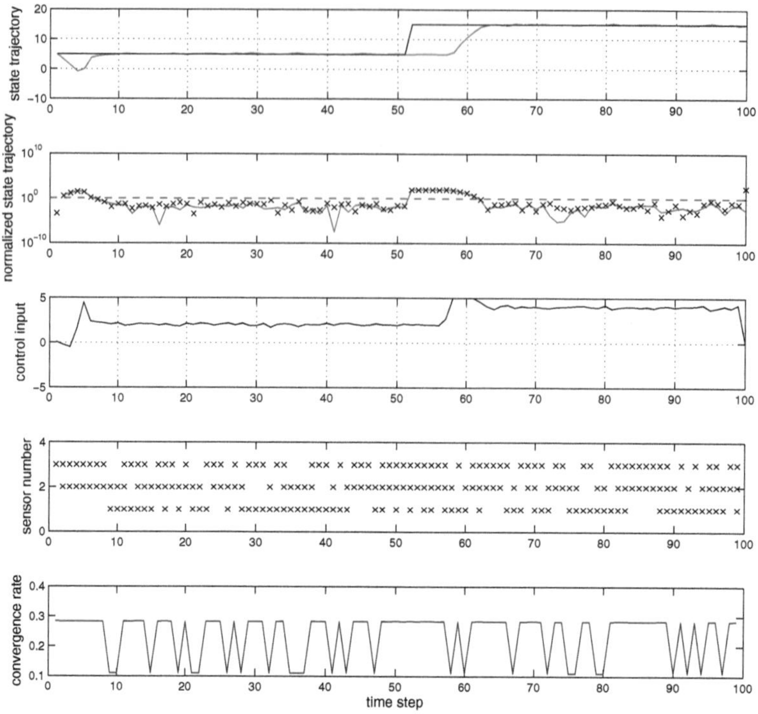

Fig. 5. Landshark cruise control with no corruption

a history of measurements is required to generate a control sequence. It remains a focus of future research to reduce this lag. A portion of the state trajectory lag is due to the saturation of the control signal, which does not affect the stability of the system despite the resilient control law being designed without a constraint on the control sequence.

The cruise controller performance in the presence of a sensor attack on sensor three is illustrated in Fig. 6. Here we observe, and consistent with previous observations, that when attacked, the measurement of sensor three is accepted less often. In this case, sensor one (known to be more noisy) is selected more often. This results in the need for, on a average, a more aggressive control law to meet the performance criteria.

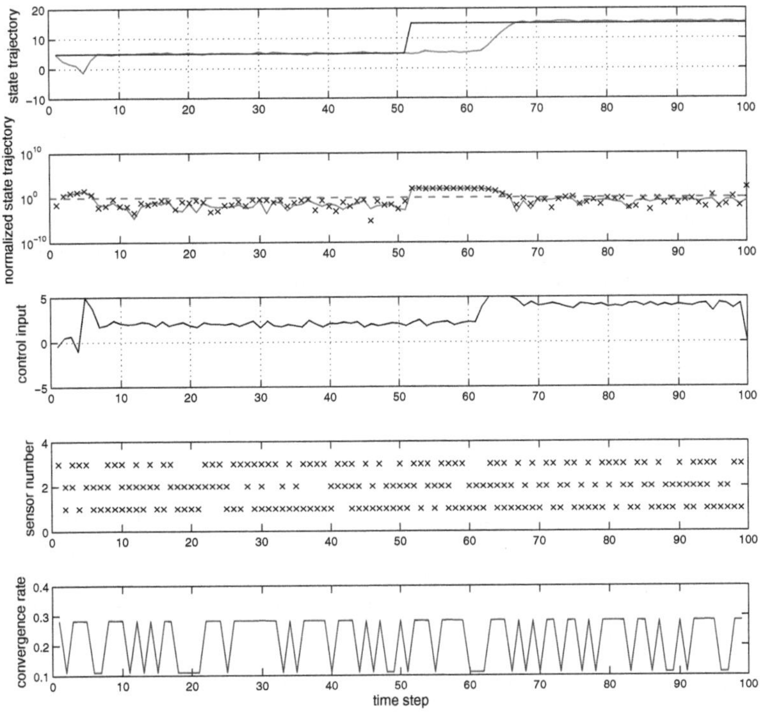

Fig. 6. Landshark cruise control with corruption

6 Discussion and Future Work

This work addresses the problem of resilient control of unknown first-order stochastic LTI systems in the presence of sensor attacks. A resilient sensor fusion strategy is introduced that minimizes the likelihood of a corrupted sensor being trusted. A time-series concatenation of secure estimates is utilized in cooperation with a desired state trajectory and previous control inputs to design a mean-stabilizing finite-horizon control sequence, invariant to the unknown system parameters. The parameter-invariant controller is designed using a frequency domain representation of an equivalent time-series representation of the system inputs to system outputs, thus allowing the design of a mean-stabilizing controller. Simulated results illustrate that the resilient parameter-invariant controller is capable of stabilizing unknown state disturbances and can perform state trajectory tracking.

Future work on this topic includes the extension of the scalar results in this paper to multi-dimensional systems with known dynamical structures but unknown parameters. A limiting factor in the application of the resilient parameter-invariant controller is the requirement (at this time) that the system-to-be-controlled behave as a linear first order system. While most stable systems can (at some time scale) be approximated by a first order system, this approximation can be very poor. As future work, we plan to evaluate the proposed control strategy on various experimental platforms to determine the robustness of the parameter-invariant control with respect to the design dependency on a first-order linear system. Additionally, further insight is needed to investigate a method to reduce (or remove) the lag and to quantify its behavior in terms of the underlying system and design parameters. Experimentation of the resilient parameter-invariant controller is planned on a robotic platform as a potentially cooperative approach with model-based approaches which assume knowledge of model parameters.

References

1. Willsky, A.: A survey of design methods for failure detection in dynamic systems. Automatica 12, 601–611 (1976)
2. Zhou, K., Doyle, J.C.: Essentials of robust control. Prentice-Hall, Englewood Cliffs (1998) ISBN: 0-13-790874-1; Qiu, L.: Automatica 38(5), 910–912 (2002)
3. Astrom, K.J., Wittenmark, B.: Adaptive Control, 2nd edn. Addison-Wesley Longman Publishing Co., Inc., Boston (1994)
4. Trees, H.L.V.: Detection, Estimation, and Modulation Theory. John Wiley & Sons, Inc., New York (1968)
5. Scharf, L.L.: Statistical Signal Processing, Detection, Estimation, and Time Series Analysis. Addison-Welsley Publishing Company Inc., Reading (1991)
6. Schenato, L., Sinopoli, B., Franceschetti, M., Poolla, K., Sastry, S.: Foundations of control and estimation over lossy networks. Proceedings of the IEEE 95(1), 163–187 (2007)
7. Gupta, A., Langbort, C., Basar, T.: Optimal control in the presence of an intelligent jammer with limited actions. In: 2010 49th IEEE Conference on Decision and Control (CDC), pp. 1096–1101 (December 2010)
8. Pasqualetti, F., Dörfler, F., Bullo, F.: Attack detection and identification in cyber-physical systems – part ii: Centralized and distributed monitor design. Technical Report arXiv:1202.6049 (February 2012)
9. Sundaram, S., Pajic, M., Hadjicostis, C.N., Mangharam, R., Pappas, G.J.: The wireless control network: Monitoring for malicious behavior. In: CDC, pp. 5979–5984 (2010)
10. Weimer, J., Kar, S., Johansson, K.H.: Distributed detection and isolation of topology attacks in power networks. In: Proceedings of the 1st International Conference on High Confidence Networked Systems, HiCoNS 2012, pp. 65–72. ACM, New York (2012)
11. Weimer, J., Ahmadi, S.A., Araujo, J., Mele, F.M., Papale, D., Shames, I., Sandberg, H., Johansson, K.H.: Active actuator fault detection and diagnostics in hvac systems. In: 4th ACM Workshop on Embedded Systems For Energy-Efficiency in Buildings (BuildSys), Toronto, Canada (2012)
12. Fawzi, H., Tabuada, P., Diggavi, S.N.: Secure estimation and control for cyber-physical systems under adversarial attacks. CoRR abs/1205.5073 (2012)

13. Pintelon, R., Guillaume, P., Rolain, Y., Schoukens, J.: Parametric identification of transfer functions in the frequency domain-a survey. IEEE Transactions on Automatic Control 39(11), 2245–2260 (1994)
14. Ljung, L. (ed.): System identification: theory for the user, 2nd edn. Prentice Hall PTR, Upper Saddle River (1999)
15. Van Den Hof, P.: Identification and control - closed-loop issues. Automatica 31(12), 1751–1770 (1995)

A Appendix: Notation and Definitions

This appendix introduces notation and definitions that prove useful in this work.

A.1 Notation

In this subsection, we illustrate the various variable notations using varying fonts and capitalization of the letter z:

- plain upper case italic fonts → constant, Z;
- plain lower case italic fonts → scalar (or function with scalar range), z;
- bold lower case italic fonts → vector (or function with vectoral range), $\boldsymbol{z}$;
- bold lower case plain fonts → vector of concatenated vectors, $\mathbf{z}$;
- bold upper case italic fonts → matrix, $\boldsymbol{Z}$;

For vectors we write $\boldsymbol{z}_i$ to denote the i-th position of $\boldsymbol{z}$ and $\boldsymbol{z}_{i:j}$ to be the sub-vector of $\boldsymbol{z}$ consisting of the i-th through j-th elements, inclusively. Similarly, For vectors of vectors we write $\mathbf{z}_i$ to denote the i-th sub-vector and $\mathbf{z}_{i:j}$ to be the sub-vector of $\mathbf{z}$ consisting of the i-th through j-th sub-vectors, inclusively. Lastly, for matrices we write $\boldsymbol{Z}_i$ to be the i-th column of $\boldsymbol{Z}$ and $\boldsymbol{Z}_{i:j}$ to be the sub-matrix consisting of the i-th through j-th columns, inclusively.

We use the notation $\Pr[x|y]$ and $\mathbb{E}[x|y]$ to denote the probability of x given y and the expected value of x given y, respectively,

A.2 Definitions

This subsection defines matrices and constants which appear in the remainder of this work. Consistent with the previous subsection's use of the letter z to illustrate various properties, we write $\boldsymbol{P}_{\boldsymbol{Z}}$ to be the projection matrix corresponding the general matrix $\boldsymbol{Z}$ and $\boldsymbol{P}_{\boldsymbol{Z}}^{\perp}$ to be the projection matrix corresponding to the null-space of $\boldsymbol{Z}$,

$$\boldsymbol{P}_{\boldsymbol{Z}} = \boldsymbol{Z}\left(\bar{\boldsymbol{Z}}^{\top}\boldsymbol{Z}\right)^{-1}\bar{\boldsymbol{Z}}^{\top} \quad \text{and} \quad \boldsymbol{P}_{\boldsymbol{Z}}^{\perp} = \boldsymbol{I} - \boldsymbol{P}_{\boldsymbol{Z}},$$

where $\bar{\boldsymbol{Z}}^{\top}$ is notation denoting the transpose of the complex conjugate of $\boldsymbol{Z}$ and $\boldsymbol{I}$ the identity matrix.

For an arbitrary positive integer $N \in \mathbb{N}^+$, $\omega = \frac{2\pi\sqrt{-1}}{N}$, $\lambda(n) = \mathbf{e}^{\omega n}$, we define the matrices $\boldsymbol{V}, \boldsymbol{\Lambda} \in \mathbb{C}^{N \times N}$ as

$$\boldsymbol{\Lambda} = \text{diag}[\lambda(0), \ldots, \lambda(N-1)]$$
$$\boldsymbol{V}_n = \frac{1}{\sqrt{N}} \left[1, \quad (\lambda(n))^1, \quad \ldots, \quad (\lambda(n))^{N-1}, \right]^\top$$

and note that $\boldsymbol{V}$ is the normalized N-point Discrete Fourier Transform (DFT) matrix, such that

$$\bar{\boldsymbol{V}}^\top \boldsymbol{V} = \boldsymbol{V} \bar{\boldsymbol{V}}^\top = \boldsymbol{I}$$

Part V

Application Domains: Robotics and Power Systems

Towards the Unification of Locomotion and Manipulation through Control Lyapunov Functions and Quadratic Programs

Aaron D. Ames and Matthew Powell*

Mechanical Engineering
Texas A&M University
College Station TX, 77840
{aames,mjpowell}@tamu.edu

Abstract. This paper presents the first steps toward unifying locomotion controllers and algorithms with whole-body control and manipulation. A theoretical framework for this unification will be given based upon quadratic programs utilizing control Lyapunov functions. In particular, we will first consider output based feedback linearization strategies for locomotion together with whole-body control methods for manipulation. We will show that these two traditionally disjoint methods are equivalent through the correct choice of controller. We will then present a method for unifying these two methodologies through the use of control Lyapunov functions presented in the form of a quadratic program. In addition, it will be shown that these controllers can be combined with force-based control to achieve locomotion and force-based manipulation in a single framework. Finally, simulation results will be presented demonstrating the validity of the proposed framework.

Keywords: Robotics, cyber-physical systems, bipedal locomotion, manipulation, control Lyapunov functions, quadratic programs.

1 Introduction

Robots provide a quintessential example of cyber-physical systems (CPSs); the ability of robots, and especially humanoid robots, to perform complex dynamic tasks requires a complete and unified understanding of dynamics, control, software and hardware, along with their interconnection and integration. In the context of the control of robotic CPSs, a variety of traditionally disparate approaches have been taken ranging from nonlinear control via input/output (IO) feedback linearization through output based control [12,34], force-based control methods [7,17,30,21,31], and whole-body control methods for manipulation [14,15,27], to name a few. Each of these methods, along with the wide variety of other methods taken in the control of robotic systems, have proven success in their domains of consideration. Yet combining these different approaches into a single unified and implementable framework remains an open problem. With

* This research is supported by NASA Grants NNX12AQ68G and NNX11AN06H.

D.C. Tarraf (ed.), *Control of Cyber-Physical Systems*,
Lecture Notes in Control and Information Sciences 449,
DOI: 10.1007/978-3-319-01159-2_12,

a view toward unification, this paper describes how locomotion controllers, as described through human-inspired control methods which utilize IO feedback linearization, can be unified with whole-body and multi-contact force control into a single coherent framework given in the form of quadratic programs utilizing control Lyapunov functions. These theoretical developments will be supported by simulation results.

We begin by reviewing the basic background necessary to introduce these results. In particular, the robotic models considered are defined, and a review of human-inspired control [1,2,3,4,22] is given. In the context of human-inspired control, we obtain outputs, or virtual constraints, that represent the desired behavior of a bipedal robot. We show how, through input/output (IO) linearization, the dynamics of these outputs can transformed into a linear system. The end result of this process is that the dynamics of the outputs can be chosen to achieve convergence of these outputs to zero at a desired rate. This implies the convergence of the robotic system to the surface defined by the zero level set of these output functions: the *zero dynamics surface*. By ensuring that this surface is invariant through impact, i.e., that we have *hybrid zero dynamics (HZD)* [12,29,34], the end result of these control methods is provably stable locomotion.

The first result of this paper is obtained by considering whole-body control methodologies, and specifically null-space control [14,15,27]. In the context of the outputs associated with locomotion, we demonstrate how Jacobians can be constructed in the case of mixed position and velocity based outputs. In addition, by projecting the dynamics of the system down to the operational space dynamics, we are able to show that the end result is dynamics for the outputs that can be utilized to achieve a linear relationship between the inputs and outputs. In more concrete terms, we establish that IO linearization and null-space control result in equivalent constructions in terms of outputs. Building on this idea, we then consider the case when manipulation tasks have been specified. A procedure for merging the locomotion and manipulation tasks is given in the context of null-space control. This method benefits from explicitly separating the locomotion tasks and the manipulation tasks so that the manipulation tasks do not affect the locomotion tasks. This is an advantage due to the fact that the manipulation tasks can therefore never destabilize the robot. Conversely, it can be a disadvantage since it does not allow for dynamic balancing between the locomotion and manipulation tasks. For example, one may want to slightly relax the tracking of the outputs associated with locomotion—as long as it does not destabilize the system—in order to better achieve a given manipulation task.

The need for a dynamic way to balance multiple tasks related to locomotion, manipulation and force control motivates the introduction of quadratic programs (QPs) that allow for this dynamic balancing coupled with the ability to add constraints on the evolution of the system, e.g., torque bounds. We begin by again considering only the locomotion task, and show how the IO representation achieved through output functions can be used to explicitly construct a control Lyapunov function (CLF) [10,28]. Importantly, this CLF results in an inequality that is linear in torque such that, when it is satisfied, convergence to the hybrid zero dynamics surface is guaranteed [5,6]. This naturally leads to the formulation of a quadratic program (QP) in terms of torque with a constraint given by the CLF associated with locomotion [11]. The strength of this representation is that additional constraints can be added to the controller such as torque bounds, moment bounds, etc., and by solving the QP the controller will naturally find the best balance between these hard physical constraints and the control objective of converging to the HZD surface.

To demonstrate the extensibility of the CLF based QP controller, we then extend the formulation to include manipulation tasks in the form of both position-based tasks and force-based tasks. In these cases, additional CLFs can be added to the QP through the form of additional constraints; these CLFs represent the manipulation objectives in the system. The advantage of this representation is that the locomotion and manipulation tasks can be dynamically balanced through the QP—the QP will naturally find the optimal balance between these control objectives. In addition, we will discuss how null-space control can be expressed in this framework. Finally, we discuss force-based tasks in the presence of multi-contact. Again, we show how these can naturally be expressed in the context of QPs and CLFs.

These formal ideas and results are demonstrated through simulation results. We begin by obtaining a stable walking gait for the lower body of a simple full-body (2D) humanoid robot. As a first step, we demonstrate how this purely lower-body controller can be embedded into the full-body robot through a QP, with the end result again being stable walking. Taking this idea even further, we show how manipulation and force tasks can be accomplished without modifying the original locomotion controller. In particular, we perform the task of holding the hand at a constant height and, more significantly, show how we can hold the hand in contact with a wall with a desired force while simultaneously walking with the locomotion controllers that were designed with no knowledge of the upper body. As a final demonstration, we show the robustness of this method through unknown rough terrain locomotion with the full-body robot.

2 Background

We begin by giving some basic terminology utilized throughout the paper. Specifically, we will introduce the basic equations for a robotic system, and show how these are converted to an affine control system (see [20,34] for additional details).

Let $\mathcal{Q}$ be the configuration space of a robot with n degrees of freedom, i.e., $n = \dim(\mathcal{Q})$, with coordinates $q \in \mathcal{Q}$. For the sake of definiteness, it may be necessary to choose $\mathcal{Q}$ to be a subset of the actual configuration space of the robot so that global coordinates can be defined[1], i.e., such that $\mathcal{Q}$ is embeddable in $\mathbb{R}^n$, or more simply $\mathcal{Q} \subset \mathbb{R}^n$. Consider the equations of motion for a robot given in the general form by the Euler-Lagrange equations:

$$D(q)\ddot{q} + H(q, \dot{q}) = Bu, \tag{1}$$

[1] Note that at various points we will assume that matrix functions have full rank; it may be necessary to carefully choose $\mathcal{Q}$ to satisfy these conditions.

where D is the inertia matrix, H is a vector containing the coriolis and gravity terms, and $B \in \mathbb{R}^{n \times n}$ is the actuation matrix which determines the way in which the torque inputs, $u \in \mathbb{R}^n$, actuate the system. Note that here, for the sake of simplicity, we assume full actuation (and hence a square actuation matrix, B).

In the context of control constructions, it is desirable to convert this system to an ODE of the form:

$$\dot{x} = f(x) + g(x)u,$$

where $x = (q, \dot{q}) \in TQ \subset \mathbb{R}^{2n}$ and

$$f(q, \dot{q}) = \begin{bmatrix} \dot{q} \\ -D^{-1}(q)H(q, \dot{q}) \end{bmatrix}, \quad g(q, \dot{q}) = \begin{bmatrix} 0 \\ D^{-1}(q)B \end{bmatrix}, \tag{2}$$

where $0 \in \mathbb{R}^{n \times n}$ is a matrix of zeros.

When modeling the bipedal robot, discrete behavior must also be considered in conjunction with these continuous dynamics. In particular, the robotic system exhibits discrete impacts when guards are reached, i.e., when contacts with the world are created or broken [13]. The end result is that the system is a hybrid system:

$$\mathscr{HC} = (\mathcal{D}, \mathcal{U}, \mathcal{S}, \Delta, (f, g)), \tag{3}$$

where $\mathcal{D}$ is the domain of the continuous dynamics, i.e., $\mathcal{D} \subset TQ \subset \mathbb{R}^{2n}$, $\mathcal{U} \subset \mathbb{R}^n$ is the set of admissible control values, $\mathcal{S}$ is the guard which determines when a discrete change in the dynamics occurs, Δ determines the discrete change in dynamics, and (f, g) is the affine control system dictating the continuous dynamics. In particular, this implies that when $(q^-, \dot{q}^-) \in \mathcal{S}$, there exists an impact of the form: $(q^-, \dot{q}^-) \mapsto (q^+, \dot{q}^+) = \Delta(q^-, \dot{q}^-)$.

3 Human-Inspired Control for Locomotion

In the context of mobility, the control algorithms utilized in this paper build off the framework of human-inspired control. While detailed algorithms can be found in [1,2,3,4,23], we will outline these methods only as they relate to the constructions presented in this paper. We note that human-inspired control builds upon the concept of human-inspired constraints that, when enforced with the proper choice of parameters, provably guarantee robotic walking. These methods are applicable both in the case of under and full actuation, and have been applied to both 2D and 3D robots to achieve walking experimentally [22,35].

Consider a *human output combination*: $Y^H = (\mathcal{Q}, y_1^H, y_2^H)$, consisting of the configuration space of a robot, $\mathcal{Q} \subset \mathbb{R}^n$, a velocity modulating output $y_1^H : \mathcal{Q} \to \mathbb{R}$, position modulating outputs $y_2^H : \mathcal{Q} \to \mathbb{R}^{n_\ell - 1}$ given by $y_2^H(q) = [y_2^H(q)_i]_{i \in O}$ with O an indexing set for y_2^H, and n_ℓ the total number of position and velocity modulation outputs. Human-inspired outputs consist of (vector) relative 1 and 2 degree output functions $y_1 : T\mathcal{Q} \to \mathbb{R}$ and $y_2 : \mathcal{Q} \to \mathbb{R}^{n_\ell - 1}$ of the form:

$$y_1(q, \dot{q}) = \frac{\partial y_1^H(q)}{\partial q}\dot{q} - v, \tag{4}$$

$$y_2(q) = y_2^H(q) - [y_{\mathrm{CWF}}(\tau(q), \alpha_i)]_{i \in O}, \tag{5}$$

where y_1 is the velocity-based output, e.g., based upon the center of mass or forward position of the hip, $v \in \mathbb{R}$ the desired velocity, and y_2 are the position based outputs, e.g., a vector of outputs including things like the angle of the stance knee, etc. In this case, $y_2^H(q)$ is the actual value of these outputs as computed from the robot, and $[y_{\mathrm{CWF}}(\tau(q), \alpha_i)]_{i \in O}$ is the desired value of these outputs as computed from the canonical walking function:

$$y_{CWF}(t, \alpha_i) = e^{-\alpha_{i,4} t}(\alpha_{i,1} \cos(\alpha_{i,2} t) + \alpha_{i,3} \sin(\alpha_{i,2} t)) + \alpha_{i,5},$$

which is simply the time-solution to a linear mass-spring damper system. Thus, we drive the relative degree 2 outputs to the behavior of a compliant system, mirroring methods in compliant-based control. Moreover, τ is a parametrization of time based upon the relative 1 degree output:

$$\tau(q) = \frac{y_1^H(q) - y_1^H(q^+)}{v}, \tag{6}$$

with $y_1^H(q^+)$ the initial value of the velocity modulating output, e.g., at the beginning of a step. It follows from these constructions that the control parameters are $v \in \mathbb{R}$ and $\alpha \in \mathbb{R}^{n_\ell - 1 \times 5}$. To provide a specific example, in the case of a 5-link 2D walking robot, $n_\ell = 5$. Therefore, there are a total of 21 control parameters; while this may appear to be a large number, the parameters are automatically determined by a human-inspired optimization and so no tuning of parameters is needed.

With the objective of driving $y_1 \to 0$ and $y_2 \to 0$, we create a linear input/output (IO) relationship [26]. More formally, the goal is to drive the dynamics of the system to the *zero dynamics surface*:

$$\mathbf{Z}_{v,\alpha} = \{(q, \dot{q}) \in TQ : y_1(q, \dot{q}) = 0,\ \ y_2(q) = 0,\ \ \dot{y}_2(q, \dot{q}) = 0\},$$

which depends on the parameters v and α. With this goal in mind, differentiating the relative degree 1 output once and differentiating the relative degree 2 output twice yields:

$$\begin{bmatrix} \dot{y}_1 \\ \ddot{y}_2 \end{bmatrix} = \underbrace{\begin{bmatrix} L_f y_1(q, \dot{q}) \\ L_f^2 y_2(q, \dot{q}) \end{bmatrix}}_{L_f} + \underbrace{\begin{bmatrix} L_g y_1(q, \dot{q}) \\ L_g L_f y_2(q, \dot{q}) \end{bmatrix}}_{A} u, \tag{7}$$

with L denoting the Lie derivative, and A the decoupling matrix [26]. Note that the dependence of L_f and A on q and $\dot{q}$ has been suppressed for notational simplicity. Traditionally, in nonlinear control, this matrix is assumed to be nonsingular, in which case, we can pick:

$$u = A^{-1}(-L_f + \mu) \tag{8}$$

for some $\mu \in \mathbb{R}^{n_\ell}$ resulting in

$$\begin{bmatrix} \dot{y}_1 \\ \ddot{y}_2 \end{bmatrix} = \mu. \tag{9}$$

Therefore, one obtains a linear relationship between input and output. As a result, one can drive the system to the surface $\mathbf{Z}_{v,\alpha}$ by, for example, choosing

$$\mu = \begin{bmatrix} -\epsilon y_1 \\ -2\epsilon \dot{y}_2 - \epsilon^2 y_2 \end{bmatrix} \tag{10}$$

for $\epsilon > 0$; here, as $\epsilon \to \infty$, the speed of convergence to $\mathbf{Z}_{v,\alpha}$ increases. In particular, it yields the output dynamics:

$$\begin{bmatrix} \dot{y}_1 \\ \dot{y}_2 \\ \ddot{y}_2 \end{bmatrix} = \underbrace{\begin{bmatrix} -\epsilon & 0 & 0 \\ 0 & 0 & I \\ 0 & -\epsilon^2 I & -\epsilon 2I \end{bmatrix}}_{F_{\mathrm{cl}}} \begin{bmatrix} y_1 \\ y_2 \\ \dot{y}_2 \end{bmatrix}, \tag{11}$$

with F_{cl} chosen so that all of the eigenvalues are located at $-\epsilon$. Note that while this choice of μ guarantees convergence, it does not achieve convergence in an optimal fashion; this motivates later constructions related to CLFs.

It is important to note that, in the context of locomotion, the impacts Δ in the hybrid model (3) that occur at foot strike must be considered when designing the control parameters v and α. In particular, the impact map Δ can throw the system away from the surface $\mathbf{Z}_{v,\alpha}$, resulting in the system being destabilized. With this consideration in mind, since we wish to allow the velocity-based relative degree 1 outputs to "jump" during impact, we consider the partial zero dynamics surface:

$$\mathbf{PZ}_{v,\alpha} = \{(q,\dot{q}) \in TQ : y_2(q) = 0,\ \ \dot{y}_2(q,\dot{q}) = 0\}.$$

This surface is termed hybrid invariant, or a hybrid zero dynamics surface, if $\Delta(\mathbf{PZ}_{v,\alpha} \cap S) \subset \mathbf{PZ}_{v,\alpha}$—that is, if the surface is invariant through impacts in the system. Creating this invariance is the basis for human-inspired optimization:

$$\begin{aligned} (v^*, \alpha^*) = \underset{(v,\alpha)\in\mathbb{R}\times\mathbb{R}^{n_\ell - 1\times 5}}{\operatorname{argmin}} \quad & \mathrm{Cost}(v,\alpha) \\ \text{s.t.} \quad & \Delta(\mathbf{PZ}_{v,\alpha} \cap S) \subset \mathbf{PZ}_{v,\alpha} \end{aligned} \tag{PHZD}$$

which automatically generates parameters v^*,α^* that ensure invariance of this surface while minimizing a cost, $\mathrm{Cost}(v,\alpha)$, often chosen based upon human data [1]. Detailed methods for constructing this optimization problem are given in [3], but the main result of human-inspired control is that it automatically generates stable periodic walking gaits.

4 Merging Locomotion and Whole-Body Control

In the context of unifying the mobility controllers with other manipulation tasks, it is possible to associate Jacobians and null-space projections [14,15,27] to the locomotion tasks. This allows for the development of a controller that will prioritize locomotion and achieve the requested manipulation tasks with the remaining degrees of freedom of the system. Before developing this idea, it is necessary to note that the constructions based upon IO linearization can be reframed in an equivalent manner in the context of null-space control.

4.1 Equivalence between IO and Null-Space Control for Mobility

Consider the outputs y_1 and y_2 constructed in the previous section (see (4) and (5)). These can be viewed as two separate tasks: one for velocity regulation, and one that drives the remaining degrees of freedom to the system to the partial hybrid zero dynamics surface $\mathbf{PZ}_{v,\alpha}$.

Define the Jacobians associated with velocity and position modulating tasks, $y_1(q,\dot{q})$ and $y_2(q)$, as:

$$J_1(q,\dot{q}) := \frac{\partial y_1(q,\dot{q})}{\partial \dot{q}}, \quad J_2(q) = \frac{\partial y_2(q)}{\partial q}.$$

Differentiating the relative 1 degree task, y_1, once and the relative 2 degree task, y_2, twice yields:

$$\begin{bmatrix} \dot{y}_1 \\ \ddot{y}_2 \end{bmatrix} = \underbrace{\begin{bmatrix} J_1(q,\dot{q}) \\ J_2(q) \end{bmatrix}}_{J_\ell} \ddot{q} + \underbrace{\begin{bmatrix} \frac{\partial y_1(q,\dot{q})}{\partial q} \\ \dot{J}_2(q,\dot{q}) \end{bmatrix}}_{\dot{J}_\ell} \dot{q}. \tag{12}$$

For simplicity of notation, we will now suppress the dependence of matrices on q and $\dot{q}$.

Through these constructions, we have a Jacobian $J_\ell \in \mathbb{R}^{n_\ell \times n}$ associated with the locomotion task. This allows for the construction of a pseudo-inverse for the locomotion task given by:

$$\bar{J}_\ell = D^{-1} J_\ell^T [J_\ell D^{-1} J_\ell^T]^{-1}.$$

This is a specific example of a right pseudo-inverse for J_ℓ and therefore satisfies: $J_\ell \bar{J}_\ell = I$. In addition, the null-space projection, N_ℓ, associated with these Jacobians is given by:

$$N_\ell^T = [I - J_\ell^T \bar{J}_\ell^T].$$

The Jacobian used to describe the locomotion task can be used to project down to the locomotion dynamics, even in the case when the full-order robot dynamics are of a much larger dimension, i.e., in the case when locomotion is described only for the lower-body, while the robot consists of both a lower and upper body. In particular, the locomotion dynamics are given by choosing:

$$u = J_\ell^T u_\ell$$

for some $u_\ell \in \mathbb{R}^{n_\ell}$, wherein it follows that, by applying the left pseudo-inverse for J_ℓ^T, given by $\bar{J}_\ell^T$, one obtains:

$$\bar{J}_\ell^T (D(q)\ddot{q} + H(q,\dot{q})) = \bar{J}_\ell^T B J_\ell^T u_\ell,$$

where, as long as the locomotion task is consistent with the actuation in the system, $\bar{J}_\ell^T B J_\ell^T$ is invertible; for example, in the case of full actuation and in the proper coordinates $B = I$ and so $\bar{J}_\ell^T B J_\ell^T = I$. Defining

$$D_\ell = (J_\ell D^{-1} J_\ell^T)^{-1}, \quad H_\ell = \bar{J}_\ell^T H - D_\ell \dot{J}_\ell \dot{q}, \quad B_\ell = \bar{J}_\ell^T B J_\ell^T, \tag{13}$$

yields

$$D_\ell \begin{bmatrix} \dot{y}_1 \\ \ddot{y}_2 \end{bmatrix} + H_\ell = B_\ell u_\ell$$

which yields dynamics for the locomotion outputs. Picking

$$u_\ell = B_\ell^{-1}(D_\ell \mu_\ell + H_\ell) \tag{14}$$

results in

$$\begin{bmatrix} \dot{y}_1 \\ \ddot{y}_2 \end{bmatrix} = \mu_\ell. \tag{15}$$

exactly as in the case of IO linearization, i.e., we arrive at exactly the same form as given in (9). Therefore, one can pick μ_ℓ exactly as in (10) to achieve the same convergence objectives. More formally, we have established the following result:

Theorem 1. *For a robotic system with dynamics expressed as* (1) *and* (2) *with outputs of the form* (4),(5)*, the control laws for IO linearization* (8) *and null-space control* (14)*:*

$$u = A^{-1}(-L_f + \mu)$$
$$u = B_\ell^{-1}(D_\ell \mu + H_\ell)$$

yield equivalent output dynamics of the form:

$$\begin{bmatrix} \dot{y}_1 \\ \ddot{y}_2 \end{bmatrix} = \mu. \tag{16}$$

4.2 Merging Locomotion with Manipulation

The advantage to the null-space formulation, as opposed to the IO approach, is that, although the two are equivalent as established by Theorem 1, in the null-space representation there is a well-defined null-space that can be utilized to ensure that any tasks performed on the upper body do not destabilize locomotion and/or balance. Let a manipulation task, or collection of manipulation tasks, be represented by a set of outputs $y_m(q) \in \mathbb{R}^{n_m}$. Defining the Jacobian for these tasks by J_m we obtain (as in the case of locomotion):

$$\ddot{y}_m = \dot{J}_m \dot{q} + J_m \ddot{q}$$

and the left pseudo-inverse for J_m^T, given by $\bar{J}_m^T$, can be again constructed in this case. This allows us to define the corresponding control law:

$$u = J_\ell^T u_\ell + N_\ell^T J_m^T u_m,$$

which guarantees that manipulation tasks do not interfere with locomotion and balance. Using, $\bar{J}_m^T$, we can again project down to the dynamics for manipulation expressed as:

$$D_m \ddot{y}_m + H_m = B_{m,\ell} u_\ell + B_{m,m} u_m,$$

where D_m and H_m are defined as in (13) with the subscripts changed from ℓ to m and

$$B_{m,m} = \bar{J}_m^T B N_\ell^T J_m^T, \quad B_{m,\ell} = \bar{J}_m^T B J_\ell^T,$$

with the locomotion controllers affecting the manipulation controllers via $B_{m,\ell}$.

Begin by assuming that the manipulation tasks are consistent with the locomotion tasks; formally, this is characterized by $\text{Null}(N_\ell^T J_m^T) = \emptyset$. In this case, $B_{m,m}$ is nonsingular and, as in the case of locomotion, we can again shape the manipulation dynamics to be any desired dynamics. To see this we can pick

$$u_m = B_{m,m}^{-1}(D_m \mu_m + H_m - B_{m,\ell} u_\ell)$$

yielding:

$$\ddot{y}_m = \mu_m.$$

If, for example, the goal was to drive $y_m \to 0$, one need only pick

$$\mu_m = -2\epsilon \dot{y}_m - \epsilon^2 y_m. \tag{17}$$

5 Implementation through Quadratic Programs

A method for achieving convergence for both the locomotion tasks and manipulation tasks was presented in the previous sections. Specifically, μ_ℓ and μ_m were chosen in (10) and (17) such that $y_\ell = (y_1, y_2) \to 0$ and $y_m \to 0$ exponentially. Yet this specific choice is in no way optimal for achieving convergence since it forces the outputs to evolve according to pre-specified dynamics that may not be consistent with the natural dynamics of the system. Therefore, we present a method for achieving convergence through control Lyapunov functions that give the desired convergence without explicitly choosing output dynamics. This not only gives optimal convergence (with respect to controllers of minimum norm), but proves much more robust to disturbances. In addition, this control methodology can be converted to a quadratic program wherein additional physical constraints can be added to the controller construction. Additional details on the mathematics behind these constructions can be found in [5,6,11].

It is important to note that there are numerous QP-based formulations of feedback control laws. A prime example is model predictive control (MPC) [8,9,19,33], but other methods include LQR-trees [32], whole-body control methods that enforce constraints through LQPs and QPs [25,24], and QPs for dynamic balancing [30,31], to name only a few. The fundamental differentiator between existing methods and the proposed method is that control objectives are represented by inequality constraints through CLFs, allowing them to be dynamically balanced with each other and with physical constraints.

5.1 Quadratic Programs for Locomotion

The IO feedback controller results in dynamics of the form given in (16). Therefore, if we define the vector $\eta = (y_1, y_2, \dot{y}_2) \in \mathbb{R}^{2n_\ell - 1}$, (16) can be equivalently written as a linear control system:

$$\dot{\eta} = \underbrace{\begin{bmatrix} 0 & 0 \\ 0 & I \\ 0 & 0 \end{bmatrix}}_{F} \eta + \underbrace{\begin{bmatrix} 1 & 0 \\ 0 & 0 \\ 0 & I \end{bmatrix}}_{G} \mu. \tag{18}$$

In the context of this control system, we can consider the continuous time algebraic Riccati equations (CARE):

$$F^T P + PF - PGG^T P + Q = 0 \tag{19}$$

for $Q = Q^T > 0$ with solution $P = P^T > 0$. One can use P to construct a *exponentially stabilizing control Lyapunov function (ES-CLF)* that can be used to stabilize the output dynamics (18) exponentially [6]. It is important to note that if we wish to exponentially stabilize at a rate $\epsilon > 0$, we would instead construct a *rapidly* exponentially stabilizing control Lyapunov function (RES-CLF) as detailed in [6]; this can be easily achieved utilizing P, but we forgo the construction in this paper for simplicity of exposition—all presented results apply equally to ES-CLFs and RES-CLFs.

Defining $V(\eta) = \eta^T P \eta$, it is easy to verify that this is a ES-CLF. In particular, it follows that

$$\dot{V}(\eta) = L_f V(\eta) + L_g V(\eta)\mu$$

with

$$\begin{aligned} L_f V(\eta) &= \eta^T (F^T P + PF)\eta, \\ L_g V(\eta) &= 2\eta^T PG. \end{aligned}$$

The goal is exponentially stabilize η to zero. In other words, we wish to find μ such that:

$$L_f V(\eta) + L_g V(\eta)\mu \leq -\gamma V(\eta)$$

for some $\gamma > 0$. In addition to simply satisfying this inequality, we could search for μ that does this in an optimal fashion:

$$m(\eta) = \operatorname{argmin}\{\|\mu\| : \psi_0(\eta) + \psi_1^T(\eta)\mu \leq 0\} \tag{20}$$

where

$$\begin{aligned} \psi_0(\eta) &= L_f V(\eta) + \gamma V(\eta), \\ \psi_1(\eta) &= L_g V(\eta)^T. \end{aligned} \tag{21}$$

The controller $m(\eta)$ that minimizes the control effort required to achieve exponential convergence is termed the *min-norm* controller [10], and can be stated in close form as:

$$m(\eta) = \begin{cases} -\frac{\psi_0(\eta)\psi_1(\eta)}{\psi_1(\eta)^T \psi_1(\eta)} & \text{if } \psi_0(\eta) > 0 \\ 0 & \text{if } \psi_0(\eta) \leq 0 \end{cases}$$

While the min-norm controller, m, can be computed in closed form, it is important to note that this closed form solution is the solution to the quadratic program (QP):

$$
\begin{aligned}
m(q,\dot{q}) = \underset{\mu \in \mathbb{R}^{n_\ell}}{\operatorname{argmin}} \quad & \mu^T \mu && (22)\\
\text{s.t.} \quad & \psi_0(q,\dot{q}) + \psi_1^T(q,\dot{q})\mu \leq 0 && \text{(CLF)}
\end{aligned}
$$

where the the optimization problem is now expressed in terms of $(q,\dot{q})$ since η is a function of $(q,\dot{q})$. The end result of solving this QP is the control law for locomotion:

$$u(q,\dot{q}) = A^{-1}(q,\dot{q})(-L_f(q,\dot{q}) + m(q,\dot{q})).$$

There are numerous advantages to this formulation of the problem, some of which will be developed throughout the rest of this section. Yet the most immediate advantage, as first discovered in [11], is that torque bounds can be directly implemented in this formulation where, as opposed to thresholding, the optimal control value that respects the torque bounds can be found. This is achieved by relaxing the constraints (CLF) and penalizing for this relaxation. In particular we consider the locomotion quadratic program (where we now suppress the dependence of functions on $(q,\dot{q})$) first formulated in [11]:

$$
\begin{aligned}
\underset{(\delta,\mu)\in\mathbb{R}^{n_\ell+1}}{\operatorname{argmin}} \quad & p\delta^2 + \mu^T\mu && \text{(L-QP)}\\
\text{s.t.} \quad & \psi_0 + \psi_1^T \mu \leq \delta && \text{(CLF)}\\
& A^{-1}(-L_f + \mu) \leq u_{\max} && \text{(Max Torque)}\\
& -A^{-1}(-L_f + \mu) \leq u_{\max} && \text{(Min Torque)}
\end{aligned}
$$

where $p > 0$ is a large value that penalizes violations of the CLF constraint, and $u_{\max}$ are maximum torque values (in vector form).

5.2 Quadratic Programs for Locomotion and Manipulation

The method for obtaining control laws through CLFs, and specifically QPs, can be easily extended to include mobility tasks, even in the case of manipulation tasks that are not necessarily consistent with the mobility tasks. Before developing this, it is necessary to discuss how mobility tasks are unified with locomotion in the context of IO control.

Given the set of outputs associated with manipulation, y_m, we can differentiate these outputs twice and combine with (7) to obtain:

$$
\begin{bmatrix} \dot{y}_1 \\ \ddot{y}_2 \\ \ddot{y}_m \end{bmatrix} = \underbrace{\begin{bmatrix} L_f y_1(q,\dot{q}) \\ L_f^2 y_2(q,\dot{q}) \\ L_f^2 y_m(q,\dot{q}) \end{bmatrix}}_{L_f} + \underbrace{\begin{bmatrix} L_g y_1(q,\dot{q}) \\ L_g L_f y_2(q,\dot{q}) \\ L_g L_f y_m(q,\dot{q}) \end{bmatrix}}_{A} u. \qquad (23)
$$

Assuming that A is invertible, we can again utilize the controller $u = A^{-1}(-L_f + \mu)$ to obtain equations of the form:

$$\dot{\eta}_\ell = F_\ell \eta_\ell + G_\ell \mu_\ell,$$
$$\dot{\eta}_m = F_m \eta_m + G_m \mu_m,$$

where $\eta_\ell = (y_1, y_2, \dot{y}_2) \in \mathbb{R}^{2n_\ell - 1}$, $\eta_m = (y_m, \dot{y}_m) \in \mathbb{R}^{2n_m}$, and $\mu = (\mu_\ell, \mu_m) \in \mathbb{R}^{n_\ell + n_m}$. For each of these linear control systems, we can construct control Lyapunov functions $V_\ell(\eta_\ell) = \eta_\ell^T P_\ell \eta_\ell$ and $V_m(\eta_m) = \eta_m^T P_m \eta_m$ by solving the CARE (19) for (F_ℓ, G_ℓ) and (F_m, G_m), respectively. This results in the CLF inequalities:

$$\psi_0^\ell + (\psi_1^\ell)^T \mu_\ell \leq 0$$
$$\psi_0^m + (\psi_1^m)^T \mu_m \leq 0$$

defined as in (20). The top inequality ensures exponential convergence of the locomotion outputs (tasks), and the bottom inequality ensures exponential convergence of the manipulation outputs (tasks).

Since the manipulation tasks may not be consistent with the locomotion tasks, A may not necessarily be invertible. Because it may not be possible to calculate u from μ as in (8), we can convert the QPs (22) and (L-QP) into QPs that are functions of u and do not require A to be inverted. By noting that $Au = -L_f + \mu$ it follows that:

$$\mu^T \mu = u^T A^T A u + 2 L_f^T A u + L_f^T L_f.$$

Therefore, the quadratic program combining locomotion and manipulation is expressed as:

$$\underset{(\delta_\ell, \delta_m, u) \in \mathbb{R}^{n+2}}{\operatorname{argmin}} \quad p_\ell \delta_\ell^2 + p_m \delta_m^2 + u^T A^T A u + 2 L_f^T A u \qquad \text{(L+M-QP)}$$
$$\text{s.t.} \quad \psi_0^\ell + (\psi_1^\ell)^T (A^\ell u + L_f) \leq \delta_\ell \qquad \text{(Locomotion CLF)}$$
$$\psi_0^m + (\psi_1^m)^T (A^m u + L_f) \leq \delta_m \qquad \text{(Manipulation CLF)}$$

where $p_\ell, p_m > 0$ are penalties for CLF violations and

$$A^\ell = \begin{bmatrix} L_g y_1(q, \dot{q}) \\ L_g L_f y_2(q, \dot{q}) \end{bmatrix}, \qquad A^m = \begin{bmatrix} L_g L_f y_m(q, \dot{q}) \end{bmatrix}.$$

There are numerous important advantages to this representation of a whole-body controller:

- It does not require A to be invertible. Therefore, conflicting (or inconsistent) locomotion and manipulation tasks can be defined in this formulation.
- In the case of inconsistent locomotion and manipulation tasks, there are two CLFs—one for locomotion and one for manipulation. These are each relaxed, so in the case of conflicting constraints where it is not possible to achieve both tasks at the same time, one can adjust the penalty values to prioritize one task over the other.

- Note that removing the relaxations ensures simultaneous exponential convergence of both tasks, but can result in infeasibility of the QP.
- Torque bounds can easily be added to this QP (as in (L-QP)). In this case, the constraints to (L+M-QP) become:

$$u \leq u_{max} \qquad \text{(Max Torque)}$$
$$-u \leq u_{max} \qquad \text{(Min Torque)}$$

- Finally, this representation of the controller allows for force-based tasks and multi-contact to be added to the QP as will be discussed in the next section.

5.3 Quadratic Programs with Null-Space Control

Null-space control methods can be easily subsumed into the whole-body QP presented in (L+M-QP). Using the notation of the previous section, we have:

$$u = J_\ell^T u_\ell + N_\ell^T J_m^T u_m.$$

Therefore, the equations of motion (1) can be written as

$$D(q)\ddot{q} + H(q,\dot{q}) = \underbrace{\begin{bmatrix} BJ_\ell^T & BN_\ell^T J_m^T \end{bmatrix}}_{\widehat{B}(q)} \underbrace{\begin{bmatrix} u_\ell \\ u_m \end{bmatrix}}_{\widehat{u}}.$$

In this case, one obtains new equations of motion:

$$\dot{x} = f(x) + \widehat{g}(x)\widehat{u}$$

where f is as originally defined in (2) and

$$\widehat{g}(q,\dot{q}) = \begin{bmatrix} 0 \\ D^{-1}(q)\widehat{B}(q) \end{bmatrix}. \tag{24}$$

This results in $\widehat{A}$ as obtained in (23), but with g replaced by $\widehat{g}$. The quadratic program (L+M-QP) can thus be utilized by replacing A with $\widehat{A}$.

It is important to note that this formulation has both advantages and disadvantages. The advantage is that it limits the manipulation tasks so that they only act in the null-space of the locomotion task thus preventing conflicts. This is also the disadvantage—it does not allow for dynamic weighting of the tasks as can be achieved through control Lyapunov functions in the case when null-space projections are not used.

6 Force-Based Multi-contact Tasks

Extending beyond locomotion and manipulation, it is necessary to consider controllers that are able to accomplish force-based multi-contact tasks. In the context of traditional nonlinear control methods such as IO linearization, the fact that this results in over-actuation requires the outputs to be explicitly chosen so that they do not conflict with the external forces being applied to and by the system. In addition, since the number of outputs is necessarily less than the degrees of actuation, a priori optimization is needed to distribute the torques [16]. This is non-ideal from both a control and implementation perspective. Therefore, motivated by existing methods for force-based control [17,31,7,21], we present a method for handling force-based multi-contact tasks directly through the QP based formulation presented in this paper. In particular, rather than first constraining the dynamics based upon holonomic constraints representing mult-contact, we consider the unconstrained dynamics and allow for the dynamics to be constrained in the QP—simultaneously, we enforce CLF associated with locomotion and manipulation. This allows for a holistic approach to control (for locomotion and manipulation), multi-contact and force-based tasks, all of which can be formulated in a single QP.

6.1 Contact Constraints

Consider a vector of holonomic constraints: $h(q) = 0$, with $h(q) \in \mathbb{R}^{n_c}$. Defining the Jacobian $J_h(q) = \frac{\partial h(q)}{\partial q}$, the holonomic constraints are enforced through constraint (or contact) forces $F \in \mathbb{R}^{n_c}$ which are enforced through the dynamics via:

$$D(q)\ddot{q} + H(q,\dot{q}) = Bu + J_h^T F, \tag{25}$$

where $J_h^T F$ projects the contact wrench into joint-space coordinates. Note that in this case, the dynamics describe the "unpinned" model, i.e., they are expressed in terms of generalized coordinates (unlike the previous case in which the dynamics were implicitly given in body (or joint) coordinates where the constraint forces where a priori assumed to be satisfied).

For the constraint forces, F, to be valid, they must satisfy the following equalities and inequalities:

$$D(q)\ddot{q} + H(q,\dot{q}) = Bu + J_h^T F, \tag{26}$$

$$\dot{J}_h\dot{q} + J_h\ddot{q} = 0, \tag{27}$$

$$\mathcal{A}(F) \geq 0, \tag{28}$$

where $\mathcal{A}(F) \in \mathbb{R}^{n_a}$ is a set of *admissibility* constraints on the reaction wrench [13] which ensure physical validity of the model, e.g. positive normal force and friction constraints.

Traditionally, in modeling the robotic system, $\ddot{q}$ in (26) is explicitly solved and substituted into (27), yielding:

$$\dot{J}_h\dot{q} + J_h D(q)^{-1}(Bu + J_h^T F - H(q,\dot{q})) = 0. \tag{29}$$

Rearranging terms gives an explicit expression for the constraint forces:

$$F = (J_h D(q)^{-1} J_h^T)^{-1}(J_h D(q)^{-1}(H(q,\dot{q}) - Bu) - \dot{J}_h\dot{q}) \tag{30}$$

which can be substituted back into (25) to yield the constrained dynamical system. This method ensures that (26) and (27) are satisfied for all control inputs u; however, there is no guarantee that either (28) is satisfied, i.e., it implicitly assumes that the constraint forces are valid.

6.2 Quadratic Program Formulation

With a view toward formulating contacts and force-based tasks in the QP framework presented, the constrained equations of motion (25) can be written as:

$$D(q)\ddot{q} + H(q,\dot{q}) = \underbrace{\begin{bmatrix} B \; J_h^T \end{bmatrix}}_{\overline{B}(q)} \underbrace{\begin{bmatrix} u \\ F \end{bmatrix}}_{\overline{u}}. \tag{31}$$

Noting that (31) takes the same general form of the Euler-Lagrange equations as (1), $\overline{A}$ can be calculated utilizing $\overline{g}$ obtained in a similar fashion to (24). The end result is the input/output relationship

$$\overline{A}\overline{u} = (-L_f + \mu). \tag{32}$$

This allows for the Contact Force QP formulation given by

$$\begin{aligned}
\underset{(\delta,\overline{u})\in\mathbb{R}^{n+n_c+1}}{\operatorname{argmin}} \quad & p\delta^2 + \overline{u}^T\overline{A}^T\overline{A}\overline{u} + 2L_f^T\overline{A}\overline{u} && \text{(CF-QP)}\\
\text{s.t.} \quad & \dot{J}_h\dot{q} + J_h D(q)^{-1}(\overline{B}\overline{u} - H(q,\dot{q})) = 0 && \text{(Constrained Dynamics)}\\
& \psi_0 + \psi_1^T(\overline{A}\overline{u} + L_f) \le \delta && \text{(CLF)}\\
& \mathcal{A}(F) \ge 0 && \text{(Contact Force)}\\
& F = F^d(t,q,\dot{q}) && \text{(Desired Force)}
\end{aligned}$$

The solution to this quadratic program, $\overline{u}^*$, is a set of actuator torques u^* and manipulator contact forces F^* that satisfy the constrained dynamic equations, (Constrained Dynamics), while guaranteeing that the relative degree one and two outputs converge exponentially (CLF), *and* the relative degree zero, i.e. force-based, task F^d is performed (Desired Force) in a way that that is consistent with the required contact forces (Contact Force).

There are some important points that should be made regarding this QP for force-based multi-contact control:

- Note that one can remove the constraint (Desired Force) if the only goal is to remain in contact with the environment.
- Compliant force control can be achieved through specific choices of F^d.
- Torque constraints can easily be added to the QP through the constraints:

$$\begin{aligned}
u &\le u_{max} && \text{(Max Torque)}\\
-u &\le u_{max} && \text{(Min Torque)}
\end{aligned}$$

- Additional CLFs can be added for additional manipulation tasks as in the case of (L+M-QP).

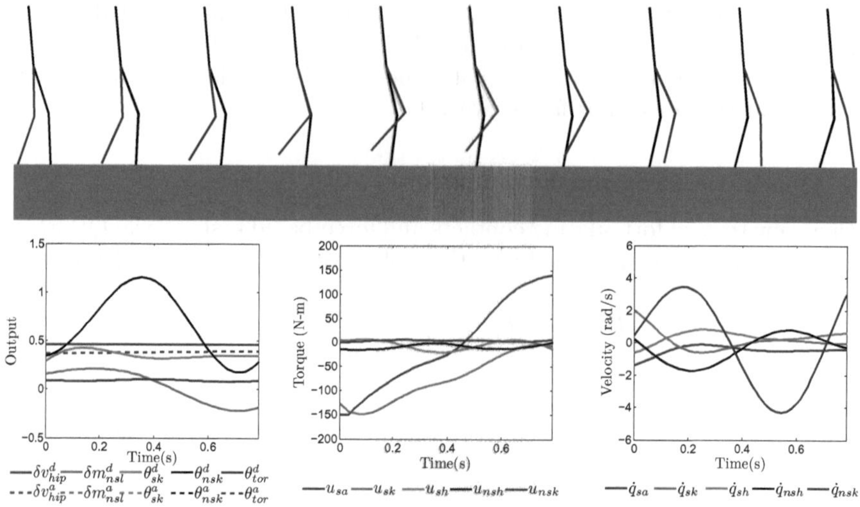

(a) Outputs (b) Torques (c) Velocities

Fig. 1. Simulation results from one step of a steady-state robotic walking gait using human-inspired control via the quadratic program (L-QP)

7 Simulation Results

To demonstrate the results presented in this paper, we will apply them through a series of simulation results of progressive complexity. In particular, we will apply the three main QPs constructed in this paper: (L-QP), (L+M-QP) and (CF-QP). We will begin with a lower-body locomotion controller implemented through (L-QP). This controller will then be embedded into a whole-body robot model (without any knowledge of this whole-body model) and coupled with a manipulation controller through (L+M-QP). The robustness of this combined controller will be tested through rough terrain locomotion, where there is no knowledge of the terrain. Finally, these preceding controllers will be combined with a force-based contact task through (CF-QP).

Lower-body locomotion: Locomotion controllers were first obtained for the lower body using human-inspired control through the methods outlined in [1,2,4]. The walking gait, and associated outputs, were then used to find a CLF through the methods given in [5]. This was then implemented through the QP (L-QP) with torque bounds of 150 Nm. The results are shown in Fig. 1. In particular, the walking gait is shown along with the actual and desired controller output profiles (a), joint torques (b) and velocities (c). Note that the relative degree two outputs are almost exactly tracked; the exception is at the beginning of the gait when it is necessary to relax the CLF condition in order to satisfy the torque bounds.

Whole-body locomotion: The locomotion controller, and associated CLF, was then embedded on the whole-body robotic model. On the upper body, outputs were chosen that keep the arms at the robot's side, i.e., in (5) the actual (position

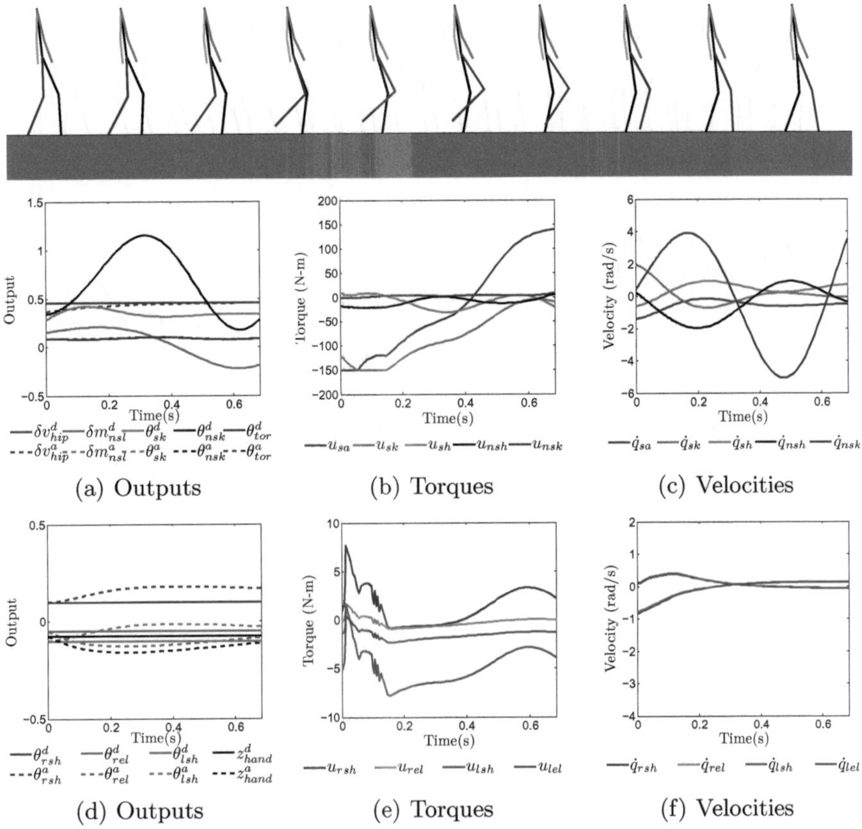

Fig. 2. Simulation results from one step of robotic walking in which a human-inspired locomotion controller is embedded in the full body of a humanoid via the quadratic program (L+M-QP)

modulating) outputs are the angles of the upper body and the desired outputs are small constant values, and the associated CLF was calculated. The controllers were then integrated through the QP (L+M-QP) in which the locomotion was given a higher priority while the manipulation task was given a low priority (through the choice of penalties p_l and p_m; that is $p_l >> p_m$). In addition, a max torque constraint of 150 Nm was enforced. The end result is that the robot walks, and the arms naturally swing to provide greater stability for the locomotion task. The results are shown in Fig. 2, including: actual and desired controller output profiles for locomotion (a), joint torques (b), and velocities (c) of the lower body; actual and desired controller output profiles for manipulation (d), joint torques (e) and velocities (f) of the upper body. In this case, slight deviations from the desired outputs are seen for the lower-body outputs while large deviations are seen for the upper-body outputs due to the prioritization of the associated CLFs.

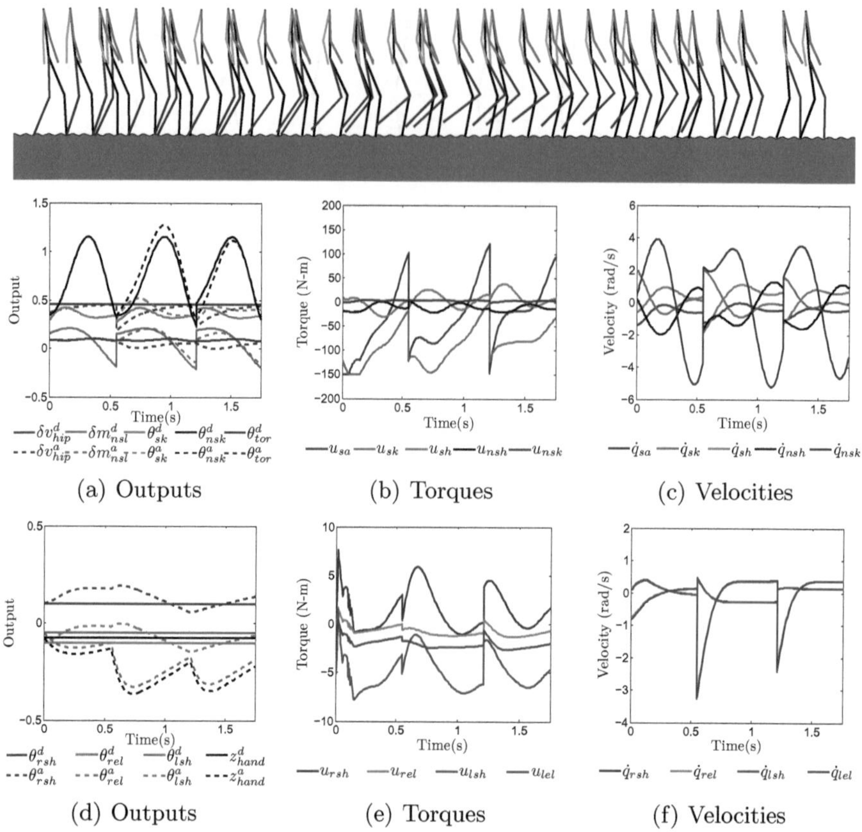

Fig. 3. Simulation results displaying robustness: three steps of robotic walking in which a human-inspired locomotion controller is embedded in the full body of a humanoid walking over—and without knowledge of—sinusoidally varying terrain (1cm peak amplitude and $\frac{2\pi}{50}$cm period)

Robustness of whole-body locomotion: To demonstrate the robustness of the control method, we consider rough terrain and solve the same QP that was considered for whole-body locomotion. In this case, the robot has no knowledge of the terrain, so the controller must dynamically compensate. The robot is able to do so by swinging its arms more (as can be seen through the drift in the upper-body outputs shown in Fig. 3(d)) to maintain stability of the locomotion task; this is all done dynamically through the QP, without the user specifying this behavior. The results of this can be seen in Fig. 3 which plots the results for three steps.

Whole-body locomotion with force-based task: As a final test of the control method, we now consider the case when we wish to locomote and perform a force-based task; in this case, we want the robot to push against the ceiling with a sinusoidal force. We achieve this by solving the QP (CF-QP) with F^d being a time-based

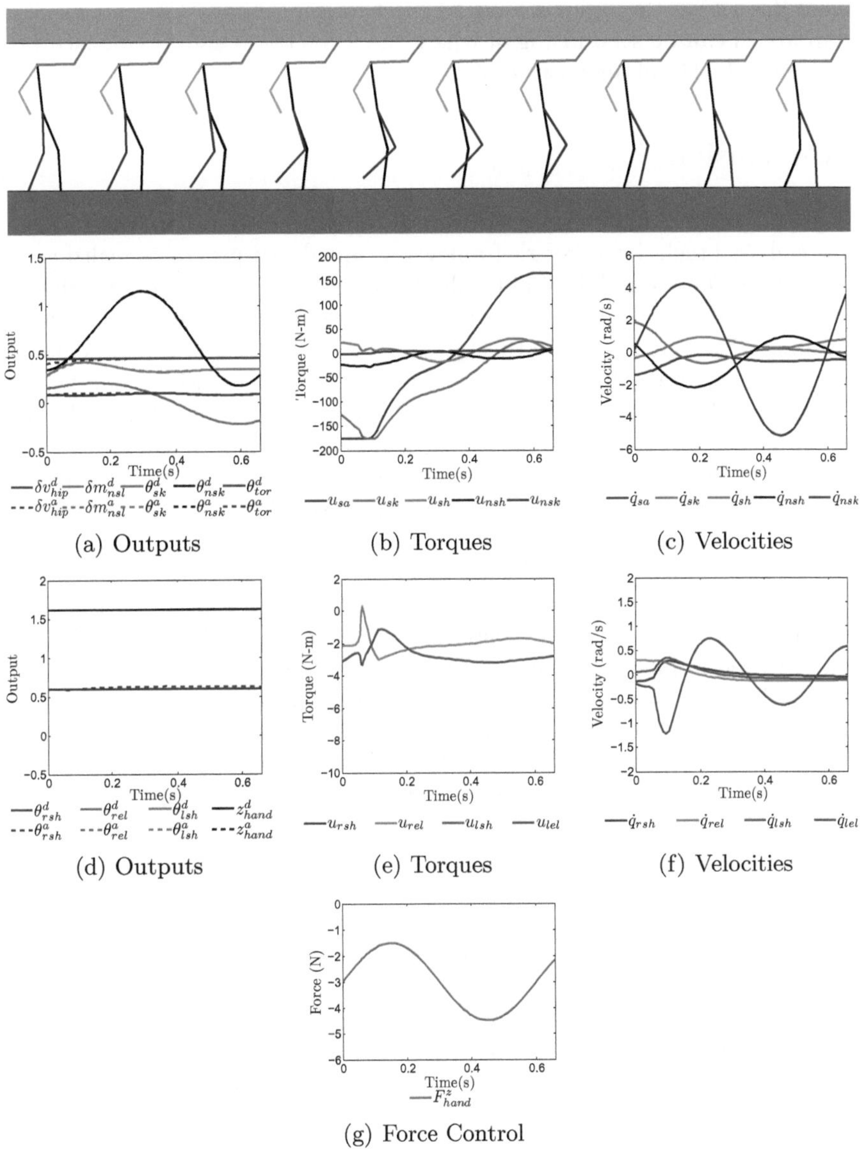

Fig. 4. Simulation results demonstrating locomotion together with force-based manipulation. The manipulation controller applies a sinusoidal force (profile shown in (g)) to the ceiling while regulating the remaining arm joints. Locomotion and manipulation controllers are combined in a single quadratic program (CF-QP) with control Lyapunov function constraints enforcing contact between the left hand and the ceiling while also specifying a desired applied force to the ceiling, strict convergence for locomotion outputs and relaxed convergence in upper body joint outputs, and constraints on the admissible joint torques.

sinusoid describing the desired force to exert on the ceiling. The end results of this approach can be seen in Fig. 4 where the locomotion and force-based tasks are shown to be simultaneously achieved.

8 Conclusion

This paper presented the first step toward unifying locomotion, manipulation and force-based tasks into a single framework—quadratic programs utilizing control Lyapunov functions. The end result was a single quadratic program that can dynamically balance all of these disparate objectives through weighted inequality constraints. The construction of this QP was motivated theoretically and demonstrated through simulation, with the end result being locomotion, manipulation and force-based control on a simplified humanoid robot. The presented results potentially have important ramifications for robotic cyber-physical systems. The QP can be implemented as a single algorithm that includes both controllers and the interaction of the robot with the physical world. This could allow for more holistic implementation of controllers on physical systems, thus permitting for a more complete understanding of their behavior and proofs of their correctness.

On both a practical and theoretic level, there are numerous areas in which to further explore the concepts presented. Practically, the CLF based QP formulation has been implemented in real-time to experimentally achieve 2D bipedal robotic walking [11] (with control rates exceeding 1 kHz utilizing embedded optimization methods [18]), and similar benchmarks have been achieved using the formalism presented in this paper for 3D walking robots. Yet, the speed of the QP depends on the feasibility of the inequality and equality constraints, so understanding the interplay between computation time of the QP and feasibility of the constraints is an interesting problem. This naturally motivates theoretic research questions related to the CLF based QP formulation. In particular, the CLF inequality constraints were relaxed to allow for solvability of the QP in the presence of hard constraints like torque bounds—yet these relaxations can result in drift in the control objectives and, for aggressive torque bounds, can result in loss of convergence. Conversely, if the relaxations are removed, guarantees on convergence can be made but the QP may become infeasible. Understanding this interplay between proofs of correctness, solvability of the QP, and speed of controllers running CLF based QPs form the basis for a variety of interesting theoretic questions. All of these questions are deeply rooted in core problems related to robotic CPSs and, therefore, promise to be fruitful areas of research.

Acknowledgements. The authors would like to thank Jessy Grizzle, Kevin Galloway and Koushil Sreenath for the many discussions on CLFs, and especially for the collaboration that motivated the constructions in this paper. We would also like to thank John Yamokoski and Nicolaus Radford for encouraging the development of methods for unifying locomotion and manipulation. Finally, we would like to thank the reviewers for their helpful comments and suggestions.

References

1. Ames, A.D.: First steps toward automatically generating bipedal robotic walking from human data. In: Kozłowski, K. (ed.) Robot Motion and Control 2011. LNICS, vol. 422, pp. 89–116. Springer, Heidelberg (2012)
2. Ames, A.D.: First steps toward underactuated human-inspired bipedal robotic walking. In: IEEE International Conference on Robotics and Automation, St. Paul, MN (2012)
3. Ames, A.D.: Human-inspired control of bipedal walking robots. To appear in the IEEE Trans. Automatic Control (2013)
4. Ames, A.D., Cousineau, E.A., Powell, M.J.: Dynamically stable robotic walking with NAO via human-inspired hybrid zero dynamics. In: Hybrid Systems: Computation and Control, Beijing (2012)
5. Ames, A.D., Galloway, K., Grizzle, J.W.: Control Lyapunov functions and hybrid zero dynamics. In: Proc. 51st IEEE Conf. Decision and Control (2012)
6. Ames, A.D., Galloway, K., Grizzle, J.W., Sreenath, K.: Rapidly exponentially stabilizing control Lyapunov runctions and hybrid zero dynamics. To appear in IEEE Trans. Automatic Control (2013)
7. Anitescu, M., Potra, F.A.: Formulating dynamic multi-rigid-body contact problems with friction as solvable linear complementarity problems. Nonlinear Dynamics 14, 231–247 (1997)
8. Bemporad, A., Morari, M.: Robust model predictive control: A survey. Robustness in Identification and Control 245, 207–226 (1999)
9. Bemporad, A., Morari, M., Dua, V., Pistikopoulos, E.N.: The explicit solution of model predictive control via multiparametric quadratic programming. In: Proceedings of the American Control Conference (2012)
10. Freeman, R.A., Kokotović, P.V.: Robust Nonlinear Control Design. Birkhäuser (1996)
11. Galloway, K., Sreenath, K., Ames, A.D., Grizzle, J.W.: Torque saturation in bipedal robotic walking through control lyapunov function based quadratic programs. CoRR, abs/1302.7314 (2013)
12. Grizzle, J.W., Abba, G., Plestan, F.: Asymptotically stable walking for biped robots: Analysis via systems with impulse effects. IEEE Transactions on Automatic control 46(1), 51–64 (2001)
13. Grizzle, J.W., Chevallereau, C., Ames, A.D., Sinnet, R.W.: 3D bipedal robotic walking: models, feedback control, and open problems. In: IFAC Symposium on Nonlinear Control Systems, Bologna (September 2010)
14. Khatib, O.: A unified approach for motion and force control of robot manipulators: The operational space formulation. IEEE Journal of Robotics and Automation 3, 43–53 (1987)
15. Khatib, O., Sentis, L., Park, J., Warren, J.: Whole-body dynamic behavior and control of human-like robots. International Journal of Humanoid Robotics 1, 29–43 (2004)
16. Kolavennu, S., Palanki, S., Cockburn, J.C.: Nonlinear control of nonsquare multivariable systems. Chemical Engineering Science 56, 2103–2110 (2001)
17. Lee, S.H., Goswami, A.: A momentum-based balance controller for humanoid robots on non-level and non-stationary ground. Autonomous Robots 33(4), 399–414 (2012)
18. Mattingley, J., Boyd, S.: Cvxgen: a code generator for embedded convex optimization. Optimization and Engineering 13(1), 1–27 (2012)

19. Mayne, D.Q., Rawlings, J.B., Rao, C.V., Scokaert, P.O.M.: Constrained model predictive control: Stability and optimality. Automatica 36, 789–814 (2000)
20. Murray, R.M., Li, Z., Sastry, S.S.: A Mathematical Introduction to Robotic Manipulation. Boca Raton (1994)
21. Oppenheimer, M.W., Doman, D.B., Bolender, M.A.: Dynamic balance force control for compliant humanoid robots. In: 14th Mediterranean Conference on Control and Automation, MED 2006, pp. 1–6 (2006)
22. Powell, M., Hereid, A., Ames, A.D.: Speed regulation in 3D robotic walking through motion transitions between human-inspired partial hybrid zero dynamics. To appear in the IEEE International Conference on Robotics and Automation (2013)
23. Powell, M.J., Zhao, H., Ames, A.D.: Motion primitives for human-inspired bipedal robotic locomotion: Walking and stair climbing. In: IEEE International Conference on Robotics and Automation, St. Paul, MN (2012)
24. Saab, L., Ramos, O.E., Keith, F., Mansard, N., Soueres, P., Fourquet, J.-Y.: Dynamic whole-body motion generation under rigid contacts and other unilateral constraints. IEEE Transactions on Robotics 29(2), 346–362 (2013)
25. Salini, J., Padois, V., Bidaud, P.: Synthesis of complex humanoid whole-body behavior: A focus on sequencing and tasks transitions. In: IEEE International Conference on Robotics and Automation (ICRA), pp. 1283–1290 (2011)
26. Sastry, S.S.: Nonlinear Systems: Analysis, Stability and Control. Springer (1999)
27. Siciliano, B., Slotine, J.J.E.: A general framework for managing multiple tasks in highly redundant robotic systems. In: Fifth International Conference on Advanced Robotics, ICAR (1991)
28. Sontag, E.: A 'universal' contruction of Artstein's theorem on nonlinear stabilization. Systems & Control Letters 13, 117–123 (1989)
29. Srinivasan, S., Raptis, I.A., Westervelt, E.R.: Low-dimensional sagittal plane model of normal human walking. ASME Journal of Biomechanical Engineering 130(5) (2008)
30. Stephens, B.J., Atkeson, C.G.: Push recovery by stepping for humanoid robots with force controlled joints. In: IEEE International Conference on Humanoid Robots (2010)
31. Stephens, B.J., Atkeson, C.G.: Dynamic balance force control for compliant humanoid robots. In: IEEE/RSJ International Conference on Intelligent Robots and Systems, IROS (2010)
32. Tedrake, R., Manchester, I.R., Tobenkin, M., Roberts, J.W.: LQR-trees: Feedback motion planning via sums of squares verification. International Journal of Robotics Research 29, 1038–1052 (2010)
33. Wang, Y., Boyd, S.: Fast model predictive control using online optimization. IEEE Transations on Control Systems Technology 18(2), 267–278 (2010)
34. Westervelt, E.R., Grizzle, J.W., Chevallereau, C., Choi, J.H., Morris, B.: Feedback Control of Dynamic Bipedal Robot Locomotion, Boca Raton (June 2007)
35. Nadubettu Yadukumar, S., Pasupuleti, M., Ames, A.D.: From formal methods to algorithmic implementation of human inspired control on bipedal robots. In: Tenth International Workshop on the Algorithmic Foundations of Robotics (WAFR), Boston, MA (2012)

Distributed Multiparty DC Power Flow Algorithm with Secure Exchange of Information

Sanja Cvijić and Marija Ilić

Electrical and Computer Engineering, Carnegie Mellon University
5000 Forbes Ave, Pittsburgh, PA 15213, USA
sanja13@andrew.cmu.edu, milic@ece.cmu.edu
http://www.ece.cmu.edu/

Abstract. This paper introduces a new distributed algorithm for computing DC power flow with secure information exchange among participating areas. The algorithm is based on the principles of the Diakoptics approach to distributed power flow execution. Power flow is computed in multiple steps through a precisely defined information exchange protocol between the coordinator and the individual areas. In the conventional Diakoptics algorithm, areas exchange the information about their generation and load with the coordinator. Our algorithm redefines the exchange of information protocol so that areas do not need to reveal their internal generation and load but only their effects on the rest of the grid. In this way, proprietary information about the exact locations and values of generation and demand are protected and cannot be uniquely deciphered. We illustrate this algorithm and the corresponding information exchange protocol on the IEEE 14-bus system.

Keywords: cyber security, distributed power flow, diakoptics, information exchange, communication.

1 Introduction

The current rate of increase in power grid size and complexity requires improved distributed power flow algorithms. Additionally, interconnected areas in a deregulated environment are strongly motivated to increase self-sufficiency and reveal the minimum amount of information necessary for reliable system operation. Consequently, distributed algorithms designed for power systems should fit the needs of competing areas while taking into consideration the sensitive nature of exchanged information.

The "DC" simplification of the full AC power flow problem is widely used in power systems analysis for market related purposes. Using "DC" assumptions, areas can asses the effect of energy transactions across areas on active power flows in order to prevent transmission line overloading. Additionally, "DC" power flow is widely used in contingency analysis for assessing active power line flows.

A distributed power flow implementation assumes that computations are divided across multiple parties who jointly solve a system-level problem through

D.C. Tarraf (ed.), *Control of Cyber-Physical Systems*,
Lecture Notes in Control and Information Sciences 449,
DOI: 10.1007/978-3-319-01159-2_13,

communication. Although the distributed execution brings high performance benefits, it also puts the grid at risk if the flow of information is intercepted. Recent trends in power grids show that the overall grid vulnerability continues to increase with emerging communication and networking protocols. However, the majority of distributed algorithms do not address the question of the security of the exchanged data nor the willingness of parties to exchange sensitive information. For economic reasons in competitive electricity markets, the question about willingness of areas to share sensitive information plays an important role. Areas tend to keep the outputs and locations of their generators private in order to protect their bidding strategies.

Various cryptographic approaches have been proposed for maintaining the privacy of sensitive data in power systems. The best-known one is the Secure Multiparty Computation (SMC) algorithm used to carry out extensive computations of a function among multiple parties [1],[2]. The SMC problem assumes that multiple parties can jointly and accurately compute a function over their inputs while keeping them private. Clearly, a similar concept is needed for computing power flow securely. Since the goal is to protect information about their own generation and load, areas in an interconnected multi-area electric system need to solve power flow jointly without revealing their inputs. The main disadvantage of general cryptographic methods is that they increase computational overhead in order to achieve privacy.

In the literature, several methods have been proposed for solving distributed power flow. In the Sixties, Diakoptics was introduced by G. Kron [3], and applied to power systems by H. Happ [4],[5]. Although, the original formulation of this algorithm was in terms of electric circuit variables, in this paper, we extend it to the "DC" power flow problem. The "DC" (or simply DC) acronym denotes an approximation of the full AC power flow problem. When analogies between electrical and DC power networks [6] are applied, it becomes clear that the original Diakoptics algorithm represents one of the first distributed power flow algorithms. However, this algorithm requires areas to communicate their internal generation and demand with the coordinator making the grid vulnerable to cyber attacks.

Recently, new contributions have been made in computing power flow in a fully decentralized way through communication with nearest neighbors only. Distributed AC power flow can be solved by installing small line and bus calculators that, through the exchange of information with nearest neighbors, can compute power flow in a distributed way [7], [8]. Regarding the communication aspect, this algorithm requires lines to exchange flow variables with adjacent buses and buses to exchange Lagrange multipliers with directly connected lines. Due to Kirchhoff's Flow Law, it should be noted that line flow variables uniquely reveal power injections into buses because all together sum to zero.

A distributed power flow algorithm based on determining the Thevinan equivalent for the exterior network and gradually correcting the equivalent power injection of the exterior system has been presented in [9]. This method is based on equivalencing which does not require areas to share their nodal injections with the coordinator. Due to the nature of equivalencing, topological and operational data are merged together requiring all computations to be repeated as network conditions change. Inability to re-use topological information, in a network in which only loading conditions change, increases the overall computational complexity of the algorithm.

As an alternative approach, we are proposing a new non-cryptographic method for solving distributed power flow in a secure manner. The idea is to find a solution to the power flow problem in a distributed way while keeping nodal injections private. The power flow problem cannot be solved in totally isolated areas with no information exchange since areas are electrically coupled. This is why some communication is required for solving power flow accurately. Computation of the effect of interconnections generally requires the exact values of generation and demand to be shared. In our proposed algorithm, the effect of interconnections can be accurately incorporated into the power flow solution by each area without explicit knowledge of nodal injections. The main idea is that areas compute how their internal injections affect the external network while the coordinator computes how the flows in the external interconnections affect power flow inside areas.

2 Algorithm for Secure Distributed Power Flow

This multi-layered algorithm originates from the Diakoptics approach explained in detail by H. Happ in [5]. Inspired by this procedure which was initially designed for electric circuits, we have formulated a coordinated distributed DC power flow algorithm and have applied it to more efficient contingency screening in [10]. For the sake of comparison, this algorithm is revisited in this paper and referred to as the Simple Distributed DC Power Flow Algorithm. As part of that algorithm, it was necessary that individual areas exchange their generation and load data with the coordinator. In this paper, areas send coded information about the effect of their generation and load onto the rest of the network. The coded information is insufficient for uniquely restoring areas' generation and demand, but sufficient for solving the multi-area power flow problem.

The goal of Diakoptics is to enable distributed execution through decomposition into two layers: one that operates on disconnected areas while neglecting interconnections with the rest of the system and another that integrates the effects of interconnections. These layers emphasize the modeling level of interest and will be referred to as *zoomed-in* and *zoomed-out* layers respectively.

At the zoomed-in layer, power flow is computed within individual areas completely disconnected from the rest of the system. Due to Kirchhoff's laws and Ohm's law, there exist couplings among all areas in the system, i.e., power generation in one area affects line flows in another. The zoomed-out layer, or the coordinating layer, quantifies the effect of interconnections by operating on a simplified representation of the original network. Finally, each area updates its power flow solution by incorporating that effect into the solution previously computed when the area was disconnected.

2.1 Mesh-to-Tree Transformation

The key idea that enables distributed execution of power flow lies in decomposing and simplifying a multi-area network. The network topology is broken down into subdivisions, or areas, and then simplified through transformation of a meshed electrical network into a spanning tree [11]. This graph-based transformation is first applied at the zoomed-in layer so that topologies of individual areas are converted into trees in parallel. Then, spanning trees are interconnected with tie-lines and transformations proceed at the zoomed-out layer until the initial network has been converted into a tree.

The motivation for converting a meshed network into a tree comes from the fact that line flows in a spanning tree have a physical interpretation. They can be understood as bilateral transactions or financial contracts that specify the amount of traded power along a contract path that the trade is supposed to take. However, physical laws that govern power grids disrupt the intended path by superposing loop flows inside basic loops in a meshed network. Additionally, due to Kirchhoff's Voltage Law, only line voltages across spanning tree branches can be considered independent while all remaining voltages can be expressed in terms of tree voltages. Therefore, the spanning tree contains the minimum information about the network. A spanning tree can be considered a financial network with trades assigned along its branches.

All network transformations obey the law of conservation of power while each network representation satisfies Kirchhoff's Flow Law (KFL), Kirchhoff's Voltage Law (KVL) and Ohm's law. A network representation, also called a *reference frame*, is fully described with a flow vector F, a voltage angle difference vector Δ and a reactance matrix X. Mapping between two reference frames (a and b) is established through a connection matrix C_{ab}. Assuming that variables in one reference frame are known, it is possible to transform them into the corresponding variables in another reference frame. Starting from the physical law equations, the resulting relationships follow:

$$F^a = C_{ab} \cdot F^b \tag{1}$$

$$\Delta_b = C_{ab}^T \cdot \Delta_a \tag{2}$$

$$X_{bb} = C_{ab}^T \cdot X_{aa} \cdot C_{ab} \tag{3}$$

The algorithm consists of five reference frames in which computations are performed in two directions, as shown in the flowchart in Figure 1. Topological transformations are performed in the *forward* direction from reference frame 1 to 5, while power flow is solved in the *backward* direction from reference frame 5 to 1.

We briefly illustrate the method on the 14-bus system in Figure 2, which has been clustered into three areas using spectral clustering with electrical distances as the measure of distance [12].

Zoomed-In Layer: After decomposing a large network into areas, the zoomed-in layer analyzes each area as if the others did not exist. Initially, it performs

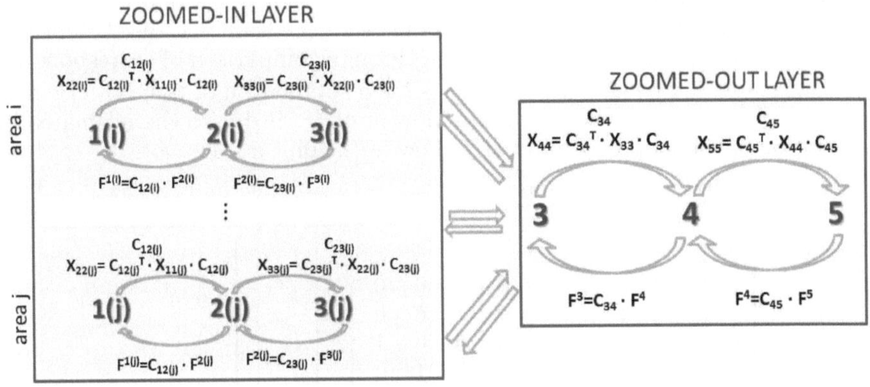

Fig. 1. Algorithm flowchart

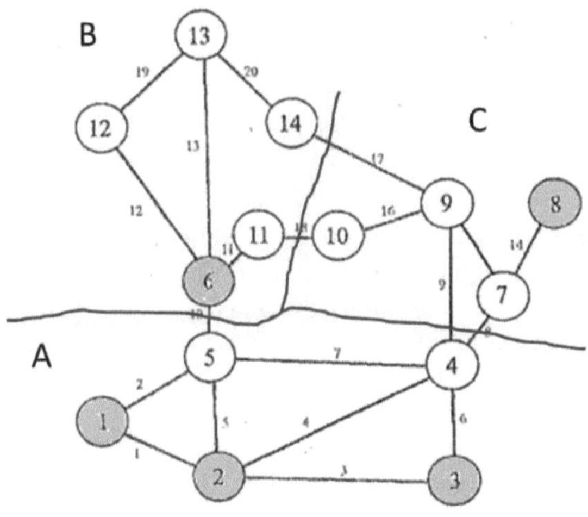

Fig. 2. Clustered 14-bus system

transformations of internal area topologies into spanning trees in a distributed way. It consists of three reference frames that gradually lead to the spanning tree representation. *Reference frame 1* represents the original meshed network which is mapped into *reference frame 2* which is composed of spanning tree branches and basic loops. The choice of a spanning tree is arbitrary; however, it uniquely determines a set of basic loops. It can be shown that, due to KVL, basic loops can be eliminated, resulting in a completely radial network representation. This representation is referred to as *reference frame 3*.

The mesh-to-tree transformation is illustrated on area A of the 14-bus example. On the zoomed-in layer, network topologies of all three areas are converted into spanning tree representations. The connection between reference frames 1 and 2 is established through a connection matrix, C_{12}, as shown in Equation (4). The C_{12} matrix defines mapping between flows in reference frames 1 and 2, depicted in Figure 3. It is determined by inspection, through comparison of corresponding flows. For example, flow f^1 through line 1-2 is a summation of tree flow $f^{1'}$ and loop flow $-f^{c1}$, where the negative sign in front of f^{c1} is due to the mismatch in the direction of the loop flow and the f^1 flow in the physical network.

$$F^{1(A)} = C_{12(A)} \cdot F^{2(A)}$$

$$\begin{bmatrix} f^1 \\ f^2 \\ f^3 \\ f^4 \\ f^5 \\ f^6 \\ f^7 \end{bmatrix} = \left[\begin{array}{cccc|ccc} 1 & 0 & 0 & 0 & -1 & 0 & 0 \\ 0 & 0 & 0 & 0 & 1 & 0 & 0 \\ 0 & 1 & 0 & 0 & 0 & 0 & 1 \\ 0 & 0 & 0 & 0 & 0 & 1 & 0 \\ 0 & 0 & 1 & 0 & -1 & -1 & -1 \\ 0 & 0 & 0 & 0 & 0 & 0 & 1 \\ 0 & 0 & 0 & 1 & 0 & 1 & 1 \end{array}\right] \cdot \begin{bmatrix} f^{1'} \\ f^{2'} \\ f^{3'} \\ f^{4'} \\ \hline f^{c1} \\ f^{c2} \\ f^{c3} \end{bmatrix} \begin{array}{l} \left.\vphantom{\begin{array}{c} f \\ f \\ f \\ f \end{array}}\right\} F^{prime} \\ \left.\vphantom{\begin{array}{c} f \\ f \\ f \end{array}}\right\} F^{ic} \end{array} \tag{4}$$

$$\Delta_{2(A)} = X_{22(A)} \cdot F^{2(A)}$$

$$\left[\frac{\Delta_{prime}}{0}\right] = \left[\begin{array}{c|c} X_{prime\_prime} & X_{prime\_ic} \\ \hline X_{ic\_prime} & X_{ic\_ic} \end{array}\right] \cdot \left[\frac{F^{prime}}{F^{ic}}\right] \tag{5}$$

In reference frame 2, the sum of phase angle differences along any loop in an electric power network is equal to zero due to KVL, Equation (5). Consequently, line flows in reference frame 2 are linearly dependent. Moreover, loop flows can be expressed as a linear combination of spanning tree flows as in Equation (6).

$$\begin{aligned} F^{ic} &= -X_{ic\_ic}^{-1} \cdot X_{ic\_prime} \cdot F^{prime} = C_{ic\_prime} \cdot F^{prime} \\ &= \sum_{i=1}^{N} c_i \cdot F^{prime} \end{aligned} \tag{6}$$

Since loop flows are dependent on spanning tree flows, they can be eliminated while the mapping between them can be saved in form of a connection matrix, C_{23}, Equation (7). The final area representation contains spanning tree flows only, F^{prime} (called F^3 from here on), depicted in Figure 4.

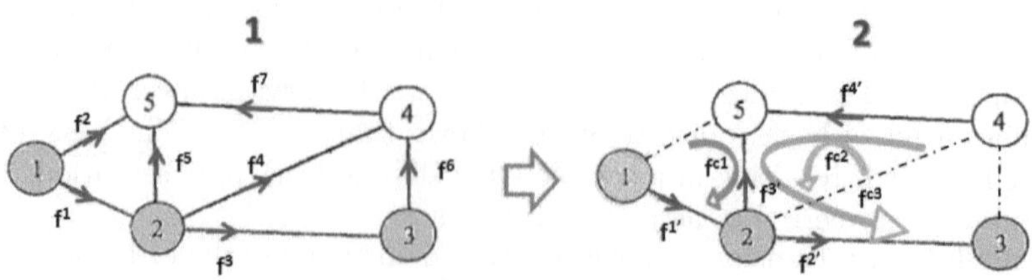

Fig. 3. Reference frame transition from 1 to 2 in area A

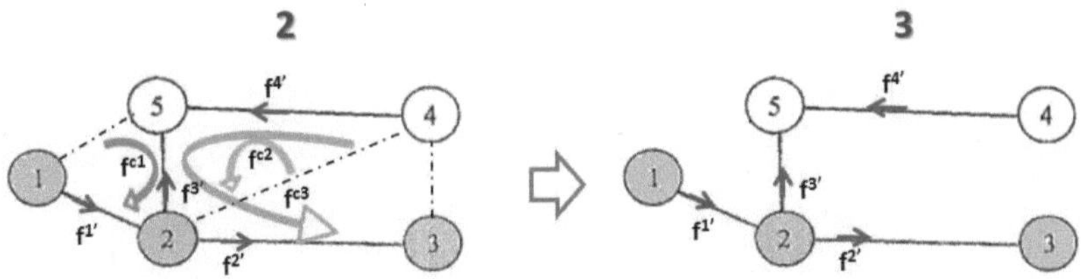

Fig. 4. Reference frame transition from 2 to 3 in area A

$$F^{2(A)} = C_{23(A)} \cdot F^{3(A)}$$
$$\begin{bmatrix} f^{1'} \\ f^{2'} \\ f^{3'} \\ f^{4'} \\ \hline f^{c1} \\ f^{c2} \\ f^{c3} \end{bmatrix} = \begin{bmatrix} 1_{4\times4} \\ \hline C_{ic\_prime} \end{bmatrix} \cdot \begin{bmatrix} f^{1'} \\ f^{2'} \\ f^{3'} \\ f^{4'} \end{bmatrix} \tag{7}$$

At this point, areas send their spanning tree representations and nodal injections to the coordinator who computes the effect of interconnections. Afterwards, areas receive their updated spanning tree flows, denoted as F^3, which contain flows due to internal injections, F^{ist}, as well as flows due to couplings with the rest of the network, EFF:

$$F^3 = F^{ist} + EFF \tag{8}$$

Once the F^3 flows are known, spanning tree flows can be mapped into physical line flows through reference frame 2 as in Equation (9).

$$\begin{matrix} F^2 = C_{23} \cdot F^3 \\ F^1 = C_{12} \cdot F^2 \end{matrix} \quad or \quad \begin{matrix} C_{13} = C_{12} \cdot C_{23} \\ F^1 = C_{13} \cdot F^3 \end{matrix} \tag{9}$$

Zoomed-Out Layer: The zoomed-out layer refers to network transformations which compute the effect of interconnections necessary for solving power flow accurately. First, tree representations of all areas are interconnected with tie-lines, as in Figure 5. The connection and reactance matrices that describe this interconnected network representation are created by stacking connection, $C_{13(i)}$, and reactance, $X_{33(i)}$, matrices of the zoomed-in layer and tie-lines on the main diagonal. The goal of the zoomed-out layer is to neglect the exact network topology inside an area while merging it into the slack bus. This is accomplished by

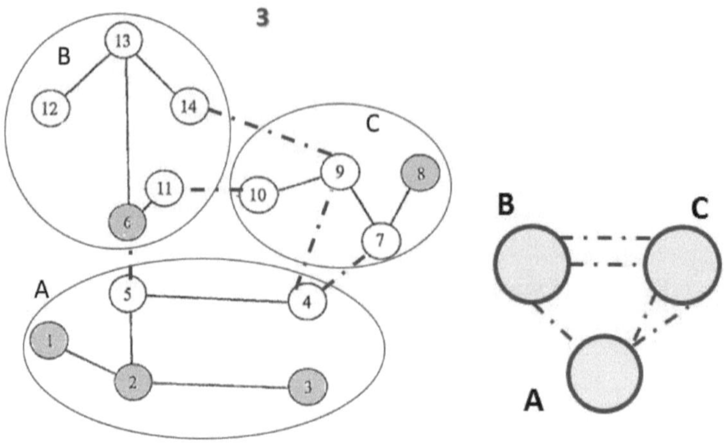

(a) Interconnected spanning trees (b) Zoomed-out view

Fig. 5. Zoomed-out representation of 14-bus system

keeping exactly one tie-line connection between slack buses of each two interconnected areas. Remaining tie-line flows are substituted with external loop flows between two areas, shown in Figure 6. This zoomed-out network representation is called *reference frame 4*. After zooming-out, the same graph-based transformations, as were used on the zoomed-in layer, are applicable for elimination of basic loops. Spanning tree branches and basic loops are identified to form a new network representation, *reference frame 5*. Starting from Ohm's law and KVL in Equation (10), external loop flows in the zoomed-out representation, F^{ec}, can be expressed as contributions of internal, F^{ist}, and external, F^{est}, spanning tree flows, Equation (11).

$$\Delta_5 = X_{55} \cdot F^5$$

$$\left[\begin{array}{c} \Delta_{ist} \\ \hline \Delta_{est} \\ \hline 0 \end{array}\right] = \left[\begin{array}{c|c|c} X_{ist\_ist} & X_{ist\_est} & X_{ist\_ec} \\ \hline X_{est\_ist} & X_{est\_est} & X_{est\_ec} \\ \hline X_{ec\_ist} & X_{ec\_est} & X_{ec\_ec} \end{array}\right] \cdot \left[\begin{array}{c} F^{ist} \\ \hline F^{est} \\ \hline F^{ec} \end{array}\right] \quad (10)$$

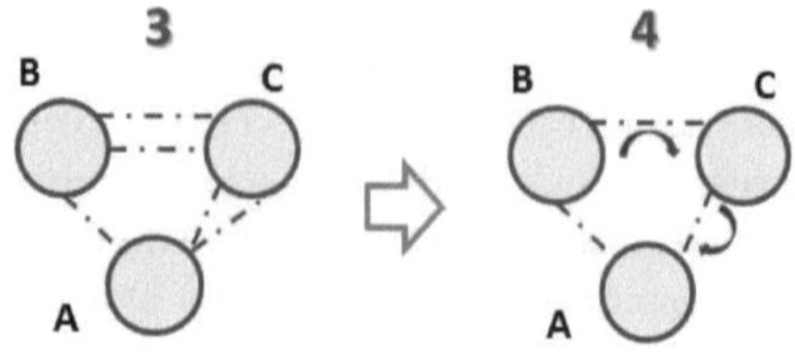

Fig. 6. Reference frame transition from 3 to 4 on the zoomed-out layer

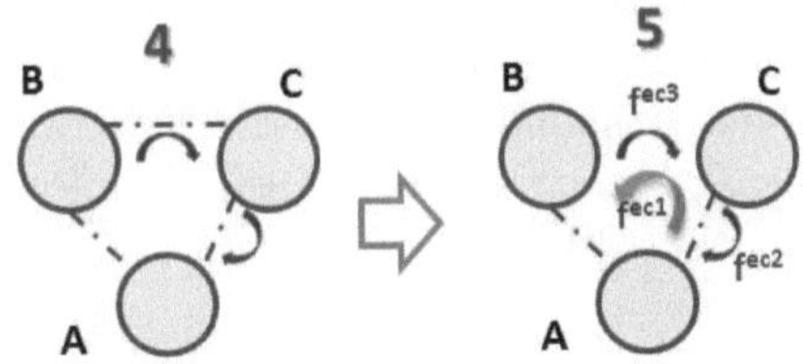

Fig. 7. Reference frame transition from 4 to 5 on the zoomed-out layer

$$F^{ec} = -X_{ec\_ec}^{-1} \cdot \left(X_{ec\_ist} \cdot F^{ist} + X_{ec\_est} \cdot F^{est}\right) \\ = C_{ec\_ist} \cdot F^{ist} + C_{ec\_est} \cdot F^{est} \tag{11}$$

Reference frame 5 is the final representation that captures internal and external spanning tree flows along with their contributions to external loop flows. Internal tree flows, F^{ist}, originate from internal nodal injections. External tree flows, F^{est}, contain information about injections into slack buses of individual areas. With reference frame 5, the topological transformations have been completed.

2.2 Simple Distributed DC Power Flow Algorithm

Once all transformations have been performed in the forward direction, power flow computation can take place in the backward direction, Figure 1. The computation of power flow starts with reference frame 5 in which all internal and external spanning tree flows have to be known. Conventionally, inputs into the power flow problem are nodal power injections. It can be shown that for a selected spanning tree, there exists a bijective mapping between specified nodal injections and spanning tree flows. Here we prove the bijective mapping in the form of two theorems.

Theorem 1: *Nodal injections uniquely determine line flows in any radial network in which KFL holds.*

In a general network with N nodes, one node, known as slack, is responsible for total power balancing while nodal injections are specified at the remaining $N-1$ nodes. The goal is to find $N-1$ spanning tree flows, F^{ist}, created by nodal injections so that KFL is satisfied. By setting up KFL balance equations for each node that is not slack, we formulate a system of $N-1$ linearly independent equations with $N-1$ unknown flow variables. This relationship can be expressed in the form of matrix T which contains only ± 1 and 0 elements, Equation (12). Since matrix T is nonsingular, there exists a unique set of spanning tree flows for the specified nodal injections, Equation (13).

$$P^{inj} = T \cdot F^{ist} \tag{12}$$

$$F^{ist} = T^{-1} \cdot P^{inj} \tag{13}$$

For example in a five-bus network with bus 1 as slack, nodal injections are specified on the left while the corresponding spanning tree flows are shown on the right in Figure 8. The corresponding T matrix that translates tree flows into injections (left) and the T^{-1} matrix that translates injections into tree flows (right) are shown in Equation (14).

$$\begin{bmatrix} p^2 \\ p^3 \\ p^4 \\ p^5 \end{bmatrix} = \begin{bmatrix} -1 & 1 & 1 & 0 \\ 0 & -1 & 0 & 0 \\ 0 & 0 & 0 & 1 \\ 0 & 0 & -1 & -1 \end{bmatrix} \cdot \begin{bmatrix} f^{A1} \\ f^{A2} \\ f^{A3} \\ f^{A4} \end{bmatrix} \quad or \quad \begin{bmatrix} f^{A1} \\ f^{A2} \\ f^{A3} \\ f^{A4} \end{bmatrix} = \begin{bmatrix} -1 & -1 & -1 & -1 \\ 0 & -1 & 0 & 0 \\ 0 & 0 & -1 & -1 \\ 0 & 0 & 1 & 0 \end{bmatrix} \cdot \begin{bmatrix} p^2 \\ p^3 \\ p^4 \\ p^5 \end{bmatrix} \tag{14}$$

Theorem 2: *Line flows uniquely determine nodal injections in any radial network in which KFL holds.*

The proof of this theorem directly follows from Equation (12).

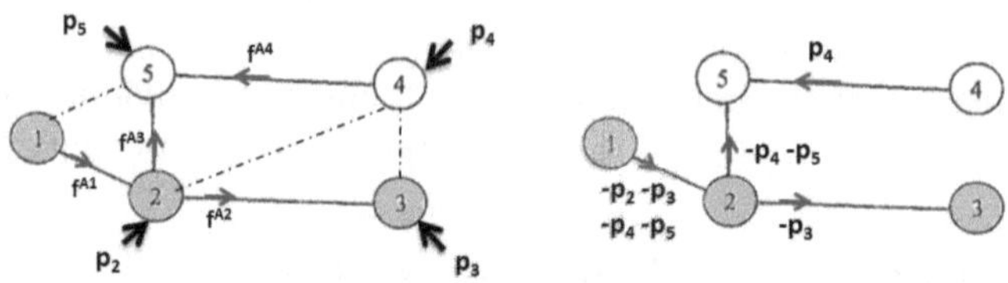

Fig. 8. Left: Nodal injections, Right: Spanning tree flows

Once nodal injections are converted into internal and external spanning tree flows using Theorem 1, power flow can be solved. Firstly, on the zoomed-out layer, external loop flows are computed from spanning tree flows using Equation (11). Secondly, flows from reference frame 5 are translated across reference frame 4 to reference frame 3 using connection matrices.

$$\begin{aligned} F^4 &= C_{45} \cdot F^5 \\ F^3 &= C_{34} \cdot F^4 \end{aligned} \quad or \quad \begin{aligned} C_{35} &= C_{34} \cdot C_{45} \\ F^3 &= C_{35} \cdot F^5 \end{aligned} \tag{15}$$

Further on, zoomed-in areas acquire updated spanning tree flows, F^3, computed on the zoomed-out layer. Finally, each area i maps these tree flows into physical line flows using its own connection matrix:

$$F^{1(i)} = C_{13(i)} \cdot F^{3(i)} \tag{16}$$

A diagram that summarizes the simple distributed power flow algorithm with the exchange of information is shown in Figure 9. In order to allow the mesh-to-tree transformation, each area needs to send their tree representation with the corresponding tree reactance matrix, $X_{33(i)}$, and all internal spanning tree flows, $F^{ist(i)}$, translated from its nodal injections. After the zoomed-out transformations, the coordinator sends updated spanning tree flows, $F^{3(i)}$, to the individual areas that then use them for computing physical line flows, $F^{1(i)}$.

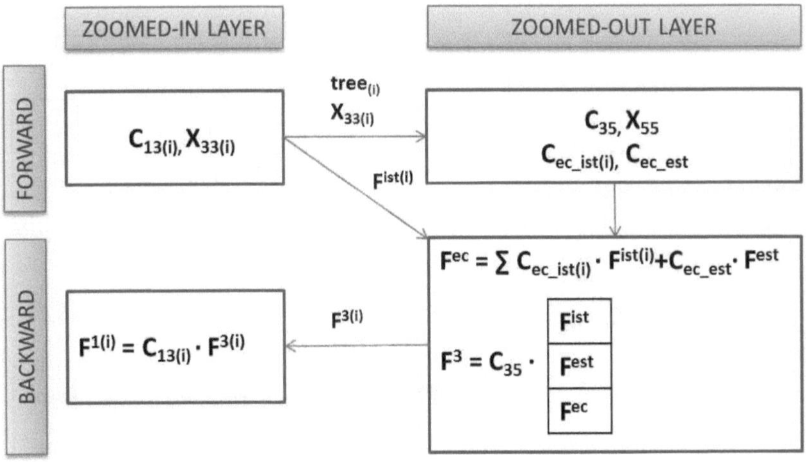

Fig. 9. Exchange of information in the simple DC power flow algorithm

2.3 Secure Distributed DC Power Flow Algorithm

In the simple distributed DC power flow formulation, all areas are required to exchange their internal spanning tree flows with the coordinator. From this exchange of information, location and outputs of each generator/load can be easily decoded based on Theorem 2. In order to overcome these security issues, we propose a new algorithm for distributed power flow execution. Compared to the previously introduced algorithm, the secure algorithm does not require areas to reveal their generation and load locations. Even if the entire flow of information has been intercepted, it is not possible to decode the locations and outputs of injection patterns of individual areas.

The key idea of this approach is based on the exchange of mutual effects between zoomed-in and zoomed-out layers. Loop flows create couplings among areas and therefore couple zoomed-in and zoomed-out layers. Areas can exchange their effects on external loop flows with the zoomed-out layer, instead of revealing their internal generation and demand. Similarly, the zoomed-out layer exchanges effects of external spanning tree flows and external loop flows on spanning tree flows inside areas, so that areas can calculate their own physical line flows. We show that it is not possible to decode information about neither internal spanning tree flows nor external spanning tree flows. This method provides an attack-resistant flow of information which successfully hides secure information.

Instead of exchanging information two-ways as in the simple distributed DC algorithm, this method requires a four-ways exchange of information, Figure 10. Starting with the zoomed-in layer, areas convert their topologies into trees and forward their tree representations to the zoomed-out layer. Further on, the zoomed-out layer interconnects spanning trees and proceeds with tree transformations in order to compute matrices C_{ec_ist} and C_{ec_est}, Equation (11). These matrices map internal and external spanning tree flows into external loop flows.

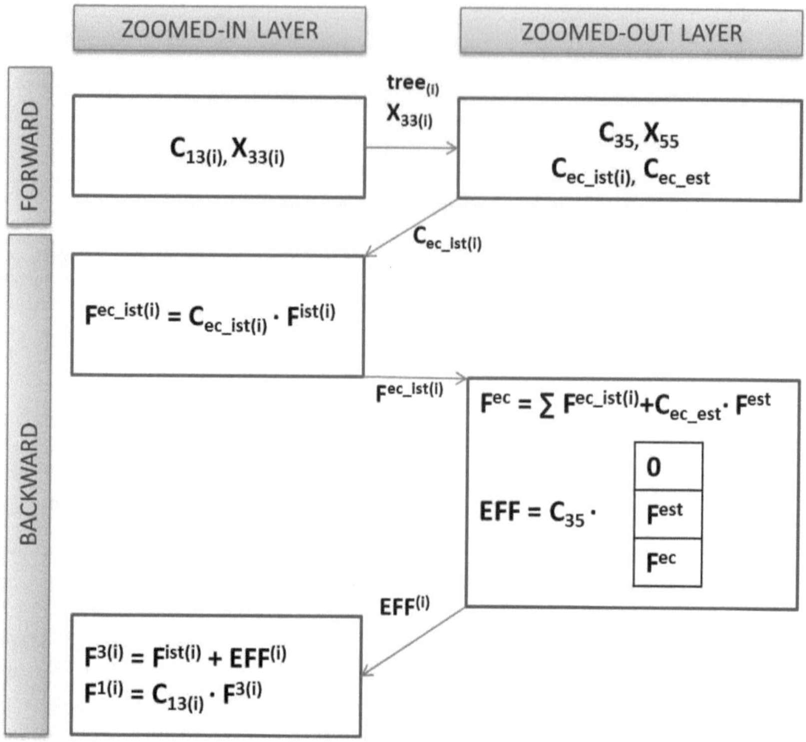

Fig. 10. Exchange of information in the secure DC power flow algorithm

The newly computed C_{ec_ist} matrix is divided into blocks corresponding to individual areas. Area i computes the effect of its internal tree flows onto external loop flows, $F^{ec\_ist(i)}$:

$$F^{ec\_ist(i)} = C_{ec\_ist(i)} \cdot F^{ist(i)} \tag{17}$$

At this point, the coordinator knows how areas affect external loop flows which enables the exact computation of external loop flows. The coordinator then sums up the effects of all areas onto external loop flows and adds them to the loop flows created by external spanning tree flows:

$$F^{ec} = \sum_i F^{ec\_ist(i)} + C_{ec\_est} \cdot F^{est} \tag{18}$$

Now, it is possible to determine the exact effect of interconnections onto individual areas. This effect is computed on the zoomed-out layer assuming that all internal spanning tree flows equal zero:

$$EFF = C_{35} \cdot \left[\begin{array}{c} 0 \\ \hline F^{est} \\ \hline F^{ec} \end{array} \right] \tag{19}$$

Individual areas receive the effect of interconnections on their tree flows which is added to their internal spanning tree flows. Then, the tree flows are mapped back to physical line flows using connection matrix $C_{13(i)}$.

3 Simulation Results

The secure distributed DC power flow algorithm is illustrated on the IEEE 14-bus example, Figure 2. Line parameters and generation and load magnitudes are given in Appendix A. Internal and external spanning tree flows are created from specified nodal injections (Table 2) based on Theorem 1, and given in Equation (20). After performing the mesh-to-tree transformation, tree representations of areas are shown in Figure 11 and the corresponding connection and reactance matrices are given in Equations (21) and (22) respectively.

$$F^{ist(A)} = \begin{bmatrix} 1.3130 \\ 0.9420 \\ 0.5540 \\ -0.4780 \end{bmatrix}, F^{ist(B)} = \begin{bmatrix} 0.0350 \\ 0.3450 \\ -0.0610 \\ 0.1490 \end{bmatrix}, F^{ist(C)} = \begin{bmatrix} -0.3850 \\ 0.3850 \\ 0.0900 \end{bmatrix}$$
$$F^{est} = \begin{bmatrix} 0.4920 \\ 0.3850 \end{bmatrix} \quad (20)$$

$$C_{13(A)} = \begin{bmatrix} 0.8383 & -0.0907 & -0.2288 & -0.0597 \\ 0.1617 & 0.0907 & 0.2288 & 0.0597 \\ -0.0271 & 0.5600 & 0.1293 & -0.0506 \\ -0.0567 & 0.2020 & 0.2706 & -0.1059 \\ -0.0778 & 0.1472 & 0.3713 & 0.0969 \\ -0.0271 & -0.4400 & 0.1293 & -0.0506 \\ -0.0839 & -0.2379 & 0.3999 & 0.8435 \end{bmatrix}$$
$$C_{13(B)} = \begin{bmatrix} 1.0000 & 0 & 0 & 0 \\ 0 & 0.2223 & -0.3411 & 0 \\ 0 & 0.7777 & 0.3411 & 0 \\ 0 & 0.2223 & 0.6589 & 0 \\ 0 & 0 & 0 & 1.0000 \end{bmatrix}$$
$$C_{13(C)} = \begin{bmatrix} 1 & 0 & 0 \\ 0 & 1 & 0 \\ 0 & 0 & 1 \end{bmatrix} \quad (21)$$

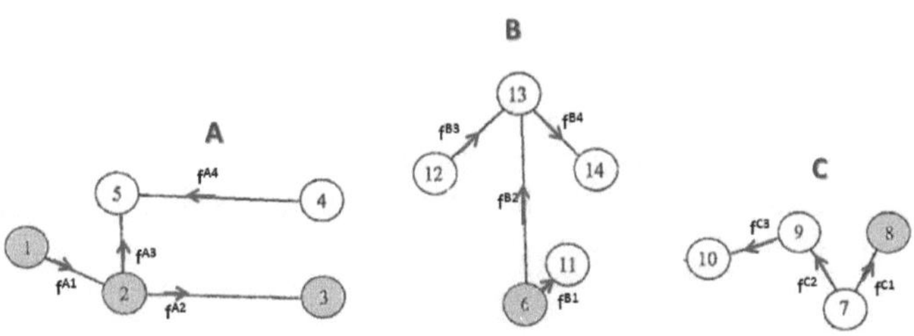

(a) Area A (b) Area B (c) Area C

Fig. 11. Spanning trees of areas on the zoomed-in layer

$$X_{33(A)} = \begin{bmatrix} 0.0496 & -0.0054 & -0.0135 & -0.0035 \\ -0.0054 & 0.1109 & 0.0256 & -0.0100 \\ -0.0135 & 0.0256 & 0.0646 & 0.0168 \\ -0.0035 & -0.0100 & 0.0168 & 0.0355 \end{bmatrix}$$

$$X_{33(B)} = \begin{bmatrix} 0.1989 & 0 & 0 & 0 \\ 0 & 0.1013 & 0.0444 & 0 \\ 0 & 0.0444 & 0.1317 & 0 \\ 0 & 0 & 0 & 0.3480 \end{bmatrix}$$

$$X_{33(C)} = \begin{bmatrix} 0.1762 & 0 & 0 \\ 0 & 0.1100 & 0 \\ 0 & 0 & 0.0845 \end{bmatrix} \quad (22)$$

The zoomed-out layer calculates contribution coefficients of internal spanning tree flows of individual areas and contribution coefficients of external spanning tree flows to external loop flows according to Equation (11):

$$C_{ec\_ist(A)} = \begin{bmatrix} -0.0045 & -0.0129 & 0.0217 & 0.0457 \\ -0.0017 & -0.0047 & 0.0079 & 0.0167 \\ -0.0027 & -0.0078 & 0.0131 & 0.0275 \end{bmatrix}$$

$$C_{ec\_ist(B)} = \begin{bmatrix} 0.1542 & 0.0519 & 0.0228 & 0.1783 \\ 0.0562 & 0.0189 & 0.0083 & 0.0650 \\ 0.2593 & -0.0535 & -0.0235 & -0.1838 \end{bmatrix}$$

$$C_{ec\_ist(C)} = \begin{bmatrix} 0 & -0.0900 & -0.0655 \\ 0 & 0.0929 & -0.0239 \\ 0 & -0.0542 & -0.1102 \end{bmatrix}$$

$$C_{ec\_est} = \begin{bmatrix} 0.3416 & -0.1997 \\ 0.1246 & 0.1661 \\ 0.2057 & -0.1202 \end{bmatrix} \quad (23)$$

Based on the tracing coefficients, individual areas can compute the effect of their internal spanning tree flows to external loop flows using Equation (17). External loop flows created by internal tree flows have magnitudes as follows

	$F^{ec\_ist(A)}$	$F^{ec\_ist(B)}$	$F^{ec\_ist(C)}$
f^{ec1}	−0.0280	0.0485	−0.0405
f^{ec2}	−0.0102	0.0177	0.0336
f^{ec3}	−0.0168	−0.0353	−0.0308

(24)

Once the zoomed-out layer collects all external loop flows created by individual areas, they are summed up together with loop flows created by external tree flows according to Equation (18). The magnitude of external loop flows are

	F^{ec}
f^{ec1}	0.0712
f^{ec2}	0.1663
f^{ec3}	−0.0280

(25)

Starting from the relationship in Equation (19), the zoomed-out layer can compute the effect of interconnections on areas' spanning tree flows:

$$EFF^{(A)} = \begin{bmatrix} 0.8770 \\ 0 \\ 0.8770 \\ -0.4562 \end{bmatrix}, EFF^{(B)} = \begin{bmatrix} 0.0280 \\ -0.0992 \\ 0 \\ -0.0992 \end{bmatrix}, EFF^{(C)} = \begin{bmatrix} 0.3850 \\ -0.0951 \\ -0.0280 \end{bmatrix} \tag{26}$$

The coordinator computes physical tie-line flows as well:

$$EFF^{tie\_lines} = \begin{bmatrix} 0.2899 \\ 0.1663 \\ 0.4208 \\ 0.0992 \\ -0.0280 \end{bmatrix} \tag{27}$$

Once areas integrate the effect of interconnections onto their internal spanning tree flows, physical line flows can be accurately computed:

$$F^{1(A)} = \begin{bmatrix} 1.4788 \\ 0.7112 \\ 0.7005 \\ 0.5523 \\ 0.4090 \\ -0.2415 \\ -0.6234 \end{bmatrix}, F^{1(B)} = \begin{bmatrix} 0.0630 \\ 0.0755 \\ 0.1703 \\ 0.0145 \\ 0.0498 \end{bmatrix}, F^{1(C)} = \begin{bmatrix} 0 \\ 0.2899 \\ 0.0620 \end{bmatrix} \tag{28}$$

4 Security Considerations

In the previous section, a new algorithm for secure distributed computation of DC power flow was introduced. The main advantage of this algorithm versus the simple distributed power flow algorithm is that it keeps internal spanning tree flows of areas private. Internal spanning tree flows uniquely map into nodal power injections using Theorem 2. Therefore, nodal power injections are kept private and it is not possible to recover them from the information exchange protocol.

The only exchanged information in the secure algorithm that incorporates internal power injections are effects of internal spanning tree flows to external loop flows, $F^{ec\_ist(i)}$. An additional information exchanged are contribution factors of internal spanning tree flows to external loop flows, $C_{ec\_ist(i)}$. The relationship between them can be stated as in Equation (17). However, the $C_{ec\_ist(i)}$ matrix is generally singular with dimensions: the number of external loop flows by the number of nodes in area i minus one. Due to its singular nature, this matrix cannot be inverted in order to decipher spanning tree flows, $F^{ist(i)}$.

5 Possible Extensions

The distributed method for solving DC power flow can be generalized into a distributed DC Optimal Power Flow (DCOPF) algorithm with secure information exchange. In paper [13], we introduced a generalized DCOPF formulation that includes wheeling loop flow charges. The key idea is that each area determines their optimal dispatch so that wheeling across third-party areas is reduced. This algorithm requires that an area exchanges only information about external loop flows created by its own dispatch. In this paper, it was shown that the loop flow information does not uniquely reveal nodal injections and therefore it can be considered secure.

Another possible extension of the algorithm is for distributed AC power flow with secure information exchange. The presented algorithm strongly relies on the linear assumption inherent in DC power flow and therefore a further extension into the non-linear AC power flow requires major breakthroughs. One potential approach, which preserves the assumption about linearity, would use the constant impedance load model. With loads modeled as constant impedance, the AC problem is linear which would make our framework fully applicable. If the load P,Q model is preserved, thesis [14] defines another approximation of the full AC load flow problem around the base case loading conditions. This approximation relies on the derivation of coefficients which relate power nodal injections to power loop flows in a linear fashion.

6 Conclusions

This paper has introduced a new method for solving distributed DC power flow with secure information exchange. We have shown how it is possible to protect information about generation and load locations and power outputs while obtaining the correct power flow solution. Although the method is non-cryptographic, the decentralized nature of the algorithm successfully hides the sensitive information. Delegating power flow computation to areas and the coordinator according to the type of information that they have access to, protects the delicate information but also allows solving power flow accurately through the exchange of intermediate results. Compared to the simple distributed power flow algorithm, the secure algorithm requires more communication for the sake of improving privacy of the data.

Acknowledgment. This work was done as a part of the 2111.004 task of the SRC ERI research initiative http://www.src.org/program/eri/. The authors would like to thank SRC Smart Grid Research Center and its member companies on their financial support which made this work possible.

References

1. Cramer, R., Damgard, I., Nielsen, J.: Multiparty Computation, an Introduction. Contemporary Cryptology, 41–87 (2005)
2. Yao, A.C.: Protocols for secure computations. In: Proceedings of the 23rd Annual Symposium on Foundations of Computer Science, SFCS 1982, pp. 160–164 (1982)
3. Kron, G.: Diakoptics: The Piecewise Solution of Large-Scale Systems. Macdonald (1963)
4. Happ, H.H.: Diakoptics and Piecewise Methods. IEEE Transactions on Power Apparatus and Systems 89 (1970)
5. Happ, H.H.: Piecewise Methods and Applications to Power Systems. John Wiley & Sons, Inc., New York (1980)
6. Ilić, M., Zaborszky, J.: Dynamics and Control of Large Electric Power Systems. John Wiley & Sons (2000)
7. Ilić, M., Hsu, A.: Toward Distributed Contingency Screening Using Line Flow Calculators and Dynamic Line Rating Units (DLRs). IEEE Transactions on Sustainable Energy 2, 37–49 (2011)
8. Hsu, A., Ilić, M.: Distributed Newton Method For Computing RealDecoupled Power Flow In Lossy Electric Energy Networks. In: North American Power Symposium (NAPS) (2012)
9. Haibo, Z., Boming, Z., Hongbin, S., Ran, A.: A new distributed power flow algorithm between multi-control-centers based on asynchronous iteration. In: International Conference on Power System Technology (2006)
10. Cvijić, S., Ilić, M.: Contingency Screening in Multi-Control Area System Using Coordinated DC Power Flow. In: ISGT Europe Manchester, UK (2011)
11. Cvijić, S., Ilić, M.: On Limits to the Graph-Theoretic Approaches in the Electric Power Systems. In: 43rd North American Power Symposium, Boston, USA (2011)
12. Cvijić, S., Ilić, M.: Optimal Clustering for Efficient Computations of Contingency Effects in Large Regional Power Systems. In: IEEE PES General Meeting (2012)
13. Cvijić, S., Ilić, M.: Area-Level Reduction of Wheeling Loop Flows in Regional Power Networks. In: ISGT Europe 2012, Berlin (2012)
14. Patel, B.C.: Synthesis of Multiarea Grid Power Systems. PhD thesis, New Jersey Institute of Technology (1979)

Appendix

A System Parameters

The IEEE 14-bus, Figure 2, is used to illustrate the proposed algorithm for distributed DC power flow. Line parameters are listed in Table 1 while nodal injections are given in Table 2.

Table 1. Line parameters

From	To	R	X	B
1	2	0.01938	0.05917	0.0528
1	5	0.05403	0.22304	0.0492
2	3	0.04699	0.19797	0.0438
2	4	0.05811	0.17632	0.034
2	5	0.05695	0.17388	0.0346
3	4	0.06701	0.17103	0.0128
4	5	0.01335	0.04211	0
4	7	0	0.20912	0
4	9	0	0.55618	0
5	6	0	0.25202	0
6	11	0.09498	0.1989	0
6	12	0.12291	0.25581	0
6	13	0.06615	0.13027	0
7	8	0	0.17615	0
7	9	0	0.11001	0
9	10	0.03181	0.0845	0
9	14	0.12711	0.27038	0
10	11	0.08205	0.19207	0
12	13	0.22092	0.19988	0
13	14	0.17093	0.34802	0

Table 2. Generation and load

Bus Number	Generation [pu]	Load [pu]
1	2.324	0
2	0.400	0.217
3	0	0.942
4	0	0.478
5	0	0.076
6	0	0.112
7	0	0
8	0	0
9	0	0.295
10	0	0.090
11	0	0.035
12	0	0.061
13	0	0.135
14	0	0.149

A Distributed Feedback Control Approach to the Optimal Reactive Power Flow Problem

Saverio Bolognani, Guido Cavraro, and Sandro Zampieri

Department of Information Engineering, University of Padova, Italy
{saverio.bolognani,guido.cavraro,zampi}@dei.unipd.it

Abstract. We consider the problem of exploiting the microgenerators connected to the low voltage or medium voltage grid in order to provide distributed reactive power compensation in the power distribution network, solving the optimal reactive power flow problem for the minimization of power distribution losses subject to voltage constraints. The proposed strategy requires that all the intelligent agents, located at the generator buses, measure their voltage and share these data with the other agents via a communication infrastructure. The agents then adjust the amount of reactive power injected into the grid according to a policy which is a specialization of duality-based methods for constrained convex optimization. Convergence of the algorithm to the configuration of minimum losses and feasible voltages is proved analytically. Simulations are provided in order to demonstrate the algorithm behavior, and the innovative feedback nature of such strategy is discussed.

Keywords: cyber-physical systems, networked control, power distribution grid, distributed control, reactive power compensation.

1 Introduction

Recent technological advances, together with environmental and economic reasons, have been motivating the deployment of small power generators in the low voltage and medium voltage power distribution grid. The availability of a large number of these generators in the distribution grid can yield relevant benefits for the network operation, which go beyond the availability of clean, inexpensive electrical power. They can be used to provide a number of ancillary services that are of great interest for the management of the grid [1,2].

We focus in particular on the problem of optimal reactive power compensation for power losses minimization and voltage support. In order to properly command the operation of these devices, the distribution network operator is required to solve an *optimal reactive power flow* (ORPF) problem. This problem is one among the typical tasks that transmission grid operators need to solve for the efficient and safe operation of the high voltage grid. Indeed, powerful solvers have been designed for the ORPF problem, and advanced optimization techniques

D.C. Tarraf (ed.), *Control of Cyber-Physical Systems*,
Lecture Notes in Control and Information Sciences 449,
DOI: 10.1007/978-3-319-01159-2_14,

have been recently specialized for this task [3,4]. However, these solvers generally assume that an accurate model of the grid is available, that all the grid buses are monitored, that loads announce their demand profiles in advance, and that generators and actuators can be dispatched on a day-ahead, hour-ahead, and real-time basis. For this reason, these solvers are in general offline and centralized, and they collect all the necessary field data, compute the optimal configuration, and dispatch the reactive power production at the generators.

These tools cannot be applied directly to the ORPF problem faced in the low voltage or medium voltage power distribution network. The main reasons are that not all the buses of the grid are monitored, individual loads are unlikely to announce they demand profile in advance, the availability of small size generators is hard to predict (being often correlated with the availability of renewable energy sources). Moreover, the grid parameters, and sometimes even the topology of the grid, are partially unknown, and generators are expected to connect and disconnect, requiring an automatic reconfiguration of the grid control infrastructure (the so called *plug and play* approach).

Different strategies have been recently proposed in order to address these issues, ranging from purely local algorithms, in which each generator is operated according to its own measurements [2], to distributed approaches that do not require any central controller, but still require measurement at all the buses of the distribution grid [5]. Only recently, algorithms that are truly scalable in the number of generators and do not require the monitoring of all the buses of the grid, have been proposed for the problem of power loss minimization (with no voltage constraints) [6,7]. While these algorithms have been designed by specializing classical nonlinear optimization algorithms to the ORPF problem, they can also be considered as *feedback* control strategies. Indeed, the key feature of these algorithms is that they require the alternation of measurement and actuation, and therefore they are inherently online algorithms. In particular, the reactive power injection of the generators is adjusted by these algorithms based on the phasorial voltage measurements that are performed at the buses where the generators are connected. The resulting closed loop system features a tight dynamic interconnection of the physical layer (the grid, the generators, the loads) with the cyber layer (where communication, computation, and decision happen). In this paper, we design a distributed feedback algorithm for the ORPF problem with voltage constraints, in which the goal is minimizing reactive power flows while ensuring that the voltage magnitude across the network is larger than a given threshold.

In Section 2, a model for the cyber-physical system of a smart power distribution grid is provided. In Section 3, the optimal reactive power flow problem is stated. An algorithm for its solution is proposed in Section 4, and its convergence is studied in Section 5. Some simulations are provided in Section 6, while Section 7 concludes the paper discussing some relevant features of the feedback nature of the proposed strategy.

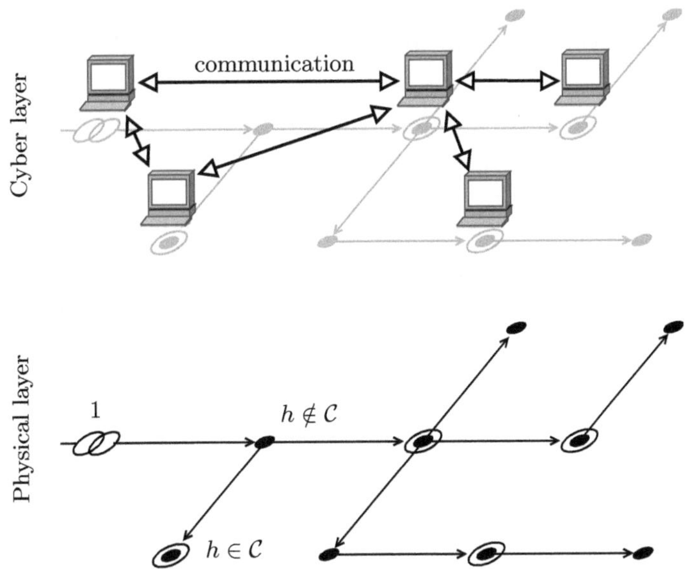

Fig. 1. A schematic representation of the the physical layer (the electric network) and the physical layer (the communication and control resources) in a smart power distribution grid. Circled nodes in the lower panel are the buses of the grid where a microgenerator is connected. Node 1 is the point of connection to the transmission grid (PCC). The other nodes are buses where loads are connected. The upper panel shows how intelligent agents in the cyber level correspond to the nodes where some sensing and actuation capabilities have been deployed, i.e. the PCC and the generator buses.

2 A Smart Power Distribution Grid

In this work, we envision a *smart* power distribution network as a cyber-physical system, in which

- the **physical layer** consists of the power distribution infrastructure, including the power lines, the loads, the microgenerators, and the point of connection to the transmission grid, while
- the **cyber layer** consists of intelligent agents, dispersed in the grid, and provided with sensing, communication, and computational capabilities.

2.1 Model for the Physical Layer

We consider a portion of the power distribution network which is populated by a number of small-size generators, together with regular loads. We model this grid via a radial directed graph $\mathcal{G}$ (i.e. a tree) in which edges represent the power lines, and the n nodes (whose set is denoted by $\mathcal{V}$) represent the buses and also the point of common coupling (PCC), i.e. the point where the distribution grid that we are considering is connected to the transmission grid (see Figure 1).

Given an edge e, we denote by $\sigma(e)$ its source node and by $\tau(e)$ its terminal node. We can therefore introduce the incidence matrix of $\mathcal{G}$, defined via its elements

$$A_{ev} = \begin{cases} -1 & \text{if } v = \sigma(e) \\ 1 & \text{if } v = \tau(e) \\ 0 & \text{otherwise.} \end{cases}$$

We introduce the following assumption on the power line impedances.

Assumption 1. *Let all the grid power line impedances have the same inductance/resistance ratio, i.e. for any edge e of the graph $\mathcal{G}$, its impedance z_e satisfies*

$$z_e = |z_e| \exp(j\theta).$$

The grid electric topology and the grid power line parameters are therefore fully described by the parameter θ and by the weighted Laplacian $L \in \mathbb{R}^{n\times n}$, defined as

$$L = A^T Z^{-1} A, \tag{1}$$

where

$$Z = \begin{bmatrix} \ddots & & \\ & |z_e| & \\ & & \ddots \end{bmatrix}$$

is the diagonal matrix of the absolute values of power line impedances.

We limit our study to the electric steady state behavior of the system, in which all voltages and currents are sinusoidal signals at the same frequency ω. We therefore assume that each signal (voltages and currents) can be represented via a complex number whose absolute value corresponds to the signal root-mean-square value, and whose phase corresponds to the phase of the signal with respect to an arbitrary common reference. In this notation, the state of a grid is therefore described by the bus voltages $u_v \in \mathbb{C}$, $v \in \mathcal{V}$.

In the following, we present a static model for the nodes of $\mathcal{G}$. We model node 1 (corresponding to the PCC) as a *slack bus*, i.e. a constant voltage generator at the nominal voltage $U_N \in \mathbb{R}$ and zero angle

$$u_1 = U_N.$$

We assume instead that every other node h is a *PQ-bus*, i.e. the complex power injected at the bus is independent from the bus voltage u_h and is equal to

$$p_h + jq_h,$$

where p_h and q_h are the injected *active* and *reactive* powers, respectively. Microgenerators fit in the PQ model (or *constant power* model) once they are commanded via a complex power reference as in [8,9]. It is also a reasonable approximation for many residential and industrial loads[1].

2.2 Model for the Cyber Layer

We assume that every generator bus, and also the PCC, correspond to an *agent* in the cyber layer (see the upper panel of Figure 1). We denote by $\mathcal{C}$ (with $|\mathcal{C}| = m$) this subset of the nodes of $\mathcal{G}$.

Each agent is provided with some computational capability, and with some sensing capability, in the form of a phasor measurement unit (i.e. a sensor that can measure voltage amplitude and angle [11]). Agents can communicate, via some communication channel that could possibly be the same power lines (via power line communication – PLC – technologies).

3 Optimal Reactive Power Flow Problem Formulation

Given the models and the definitions introduced above, we formulate the ORPF problem in the following form,

$$\begin{aligned} \min_{q_h, h \in \mathcal{C} \setminus \{1\}} \quad & J_{\text{losses}} \\ \text{subject to} \quad & |u_h| \geq U_{\min}, \quad \forall h \in \mathcal{C}, \\ & |u_h| \leq U_{\max}, \quad \forall h \in \mathcal{C}, \end{aligned} \tag{2}$$

where J_{losses} are the power distribution losses on the lines, and $U_{\min}$ is a given lower bound for bus voltage magnitudes.

Notice that the decision variables in the optimization problem (2) are the reactive power injections at the microgenerators, which are the only physical devices that we aim to control via the proposed feedback algorithm. The reactive power flowing from the transmission grid into the distribution grid via the PCC (which we modeled as a slack bus) is not part of the set of decision variables, because it automatically adjusts in order to ensure that power balance is satisfied at any time in this portion of the distribution grid.

Notice moreover that the voltage constraints are defined only on the nodes where we deployed sensing and actuation devices. Given the fact that the remaining nodes (corresponding to loads) are unmonitored, we cannot aim to design any algorithm that can guarantee the satisfaction of operational constraints at such nodes. However, in the case in which the voltage magnitude of some nodes is of particular interest, the network operator has the flexibility of deploying agents (i.e. PMU sensing units and the corresponding communication devices) also in those nodes, and include them in the algorithm in order to guarantee voltage feasibility also in those buses. Alternatively, given *a priori* bounds on the

[1] More general models could be considered, namely the *exponential model* and the *ZIP* model, similarly to what has been done in [10]. The following analysis would remain exactly the same, at the cost of a slightly more complex notation.

maximum power demand of the loads, the network operator can infer worst-case-type bounds on the maximum voltage drop that can occur between unmonitored nodes and the closest agent, and therefore increase the voltage bound $U_{\min}$ accordingly (possibly also agent-wise). Both these solutions (ad-hoc deployment of the agents, and worst-case guarantees for the voltage on unmonitored nodes) are implementation issues that we are not addressing in this paper, but that can be easily included on top of the approach that we describe here.

The ultimate goal of ORPF strategies is therefore the minimization of reactive power flows across the power grid (by injecting reactive power as close at possible to the buses that need it) while at the same time ensuring that the bus voltages (which are a function also of reactive power flows) is kept inside a given range, in order to guarantee more reliable and robust operation of the grid (see voltage stability issues in [12]).

4 Proposed ORPF Algorithm

In order to formally describe the algorithm, we need the following definitions.

Definition 1 (Path). *Let $h, k \in \mathcal{V}$ be two nodes of the graph $\mathcal{G}$. The path $\mathcal{P}_{hk} = (v_1, \dots, v_\ell)$ is the sequence of nodes, without repetitions, that satisfies*

- $v_1 = h$
- $v_\ell = k$
- *for each $i = 1, \dots, \ell - 1$, the nodes v_i and v_{i+1} are connected by an edge.*

Notice that, as the distribution grid topology is radial, there is only one path connecting a pair of nodes $h, k \in \mathcal{V}$.

Definition 2 (Neighbors in the cyber layer). *Let $h \in \mathcal{C}$. The set of nodes that are neighbors of h in the cyber layer, denoted as $\mathcal{N}(h)$, is the subset of $\mathcal{C}$ defined as*

$$\mathcal{N}(h) = \{k \in \mathcal{C} \mid \mathcal{P}_{hk} \cap \mathcal{C} = \{h, k\}\}.$$

Figure 2 gives an example of such set. We assume that every agent $h \in \mathcal{C}$ knows its set of neighbors $\mathcal{N}(h)$, and can communicate with them. Notice that this architecture can be constructed by each agent in a distributed way, for example by exploiting the PLC channel (as suggested for example in [13]). This allows also a plug-and-play reconfiguration of such architecture when new agents are connected to the grid.

It is also assumed that each agent $h \in \mathcal{C}$ has a local knowledge of the grid electric parameters, and in particular of the following parameter.

Definition 3 (G-parameters). *For each pair $h, k \in \mathcal{C}$, let us define the parameter*

$$G_{hk} = |i_h| \Big|_{\substack{u_k = 1 \\ u_\ell = 0,\ \ell \in \mathcal{C} \setminus \{k\} \\ i_\ell = 0,\ \ell \notin \mathcal{C}}} . \tag{3}$$

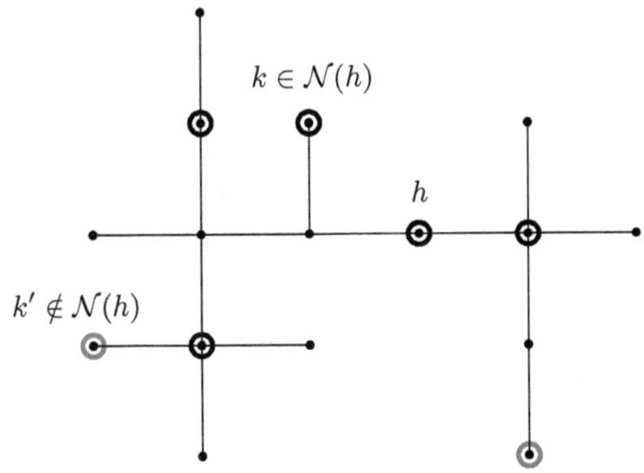

Fig. 2. An example of neighbor nodes in the cyber layer. Circled nodes (both gray and black) are nodes in $\mathcal{C}$. Nodes circled in black belong to the set $\mathcal{N}(h) \subset \mathcal{C}$. Node circled in gray are agents which do not belong to the set of neighbors of h. For each agent $k \in \mathcal{N}(h)$, the path that connects h to k does not include any other agent besides h and k themselves.

i.e. the current that would be injected at node h if

- *node k was replaced with a unitary voltage generator;*
- *all other nodes in $\mathcal{C}$ (the other agents) were replaced by short circuits;*
- *all nodes not in $\mathcal{C}$ (load buses) were replaced by open circuits.*

Notice that the parameters G_{hk} depend only on the grid electric topology, and that

$$G_{hk} \neq 0 \quad \text{if and only if} \quad k \in \mathcal{N}(h). \tag{4}$$

Figure 3 gives a representation of this definition. Notice that, in the special case in which the paths from h to its neighbors are all disjoint paths, then $G_{hk} = 1/|Z_{hk}|$, where Z_{hk} is the impedance of the electric path connecting h to k. The G-parameters could also be defined from algebraic operations on the Laplacian of the graph $\mathcal{G}$ (Schur complement), corresponding to a node elimination procedure on the electrical network, as discussed in detail in [14].

As suggested in [13], these parameters can be estimated in an initialization phase via some ranging technologies over the PLC channel. Alternatively, this limited amount of knowledge of the grid topology can be stored in the agents at the deployment time. Finally, the same kind of information can be inferred by specializing the procedures that use the extended capabilities of the generator power inverters for online grid sensing and impedance estimation [15,16].

Given the above definitions, we can state the proposed algorithm to solve the optimization problem (2). We will show in Section 5 how the algorithm is inspired by a dual decomposition approach [17] to (2). While problem (2) might not be convex in general, we rely on the results presented in [18] which show that zero duality gap holds for the ORPF problems, under some conditions that are

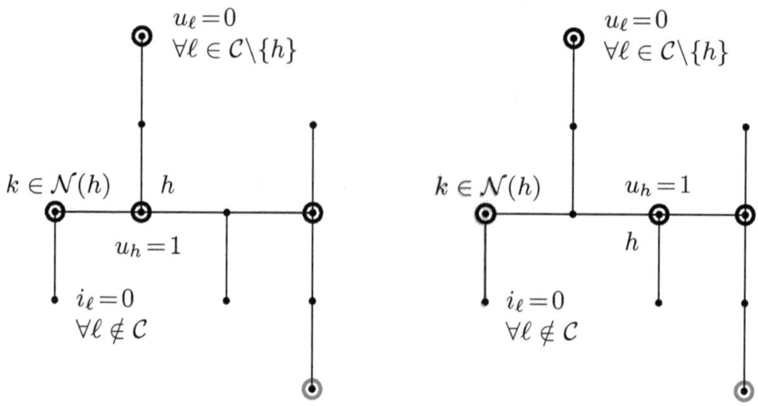

Fig. 3. A representation of how the elements G_{kh} are defined. Notice that in the configuration of the left panel, as the paths from h to its neighbors $k \in \mathcal{N}(h)$ do not share any edge, the gains G_{kh} corresponds to the absolute value of the path admittances $1/|Z_{kh}|$.

commonly verified in practice and in particular in radial networks like the ones that we are considering. We will also show that, introducing an approximate model for the grid, convergence of the algorithm can be studied analytically. Finally, in Section 6, we will validate the proposed algorithm via simulations, introducing all the non-idealities that have been neglected in the analytic study of the algorithm convergence.

ORPF Algorithm. *Let all agents store two auxiliary scalar variables λ_h^+ and λ_h^-. Let γ be a positive scalar parameter, and let θ be the impedance angle defined in Assumption 1. At every synchronous iteration of the algorithm, each agent $h \in C$ executes the following operations:*

- *measures the voltage $u_h = |u_h| \exp(j\angle u_h)$;*
- *gathers the measurements $\{u_k, k \in \mathcal{N}(h)\}$ from its neighbors;*
- *updates the auxiliary variables λ_h^+ and λ_h^- as*

$$\lambda_h^+ \leftarrow \left[\lambda_h^+ + \gamma \left(U_{min}^2 - |u_h|^2\right)\right]_+, \tag{5}$$

$$\lambda_h^- \leftarrow \left[\lambda_h^- - \gamma \left(U_{max}^2 - |u_h|^2\right)\right]_+, \tag{6}$$

  where $[\cdot]_+$ stands for the operation of projection on the positive orthant;
- *updates the injected reactive power q_h as*

$$q_h \leftarrow q_h + 2\sin\theta(\lambda_h^+ - \lambda_h^-) + \sum_{k \in \mathcal{N}(h)} G_{hk}|u_h||u_k| \sin(\angle u_k - \angle u_h - \theta). \tag{7}$$

Notice that the agent located at the PCC does not need to perform the updates (5) and (7), because the PCC is a constant-voltage slack bus, and therefore the reactive power injected by the PCC automatically results from the power balance in the grid.

5 Convergence Analysis

Before presenting the main result about the conditions that guarantee convergence of the proposed method, we show how this algorithm derives from a dual decomposition of the original problem and from the specialization of the *dual ascent methods* [17] to this specific system.

Let u and u' be the vectors obtained by stacking the voltages $u_h, h \in \mathcal{C}$ and $u_k, k \in \mathcal{V} \backslash \mathcal{C}$, respectively. In the same way, let us define i, i', p, p', q, q' as the vectors of injected currents, active power, and reactive power, respectively.

Given Assumption 1, and given the weighted Laplacian L defined in (1), the following grid equation is satisfied

$$\exp(-j\theta) L \begin{bmatrix} u \\ u' \end{bmatrix} = \begin{bmatrix} i \\ i' \end{bmatrix}.$$

It is also possible to construct a positive semi-definite symmetric pseudoinverse $X \in \mathbb{R}^{n \times n}$ (see [10]) such that

$$\left(I - \mathbb{1} e_1^T\right) \begin{bmatrix} u \\ u' \end{bmatrix} = \exp(j\theta) X \begin{bmatrix} i \\ i' \end{bmatrix}, \tag{8}$$

where $\mathbb{1}$ is a vector all ones, and e_h is a vector which is valued 1 in position h, and 0 everywhere (i.e., is the h-th canonical basis vector).

The matrix X has some notable properties, including the fact that

$$(e_h - e_k)^T X (e_h - e_k) = |Z_{hk}|, \quad h, k \in \mathcal{V} \tag{9}$$

$$X e_1 = 0, \tag{10}$$

and the fact that

$$X_{hh} \geq X_{hk} \geq 0 \quad h, k \in \mathcal{V}. \tag{11}$$

With the same partitioning as before, we can also partition X into blocks, as

$$X = \begin{bmatrix} M & N \\ N^T & Q \end{bmatrix}. \tag{12}$$

This notation allows us to introduce the approximate model for the power flows in the grid proposed in [10], where a rigorous derivation and analysis of the approximation error is also provided. According to this model

- the bus voltages of nodes in $\mathcal{C}$ can be expressed as

$$u = U_N \mathbb{1} + \frac{\exp(j\theta)}{U_N} \left[M \ N\right] \begin{bmatrix} p - jq \\ p' - jq' \end{bmatrix} + o\left(\frac{1}{U_N^2}\right); \tag{13}$$

- reactive power injections satisfy, at every time, the balance

$$\mathbb{1}^T \begin{bmatrix} q \\ q' \end{bmatrix} = o\left(\frac{1}{U_N}\right); \tag{14}$$

– the problem of optimal reactive power flow for losses minimization is equivalent to the problem of minimizing

$$J = q^T \frac{M}{2} q + q^T N q' + o\left(\frac{1}{U_N}\right). \tag{15}$$

This also allows us to express the squared voltage magnitudes $|u_h|^2$ via (13), obtaining

$$\begin{bmatrix} \vdots \\ |u_h|^2 \\ \vdots \end{bmatrix} = U_N^2 \mathbb{1} + 2\,\mathrm{Re}\left(\exp(j\theta)\,[M\ N] \begin{bmatrix} p - jq \\ p' - jq' \end{bmatrix}\right) + o\left(\frac{1}{U_N}\right). \tag{16}$$

The proposed approximation is based on the fact that the grid operating point, in its regular regime, is characterized by a relatively high nominal voltage compared to the voltage drops across the power lines, and by relatively small power distribution losses, compared to the power delivered to the loads. Notice that similar approximations have been used before in the literature for the problem of estimating power flows on the power lines (see among the others [19,20] and references therein). It also shares some similarities with the DC power flow model [21, Chapter 3], extending it to the case in which lines are not purely inductive but also resistive (which is crucial in power distribution networks). In the analysis of the algorithm convergence, we will neglect the infinitesimal terms in both (15) and (16).

In the following, we will consider only the lower bound constraint in the optimization problem (2). We will therefore need only one of the two set of auxiliary variables, i.e. λ_h^+ (which we will rename λ_h in order to simplify the notation). The upper voltage constraint can be included following exactly the same lines that will be presented hereafter for the lower voltage constraint, with no additional complication to the design. Notice that the two constraints $|u_h| \geq U_{\min}$ and $|u_h| \leq U_{\max}$ at any node h will never be violated at the same time, and therefore only one of the two additional variables (Lagrange multipliers) λ_h^+ and λ_h^- will be different from zero in practice.

We then consider the following problem in the decision variables q

$$\min_{\mathbb{1}^T q = c} \quad q^T \frac{M}{2} q + q^T N q' \tag{17a}$$

$$\text{subject to} \quad U_{\min}^2 \mathbb{1} - v(q) \leq 0, \tag{17b}$$

where $c = -\mathbb{1}^T q'$ and where $v(q)$ is defined as

$$v(q) = U_N^2 \mathbb{1} + 2\,\mathrm{Re}\left(\exp(j\theta)\,[M\ N] \begin{bmatrix} p - jq \\ p' - jq' \end{bmatrix}\right). \tag{18}$$

Based on the proposed approximated model, and in particular by plugging (13) into (7), we can also rewrite (via some algebraic manipulations) the optimization algorithm presented in the previous section as a m-dimensional discrete time system, with the following update equations.

$$q(t+1) = q(t) - G\begin{bmatrix}M & N\end{bmatrix}\begin{bmatrix}q(t)\\ q'\end{bmatrix} + 2\sin\theta\lambda(t) + k(q(t),\lambda(t))e_1 \tag{19a}$$

$$\lambda(t+1) = \left[\lambda(t) + \gamma\left(U_{\min}^2 \mathbb{1} - v(q(t+1))\right)\right]_+ \tag{19b}$$

where G is the matrix whose elements G_{hk} have been defined in Definition 3, and k is a scalar that guarantees that constraint (14) is met at every iteration:

$$k(q(t),\lambda(t)) = \mathbb{1}^T G\begin{bmatrix}M & N\end{bmatrix}\begin{bmatrix}q(t)\\ q'\end{bmatrix} - 2\sin\theta\mathbb{1}^T\lambda(t).$$

Notice that the term ke_1 models the fact that, when the system is actuated at the end of every algorithm iteration, the reactive power injected by the slack bus 1 (the PCC) automatically balances the variations in the reactive power injection that have been commanded to the generators.

In the following, we show how the algorithm (19) is a specialization of the dual ascent algorithms for the solution of the optimization problem (17). The Lagrangian of the problem (17) is

$$\mathcal{L}(q,\lambda) = q^T\frac{M}{2}q + q^T N q' + \lambda^T\left(U_{\min}^2\mathbb{1} - v(q)\right) \tag{20}$$

where λ is the Lagrangian multiplier (dual variables).

A dual ascent algorithm consists in the iterative execution of the following alternated steps

1. minimization of the Lagrangian with respect to the primal variables q

$$q(t+1) = \arg\min_{\mathbb{1}^T q = c} \mathcal{L}(q(t),\lambda(t)), \tag{21}$$

2. dual gradient ascent step on the dual variables

$$\lambda(t+1) = \left[\lambda(t) + \gamma\frac{\partial\mathcal{L}(q(t+1),\lambda(t)}{\partial\lambda}\right]_+. \tag{22}$$

The partial derivative of the Lagrangian with respect to q results to be, by inspecting (20) and (18),

$$\frac{\partial\mathcal{L}}{\partial q} = Mq + Nq' - \frac{\partial v(q)}{\partial q}\lambda = Mq + Nq' - 2\sin\theta M\lambda.$$

In order to show that (19a) is indeed the primal step that minimizes the Lagrangian with respect to the primal variable q, we need the following technical lemma.

Lemma 1. *Let G be defined element-wise as in Definition 3, and let M be defined as in* (12). *Then*

$$MG = I - \mathbb{1}e_1^T.$$

Proof. From the definition of G, and by adopting the vector notation presented in this section, we have that, given Assumption 1 and when $i' = 0$,

$$i = \exp(-j\theta)Gu.$$

Then, by using (8), we have that

$$(I - \mathbb{1}e_1^T)\begin{bmatrix} u \\ u' \end{bmatrix} = \exp(j\theta)\begin{bmatrix} M & N \\ N^T & Q \end{bmatrix}\begin{bmatrix} \exp(-j\theta)Gu \\ 0 \end{bmatrix},$$

and thus the conclusion. □

By evaluating $\frac{\partial \mathcal{L}}{\partial q}$ in $q(t+1)$ defined as is (19a), and by using the result of Lemma 1, we have

$$\begin{aligned}\frac{\partial \mathcal{L}(q(t+1), \lambda(t))}{\partial q} &= Mq(t+1) + Nq' - 2\sin\theta M\lambda(t) \\ &= Mq(t) - MG\begin{bmatrix} M & N \end{bmatrix}\begin{bmatrix} q(t) \\ q' \end{bmatrix} + Nq' \\ &= \mathbb{1}e_1^T Nq',\end{aligned}$$

which is orthogonal to the feasible set $\mathbb{1}^T q = c$, and therefore proves that $q(t+1)$ solves (21).

In order to prove that (19b) is the dual ascent step described in (22), it is enough to evaluate $\frac{\partial \mathcal{L}}{\partial \lambda}$ to see that

$$\frac{\partial \mathcal{L}(q(t+1), \lambda(t))}{\partial \lambda} = U_{\min}^2 \mathbb{1} - v(q(t+1)),$$

in accordance to the well known result in dual decomposition that the dual ascent direction is given by the constraint violation.

We can then state the following convergence result.

Theorem 1. *Consider the algorithm described in* (19) *for the optimization problem* (17)*, which is an approximated description of the algorithm presented in Section 4 for the optimization problem described in Section 3. The algorithm converges if*

$$\gamma \leq \frac{1}{4\sin^2\theta Dm},$$

where m is the cardinality of $\mathcal{C}$ (the number of generator buses plus one) and $D = \max_h |Z_{1h}|$ is the maximum electric distance of a generator bus from the PCC.

The proof is provided in the Appendix.

6 Simulations

The algorithm has been tested on the testbed IEEE 37 [22], which is an actual portion of power distribution network located in California. The load buses are a blend of constant-power, constant-current, and constant-impedance loads, with a total power demand of almost 2 MW of active power and 1 MVAR of reactive power (see [22] for the testbed data). The length of the power lines range from a minimum of 25 meters to a maximum of almost 600 meters. The impedance of the power lines differs from edge to edge (for example, resistance ranges from 0.182 Ω/km to 1.305 Ω/km). However, the inductance/resistance ratio exhibits a smaller variation, ranging from $\angle z_e = 0.47$ to $\angle z_e = 0.59$. This justifies Assumption 1, in which we claimed that $\angle z_e$ can be considered constant across the network.

The lower and upper bounds for voltage magnitudes has been set to 96% and 104% of the nominal voltage $U_N = 4800$, respectively. The algorithm presented in Section 4 has been simulated on a nonlinear exact solver of the grid. None of the assumptions that have been considered during the design of the algorithm (constant power loads, constant line impedance angle $\angle z_e$, linearized power flow equations) has been used in these simulations, being only a tool for the design of the algorithm and for the study of the algorithm's convergence.

When the grid is operated according to the testbed data, all voltages are above the threshold. In this configuration, the proposed algorithm is capable of reducing the power distribution losses practically to the minimum. Indeed we have that

Power distribution losses	
With no optimization	47.164 KW
With the proposed algorithm	38.309 KW
With numerical nonlinear optimizer	37.931 KW

and therefore the proposed optimization algorithm can achieve more than 95% of the potential loss reduction.

In order to evaluate the performance of the algorithm when voltage constraints are active, the active power demand of two nodes of the grid has been increased step-wise (see Figure 4). While this variation in the active power demand does not have effect on the optimal reactive power configuration (up to second-order effects), the increased load causes a drop in the voltage magnitudes and an increase in power distribution losses up to 57.5 kW. At this point voltage constraints are not satisfied for one of the agents. The algorithm then drives the system to a new optimal configuration that guarantees satisfaction of all voltage constraints, at the cost of slightly larger losses (from 57.5 kW to 57.7 kW).

Additional simulations, including the case of time varying generation profiles and intermittent loads, are available in [23].

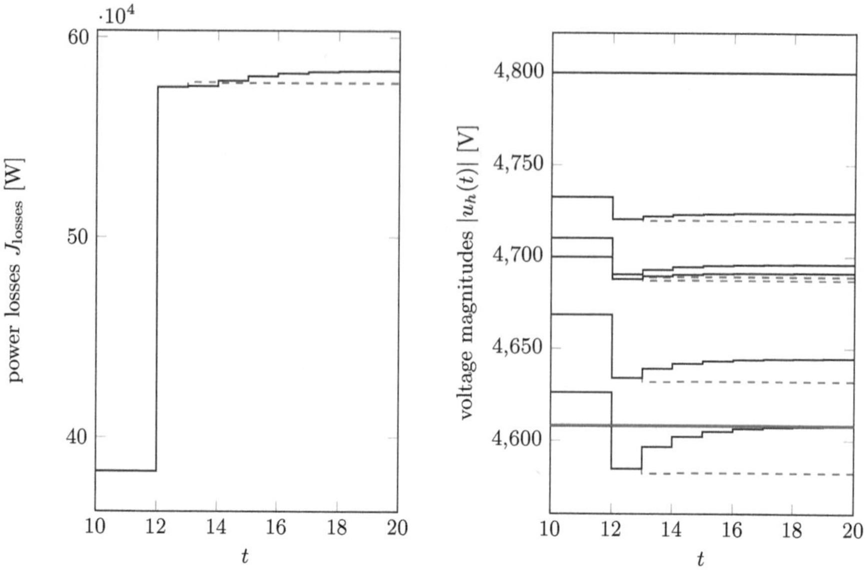

Fig. 4. The left panel represents total power losses, while the right panel represents the voltage magnitudes at the buses where generators are connected. The solid line represents the behavior of the proposed algorithm, with the parameter γ equal to 0.1 (one half of the bound provided by Theorem 1). The dashed line, on the other hand, represent the behavior of the loss minimization algorithm if no voltage constraints are enforced. The red thick line in the right panel represents the desired voltage bound.

7 Conclusion

In this paper we proposed a distributed algorithm for the problem of optimal reactive power flow in a smart power distribution grid, that is based on a *feedback strategy*, in the sense that it requires the *interleaving of actuation and measurement*, and that the control action is a function of real time data collected from the agents. Figure 5 provides a block diagram representation of the proposed algorithm. The two feedback functions K_1 and K_2 (both functions of the measured voltages) are defined element-wise as

$$[K_1(u)]_h = \sum_{k \in \mathcal{N}(h)} G_{hk}|u_h||u_k| \sin(\angle u_k - \angle u_h - \theta) \tag{23}$$

and

$$[K_2(u)]_h = U_{\min}^2 - |u_h|^2. \tag{24}$$

This interpretation of a dual-ascent optimization algorithm as a control feedback loop with memory, resembles what has been recently done in [24], and allows to state some final remarks.

By adopting a feedback strategy on the measured voltages, the active power injections in the grid ($p_h, h \in \mathcal{V}$) and the reactive power injection of the loads ($q_h, h \in \mathcal{V}\backslash\mathcal{C}$) can be considered as *disturbances* for the control system. This means that these quantities do not need to be known to the agents: in some sense, the agents are implicitly inferring this information from the voltage measurements performed on the grid. This feature differentiates the proposed algorithm from basically all the ORPF algorithms available in the power system literature, with the exception of some works, like [20], where however the feedback is only local, with no communication between the agents, and of [6] and [7]. Moreover, because of this feedback strategy, the controller (or optimizer) does not need to solve any model of the grid in order to find the optimal solution. While a model of the grid has been used in the design of the algorithm, the online controller does not need to know the grid parameter and to solve the nonlinear equations that are generally a critical issue in offline ORPF solvers. On the contrary, the computational effort required for the execution of the proposed algorithm is minimal. These features are extremely interesting for the scenario of low voltage or medium voltage power distribution networks, where real time measurement of the loads is usually not available, the grid parameters are only partially unknown, and many buses are unmonitored.

Another feature that becomes apparent from this feedback interpretation, is the guarantee that the algorithm can provide regarding the eventual satisfaction of the voltage constraints. Because the output of the function $K_2(u)$ defined in (24) is integrated, it is guaranteed that, if the algorithm converges, then $|u_h| \geq U_{\min}$ for every node $h \in \mathcal{C}$. This is true independently of the choice of the parameters γ and also of the coefficients G_{hk}.

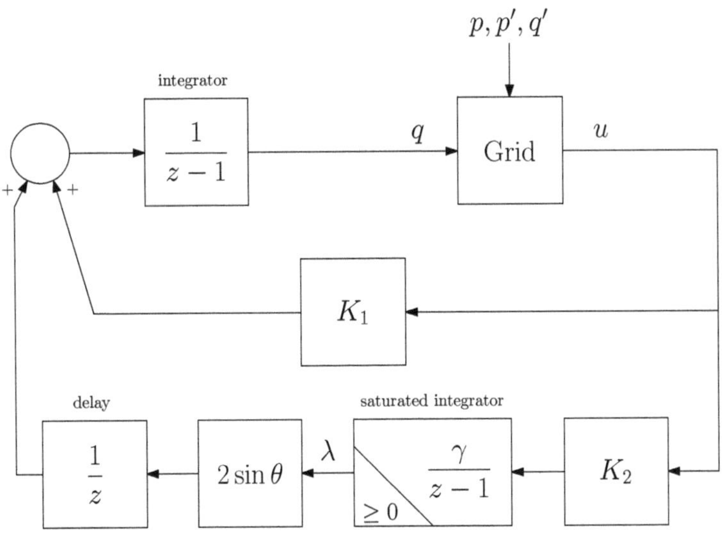

Fig. 5. A block diagram representation of the algorithm proposed in Section 4. The two feedback functions K_1 and K_2 are defined element-wise in (23) and (24).

Finally, a control-theoretic approach to the problem of optimal power flow enables a number of analyses on the performance of the closed loop system that are generally overlooked when tackling the problem with the tools of nonlinear optimization. Examples are L_2-like metrics for the resulting losses in a time-varying scenario (see for example the preliminary results in [25]), robustness to measurement noise and parametric uncertainty, stability margin against communication delays. These analyses, still not investigated, are also of interest for the design of the cyber architecture that has to support this and other real time algorithms, because they can provide specifications for the communication channels, the communication protocols, and the computational resources that need to be deployed in a smart distribution grid.

A Proof of Theorem 1

As shown in Section 5, the algorithm (19) is a dual-ascent algorithm for the solution of the constrained quadratic problem (17). As the expression (18) for the voltage constraint (17b) is an affine function of the decision variables, strong duality holds. We thus have zero duality gap and therefore if the dual ascent algorithm converges, it converges to the optimal solution of the problem.

In order to characterize converge of the algorithm via a condition on the parameter γ, we need to define the two quantities

$$x(t) = q(t) - q^* \quad \text{and} \quad y(t) = \lambda(t) - \lambda^*,$$

where q^* and λ^* are the optimal value of the primal and dual variables, respectively. Notice that, because of the constraint $\mathbb{1}^T q = c$ in (17), we have that

$$\mathbb{1}^T x(t) = 0, \quad \forall t.$$

The following two necessary conditions descend from the Uzawa saddle point theorem [26]:

$$\frac{\partial \mathcal{L}(q^*, \lambda^*)}{\partial q} = \begin{bmatrix} M & N \end{bmatrix} \begin{bmatrix} q^* \\ q' \end{bmatrix} - 2 \sin\theta M \lambda^* = \alpha \mathbb{1} \tag{25}$$

for some α, and

$$\frac{\partial \mathcal{L}(q^*, \lambda^*)}{\partial \lambda_h} = U_{\min}^2 \mathbb{1} - v(q^*) \leq 0 \tag{26}$$

with

$$\frac{\partial \mathcal{L}(q^*, \lambda^*)}{\partial \lambda_h} = U_{\min}^2 \mathbb{1} - v(q^*) < 0 \quad \Leftrightarrow \quad \lambda_h^* = 0. \tag{27}$$

The update for $x(t)$ is, given (19a),

$$\begin{aligned} x(t+1) &= q(t+1) - q^* \\ &= x(t) - G \begin{bmatrix} M & N \end{bmatrix} \begin{bmatrix} q^* + x(t) \\ q' \end{bmatrix} + 2 \sin\theta \lambda(t) + k e_1 \\ &= (I - GM) x(t) - G \begin{bmatrix} M & N \end{bmatrix} \begin{bmatrix} q^* \\ q' \end{bmatrix} + 2 \sin\theta \lambda(t) + k e_1, \end{aligned}$$

which, by using (25) and the fact that $G\mathbb{1} = 0$ becomes

$$\begin{aligned} x(t+1) &= (I - GM)x(t) - 2\sin\theta GM\lambda^* + 2\sin\theta\lambda(t) + ke_1 \\ &= e_1\mathbb{1}^T x(t) - 2\sin\theta(I - e_1\mathbb{1}^T)\lambda^* + 2\sin\theta\lambda(t) + ke_1 \\ &= 2\sin\theta y(t) + k'e_1, \end{aligned}$$

where $k' = 2\sin\theta\mathbb{1}^T\lambda^* + k$ and where we used Lemma 1 and the fact that $\mathbb{1}^T x(t) = 0$.

We then consider the update equation for $y(t)$. By using the fact that, according to (18),

$$v(q(t)) = v(q^*) + 2\sin\theta Mx(t),$$

and according to (19b), we have that

$$\begin{aligned} y(t+1) &= \lambda(t+1) - \lambda^* \\ &= \left[\lambda(t) + \gamma\left(U_{\min}^2\mathbb{1} - v(q(t+1))\right)\right]_+ - \lambda^* \\ &= \left[y(t) + \lambda^{\mathrm{opt}} + \gamma(U_{\min}^2\mathbb{1} - v(q^*)) - 2\gamma\sin\theta Mx(t+1)\right]_+ - \lambda^*. \end{aligned}$$

By using the fact that $\lambda^* \geq 0$ together with (26) and (27) we have that

$$\lambda^* = [\lambda^*]_+ = [\lambda^* + \gamma(U_{\min}^2\mathbb{1} - v(q^*))]_+$$

Therefore, by plugging in the expression for $x(t+1)$ and by using the fact that $Me_1 = 0$, we have

$$\begin{aligned} y(t+1) = \left[(I - 4\gamma\sin^2\theta M)y(t) + \lambda^* + \gamma(U_{\min}^2\mathbb{1} - v(q^*))\right]_+ \\ - [\lambda^* + \gamma(U_{\min}^2\mathbb{1} - v(q^*))]_+. \end{aligned}$$

Then, by using the fact that

$$\|a_+ - b_+\| \leq \|a - b\|,$$

we have that

$$\|y(t+1)\| \leq \left\|(I - 4\gamma\sin^2\theta M)y(t)\right\|,$$

and therefore $y(t)$ converges to zero if

$$\gamma \leq \frac{1}{4\sin^2\theta\rho(M)},$$

where $\rho(M)$ is the spectral radius of M.

Finally, because of (11), we have that

$$M_{hh} \geq M_{hk} \geq 0 \quad \forall h, k,$$

and thus

$$\rho(M) = \max_{\|v\|=1} \sqrt{\sum_h \left(\sum_k M_{hk} v_k\right)^2} \leq \max_{\|v\|=1} \sqrt{\sum_h \left(M_{hh} \sum_k v_k\right)^2}$$
$$\leq \sqrt{\sum_h \left(M_{hh}\sqrt{m}\right)^2} \leq \sqrt{\sum_h \left(\max_h M_{hh}\sqrt{m}\right)^2} = \max_h M_{hh} m.$$

By using the fact that, via (9), $M_{hh} = |Z_{1h}|$, we then finally have the sufficient condition for convergence

$$\gamma \leq \frac{1}{4 \sin^2 \theta \max_h |Z_{1h}| m}.$$

□

References

1. Katiraei, F., Iravani, M.R.: Power management strategies for a microgrid with multiple distributed generation units. IEEE Trans. Power Syst. 21(4), 1821–1831 (2006)
2. Prodanovic, M., De Brabandere, K., Van den Keybus, J., Green, T., Driesen, J.: Harmonic and reactive power compensation as ancillary services in inverter-based distributed generation. IET Gener. Transm. Distrib. 1(3), 432–438 (2007)
3. Zhao, B., Guo, C.X., Cao, Y.J.: A multiagent-based particle swarm optimization approach for optimal reactive power dispatch. IEEE Trans. Power Syst. 20(2), 1070–1078 (2005)
4. Lavaei, J., Rantzer, A., Low, S.H.: Power flow optimization using positive quadratic programming. In: Proc. 18th IFAC World Congr. (2011)
5. Lam, A.Y.S., Zhang, B., Dominiguez-Garcia, A., Tse, D.: Optimal distributed voltage regulation in power distribution networks. arXiv [math.OC] 1204.5226 (2012)
6. Tenti, P., Costabeber, A., Mattavelli, P., Trombetti, D.: Distribution loss minimization by token ring control of power electronic interfaces in residential microgrids. IEEE Trans. Ind. Electron 59(10), 3817–3826 (2012)
7. Bolognani, S., Zampieri, S.: Distributed control for optimal reactive power compensation in smart microgrids. In: Proc. 50th IEEE Conf. on Decision and Control and European Control Conf. (CDC-ECC 2011), Orlando, FL (2011)
8. Green, T.C., Prodanović, M.: Control of inverter-based micro-grids. Electr. Pow. Syst. Res. 77(9), 1204–1213 (2007)
9. Lopes, J.A., Moreira, C.L., Madureira, A.G.: Defining control strategies for microgrids islanded operation. IEEE Trans. Power Syst. 21(2), 916–924 (2006)
10. Bolognani, S., Zampieri, S.: A distributed control strategy for reactive power compensation in smart microgrids. IEEE Transactions on Automatic Control 56(11) (November 2013)
11. Phadke, A.G.: Synchronized phasor measurements in power systems. IEEE Comput. Appl. Power 6(2), 10–15 (1993)
12. Kundur, P.: Power system stability and control. McGraw-Hill (1994)

13. Costabeber, A., Erseghe, T., Tenti, P., Tomasin, S., Mattavelli, P.: Optimization of micro-grid operation by dynamic grid mapping and token ring control. In: Proc. 14th European Conf. on Power Electronics and Applications (EPE), Birmingham, UK (2011)
14. Dorfler, F., Bullo, F.: Kron reduction of graphs with applications to electrical networks. IEEE Transactions on Circuits and Systems I 60, 150–163 (2013)
15. Ciobotaru, M., Teodorescu, R., Rodriguez, P., Timbus, A., Blaabjerg, F.: Online grid impedance estimation for single-phase grid-connected systems using PQ variations. In: Proc. 38th IEEE Power Electronics Specialists Conf., PESC (2007)
16. Ciobotaru, M., Teodorescu, R., Blaabjerg, F.: On-line grid impedance estimation based on harmonic injection for grid-connected PV inverter. In: Proceedings of the IEEE International Symposium on Industrial Electronics (ISIE), pp. 2437–2442 (2007)
17. Bertsekas, D.P.: Nonlinear programming, 2nd edn. Athena Scientific, Belmont (1999)
18. Lavaei, J., Low, S.H.: Zero duality gap in optimal power flow problem. IEEE Trans. Power Syst. (2011)
19. Baran, M.E., Wu, F.F.: Optimal sizing of capacitors placed on a radial distribution system. IEEE Trans. Power Del. 4, 735–743 (1989)
20. Turitsyn, K., Šulc, P., Backhaus, S., Chertkov, M.: Options for control of reactive power by distributed photovoltaic generators. Proc. IEEE 99(6), 1063–1073 (2011)
21. Gómez-Expósito, A., Conejo, A.J., Cañizares, C.: Electric energy systems. Analysis and operation. CRC Press (2009)
22. Kersting, W.H.: Radial distribution test feeders. In: IEEE Power Engineering Society Winter Meeting, vol. 2, pp. 908–912 (2001)
23. Bolognani, S., Cavraro, G., Carli, R., Zampieri, S.: A distributed feedback control strategy for optimal reactive power flow with voltage constraints. arXiv:1303.7173 [math.OC] (2013)
24. Wang, J., Elia, N.: A control perspective for centralized and distributed convex optimization. In: Proceedings of the 50th IEEE Conference on Decision and Control (CDC), Orlando, FL, USA, pp. 3800–3805 (2011)
25. Bolognani, S., Zampieri, S.: A distributed optimal control approach to dynamic reactive power compensation. In: Proc. 51st IEEE Conf. on Decision and Control, CDC (2012)
26. Uzawa, H.: The Kuhn-Tucker theorem in concave programming. Studies in linear and nonlinear programming, pp. 32–37. Stanford University Press (1958)

Part VI
Control of Multi-agent Systems

Optimal Control of Multi-agent Persistent Monitoring Systems with Performance Constraints

Christos G. Cassandras and Xuchao Lin*

Division of Systems Engineering
and Center for Information and Systems Engineering
Boston University, Boston, MA
{cgc,mmxclin}@bu.edu

Abstract. We present an optimal control framework for persistent monitoring problems where the objective is to control the movement of multiple cooperating agents to minimize an uncertainty metric in a given mission space, while seeking to maintain some upper bound constraints on uncertainty values. In a one-dimensional mission space, we show that the optimal solution is for each agent to move at maximal speed from one switching point to the next, possibly waiting some time at each point before reversing its direction. Thus, the solution is reduced to a simpler parametric optimization problem: determining a sequence of switching locations and associated waiting times at these switching points for each agent. This amounts to a hybrid system which we analyze using Infinitesimal Perturbation Analysis (IPA) to obtain a complete on-line solution through a gradient-based algorithm. We also show that the solution is robust with respect to the uncertainty model used.

Keywords: Persistent Monitoring, Optimal Control, Hybrid System, Infinitesimal Perturbation Analysis.

1 Introduction

Autonomous cooperating agents may be used to perform tasks such as coverage control [1, 2], surveillance [3, 4] and environmental sampling [5, 6]. *Persistent monitoring* (or surveillance) arises in a large dynamically changing environment which cannot be fully covered by a stationary team of available agents. Thus, persistent monitoring differs from traditional coverage tasks due to the perpetual need to cover a changing environment, i.e., all areas of the mission space must be visited infinitely often. The main challenge in designing control strategies in this case is in balancing the presence of agents in the changing environment so that it is covered over time optimally (in some well-defined sense) while still satisfying sensing and motion constraints.

Control and motion planning for agents performing persistent monitoring tasks have been studied in the literature, e.g., [1], [7–11]. A one-dimensional persistent monitoring problem is studied in [12, 13] where the objective is to minimize a refresh time metric

* The authors' work is supported in part by NSF under Grants EFRI-0735974 and CNS-1239021, by AFOSR under grant FA9550-12-1-0113, by ONR under grant N00014-09-1-1051 and by ARO under grant W911NF-11-1-0227.

D.C. Tarraf (ed.), *Control of Cyber-Physical Systems*,
Lecture Notes in Control and Information Sciences 449,
DOI: 10.1007/978-3-319-01159-2_15,

for sensors. In [11, 14], we addressed the persistent monitoring problem by proposing an *optimal control* framework to drive multiple agents so as to minimize a metric of uncertainty over the environment. This metric is a function of both space and time defined so that uncertainty at a point grows if it is not covered by any agent sensors. To model sensor coverage, we define a probability of detecting events at each point of the mission space by agent sensors. Thus, the uncertainty of the environment decreases with a rate proportional to the event detection probability, i.e., the higher the sensing effectiveness is, the faster the uncertainty is reduced. Our ultimate goal is to optimally control a team of cooperating agents in a two or three-dimensional environment. Our contribution here is to extend the results of [11] and [14], where a single agent or multiple agents were considered moving in a one-dimensional mission space described by an interval $[0,L] \subset \mathbb{R}$, by studying the persistent monitoring problem with performance constraints imposed as thresholds on the uncertainty metric considered. Specifically, the objective function is modified to incur an additional cost when such thresholds are exceeded.

We show that, as in our earlier work, the optimal control problem can be reduced to a *parametric* optimization problem. In particular, the optimal trajectory of each agent is to move at full speed until it reaches some switching point, dwell on the switching point for some time (possibly zero), and then switch directions. In addition, we prove that all agents should never reach the end points of the mission space $[0,L]$. Thus, each agent's optimal trajectory is fully described by a set of switching points $\{\theta_1,\ldots,\theta_K\}$ and associated waiting times at these points, $\{w_1,\ldots,w_K\}$. As a result, we show that the behavior of the agents operating under optimal control is described by a hybrid system. This allows us to make use of generalized Infinitesimal Perturbation Analysis (IPA), as presented in [15], [16], to determine gradients of the objective function with respect to these parameters and subsequently obtain optimal switching locations and waiting times that fully characterize an optimal solution. When adding uncertainty thresholds, we show that only the event time instants at which the uncertainty values exceed or go below these upper bound constraints influence the evaluation of the gradient of the objective function with respect to the switching locations and waiting times. The IPA approach also allows us to exploit its inherent robustness properties and readily extend this solution approach to a *stochastic* uncertainty model.

The remainder of the chapter is organized as follows. In Section 2, the persistent monitoring problem with performance constraints is formulated. In Section 3, we review the optimal control solution of the problem for the case where no performance constraints are present and in Section 4 we provide the solution approach when such constraints are included in the form of penalizing time intervals during which they are violated. We include numerical examples illustrating the solution approach and providing some comparisons between the two cases. We conclude with Section 5 where we outline ongoing research for a similar approach in the case of two-dimensional mission spaces.

2 Persistent Monitoring Problem Formulation

Following the setting in [11] and [14], we consider N mobile agents moving in a one-dimensional mission space of length L, for simplicity taken to be an interval $[0,L] \subset \mathbb{R}$.

Let the position of the agents at time t be $s_n(t) \in [0,L]$, $n = 1,\ldots,N$, following the dynamics:

$$\dot{s}_n(t) = u_n(t) \tag{1}$$

i.e., we assume that each agent can control its direction and speed (as we will see, our analysis still holds for more complex agent dynamics). Without loss of generality, after some rescaling with the size of the mission space L, we further assume that the speed is constrained by $|u_n(t)| \leq 1$, $n = 1,\ldots,N$. For the sake of generality, we include the additional constraint:

$$a \leq s_n(t) \leq b, \ a \geq 0, \ b \leq L \tag{2}$$

over all t to allow for mission spaces where the agents may not reach the end points of $[0,L]$, possibly due to the presence of obstacles. Finally, an additional constraint may be imposed if we assume that the agents are initially located so that $s_n(0) < s_{n+1}(0)$, $n = 1,\ldots,N-1$, and we wish to prevent them from subsequently crossing each other over all t:

$$s_n(t) - s_{n+1}(t) \leq 0 \tag{3}$$

We associate with every point $x \in [0,L]$ a function $p_n(x,s_n)$ that measures the probability that an event at location x is detected by agent n. We also assume that $p_n(x,s_n) = 1$ if $x = s_n$, and that $p_n(x,s_n)$ is monotonically nonincreasing in the distance $|x - s_n|$ between x and s_n, thus capturing the reduced effectiveness of a sensor over its range which we consider to be finite and denoted by r_n (this is the same as the concept of "sensor footprint" found in the robotics literature.) Therefore, we set $p_n(x,s_n) = 0$ when $|x - s_n| > r_n$. Although our analysis is not affected by the precise sensing model $p_n(x,s_n)$, we will limit ourselves to a linear decay model as follows:

$$p_n(x,s_n) = \begin{cases} 1 - \frac{|x-s_n|}{r_n}, & \text{if } |x - s_n| \leq r_n \\ 0, & \text{if } |x - s_n| > r_n \end{cases} \tag{4}$$

Next, consider a set of points $\{\alpha_i\}$, $i = 1,\ldots,M$, $\alpha_i \in [0,L]$, and associate a time-varying measure of uncertainty with each point α_i, which we denote by $R_i(t)$. Without loss of generality, we assume $0 \leq \alpha_1 \leq \cdots \leq \alpha_M \leq L$ and, to simplify notation, we set $p_{n,i}(s_n(t)) \equiv p_n(\alpha_i, s_n(t))$. This set may be selected to contain points of interest in the environment, or sampled points from the mission space. Alternatively, we may consider a partition of $[0,L]$ into M intervals whose center points are $\alpha_i = \frac{(2i-1)L}{2M}$, $i = 1,\ldots,M$. We can then set $p_n(x,s_n(t)) = p_{n,i}(s_n(t))$ for all $x \in [\alpha_i - \frac{L}{2M}, \alpha_i + \frac{L}{2M}]$. Therefore, the joint probability of detecting an event at location $x \in [\alpha_i - \frac{L}{2M}, \alpha_i + \frac{L}{2M}]$ by all the N agents simultaneously (assuming detection independence) is $P_i(\mathbf{s}(t)) = 1 - \prod_{n=1}^{N} [1 - p_{n,i}(s_n(t))]$, where we set $\mathbf{s}(t) = [s_1(t),\ldots,s_N(t)]^{\mathrm{T}}$. We define uncertainty functions $R_i(t)$ associated with the intervals $[\alpha_i - \frac{L}{2M}, \alpha_i + \frac{L}{2M}]$, $i = 1,\ldots,M$, so that they have the following properties: (*i*) $R_i(t)$ increases with a prespecified rate A_i if $P_i(\mathbf{s}(t)) = 0$, (*ii*) $R_i(t)$ decreases with a fixed rate B if $P_i(\mathbf{s}(t)) = 1$ and (*iii*) $R_i(t) \geq 0$ for all t. It is then natural to model uncertainty so that its decrease is proportional to the probability of detection. In particular, we model the dynamics of $R_i(t)$, $i = 1,\ldots,M$, as follows:

$$\dot{R}_i(t) = \begin{cases} 0 & \text{if } R_i(t) = 0,\ A_i \le BP_i(\mathbf{s}(t)) \\ A_i - BP_i(\mathbf{s}(t)) & \text{otherwise} \end{cases} \tag{5}$$

where we assume that initial conditions $R_i(0)$, $i = 1,\ldots,M$, are given and that $B > A_i > 0$ (thus, the uncertainty strictly decreases when there is perfect sensing $P_i(\mathbf{s}(t)) = 1$.)

As described in [11] and [14], persistent monitoring can be viewed as a polling system with agents as "servers" and each point α_i (equivalently, ith interval in $[0,L]$) associated with a "virtual queue" where uncertainty accumulates with inflow rate A_i. The service rate of this queue is time-varying and given by $BP_i(\mathbf{s}(t))$, controllable through the agent position at time t.

The goal of the optimal persistent monitoring problem we consider is to control the movement of the N agents through $u_n(t)$ in (1) so that the cumulative uncertainty over all sensing points $\{\alpha_i\}$, $i = 1,\ldots,M$ is minimized over a fixed time horizon T. In addition, we would like to impose a constraint $R_i(t) \le R_i^{\max}$, for some or all $i \in \{1\ldots,M\}$, so as to provide upper bound uncertainty guarantees. Instead of such hard constraints, however, we proceed by imposing an additional cost proportional to the current uncertainty value at i, $R_i(t)$, whenever it exceeds the predefined threshold $R_i^{\max}$. Thus, setting $\mathbf{u}(t) = [u_1(t),\ldots,u_N(t)]^{\mathrm{T}}$ we aim to solve the following optimal control problem **P1**:

$$\min_{\mathbf{u}(t)} J = \frac{1}{T}\int_0^T \sum_{i=1}^{M} [R_i(t) + \beta_i R_i(t)\mathbf{1}[R_i(t) > R_i^{\max}]]dt \tag{6}$$

subject to the agent dynamics (1), uncertainty dynamics (5), control constraint $|u_n(t)| \le 1$, $t \in [0,T]$, and state constraints (2), $t \in [0,T]$. In (6), β_i is a non-negative constant representing the desired penalty at i when $R_i(t) \le R_i^{\max}$ is violated and $\mathbf{1}[\cdot]$ is the indication function:

$$\mathbf{1}[R_i(t) > R_i^{\max}] = \begin{cases} 1 & \text{if } R_i(t) > R_i^{\max} \\ 0 & \text{otherwise} \end{cases}$$

Note that we require $a \le r_n$ and $b \ge L - r_m$, for at least some $n,m = 1,\ldots,N$; this is to ensure that there are no points in $[0,L]$ which can never be sensed, i.e., any i such that $\alpha_i < a - r_n$ or $\alpha_i > b + r_m$ would always lie outside any agent's sensing range. We will omit the additional constraint (3) from our initial analysis, but we will show that, when it is included, the optimal solution never allows it to be active.

3 Optimal Control Solution

We first characterize the optimal control solution of problem **P1** when $\beta_i = 0$ for all $i = 1,\ldots,M$ and then extend the results to the more general form where $\beta_i \ge 0$. Note that when $\beta_i = 0$ the objective function (6) becomes

$$\min_{\mathbf{u}(t)} J = \frac{1}{T}\int_0^T \sum_{i=1}^{M} R_i(t)dt \tag{7}$$

We have shown in [14] that this simplified problem can be reduced to a parametric optimization problem and we review the main elements of our analysis in what follows. We define the state vector $\mathbf{x}(t) = [s_1(t),\ldots,s_N(t),R_1(t),\ldots,R_M(t)]^{\mathrm{T}}$ and the associated costate vector $\lambda(t) = [\lambda_{s_1}(t),\ldots,\lambda_{s_N}(t),\lambda_1(t),\ldots,\lambda_M(t)]^{\mathrm{T}}$. In view of the discontinuity in the dynamics of $R_i(t)$ in (5), the optimal state trajectory may contain a boundary arc when $R_i(t) = 0$ for any i; otherwise, the state evolves in an interior arc. We first analyze the system operating in such an interior arc and omit the constraint (2). Using (1) and (5), the Hamiltonian is

$$H(\mathbf{x},\lambda,\mathbf{u}) = \sum_{i=1}^{M} R_i(t) + \sum_{n=1}^{N} \lambda_{s_n}(t)\, u_n(t) + \sum_{i=1}^{M} \lambda_i(t)\, \dot{R}_i(t) \tag{8}$$

and the costate equations $\dot{\lambda} = -\frac{\partial H}{\partial x}$ are

$$\dot{\lambda}_i(t) = -\frac{\partial H}{\partial R_i(t)} = -1, \quad i = 1,\ldots,M \tag{9}$$

$$\begin{aligned}\dot{\lambda}_{s_n}(t) &= -\frac{\partial H}{\partial s_n(t)} = -\frac{B}{r_n} \sum_{i\in F_n^-(t)} \lambda_i(t) \prod_{d\neq n} \left[1 - p_{d,i}(s_d(t))\right] \\ &\quad + \frac{B}{r_n} \sum_{i\in F_n^+(t)} \lambda_i(t) \prod_{d\neq n} \left[1 - p_{d,i}(s_d(t))\right]\end{aligned} \tag{10}$$

where we have used (4) and define the sets $F_n^-(t) = \{i : s_n(t) - r_n \leq \alpha_i \leq s_n(t)\}$ and $F_n^+(t) = \{i : s_n(t) < \alpha_i \leq s_n(t) + r_n\}$ for $n = 1,\ldots,N$. Note that $F_n^-(t)$, $F_n^+(t)$ identify all points α_i to the left and right of $s_n(t)$ respectively that are within agent n's sensing range. Since we impose no terminal state constraints, the boundary conditions are $\lambda_i(T) = 0,\ i = 1,\ldots,M$ and $\lambda_{s_n}(T) = 0,\ n = 1,...,N$. Applying the Pontryagin minimum principle to (8) with $\mathbf{u}^\star(t)$, $t \in [0,T)$, denoting an optimal control, we have $H(\mathbf{x}^\star,\lambda^\star,\mathbf{u}^\star) = \min_{u_n\in[-1,1],\, n=1,\ldots,N} H(\mathbf{x},\lambda,\mathbf{u})$ and it is immediately obvious that it is necessary for an optimal control to satisfy:

$$u_n^\star(t) = \begin{cases} 1 & \text{if } \lambda_{s_n}(t) < 0 \\ -1 & \text{if } \lambda_{s_n}(t) > 0 \end{cases} \tag{11}$$

This condition excludes the possibility that $\lambda_{s_n}(t) = 0$ over some finite singular intervals [17]. We will show that if $s_n(t) = a > 0$ or $s_n(t) = b < L$, then $\lambda_{s_n}(t) = 0$ for some $n \in \{1,\ldots,N\}$ may in fact exist for some finite arc; otherwise $\lambda_{s_n}(t) = 0$ can arise only when $u_n(t) = 0$.

By examining the boundary condition $\lambda_i(T) = 0$ and the discontinuity of $\lambda_i(t)$ when the condition $R_i(t) = 0$ is satisfied, we have shown in [11] that $\lambda_i(t) \geq 0$, $i = 1,\ldots,M$, $t \in [0,T]$, with equality holding only if $t = T$, or $t = t_0^-$ with $R_i(t_0) = 0$, $R_i(t') > 0$, where $t' \in [t_0 - \delta, t_0)$, $\delta > 0$. The actual evaluation of the costate vector over the interval $[0,T]$ requires solving (10), which in turn involves the determination of all points where the state variables $R_i(t)$ reach their minimum feasible values $R_i(t) = 0$, $i = 1,\ldots,M$. This generally involves the solution of a two-point-boundary-value problem. However, our analysis thus far has already established the structure of the optimal control (11) which we have seen to remain unaffected by the presence of boundary arcs when $R_i(t) = 0$ for one or more $i = 1,\ldots,M$. We will next present some additional structural properties of an optimal trajectory, based on which one can see that it is fully characterized by a set of non-negative scalar parameters. Determining the values of these parameters is a much simpler problem that does not require the solution of a two-point-boundary-value problem.

Let us turn our attention to the constraints $s_n(t) \geq a$ and $s_n(t) \leq b$ and consider first the case where $a = 0$, $b = L$, i.e., the agents can move over the entire $[0,L]$. We shall make use of the following technical condition:

Assumption 1: For any $n = 1,\ldots,N$, $i = 1,\ldots,M$, $t \in (0,T)$, and any $\varepsilon > 0$, if $s_n(t) = 0$, $s_n(t-\varepsilon) > 0$, then either $R_i(\tau) > 0$ for all $\tau \in [t-\varepsilon,t]$ or $R_i(\tau) = 0$ for all $\tau \in [t-\varepsilon,t]$; if $s_n(t) = L, s_n(t-\varepsilon) < L$, then either $R_i(\tau) > 0$ for all $\tau \in [t-\varepsilon,t]$ or $R_i(\tau) = 0$ for all $\tau \in [t-\varepsilon,t]$.

This condition excludes the case where an agent reaches an end point of the mission space at the exact same time that any one of the uncertainty functions reaches its minimal value of zero. Proposition 1 below asserts that neither of the constraints $s_n(t) \geq 0$ and $s_n(t) \leq L$ can become active on an optimal trajectory if $a = 0$, $b = L$. On the other hand, when $a > 0$ and (or) $b < L$, Proposition 2 states that an agent may dwell at the points a or b before switching its direction of motion. Proofs of Propositions 1-4 may be found in [14].

Proposition 1. *Under Assumption 1, if $a = 0$, $b = L$, then on an optimal trajectory: $s_n^\star(t) \neq 0$ and $s_n^\star(t) \neq L$ for all $t \in (0,T)$, $n \in \{1,\ldots,N\}$.*

Proposition 2. *If $a > 0$ and (or) $b < L$, then on an optimal trajectory there exist finite length intervals $[t_0,t_1]$ such that $s_n^\star(t) = a$ and (or) $s_n^\star(t) = b$, for some $n \in \{1,\ldots,N\}$, $t \in [t_0,t_1]$, $0 \leq t_0 < t_1 \leq T$.*

The next result establishes that on an optimal trajectory, every agent either moves at full speed or is at rest.

Proposition 3. *On an optimal trajectory, either $u_n^\star(t) = \pm 1$ if $\lambda_{s_n}^\star(t) \neq 0$, or $u_n^\star(t) = 0$ if $\lambda_{s_n}^\star(t) = 0$ for $t \in [0,T]$, $n = 1,\ldots,N$.*

Next, we consider the case where the additional state constraint (3) is included. Proposition 4 states that this constraint is never active on an optimal trajectory, i.e., agents reverse their direction before making contact with any other agent.

Proposition 4. *If the constraint (3) is included in problem **P1**, then on an optimal trajectory, $s_n^\star(t) \neq s_{n+1}^\star(t)$ for $t \in (0,T]$, $n = 1,\ldots,N-1$.*

Based on this analysis, the optimal control $u_n^\star(t)$ depends entirely on the sign of $\lambda_{s_n}^\star(t)$ and, in light of Propositions 1-3, the solution of the problem reduces to determining: (*i*) *switching points* in $[0,L]$ where an agent switches from $u_n^\star(t) = \pm 1$ to either ∓ 1 or 0; or from $u_n^\star(t) = 0$ to either ± 1, and (*ii*) if an agent switches from $u_n^\star(t) = \pm 1$ to 0, *waiting times* until the agent switches back to a speed $u_n^\star(t) = \pm 1$. In other words, the full solution is characterized by two parameter vectors for each agent n: $\theta_n = [\theta_{n,1},\ldots,\theta_{n,\Gamma_n}]^{\mathrm{T}}$ and $w_n = [w_{n,1}\ldots,w_{n,\Gamma_n}]^{\mathrm{T}}$, where $\theta_{n,\xi} \in (0,L)$ denotes the ξth location where agent n

changes its speed from ± 1 to 0 and $w_{n,\xi} \geq 0$ denotes the time (which is possibly null) that agent n dwells on $\theta_{n,\xi}$. Note that Γ_n is generally not known a priori and depends on the time horizon T. In addition, we always assume that agent n reverses its velocity direction after leaving the switching point $\theta_{n,\xi}$ with respect to the one it had when reaching $\theta_{n,\xi}$. This seemingly excludes the possibility of an agent's control following a sequence $1,0,1$ or $-1,0,-1$. However, these two motion behaviors can be captured as two adjacent switching points approaching each other: when $\left|\theta_{n,\xi} - \theta_{n,\xi+1}\right| \to 0$, the agent control follows the sequence $1,0,1$ or $-1,0,-1$, and the waiting time associated with $u_n^\star(t) = 0$ is $w_{n,\xi} + w_{n,\xi+1}$.

For simplicity, we will assume that $s_n(0) = 0$, so that it follows from Proposition 1 that $u_n^\star(0) = 1$, $n = 1,\ldots,N$. Therefore, $\theta_{n,1}$ corresponds to the optimal control switching from 1 to 0. Furthermore, $\theta_{n,\xi}$ with ξ odd (even) always corresponds to $u_n^\star(t)$ switching from 1 to 0 (-1 to 0.) Thus, we have the following constraints on the switching locations for all $\xi = 2,\ldots,\Gamma_n$:

$$\begin{cases} \theta_{n,\xi} \leq \theta_{n,\xi-1}, & \text{if } \xi \text{ is even} \\ \theta_{n,\xi} \geq \theta_{n,\xi-1}, & \text{if } \xi \text{ is odd.} \end{cases} \tag{12}$$

It is now clear that the behavior of each agent under the optimal control policy is that of a *hybrid system* whose dynamics undergo switches when $u_n^\star(t)$ changes from ± 1 to 0 and from 0 to ∓ 1 or when $R_i(t)$ reaches or leaves the boundary value $R_i = 0$. As a result, we are faced with a parametric optimization problem for a system with hybrid dynamics. This is a setting where one can apply the generalized theory of Infinitesimal Perturbation Analysis (IPA) in [15], [16] to conveniently obtain the gradient of the objective function J in (7) with respect to the vectors θ and w, and therefore, determine (generally, locally) optimal vectors $\theta^\star$ and $w^\star$ through a gradient-based optimization approach. Note that this is done on line, i.e., the gradient is evaluated by observing a trajectory with given θ and w over $[0,T]$ based on which θ and w are adjusted until convergence is attained using standard gradient-based algorithms.

Regarding the agent dynamics in (1), as pointed out in [14], they can be replaced by a more general model of the form $\dot{s}_n(t) = g_n(s_n) + b_n u_n(t)$ under which it is easy to see that (11) still holds. The difference lies in (10) which would involve a dependence on $\frac{dg_n(s_n)}{ds_n}$ and further complicate the associated two-point-boundary-value problem. However, since the optimal solution is still defined by parameter vectors $\theta_n = [\theta_{n,1},\ldots,\theta_{n,\Gamma_n}]^{\mathrm{T}}$ and $w_n = [w_{n,1}\ldots,w_{n,\Gamma_n}]^{\mathrm{T}}$ for each agent n, we can still apply the IPA approach presented in the next section.

3.1 Infinitesimal Perturbation Analysis (IPA)

We have established that the nth agent's movement can be parameterized through $\theta_n = [\theta_{n,1},\ldots,\theta_{n,\Gamma_n}]^{\mathrm{T}}$ and $w_n = [w_{n,1}\ldots,w_{n,\Gamma_n}]^{\mathrm{T}}$ where $\theta_{n,\xi}$ is the ξth control switching point and $w_{n,\xi}$ is the waiting time for this agent at the ξth switching point. Therefore, the solution of problem **P1** reduces to the determination of optimal parameter vectors $\theta_n^\star$ and $w_n^\star$, $n = 1,\ldots,N$. As we pointed out, the agent's optimal behavior defines a hybrid system, and the switching locations translate to switching times between particular modes of this system. This is similar to switching-time optimization problems, e.g., [18], except that we can only control a subset of mode switching times. We make use of IPA in part to exploit robustness properties that the resulting gradients possess [19]; specifically, we will show that they do not depend on the uncertainty model parameters A_i, $i = 1,\dots,M$, and may therefore be used without any detailed knowledge of how uncertainty affects the mission space. In what follows, we use a standard definition of a hybrid automaton (e.g., see [20]) as the formalism to model the system described above.

IPA Review. The purpose of IPA [15] is to study the behavior of a hybrid system state as a function of a parameter vector $\theta \in \Theta$ for a given compact, convex set $\Theta \subset \mathbb{R}^l$. Let $\{\tau_k(\theta)\}$, $k = 1,\dots,K$, denote the occurrence times of all events in the state trajectory. For convenience, we set $\tau_0 = 0$ and $\tau_{K+1} = T$. Over an interval $[\tau_k(\theta), \tau_{k+1}(\theta))$, the system is at some mode during which the time-driven state satisfies $\dot{x} = f_k(x,\theta,t)$. An event at τ_k is classified as (i) *Exogenous* if it causes a discrete state transition independent of θ and satisfies $\frac{d\tau_k}{d\theta} = 0$; (ii) *Endogenous*, if there exists a continuously differentiable function $g_k : \mathbb{R}^n \times \Theta \to \mathbb{R}$ such that $\tau_k = \min\{t > \tau_{k-1} : g_k(x(\theta,t),\theta) = 0\}$; and (iii) *Induced* if it is triggered by the occurrence of another event at time $\tau_m \leq \tau_k$. IPA specifies how changes in θ influence the state $x(\theta,t)$ and the event times $\tau_k(\theta)$ and, ultimately, how they influence interesting performance metrics which are generally expressed in terms of these variables.

Given $\theta = [\theta_1,\dots,\theta_\Gamma]^{\mathrm{T}}$, we use the Jacobian matrix notation: $x'(t) \equiv \frac{\partial x(\theta,t)}{\partial\theta}$, $\tau_k' \equiv \frac{\partial\tau_k(\theta)}{\partial\theta}$, $k = 1,\dots,K$, for all state and event time derivatives. It is shown in [15] that $x'(t)$ satisfies:

$$\frac{d}{dt}x'(t) = \frac{\partial f_k(t)}{\partial x}x'(t) + \frac{\partial f_k(t)}{\partial\theta} \tag{13}$$

for $t \in [\tau_k, \tau_{k+1})$ with boundary condition:

$$x'(\tau_k^+) = x'(\tau_k^-) + \left[f_{k-1}(\tau_k^-) - f_k(\tau_k^+)\right]\tau_k' \tag{14}$$

for $k = 0,\dots,K$. In addition, in (14), the gradient vector for each τ_k is $\tau_k' = 0$ if the event at τ_k is exogenous and

$$\tau_k' = -\left[\frac{\partial g_k}{\partial x}f_k(\tau_k^-)\right]^{-1}\left(\frac{\partial g_k}{\partial\theta} + \frac{\partial g_k}{\partial x}x'(\tau_k^-)\right) \tag{15}$$

if the event at τ_k is endogenous (i.e., $g_k(x(\theta,\tau_k),\theta) = 0$) and defined as long as $\frac{\partial g_k}{\partial x}f_k(\tau_k^-) \neq 0$.

We now include both parameter vectors $\theta_n = [\theta_{n,1},\dots,\theta_{n,\Gamma_n}]^{\mathrm{T}}$ and $w_n = [w_{n,1},\dots w_{n,\Gamma_n}]^{\mathrm{T}}$ for each agent n and, for notational simplicity, concatenate them to construct $\theta = [\theta_1,\dots,\theta_N]^{\mathrm{T}}$ and $w = [w_1,\dots,w_N]^{\mathrm{T}}$. The solution of problem **P1** reduces to the determination of optimal parameter vectors $\theta^\star$ and $w^\star$ and we will use IPA to evaluate $\nabla J(\theta,w) = [\frac{dJ(\theta,w)}{d\theta}\,\frac{dJ(\theta,w)}{dw}]^{\mathrm{T}}$. It depends on $\nabla R_i(t) = \left[\frac{\partial R_i(t)}{\partial\theta}\,\frac{\partial R_i(t)}{\partial w}\right]^{\mathrm{T}}$ and the event times τ_k, $k = 1,\dots,K$, observed on a trajectory over $[0,T]$ with given θ and w.

IPA Equations. We begin by recalling the dynamics of $R_i(t)$ in (5) which depend on the relative positions of all agents with respect to α_i and change at time instants τ_k such that either $R_i(\tau_k) = 0$ with $R_i(\tau_k^-) > 0$ or $A_i > BP_i(\mathbf{s}(\tau_k))$ with $R_i(\tau_k^-) = 0$. Moreover, using (1) and our previous analysis, the dynamics of $s_n(t)$, $n = 1,\dots,N$, in an optimal trajectory can be expressed as follows. Define $\Theta_{n,\xi} = (\theta_{n,\xi-1}, \theta_{n,\xi})$ if ξ is odd and $\Theta_{n,\xi} = (\theta_{n,\xi}, \theta_{n,\xi-1})$ if ξ is even to be the ξth interval between successive switching points for any $n = 1, \ldots, N$, where $\theta_{n,0} = s_n(0)$. Then, for $\xi = 1, 2, \ldots$,

$$\dot{s}_n(t) = \begin{cases} 1 & s_n(t) \in \Theta_{n,\xi},\ \xi \text{ odd} \\ -1 & s_n(t) \in \Theta_{n,\xi},\ \xi \text{ even} \\ 0 & \text{otherwise} \end{cases} \tag{16}$$

where transitions for $s_n(t)$ from ± 1 to ∓ 1 are incorporated by treating them as cases where $w_{n,\xi} = 0$, i.e., no dwelling at a switching point $\theta_{n,\xi}$ (in which case $\dot{s}_n(t) = 0$.) We can now concentrate on all events causing switches either in the dynamics of any $R_i(t)$, $i = 1, \ldots, M$, or the dynamics of any $s_n(t)$, $n = 1, \ldots, N$. From (14), any other event at some time τ_k in this hybrid system cannot modify the values of $\nabla R_i(t) = \left[\frac{\partial R_i(t)}{\partial \theta}\ \frac{\partial R_i(t)}{\partial w}\right]^{\mathrm{T}}$ or $\nabla s_n(t) = \left[\frac{\partial s_n(t)}{\partial \theta_n}\ \frac{\partial s_n(t)}{\partial w_n}\right]^{\mathrm{T}}$ at $t = \tau_k$.

First, applying (13) to $s_n(t)$ with $f_k(t) = 1$, -1 or 0 due to (16), the solution yields

$$\nabla s_n(t) = \nabla s_n(\tau_k^+), \text{ for } t \in [\tau_k, \tau_{k+1}) \tag{17}$$

for all $k = 1, \ldots, K$, $n = 1, \ldots, N$. Similarly, applying (13) to $R_i(t)$ and using (5) gives:

$$\begin{bmatrix} \frac{\partial R_i}{\partial \theta_{n,\xi}}(t) \\ \frac{\partial R_i}{\partial w_{n,\xi}}(t) \end{bmatrix} = \begin{bmatrix} \frac{\partial R_i}{\partial \theta_{n,\xi}}(\tau_k^+) \\ \frac{\partial R_i}{\partial w_{n,\xi}}(\tau_k^+) \end{bmatrix} - \begin{cases} 0 & \text{if } R_i(t) = 0,\ A_i < BP_i(\mathbf{s}(t)) \\ G \begin{bmatrix} \frac{\partial s_n(\tau_k^+)}{\partial \theta_{n,\xi}} \\ \frac{\partial s_n(\tau_k^+)}{\partial w_{n,\xi}} \end{bmatrix} & \text{otherwise} \end{cases} \tag{18}$$

where $G = B \prod_{d \neq n} (1 - p_i(s_d(t))) \left(\frac{\partial p_i(s_n)}{\partial s_n}\right)(t - \tau_k)$. Thus, it remains to determine the components of $\nabla s_n(\tau_k^+)$ and $\nabla R_i(\tau_k^+)$ in (17) using (14). This involves the event time gradient vectors $\nabla \tau_k = \left[\frac{\partial \tau_k}{\partial \theta}\ \frac{\partial \tau_k}{\partial w}\right]^{\mathrm{T}}$ for $k = 1, \ldots, K$, which will be determined through (15). There are three possible cases:

Case 1: An event at time τ_k such that $\dot{R}_i(t)$ switches from $\dot{R}_i(t) = 0$ to $\dot{R}_i(t) = A_i - BP_i(\mathbf{s}(t))$. In this case, it is easy to see that the dynamics of both $s_n(t)$ and $R_i(t)$ are continuous, so that $f_{k-1}(\tau_k^-) = f_k(\tau_k^+)$ in (14) applied to $s_n(t)$ and $R_i(t)$, $i = 1, \ldots, M$, $n = 1, \ldots, N$, and we get

$$\nabla s_n(\tau_k^+) = \nabla s_n(\tau_k^-),\ n = 1, \ldots, N \tag{19}$$

$$\nabla R_i(\tau_k^+) = \nabla R_i(\tau_k^-),\ i = 1, \ldots, M \tag{20}$$

Case 2: An event at time τ_k such that $\dot{R}_i(t)$ switches from $\dot{R}_i(t) = A_i - BP_i(\mathbf{s}(t))$ to $\dot{R}_i(t) = 0$, i.e., $R_i(\tau_k)$ becomes zero. In this case, we need to first evaluate $\nabla \tau_k$ from (15) in order to determine $\nabla R_i(\tau_k^+)$ through (14). Observing that this event is endogenous, (15) applies with $g_k = R_i = 0$ and we get

$$\nabla \tau_k = -\frac{\nabla R_i(\tau_k^-)}{A_i(\tau_k^-) - BP_i(\mathbf{s}(\tau_k^-))} \tag{21}$$

It follows from (14) that

$$\nabla R_i(\tau_k^+) = \nabla R_i(\tau_k^-) - \frac{[A_i(\tau_k^-) - BP_i(\mathbf{s}(t))]\nabla R_i(\tau_k^-)}{A_i(\tau_k^-) - BP_i(\tau_k^-)} = 0 \tag{22}$$

Case 3: An event at time τ_k such that the dynamics of $s_n(t)$ switch from ± 1 to 0, or from 0 to ± 1. (20) holds since the the dynamics of $R_i(t)$ are continuous. However, determining $\nabla s_n(\tau_k^+)$ is more elaborate and requires us to consider its components separately, first $\frac{\partial s_n(\tau_k^+)}{\partial \theta_n}$ and then $\frac{\partial s_n(\tau_k^+)}{\partial w_n}$.

Case 3.1: Evaluation of $\frac{\partial s_n(\tau_k^+)}{\partial \theta_n}$.

Case 3.1.1: An event at time τ_k such that the dynamics of $s_n(t)$ in (16) switch from ± 1 to 0. This is an endogenous event and (15) applies with $g_k = s_n - \theta_{n,\xi} = 0$ for some $\xi = 1, \ldots, \Gamma_n$ and we have $\frac{\partial \tau_k}{\partial \theta_{n,\xi}} = \frac{1 - \frac{\partial s_n}{\partial \theta_{n,\xi}}(\tau_k^-)}{u_n(\tau_k^-)}$. Using (14) and omitting details (which may be found in [14]), we get:

$$\frac{\partial s_n}{\partial \theta_{n,j}}(\tau_k^+) = \begin{cases} 0, \text{ if } j \neq \xi \\ 1, \text{ if } j = \xi \end{cases} \tag{23}$$

Case 3.1.2: An event at time τ_k such that the dynamics of $s_n(t)$ in (16) switch from 0 to ± 1. This is an induced event since it is triggered by the occurrence of some other endogenous event when the agent switches from ± 1 to 0. Suppose the agent starts from an initial position $s_n(0) = a$ with $u_n(0) = 1$ and τ_k is the time the agent switches from the 0 to ± 1 at $\theta_{n,\xi}$. If $\theta_{n,\xi}$ is such that $u_n(\tau_k^+) = 1$, then ξ is even and τ_k can be calculated as follows:

$$\begin{aligned} \tau_k &= (\theta_{n,1} - a) + w_{n,1} + \ldots + (\theta_{n,\xi-1} - \theta_{n,\xi}) + w_{n,\xi} \\ &= 2\left(\sum_{v=1,\, v \text{ odd}}^{\xi-1} \theta_{n,v} - \sum_{v=2,\, v \text{ even}}^{\xi-2} \theta_{n,v} \right) + \sum_{v=1}^{\xi} w_{n,v} - \theta_{n,\xi} \end{aligned} \tag{24}$$

Similarly, if $\theta_{n,\xi}$ is the switching point such that $u_n(\tau_k^+) = -1$, then ξ is odd and we get:

$$\tau_k = 2\left(\sum_{v=1,\, v \text{ odd}}^{\xi-2} \theta_{n,v} - \sum_{v=2,\, v \text{ even}}^{\xi-1} \theta_{n,v} \right) + \sum_{v=1}^{\xi} w_{n,v} + \theta_{n,\xi} \tag{25}$$

We can then directly obtain $\frac{\partial \tau_k}{\partial \theta_{n,\xi}} = -sgn(u(\tau_k^+))$ and (14), omitting details, gives:

$$\frac{\partial s_n}{\partial \theta_{n,j}}(\tau_k^+) = \frac{\partial s_n}{\partial \theta_{n,j}}(\tau_k^-) + 2 \cdot (-1)^j u_n(\tau_k^+) \tag{26}$$

Case 3.2: Evaluation of $\frac{\partial s_n(\tau_k^+)}{\partial w_n}$.

Case 3.2.1: An event at time τ_k such that the dynamics of $s_n(t)$ in (16) switch from ± 1 to 0 or from 0 to ± 1. Using similar arguments and omitting details we get:

$$\frac{\partial s_n}{\partial w_{n,j}}(\tau_k^+) = \frac{\partial s_n}{\partial w_{n,j}}(\tau_k^-) - \frac{[u_n(\tau_k^-) - 0] \frac{\partial s_n}{\partial w_{n,j}}(\tau_k^-)}{u_n(\tau_k^-)} = 0 \tag{27}$$

Case 3.2.2: An event that the dynamics of $s_n(t)$ in (16) switch from 0 to ± 1 at time τ_k. Omitting details (which may be found in [14]), we have

$$\frac{\partial s_n}{\partial w_{n,j}}(\tau_k^+) = \begin{cases} 0, & \text{if } u_n(\tau_k^-) = \pm 1,\ u_n(\tau_k^+) = 0 \\ \mp 1, & \text{if } u_n(\tau_k^-) = 0,\ u_n(\tau_k^+) = \pm 1 \end{cases} \tag{28}$$

Finally, note that $\frac{\partial s_n}{\partial w_{n,\xi}}(t) = 0, t \in [0, \tau_k)$, since the position of agent n is not affected by $w_{n,\xi}$ prior to such an event.

Objective Function Gradient Evaluation. Based on our analysis, the objective function (7) in problem **P1** can now be written as $J(\theta, w)$, a function of θ and w, instead of $\mathbf{u}(t)$ and we can rewrite it as $J(\theta, w) = \frac{1}{T}\sum_{i=1}^{M}\sum_{k=0}^{K}\int_{\tau_k(\theta,w)}^{\tau_{k+1}(\theta,w)} R_i(t, \theta, w)\,dt$ where we have explicitly indicated the dependence on θ and w. We then obtain

$$\nabla J(\theta) = \frac{1}{T}\sum_{i=1}^{M}\sum_{k=0}^{K}\left(\int_{\tau_k}^{\tau_{k+1}} \nabla R_i(t)\,dt + R_i(\tau_{k+1})\nabla\tau_{k+1} - R_i(\tau_k)\nabla\tau_k\right)$$

Observing the cancelation of all terms of the form $R_i(\tau_k)\nabla\tau_k$ for all k (with $\tau_0 = 0$, $\tau_{K+1} = T$ fixed), we finally get $\nabla J(\theta, w) = [\frac{dJ(\theta,w)}{d\theta}\ \frac{dJ(\theta,w)}{dw}]^{\mathrm{T}}$ as

$$\nabla J(\theta, w) = \frac{1}{T}\sum_{i=1}^{M}\sum_{k=0}^{K}\int_{\tau_k(\theta,w)}^{\tau_{k+1}(\theta,w)} \nabla R_i(t)\,dt \tag{29}$$

This depends entirely on $\nabla R_i(t)$, which is obtained from (18) and the event times τ_k, $k = 1, \ldots, K$, given initial conditions $s_n(0) = a$ for $n = 1, \ldots, N$, and $R_i(0)$ for $i = 1, \ldots, M$. In (18), $\frac{\partial R_i}{\partial \theta_{n,\xi}}(\tau_k^+)$ is obtained through (20) and (22), whereas $\frac{\partial s_n(\tau_k^+)}{\partial \theta_{n,\xi}}$ is obtained through (17), (19), (23), and (26). As for $\frac{\partial R_i}{\partial w_{n,\xi}}(\tau_k^+)$ in (18), it is again obtained through (20) and (22), whereas $\frac{\partial s_n(\tau_k^+)}{\partial w_{n,\xi}}$ is obtained through (19), and (28).

Robustness Property. Observe that the evaluation of $\nabla R_i(t)$, hence $\nabla J(\theta, w)$, is independent of A_i, $i = 1, \ldots, M$. In fact, the dependence of $\nabla R_i(t)$ on A_i manifests itself through the event times τ_k, $k = 1, \ldots, K$, that do affect this evaluation, but they, unlike A_i which may be unknown, are directly observable during the gradient evaluation process. Thus, the IPA approach possesses an inherent *robustness* property: there is no need to explicitly model how uncertainty affects $R_i(t)$ in (5). Consequently, we may treat A_i as unknown without affecting the solution approach. We may also allow this uncertainty to be modeled through random processes $\{A_i(t)\}$, $i = 1, \ldots, M$; in this case, however, the result of Prop. 3 no longer applies without some conditions on the statistical characteristics of $\{A_i(t)\}$ and the resulting $\nabla J(\theta, w)$ is an estimate of a stochastic gradient.

3.2 Objective Function Optimization

In view of the objective function gradient in (29) evaluated using (18) and the IPA equations given above, we can now seek to obtain $\theta^\star$ and $w^\star$ minimizing $J(\theta, w)$ through a standard gradient-based optimization scheme of the form

$$[\theta^{l+1} w^{l+1}]^{\mathrm{T}} = [\theta^{l} w^{l}]^{\mathrm{T}} - [\eta_\theta \ \eta_w] \tilde{\nabla} J(\theta^l, w^l) \tag{30}$$

where $\{\eta_\theta^l\}, \{\eta_w^l\}$ are appropriate step size sequences and $\tilde{\nabla} J(\theta^l, w^l)$ is the projection of the gradient $\nabla J(\theta^l, w^l)$ onto the feasible set (satisfying (12), $a \leq \theta^{l+1} \leq b$, and $w^l \geq 0$). The algorithm terminates when $|\tilde{\nabla} J(\theta, w)| < \varepsilon$ (for some fixed ε) leading to $\theta^\star$ and $w^\star$ and is summarized in Algorithm 1 where we have adopted the Armijo method in step-size selection (see [21]) for $\{[\eta_\theta^l \ \eta_w^l]\}$.

One of the unusual features in (30) is the fact that the dimension $\Gamma_n^\star$ of $\theta_n^\star$ and $w_n^\star$ is a priori unknown (it depends on T). Thus, the algorithm must implicitly determine this value along with $\theta_n^\star$ and $w_n^\star$. We determine an approximate upper bound for Γ_n as follows. First, we set the initial switching points such that the distance between $\theta_{n,\xi}$ for ξ odd and even is 2σ, where $\sigma > 0$ is an arbitrarily small number, $n = 1, \ldots, N$. In addition, set $w_n = \mathbf{0}$. Then, T must satisfy

$$\theta_{n,1} - s_n(0) + 2\sigma(\Gamma_n - 1) \leq T \leq \theta_{n,1} - s_n(0) + 2\sigma\Gamma_n \tag{31}$$

σ is selected sufficiently small while ensuring that the algorithm can be executed sufficiently fast. When Algorithm 1 stops, there exists $\zeta_n < \Gamma_n$, such that θ_{n,ζ_n} is the last switching point agent n can reach within $(0, T]$. Observe that there generally exist ξ such that $\zeta_n < \xi \leq \Gamma_n$ which correspond to points $\theta_{n,\xi}$ that agent n cannot reach within $(0, T]$; the associated derivatives of the cost with respect to such $\theta_{n,\xi}$ are 0, since perturbations in these $\theta_{n,\xi}$ will not affect $s_n(t)$, $t \in (0, T]$ and thus the cost $J(\theta, w)$. Several numerical examples of our approach applied to single and multiple agent persistent monitoring problems may be found in [14].

Algorithm 1. IPA-based optimization algorithm

Pick $\sigma > 0$ and $\varepsilon > 0$.
Define $D_n = a + \frac{2n-1}{2N}(b-a), n = 1, \ldots, N$, and set $\begin{cases} \theta_{n,\xi} = D_n - \sigma \text{ if } \xi \text{ even} \\ \theta_{n,\xi} = D_n + \sigma \text{ if } \xi \text{ odd} \end{cases}$.
  Set $w = [w_1, \ldots, w_N] = 0.$, where $w_n = [w_{n,1}, \ldots, w_{n,\xi_n}]$ and $\Gamma_n = \lceil \frac{1}{2\sigma}[T - \theta_{n,1} + s_n(0)] \rceil$
repeat
  Compute $s_n(t), t \in [0, T]$ using $s_n(0)$, (11), θ and w for $n = 1, \ldots, N$
  Compute $\tilde{\nabla} J(\theta, w)$ and update θ, w through (30)
until $|\tilde{\nabla} J(\theta, w)| < \varepsilon$
Set $\theta_n^\star = [\theta_{n,1}^\star, \ldots, \theta_{n,\zeta_n}^\star]$ and $w_n^\star = [w_{n,1}^\star, \ldots, w_{n,\zeta_n}^\star]$, where ζ_n is the index of θ_{n,ζ_n}, which is the last switching point agent n can reach within $(0, T]$, $n = 1, \ldots, N$

---

4 Optimal Control Solution with Performance Constraints

We now consider the case where $\beta_i > 0$ for at least some $i \in 1, \ldots, M$ in (6). Following the Hamiltonian analysis presented earlier, the optimal control $u_n^\star(t)$ still only depends on the sign of $\lambda_{s_n}^\star(t)$. It is easy to verify that the proofs of Prop. 1, 2 and 4 still hold in this case, so that, under optimal control, agents never reach end points of the mission space and if $a > 0$ and (or) $b < L$, they may remain at points a and b for finite time interval. However, the proof of Proposition 3 in [14] needs to be modified so as to accommodate the penalty term added to the cost function; this changes the Hamiltonian expression, as well as the dynamics of $\lambda_i(t)$ in (9). We state the modified proposition below and provide a proof.

Proposition 5. *On an optimal trajectory with $\beta_i > 0$ for at least some $i \in 1,\dots,M$, either $u_n^\star(t) = \pm 1$ if $\lambda_{s_n}^\star(t) \neq 0$, or $u_n^\star(t) = 0$ if $\lambda_{s_n}^\star(t) = 0$ for $t \in [0,T]$, $n = 1,\dots,N$.*

Proof. When $\beta_i > 0$ for at least some $i \in \{1\dots,M\}$ in (6), if we analyze the system operating in an interior arc and omit the constraint (2), the Hamiltonian in (8) becomes

$$H(\mathbf{x},\lambda,\mathbf{u}) = \sum_{i=1}^{M} R_i(t)\left[1+\beta_i \mathbf{1}\left[R_i(t) > R_i^{\max}\right]\right] + \sum_{n=1}^{N} \lambda_{s_n}(t)\, u_n(t) + \sum_{i=1}^{M} \lambda_i(t)\, \dot{R}_i(t) \quad (32)$$

For simplicity, we set $I_i(R_i(t)) = R_i(t)\left[1+\beta_i \mathbf{1}\left[R_i(t) > R_i^{\max}\right]\right]$ and (32) becomes

$$H(\mathbf{x},\lambda,\mathbf{u}) = \sum_{i=1}^{M} I_i(R_i(t)) + \sum_{n=1}^{N} \lambda_{s_n}(t)\, u_n(t) + \sum_{i=1}^{M} \lambda_i(t)\, \dot{R}_i(t) \quad (33)$$

The costate equations $\dot{\lambda} = -\frac{\partial H}{\partial x}$ give the dynamics of $\lambda_i(t)$ as

$$\begin{aligned} \dot{\lambda}_i(t) &= -\frac{\partial H}{\partial R_i(t)} = -\frac{\partial H}{\partial I_i(t)} \frac{\partial I_i(t)}{\partial R_i(t)} \\ &= -\frac{\partial I_i(t)}{\partial R_i(t)}, i = 1,\dots,M \end{aligned} \quad (34)$$

When $\lambda_{s_n}^\star(t) \neq 0$, we have shown in (11) that $u_n^\star(t) = \pm 1$, depending on the sign of $\lambda_{s_n}^\star(t)$. Thus, it remains to consider the case $\lambda_{s_n}^\star(t) = 0$ for some $t \in [t_1,t_2]$, where $0 \le t_1 < t_2 \le T$. Since the state is in a singular arc, $\lambda_{s_n}^\star(t)$ does not provide information about $u_n^\star(t)$. On the other hand, the Hamiltonian in (33) is not an explicit function of time, therefore, setting $H(\mathbf{x}^\star,\lambda^\star,\mathbf{u}^\star) \equiv H^\star$, we have $\frac{dH^\star}{dt} = 0$, which gives

$$\begin{aligned} \frac{dH^\star}{dt} = \sum_{i=1}^{M} \frac{\partial I_i(t)}{\partial R_i(t)} \dot{R}_i^\star(t) + \sum_{n=1}^{N} \dot{\lambda}_{s_n}^\star(t)\, u_n^\star(t) + \sum_{n=1}^{N} \lambda_{s_n}^\star(t)\, \dot{u}_n^\star(t) + \sum_{i=1}^{M} \dot{\lambda}_i^\star(t)\, \dot{R}_i^\star(t) \\ + \sum_{i=1}^{M} \lambda_i^\star(t)\, \ddot{R}_i^\star(t) = 0 \end{aligned} \quad (35)$$

Define $S(t) = \{n | \lambda_{s_n}(t) = 0, n = 1,\dots,N\}$ as the set of indices of agents that are in a singular arc and $\bar{S}(t) = \{n | \lambda_{s_n}(t) \neq 0, n = 1,\dots,N\}$ as the set of indices of all other agents. Thus, $\lambda_{s_n}^\star(t) = 0$, $\dot{\lambda}_{s_n}^\star(t) = 0$ for $t \in [t_1,t_2], n \in S(t)$. In addition, agents move with constant full speed, either 1 or -1, so that $\dot{u}_n^\star(t) = 0$, $n \in \bar{S}(t)$. Then, (35) becomes

$$\frac{dH^\star}{dt} = \sum_{i=1}^{M} \left[\frac{\partial I_i(t)}{\partial R_i(t)} + \dot{\lambda}_i^\star(t)\right] \dot{R}_i^\star(t) + \sum_{n \in \bar{S}(t)} \dot{\lambda}_{s_n}^\star(t)\, u_n^\star(t) + \sum_{i=1}^{M} \lambda_i^\star(t)\, \ddot{R}_i^\star(t) = 0$$

From (34), $\dot{\lambda}_i^\star(t) = -\frac{\partial I_i(t)}{\partial R_i(t)}$, $i = 1,\ldots,M$, so $\frac{\partial I_i(t)}{\partial R_i(t)} + \dot{\lambda}_i^\star(t) = 0$, leaving only the last two terms above. Note that $\dot{\lambda}_{s_n}^\star(t) = -\frac{\partial H^\star}{\partial s_n^\star(t)}$ and writing $\ddot{R}_i^\star(t) = \frac{d\dot{R}_i^\star(t)}{dt}$ we get:

$$-\sum_{n\in\bar{S}(t)} u_n^\star(t)\frac{\partial H^\star}{\partial s_n^\star(t)} + \sum_{i=1,R_i\neq 0}^{M} \lambda_i^\star(t)\frac{d\dot{R}_i^\star(t)}{dt} = 0$$

The remainder of the proof is the same as given in [14] and is included here for the sake of completeness. Recall from (5) that when $R_i(t) \neq 0$, we have $\dot{R}_i(t) = A_i - B[1 - \prod_{n=1}^{N}[1-p_i(s_n(t))]]$, so that

$$\frac{\partial H^\star}{\partial s_n^\star(t)} = -B\sum_{i=1,R_i\neq 0}^{M}\lambda_i^\star(t)\frac{\partial p_i(s_n^\star(t))}{\partial s_n^\star(t)}\prod_{d\neq n}^{N}(1-p_i(s_d^\star(t))) \tag{36}$$

$$\frac{d\dot{R}_i^\star(t)}{dt} = -B\sum_{n=1}^{N}u_n^\star(t)\frac{\partial p_i(s_n^\star(t))}{\partial s_n^\star(t)}\prod_{d\neq n}^{N}(1-p_i(s_d^\star(t))) \tag{37}$$

which results in

$$\begin{aligned}
&B\sum_{\substack{i=1\\R_i\neq 0}}^{M}\lambda_i^\star(t)\sum_{n\in\bar{S}(t)}u_n^\star(t)\frac{\partial p_i(s_n^\star(t))}{\partial s_n^\star(t)}\prod_{d\neq n}^{N}(1-p_i(s_d^\star(t)))\\
&-B\sum_{\substack{i=1\\R_i\neq 0}}^{M}\lambda_i^\star(t)\sum_{n=1}^{N}u_n^\star(t)\frac{\partial p_i(s_n^\star(t))}{\partial s_n^\star(t)}\prod_{d\neq n}^{N}(1-p_i(s_d^\star(t)))\\
&= -B\sum_{\substack{i=1\\R_i\neq 0}}^{M}\lambda_i^\star(t)\sum_{n\in S(t)}u_n^\star(t)\frac{\partial p_i(s_n^\star(t))}{\partial s_n^\star(t)}\times\prod_{d\neq n}^{N}(1-p_i(s_d^\star(t))) = 0
\end{aligned} \tag{38}$$

Note that $\frac{\partial p_i(s_1^\star(t))}{\partial s_n^\star(t)} = \pm\frac{1}{r_1}$ or 0, depending on the relative position of $s_1^\star(t)$ with respect to α_i. Moreover, (38) is invariant to M or the precise way in which the mission space $[0,L]$ is partitioned, which implies that

$$\lambda_i^\star(t)\sum_{n\in S(t)}u_n^\star(t)\frac{\partial p_i(s_n^\star(t))}{\partial s_n^\star(t)}\prod_{d\neq n}^{N}(1-p_i(s_d^\star(t))) = 0$$

for all $i = 1,\ldots,M, t\in[t_1,t_2]$. Since $\dot{\lambda}_i^\star(t) = -1, i = 1,\ldots,M$, it is clear that to satisfy this equality we must have $u_n^\star(t) = 0$ for all $t\in[t_1,t_2], n\in S(t)$. In conclusion, in a singular arc with $\lambda_{s_n}^\star(t) = 0$ for some $n\in\{1,\ldots,N\}$, the optimal control is $u_n^\star(t) = 0$. ■

This proof establishes the fact that on an optimal control trajectory, every agent either moves at full speed or is at rest. Since Propositions 1, 2, 4 and 5 hold, the solution of the problem with $\beta_i > 0$ for at least some $i\in\{1\ldots,M\}$ again reduces to determining the switching points $\theta = [\theta_1,\ldots,\theta_N]^{\mathrm{T}}$ and the waiting time on those switching points $w =$

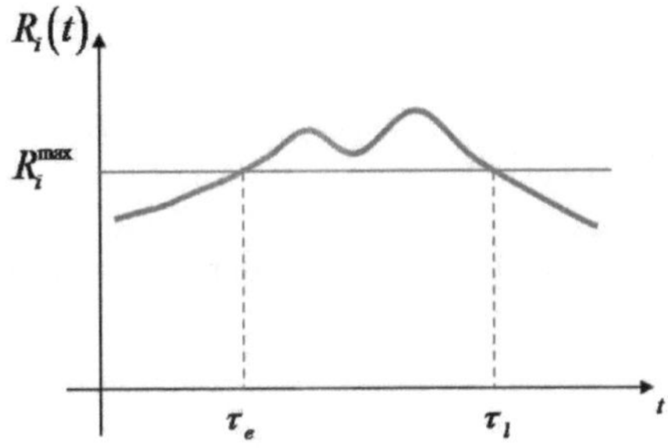

Fig. 1. τ_e is the event time when the system enters a region violating the constraint $R_i(t) \leq R_i^{\max}$ and τ_l is the event time when the system leaves this region

$[w_1, \ldots, w_N]^{\mathrm{T}}$. We obtain $\theta^\star$ and $w^\star$ minimizing $J(\theta, w)$ through a standard gradient-based optimization method using IPA to evaluate $\nabla J(\theta, w) = [\frac{dJ(\theta,w)}{d\theta} \frac{dJ(\theta,w)}{dw}]^{\mathrm{T}}$ in (39). Note that this gradient only depends on $\nabla R_i(t) = \left[\frac{\partial R_i(t)}{\partial \theta} \frac{\partial R_i(t)}{\partial w}\right]^{\mathrm{T}}$ and the event times τ_k, $k = 1, \ldots, K$, observed on a trajectory over $[0, T]$. However, we now have additional events which correspond to (*i*) the uncertainty value $R_i(t)$ exceeding the threshold value $R_i^{\max}$ for some $i \in \{1 \ldots, M\}$, and (*ii*) the uncertainty value $R_i(t)$ dropping below the threshold value $R_i^{\max}$ for the same $i \in \{1 \ldots, M\}$. Next we show how these events affect the objective function gradient evaluation.

4.1 Objective Function Gradient Evaluation

Based on our analysis, we are still interested in minimizing the objective function $J(\theta, w)$ in (6) with respect to θ and w and can obtain $\nabla J(\theta, w) = [\frac{dJ(\theta,w)}{d\theta} \frac{dJ(\theta,w)}{dw}]^{\mathrm{T}}$ as the gradient of

$$J(\theta, w) = \frac{1}{T} \sum_{i=1}^{M} \sum_{k=0}^{K} \int_{\tau_k(\theta,w)}^{\tau_{k+1}(\theta,w)} [R_i(t) + \beta_i R_i(t) \mathbf{1}[R_i(t) > R_i^{\max}]] dt$$

As illustrated in Fig. 1, suppose τ_e is the time instant when some $R_i(t)$ enters a region violating the constraint $R_i(t) \leq R_i^{\max}$ and τ_l is the first time instant the same $R_i(t)$ leaves this region after entering it at time τ_e. To generalize, let $[\tau_{e_i^j}, \tau_{l_i^j})$ be the jth such interval, i.e., $R_i(\tau_{e_i^j}^-) \leq R_i^{\max}$, $R_i(t) > R_i^{\max}$ for all $t \in (\tau_{e_i^j}, \tau_{l_i^j})$, and $R_i(\tau_{l_i^j}^+) \leq R_i^{\max}$. Then, define

$$\Phi_i = \{(e_i^j, l_i^j),\ j = 1, \ldots, \phi_i\}$$

to be the set of event index pairs with the starting and ending points of such intervals, where ϕ_i is the number of such intervals where $R_i(t) > R_i^{\max}$ (noting that ϕ_i may be zero), and

$$\mathscr{E}_i = \{e : e = e_i^j \text{ for some } j = 1, \ldots, \phi_i\}$$
$$\mathscr{L}_i = \{l : l = l_i^j \text{ for some } j = 1, \ldots, \phi_i\}$$

Then, we can the rewrite $J(\theta, w)$ above as

$$J(\theta,w) = \frac{1}{T}\sum_{i=1}^{M}\left[\sum_{k=0}^{K}\int_{\tau_k}^{\tau_{k+1}} R_i(t)dt + \sum_{j=1}^{\phi_i}\int_{\tau_{e_i^j}}^{\tau_{l_i^j}} \beta_i R_i(t)\,dt\right]$$

Taking derivatives with respect to θ and w and observing the cancelation of terms of the form $R_i(\tau_k)\nabla\tau_k$, we finally get

$$\nabla J(\theta,w) = \frac{1}{T}\sum_{i=1}^{M}\left[\sum_{k=0}^{K}\int_{\tau_k}^{\tau_{k+1}} \nabla R_i(t)\left(1+\beta_i \mathbf{1}\left[R_i(t) > R_i^{\max}\right]\right)dt + \beta_i R_i^{\max}\left(\sum_{l\in\mathscr{L}_i}\nabla\tau_l - \sum_{e\in\mathscr{E}_i}\nabla\tau_e\right)\right] \tag{39}$$

The evaluation of $\nabla J(\theta,w)$ therefore depends on $\nabla R_i(t)$, $\nabla\tau_e$, $e\in\mathscr{E}_i$, and $\nabla\tau_l$, $l\in\mathscr{L}_i$, for all $i=1,\ldots,M$. We have already provided the details for calculating $\nabla R_i(t)$ in Section. 3 and note that this gradient is not affected by events at τ_e and τ_l since all agent dynamics and uncertainty dynamics remain unaffected in applying (14). Therefore, we are only left with the additional evaluation of $\nabla\tau_e$ and $\nabla\tau_l$. They are both determined by the movement of agents, hence the controllable parameters θ and w, so entering and leaving the region defined by $R_i(t) > R_i^{\max}$ are both endogenous events. The associated guard conditions involved in (15) are both $R_i - R_i^{\max} = 0$. Thus, from (15), similar to (21) we have,

$$\nabla\tau_k = -\frac{\nabla R_i\left(\tau_k^-\right)}{A_i - BP_i\left(s\left(\tau_k^-\right)\right)}, \quad \text{for } k = e \text{ or } l \tag{40}$$

Compared to the case without any constraints, the evaluation of the gradient of the cost function with respect to θ and w has one more term: $\beta_i R_i^{\max}(\sum_l \nabla\tau_l - \sum_e \nabla\tau_e)$, which only requires the calculation of $\nabla\tau_k$ using (40) for the time instant when some $R_i(t)$ enters or leaves the region defined by $R_i(t) > R_i^{\max}$.

4.2 Numerical Examples

We present two examples of persistent monitoring problems with and without the $R_i^{\max}$ constraint respectively so as to compare them and evaluate the effect of the constraints. In both cases, we use two agents and parameters $L=40$, $M=41$, $\alpha_1=0$, $\alpha_M=40$. Also, the remaining sampling points are evenly spaced over $[0,40]$. The sensing range is set to $r=4$, the initial values of the uncertainty are $R_i(0)=2$ for all i, and the time horizon is $T=200$. We select $B=6$ and $A_i=0.2$ for all i except $A_{20}=0.4$, which means that the middle sampling point has uncertainty increasing twice as fast as all other points. We set the threshold constraint $R_i^{\max}=4$ for $i=0,1,...,40$. Figure 2 is an example with $\beta_i=0$, $i=0,\ldots,40$, and it is equivalent to the problem without any constraint. Figure 3 shows an example with $\beta_i=10$, $i=0,\ldots,40$.

On the left sides of Fig. 2 and Fig. 3, the top plot shows the cost $J(\theta^l, w^l)$ as a function of iteration number $l=1,2,\ldots$, while the bottom plot shows the optimal agent trajectories. On the right sides of Fig. 2 and Fig. 3, nine sampling points are evenly

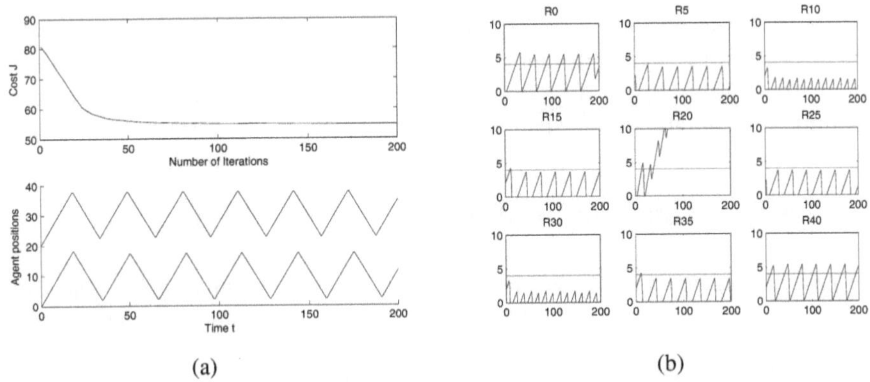

Fig. 2. Two agent example for $\beta_i = 0$, $i = 0, 1, \ldots, 40$. Left top plot: cost $J(\theta^l, w^l)$ as a function of iteration number with $J_{\beta=0}(\theta^\star_{\beta=0}, w^\star_{\beta=0}) = 54.59$; left bottom plot shows the optimal agent trajectories. Right plot: Selected $R_i(t)$ vs. time. Red lines are the $R_i^{\max}$ constraint.

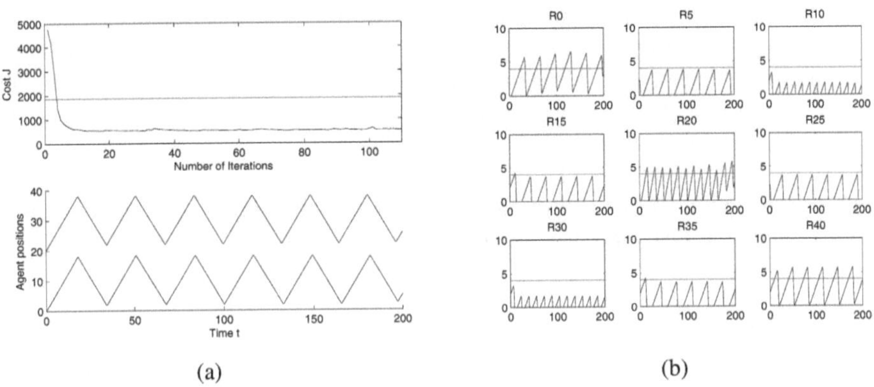

Fig. 3. Two agent example for $\beta_i = 10$, $i = 0, 1, \ldots, 40$. Left top plot: cost $J(\theta^l, w^l)$ as a function of iteration number with $J^\star_{\beta=10}(\theta^\star_{\beta=10}, w^\star_{\beta=10}) = 530.1$, red line $J_{\beta=10}(\theta^\star_{\beta=0}, w^\star_{\beta=0}) = 1861$ represents the cost using the agent trajectories with $\beta_i = 0$ and cost function with $\beta_i = 10$, $i = 0, \ldots, 40$; left bottom plot shows the optimal agent trajectories. Right plot: Seleccted $R_i(t)$ vs. time. Red lines are the $R_i^{\max}$ constraint.

selected and their corresponding $R_i(t)$ is shown. The horizontal red line in each such plot shows $R_i^{\max}$. If we compare the 9 selected $R_i(t)$ between these two examples, the effect of increasing β_i from $\beta_i = 0$ to $\beta_i = 10$, $i = 0, \ldots, 40$, is obvious in the sense that R_{20} in Fig. 2 increases to unacceptably high values beyond $R_i^{\max}$, whereas it is maintained to acceptable levels around $R_i^{\max}$ in Fig. 3. Comparing the optimal agent trajectories shown in Fig. 2 and Fig. 3, we can see that, when $\beta_i = 10$ for all i, the agents sacrifice coverage at the end points in order to maintain the middle point uncertainty R_{20} low.

In addition, if we apply the optimal agent trajectory with $\beta_i = 0$ to the cost function with $\beta_i = 10$, we obtain the cost $J_{\beta=10}(\theta^\star_{\beta=0}, w^\star_{\beta=0}) = 1861$, which is shown as the horizontal red line in the left plot of Fig. 3, while the actual optimal cost with $\beta_i = 10$, $i = 0,\ldots,40$, is $J^\star_{\beta=10}(\theta^\star_{\beta=10}, w^\star_{\beta=10}) = 530.1$. This is a significant cost reduction resulting from our algorithm which accounts for the desired performance constraints.

5 Conclusion and Future Work

We have formulated an optimal persistent monitoring problem with the objective of controlling the movement of multiple cooperating agents to minimize an uncertainty metric in a given mission space with some upper bound constraints for the uncertainty values in the mission space. In a one-dimensional mission space, as in the case of no constraints analyzed in [14], we have shown that the optimal solution is reduced to the determination of two parameter vectors for each agent: a sequence of switching locations and associated waiting times at these switching points. We have used Infinitesimal Perturbation Analysis (IPA) to evaluate sensitivities of the objective function with respect to all parameters and, therefore, obtain a complete on-line (locally optimal) solution through a gradient-based algorithm. Our work to date has established the basis for extending this optimal control approach to a two-dimensional mission space. In this case, similar simple solutions can no longer be derived. An alternative is to optimally assign each agent a linear trajectory, motivated by the one-dimensional analysis. We prove in [22], however, that elliptical trajectories outperform linear ones. Therefore, by parametrically describing an elliptical trajectory, we formulate and solve instead a parametric optimization problem in which we seek to determine such trajectories optimizing an uncertainty metric over a two-dimenisonal mission space.

References

1. Rekleitis, I., Lee-Shue, V., New, A., Choset, H.: Limited communication, multi-robot team based coverage. In: IEEE Intl. Conf. on Robotics and Automation, vol. 4, pp. 3462–3468 (2004)
2. Cortes, J., Martinez, S., Karatas, T., Bullo, F.: Coverage control for mobile sensing networks. IEEE Trans. on Robotics and Automation 20(2), 243–255 (2004)
3. Girard, A., Howell, A., Hedrick, J.: Border patrol and surveillance missions using multiple unmanned air vehicles. In: 43rd IEEE Conf. Decision and Control, vol. 1, pp. 620–625 (2005)
4. Grocholsky, B., Keller, J., Kumar, V., Pappas, G.: Cooperative air and ground surveillance. IEEE Robotics & Automation Magazine 13(3), 16–25 (2006)
5. Smith, R., Mac Schwager, S., Rus, D., Sukhatme, G.: Persistent ocean monitoring with underwater gliders: Towards accurate reconstruction of dynamic ocean processes. In: IEEE Conf. on Robotics and Automation, pp. 1517–1524 (2011)
6. Paley, D., Zhang, F., Leonard, N.: Cooperative control for ocean sampling: The glider coordinated control system. IEEE Trans. on Control Systems Technology 16(4), 735–744 (2008)
7. Smith, S.L., Schwager, M., Rus, D.: Persistent monitoring of changing environments using robots with limited range sensing. In: IEEE International Conf. Robotics and Automation, pp. 5448–5455 (2011)

8. Nigam, N., Kroo, I.: Persistent surveillance using multiple unmanned air vehicles. In: IEEE Aerospace Conf., pp. 1–14 (2008)
9. Hokayem, P., Stipanovic, D., Spong, M.: On persistent coverage control. In: 46th IEEE Conf. Decision and Control, pp. 6130–6135 (2008)
10. Elmaliach, Y., Agmon, N., Kaminka, G.: Multi-robot area patrol under frequency constraints. In: IEEE Intl. Conf. on Robotics and Automation, pp. 385–390 (2007)
11. Cassandras, C.G., Ding, X.C., Lin, X.: An optimal control approach for the persistent monitoring problem. In: Proc. of 50th IEEE Conf. Decision and Control, pp. 2907–2912 (2011)
12. Pasqualetti, F., Franchi, A., Bullo, F.: On cooperative patrolling: Optimal trajectories, complexity analysis, and approximation algorithms. IEEE Transactions on Robotics 28, 592–606 (2012)
13. Spindler, M., Pasqualetti, F., Bullo, F.: Distributed multi-camera synchronization for smart-intruder detection. In: American Control Conference, pp. 5120–5125 (2012)
14. Cassandras, C.G., Lin, X., Ding, X.C.: An optimal control approach to the multi-agent persistent monitoring problem. IEEE Trans. on Automatic Control 58(4), 947–961 (2013)
15. Cassandras, C.G., Wardi, Y., Panayiotou, C.G., Yao, C.: Perturbation analysis and optimization of stochastic hybrid systems. European Journal of Control 16(6), 642–664 (2010)
16. Wardi, Y., Adams, R., Melamed, B.: A unified approach to infinitesimal perturbation analysis in stochastic flow models: the single-stage case. IEEE Trans. on Automatic Control 55(1), 89–103 (2009)
17. Bryson, A., Ho, Y.: Applied optimal control. Wiley, N.Y (1975)
18. Xu, X., Antsaklis, P.: Optimal control of switched systems based on parameterization of the switching instants. IEEE Trans. on Automatic Control 49(1), 2–16 (2004)
19. Yao, C., Cassandras, C.G.: Perturbation analysis of stochastic hybrid systems and applications to resource contention games. Frontiers of Electr. and Electr. Engin. in China 6(3), 453–467 (2011)
20. Cassandras, C.G., Lygeros, J.E.: Stochastic hybrid systems. Taylor and Francis (2007)
21. Polak, E.: Optimization: algorithms and consistent approximations. Springer (1997)
22. Lin, X., Cassandras, C.G.: An optimal control approach to the multi-agent persistent monitoring problem in two-dimensional spaces. In: 52nd IEEE Conf. Decision and Control (submitted, 2013), `http://arxiv.org/submit/694747/view`

Resilient Distributed Control of Multi-agent Cyber-Physical Systems

Quanyan Zhu[1], Linda Bushnell[2], and Tamer Başar[1,*]

[1] Coordinated Science Laboratory
University of Illinois at Urbana-Champaign
1308 W. Main St., Urbana, IL
{zhu31,basar1}@illinois.edu
[2] Networked Control Systems Lab
EE Dept., University of Washington
Seattle, WA, 98195, USA
lb2@uw.edu

Abstract. Multi-agent cyber-physical systems (CPSs) are ubiquitous in modern infrastructure systems, including the future smart grid, transportation networks, and public health systems. Security of these systems are critical for normal operation of our society. In this paper, we focus on physical layer resilient control of these systems subject to cyber attacks and malicious behaviors of physical agents. We establish a cross-layer system model for the investigation of cross-layer coupling and performance interdependencies for CPSs. In addition, we study a two-system synchronization problem in which one is a malicious agent who intends to mislead the entire system behavior through physical layer interactions. Feedback Nash equilibrium is used as the solution concept for the distributed control in the multi-agent system environment. We corroborate our results with numerical examples, which show the performance interdependencies between two CPSs through cyber and physical interactions.

Keywords: Cyber-Physical Systems, Network Security, Differential Games, Multi-Resolution Games, Games-in-Games, Coupled Riccati Differential Equations, Secure Control, Resilient Control Systems.

1 Introduction

Recent years have witnessed increasing integration of information technologies into modern critical infrastructures including energy systems, transportation systems and public health. The technological advancement has also brought many challenges for understanding the efficient and reliable integration of cyber and physical components of the system. Security is one of the major concerns of such

* Research was supported in part by an AFSOR MURI Grant (FA9550-10-1-0573), and in part by an NSA Grant through the Information Trust Institute at the University of Illinois.

DOI: 10.1007/978-3-319-01159-2_16,

cyber-physical systems (CPSs). With the migration from a closed network to an open and public network, adversaries can take advantage of vulnerabilities existing in cyber world in order to compromise or inflict damages on the physical system. To protect these systems, it is imperative for us to design defense schemes both at the cyber and physical layers of the system to provide security mechanisms for reliable operations.

Modern systems are increasingly complex because of multi-layer system integrations, which lead to "systems of systems". Moreover, the complexity also comes from the large scale of the system, composed of a large number of interacting distributed systems or agents that are coordinated or controlled to accomplish a certain task in a decentralized manner. Illustrated in Fig. 1, multiple CPSs are interconnected with each system autonomously controlling itself and reacting to the environment as well as cyber or physical signals of other systems. The multi-agent system architecture renders it difficult to study the security issues of such multi-agent CPSs using conventional methods. Instead, it is important to establish new frameworks for understanding the system security interdependencies. The vulnerability of the cyber component of one system can lead to insecurity of physical components of another system. Similarly, it is also possible that the physical compromise of one system can become the cyber vulnerability of another connected system.

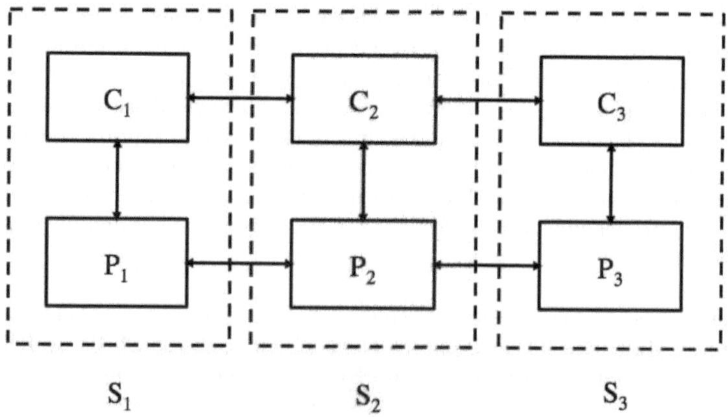

Fig. 1. Three interconnected CPSs: Each system $S_j, j = 1, 2, 3$, is composed of its cyber system C_j and physical system P_j. Three CPSs have interactions at both the physical layer and the cyber layer.

Game theory provides systematic modeling and computational tools to address these issues. Recent literature has seen a surge of interest in applying game-theoretic methods to understand cyber security and secure control systems [2,15]. In [3], static and dynamic game frameworks have been used to design equilibrium revocation strategies for defending sensor networks from node capturing and cloning attacks. In [4], a stochastic game has been used to model the strategic interactions between an intrusion detection system (IDS) and a malicious intruder, and the authors have used online reinforcement learning methods to provide data-driven defense polices for dynamic IDS configurations.

Most of current literature have focused on problems residing at either cyber or physical components of the system. To address the system integration problem, it is essential and also inevitable to establish system models that allow detailed investigations of cross-layer coupling and performance interdependencies for CPSs. In [4, 5], we have proposed a stochastic hybrid system model, where each mode represents the condition under which physical dynamical systems evolve, and the system switches from one mode to another, depending on cyber attacks and security policies. The robust control design for the physical system against noise and disturbances is strongly coupled with the cyber defense mechanism design against cyber attacks. It has been shown that under the linear-quadratic robust control system paradigm as in [6] and the stochastic game modeling of cyber security systems as in [7, 8], the design of the CPS results in a set of coupled equations to be solved for achieving resilient and robust control of the system. This work has provided us fundamental and rich concepts in designing optimal cross-layer CPSs. The cyber system model can be further extended by including detailed models for describing attacks on cyber components based on recently developed games-in-games principle for multi-resolution games [9, 10], while the physical system model can also be extended to investigate multi-agent CPSs for understanding the multi-system interdependencies.

The goal of this paper is to focus on the latter part of the extension. We first discuss a general framework for designing distributed control schemes for multi-agent CPSs, and then establish a stochastic hybrid differential game model for studying the impact of a malicious physical system on the physical dynamics of other systems. In particular, we study a two-person synchronization problem where S_1 aims to achieve synchrony with S_2, while S_2 intends to mislead S_1 to an unfavorable system state. We provide a set of coupled Riccati differential equations to characterize the feedback Nash equilibrium solution of an N-person stochastic hybrid differential game. This investigation provides an initial step toward addressing more complex scenarios where cyber systems can be reconfigured in response to physical systems. We also see that this work serves as the inner-most game within the games-in-games framework for large scale hierarchical systems.

The paper is organized as follows. In Section 2, we discuss related work to our problem. In Section 3, we present the general system model for multi-agent cyber-physical systems. In Section 4, we study the feedback Nash equilibrium strategies for a two-person game problem. Section 5 provides numerical examples to illustrate the equilibrium solutions, and we conclude in Section 6.

2 Related Works

Our work falls into many different research areas in the literature. Our system model for describing CPSs is based on the continuous-time Markov jump linear systems, which has been widely studied in the literature in [11–13]. In [11], zero-sum differential game frameworks are used to study H-infinity robust control of Markovian jump systems. [13] has studied the minimax control of randomly switching systems under sampled state information. In this work, we build our multi-agent system model based on systems of similar structures, and investigate distributed control using N-person nonzero-sum differential games.

This work focuses on the physical component under the larger framework of resilient control established in [4, 5]. With the parameters of the cyber components fixed, we investigate the control of multi-agent systems at different system modes. This includes the design of distributed controllers at critical systems states, which allows to provide certain level of system performance after cyber attacks.

Adversarial behaviors reside at multiple layers of the system. [14] has proposed a hierarchical security framework for CPS, in particular for the emerging smart grid, and discussed security issues at the control, communications, and information management layers of the system. The goal of resilient control of CPS is to adopt first a divide-and-conquer approach and then integrate the layer-specific solutions together as the system-level solution. Following this methodology, this work considers malicious behaviors at the physical layer, where some agents intend to mislead or inflict damage on the agents through physical interactions. Solution to this problem can be interfaced with solutions from the other layers, such as those in [15, 16], through recently developed games-in-games principle for multi-resolution games [9, 10].

3 System Model for Multi-agent CPSs

In this section, we present a general system model for describing the interactions between multi-agent CPSs. Let $\mathcal{N} = \{1, 2, \cdots, N\}$ be the index set, and $\mathcal{S} = \{S_1, S_2, \cdots, S_N\}$ the set of N interconnected CPSs. Each system $S_j \in \mathcal{S}$ is composed of a cyber system C_j and a physical system P_j. We let $\mathcal{C} = \{C_1, C_2, \cdots, C_N\}$ be the N cyber systems associated with $\mathcal{S}$, and $\mathcal{P} = \{P_1, P_2, \cdots, P_N\}$ be the set of N physical dynamical systems. The cyber systems are often described by graphical models and their modeling can be found in many recent literature on decision and control at cyber-level of CPS [9, 15], while the physical systems are often modeled through dynamical systems governed by physical laws and constraints. The focus of this paper will be on the interconnected physical systems $\mathcal{P}$ and the impact of cyber systems $\mathcal{C}$ on the performance of $\mathcal{P}$. The multi-agent interactions between N systems can be represented by two graphs. One is $\mathcal{G}_P := \langle \mathcal{N}, \mathcal{E}_P \rangle$ that represents the cyber relations among N systems. Such relations can exist at multiple levels within the cyber system, ranging from communication links between two work stations [16] to security interdependencies through economic investment [17]. The other graph is denoted by $\mathcal{G}_C := \langle \mathcal{N}, \mathcal{E}_C \rangle$, which captures the interconnections between physical systems. It can represent the underlying information flow of sensing or actuation signals. In Fig. 1, both cyber and physical parts of the systems are interconnected in the same fashion, i.e., $\mathcal{E}_C = \mathcal{E}_P := \{(1, 2), (2, 3)\}$.

We describe the dynamics of each CPS S_j by a continuous-time Markov jump linear system as follows:

$$\dot{x}_j = A_j(t, \theta_j(t))x_j + B_j(t, \theta_j(t))u_j; \quad x_j(t_0) = x_{j,0}, \tag{1}$$

where $x_j \in \mathbb{R}^{n_j}$ is the n_j-dimensional system state vector of system $P_j \in \mathcal{P}$; $u_j \in \mathbb{R}^{p_j}$ is the p_j-dimensional control input determined by P_j; θ_j is a finite state Markov chain defined on the state space $\Theta_j = \{\theta_{j,1}, \theta_{j,2}, \cdots, \theta_{j,M}\}$ with a positive initial distribution $\pi_j^0 := [\pi_{j,1}^0, \pi_{j,2}^0, \cdots, \pi_{j,M}^0]$ and the infinitesimal generator matrix $\Lambda_j = (\lambda_{i,i'}(t))_{i\in\Theta_j, i'\in\Theta_j}$, such that $\lambda_{ii'} \geq 0$ for $i \neq i'$ and

$$\mathbb{P}\{\theta_j(t+h) = i' | \theta_j(t) = i\} = \begin{cases} \lambda_{ii'}h + o(h), & i' \neq i \\ 1 + \lambda_{ii}h + o(h), & i' = i \end{cases}.$$

The system states x_j and inputs u_j each belong to appropriate Hilbert spaces $\mathcal{H}_{x_j}$ and $\mathcal{H}_{u_j}$, respectively, defined on the time interval $[0, t_f]$. The system P_j is stochastic due to the switching between different modes or forms governed by Λ_j. Here, we assume that $A_j(t, i) \in \mathbb{R}^{n_j \times n_j}$ and $B_j(t, i) \in \mathbb{R}^{n_j \times p_j}$ are piecewise continuous in t for each $i \in \Theta$.

Note that the process θ_j in stochastic hybrid dynamics (1) captures structural changes of the physical system caused by successful cyber attacks, while x_j models the evolution of physical states of P_j. For example, the attack on circuit breakers in energy systems will change the system from being in normal mode to restorative or emergency mode, where partial load is lost or power flow constraint is violated. It is important to design contingent voltage or frequency control strategies in response to mode changes at the physical layer. However, it is also necessary to take appropriate cyber control actions to restore the system to its normal state [14, 18].

For each system S_j, input u_j directly controls the physical state x_j, while the defense in the cyber domain determines the rate matrix Λ. In this paper, we assume that Λ is given and find distributed control strategies of each system when centralized coordination is not possible. In particular, we study the case where malicious behaviors are present in the physical component of the system. These adversarial effects can be caused by physical compromise of a normal system or manual placement of malicious agents into the network, which is in the same spirit as the node capturing and cloning attacks in sensor networks [3]. In addition, the malicious behavior can also be induced through cyber attacks, where normal system behavior can be altered by Stuxnet-like worms through the enterprise and process control networks [19].

The goal of each system S_j can be captured by the performance index L_j given as:

$$L_j(x_j, u_j; \theta_j, t_0) = q_j^f(x(t_f); \theta(t_f)) + \int_{t_0}^{t_f} g_j(t, x_j(t), u_j(t); \theta(t))dt,$$

where $x(t) = [x_1^T(t), x_2^T(t), \cdots, x_N^T(t)]'$, $\theta(t) = [\theta_1, \theta_2, \cdot, \theta_N]^T$, q_j^f is continuous in x, and g_j is jointly continuous in (t, x_j, u_j). We consider the feedback perfect-state measurement information structure for all systems and design controller in the form of

$$u_j(t) = \gamma_j(t, x(t); \theta(t)), \tag{2}$$

where γ_j is an admissible feedback control strategy, piecewise continuous in its first argument, and Lipschitz continuous in its second argument. We denote the class of all such control strategies by $\mathcal{M}_j^{\text{FB}}$. Note that each P_j can observe the state information of other systems through interconnection graph $\mathcal{G}$. Hence in general, the control and its performance index will be dependent on the aggregate state vector x.

Under the above assumptions and with controls picked as in (2), the system (1) admits a well-defined solution, which will induce corresponding "open-loop" representations of γ_j. By taking the expected value of the resulting stochastic cost L_j over the statistics of θ will lead to the average cost corresponding to the inputs generated by γ_j, which we write as:

$$J_j(\gamma_j, \gamma_{-j}; t_0) = \mathbb{E}_\theta\{L_j(x_j, u_j; \theta)\}, \tag{3}$$

where $\gamma_{-j} := \{\gamma_1, \cdots, \gamma_{j-1}, \gamma_{j+1}, \cdots, \gamma_N\}$ denotes the set of control strategies other than γ_j.

Since each system computes its own optimal control, the objective of each system S_j is to minimize the cost J_j over all its own feedback control policies:

$$\min_{\gamma_j \in \mathcal{M}_j^{\text{FB}}} J_j(\gamma_j, \gamma_{-j}; t_0). \tag{4}$$

This will lead to an $N-$person differential game model with each system solving (4), and its solution is characterized by feedback Nash equilibrium (FBNE) defined as follows.

Definition 1 (Feedback Nash Equilibrium, [1]). *The strategy profile $(\gamma_1^*, \gamma_2^*, \cdots, \gamma_N^*)$ is a feedback Nash equilibrium (FBNE) for the N-person stochastic differential game described by (1) and (3) if for all $j \in \mathcal{N}$ and $\gamma_j \in \mathcal{M}_j^{FB}$,*

$$J_j(\gamma_j^*, \gamma_{-j}^*; t_0) \leq J_j(\gamma_j, \gamma_{-j}^*; t_0).$$

In addition, the equilibrium strategies are strongly time-consistent if for all $j \in \mathcal{N}$, $t \in [t_0, t_f)$, and $\gamma_j \in \mathcal{M}_j^{FB}$,

$$J_j(\gamma_j^*, \gamma_{-j}^*; t) \leq J_j(\gamma_j, \gamma_{-j}^*; t).$$

4 Two-System Problem

In this section, we consider the case of two interconnected systems $S_1 = \langle C_1, P_1 \rangle$, and $S_2 = \langle C_2, P_2 \rangle$. Each system $S_j, j \in \mathcal{N}$, has two modes. We let $\theta_{j,1}$ refer to the normal or safe operation state, while $\theta_{j,2}$ refers to the compromised state after the success of cyber attacks. The goal of S_1 is to achieve physical synchronization with S_2 and its finite-horizon cost function (3) can be rewritten as

$$J_1 = \mathbb{E}\left\{\frac{1}{2}|x_1(t_f) - x_2(t_f)|^2_{Q_1^f(\theta(t_f))} + \frac{1}{2}\int_0^{t_f}\left(|x_1(t) - x_2(t)|^2_{Q_1(t,\theta(t))}\right.\right.$$
$$\left.\left. + |u_1(t)|^2_{R_1(t,\theta(t))}\right) dt\right\}, \quad (5)$$

The goal of S_2 is a malicious system which intends to mislead the state of S_1 to its preferred state $\bar{x}_2 \in \mathbb{R}$. Hence its associated performance index is described by

$$J_2 = \mathbb{E}\left\{\frac{1}{2}|x_2(t_f) - \bar{x}_2|^2_{Q_2^f(\theta(t_f))} + \frac{1}{2}\int_0^{t_f}\left(\alpha|x_1(t) - x_2(t)|^2_{Q_2(t,\theta(t))}\right.\right.$$
$$\left.\left. + (1-\alpha)|x_2(t) - \bar{x}_2|^2_{Q_2(t,\theta(t))} + |u_2(t)|^2_{R_2(t,\theta(t))}\right) dt\right\}, \quad (6)$$

where $\alpha \in (0,1)$ is a weighting parameter. In (5) and (6), $Q_j^f(\cdot), Q_j(\cdot,\cdot), j = 1,2,$ are non-negative definite matrices of proper dimensions, and $R_j(\cdot,\cdot)$ is positive definite. In addition, we assume that $Q_j(t,i), R_j(t,i)$ are piece-wise continuous in t for each $i \in \Theta$.

We assume that each system S_j has perfect observation of its own physical states x_j and system mode θ_j as well as the state and the mode of the other system. Hence we have $\mathcal{E}_C = \mathcal{E}_P := \{(1,2)\}$. Let the aggregate state and mode vectors be given by $x := [x_1^T, x_2^T]^T \in \mathbb{R}^2$ and $\theta := [\theta_1, \theta_2]^T \in \Theta := \Theta_1 \times \Theta_2$. The control input u_j of P_j is generated by a feedback strategy γ_j^F, according to (2).

4.1 Feedback Nash Equilibrium Solution

In this subsection, we characterize the feedback Nash equilibrium of the game associated with (1), (5), (6), and (2). The evolution of aggregated system $\mathcal{S}$ is described by

$$\dot{x} = \tilde{A}(t,\theta(t))x + \sum_{j=1}^{N}\tilde{B}_j(t,\theta(t))u_j, \quad (7)$$

where the system parameters $\tilde{A} \in \mathbb{R}^{(n_1+n_2)\times(n_1+n_2)}$ and $\tilde{B}_1 \in \mathbb{R}^{(n_1+n_2)\times n_1}, \tilde{B}_2 \in \mathbb{R}^{(n_1+n_2)\times n_2}$ are given by

$$\tilde{A}(t,\theta(t)) = \begin{bmatrix} A_1(t,\theta(t)) & 0_{n_1\times n_2} \\ 0_{n_2\times n_1} & A_2(t,\theta(t)) \end{bmatrix},$$

$$\tilde{B}_1(t,\theta(t)) = \begin{bmatrix} B_1 \\ 0_{n_2\times n_1} \end{bmatrix} \text{ and } \tilde{B}_2(t,\theta(t)) = \begin{bmatrix} 0_{n_1\times n_2} \\ B_2 \end{bmatrix}.$$

In addition, we define the following quantities:

$$\tilde{Q}_1(t,\theta(t)) := \begin{bmatrix} Q_1(t,\theta(t)) & -Q_1(t,\theta(t)) \\ -Q_1(t,\theta(t)) & Q_1(t,\theta(t)) \end{bmatrix},$$

$$\tilde{Q}_2(t,\theta(t)) := \begin{bmatrix} \alpha Q_2(t,\theta(t)) & -\alpha Q_2(t,\theta(t)) \\ -\alpha Q_2(t,\theta(t)) & Q_2(t,\theta(t)) \end{bmatrix},$$

$$\tilde{Q}_1^f(\theta(t)) = \begin{bmatrix} Q_1^f(\theta(t)) & -Q_1^f(\theta(t)) \\ -Q_1^f(\theta(t)) & Q_1^f(\theta(t)) \end{bmatrix},$$

$$\tilde{Q}_2^f(\theta(t)) = \begin{bmatrix} 0_{n_1\times n_1} & 0_{n_1\times n_2} \\ 0_{n_2\times n_1} & Q_2^f(\theta(t)) \end{bmatrix}, \quad \text{and}$$

$$p_1^T(t,\theta(t)) = 0_{1\times(n_1+n_2)}, \quad p_2^T(t,\theta(t)) = \begin{bmatrix} 0_{1\times n_1} & (1-\alpha)\bar{x}_2^T(Q_2^T + Q_2) \end{bmatrix}$$

We can rewrite (5) and (6) into the following equivalent cost functions:

$$\tilde{J}_1 = \mathbb{E}\left\{\frac{1}{2}x^T\tilde{Q}_1^f(\theta(t_f))x + \frac{1}{2}\int_0^{t_f}\left(x^T\tilde{Q}_1(t,\theta(t))x \right.\right. \\ \left.\left. -p_1^T(t,\theta(t))x + u_1^T R_1(t,\theta(t))u_1\right)dt\right\}, \tag{8}$$

$$\tilde{J}_2 = \mathbb{E}\left\{\frac{1}{2}x^T\tilde{Q}_2^f(\theta(t_f))x + \frac{1}{2}\int_0^{t_f}\left(x^T\tilde{Q}_2(t,\theta(t))x \right.\right. \\ \left.\left. -p_2^T(t,\theta(t))x + u_2^T R_2(t,\theta(t))u_2\right)dt\right\}, \tag{9}$$

Note that the direct coupling between two systems in this problem comes from the cost function while the indirect coupling enters in the dynamics though the feedback control strategies based on the information flow topology $\mathcal{G}_C, \mathcal{G}_P$. Following [20], to characterize the equilibrium solution, we let value functions V_j take the form of

$$V_j(t,x,\theta(t)) = \frac{1}{2}x^T Z_j(t,\theta(t))x + c_j^T(t,\theta(t))x + \eta_j(t,\theta(t)). \tag{10}$$

In addition, denote by $Z_j^l(t) := Z_j(t,\theta(t))$, $c_j^l(t) := c_j(t,\theta(t)), p_j^l(t) := p_j(t,\theta(t))$, $\eta_j^l(t) := \eta_j(t,\theta(t))$, when $\theta(t) = l, l \in \Theta$.

Theorem 1. *For the N-person stochastic differential game described above, let there exist a set of matrix valued functions $Z_j^l(t) \geq 0, j \in \mathcal{N}, l \in \Theta$, satisfying the following N coupled matrix Riccati differential equations:*

$$\dot{Z}_j^l + Z_j^l F_j^l + (F_j^l)^T Z_j^l + \tilde{Q}_j^l + Z_j^l\tilde{B}_j^l(R_j^l)^{-1}(\tilde{B}_j^l)^T Z_j^l + \sum_{l'\in\Theta}\lambda_{ll'}Z_j^{l'} = 0, \tag{11}$$

$$Z_j^l(t_f) = \tilde{Q}_j^f, \quad l \in \Theta \tag{12}$$

where

$$F_j^l := A^l - \sum_{j'=1,2}\tilde{B}_{j'}^l(R_{j'}^l)^{-1}(\tilde{B}_{j'}^l)^T Z_{j'}^l.$$

Then, the differential game admits a FBNE solution, affine in the current value of the aggregate state $x(t)$, given by

$$\begin{aligned} u_j^*(t) &= \gamma_j^*(t, x(t), \theta(t)) \\ &= -R_j^{-1}(t, \theta(t))\tilde{B}_j^T(t, \theta(t))[Z_j(t, \theta(t))x(t) + c_j(t, \theta(t))], \quad j \in \mathcal{N}, \end{aligned} \tag{13}$$

where $c_j(t, \theta(t))$ are obtained as the unique solution of the coupled linear differential equations:

$$\dot{c}_j^l + (F_j^l)^T c_j^l - Z_j^l \left\{ \sum_{j' \neq j} \tilde{B}_{j'}^l (R_{j'}^l)^{-1} (\tilde{B}_{j'}^l)^T c_{j'}^l \right\} - \frac{1}{2} p_j^l + \sum_{l' \in \Theta} \lambda_{ll'} c_j^{l'} = 0 \tag{14}$$

$$c_1^l(t_f) = 0, \quad l \in \Theta \tag{15}$$

$$c_2^l(t_f) = -\frac{1}{2} \bar{x}_2^T ((Q^f)^T + Q^f), \quad l \in \Theta \tag{16}$$

The corresponding values of the cost functionals associated with each mode are

$$V_j^l(0, x_0) = \frac{1}{2} x_0^T Z_i^l(0) x_0 + (c_j^l)^T x_0^T + \eta_j^l(0),$$

where $\eta_j^l,\ j \in \mathcal{N},\ l \in \Theta$ are obtained from

$$\begin{aligned} &\dot{\eta}_j^l - (c_j^l)^T \left\{ \sum_{j'=1,2} \tilde{B}_{j'}^l (R_{j'}^l)^{-1} (\tilde{B}_{j'}^l)^T c_{j'}^l \right\} \\ &+ \frac{1}{2} (c_j^l)^T \tilde{B}_j^l (R_j^l)^{-1} (\tilde{B}_j^l)^T c_j^l + \sum_{l' \in \Theta} \lambda_{ll'} \eta_j^{l'} = 0, \end{aligned} \tag{17}$$

$$\eta_j^l(t_f) = 0, \quad l \in \Theta \tag{18}$$

Proof (Sketch of Proof). With γ_2^* fixed, the sufficient condition for the feedback strategy γ_1^* to be optimal is that the cost-to-go function $V_1(t, x, \theta(t))$ satisfies the following partial differential equations [20]:

$$\begin{aligned} \min_{u_1} \Bigg\{ &\frac{\partial}{\partial t} V_1(t, x, i) + \frac{\partial}{\partial x} V_1(t, x, i) \cdot \Big(\tilde{A}(t, i)x + \tilde{B}_1(t, i)u_1 \\ &+ \tilde{B}_2(t, i)\gamma_2^*(t, x, i) \Big) + \sum_{i' \in \Theta} \lambda_{ii'} V_1(t, x, i') \Bigg\} = 0. \end{aligned} \tag{19}$$

Likewise, with γ_1^* fixed, the sufficient condition for γ_2^* is that $V_2(t, x, \theta(t))$ satisfies

$$\begin{aligned} \min_{u_2} \Bigg\{ &\frac{\partial}{\partial t} V_2(t, x, i) + \frac{\partial}{\partial x} V_2(t, x, i) \cdot \Big(\tilde{A}(t, i)x + \tilde{B}_1(t, i)\gamma_1^* \\ &+ \tilde{B}_1(t, i)u_1 \Big) + \sum_{i' \in \Theta} \lambda_{ii'} V_2(t, x, i') \Bigg\} = 0. \end{aligned} \tag{20}$$

The results follow from using (19) and (20) and the value function in the form of (10).

Note that the FBNE obtained above is also strongly-time consistent. The equilibrium control strategies retain the same form for any initial conditions of the game. This leads to a robust and optimal solution in case of disturbances and errors.

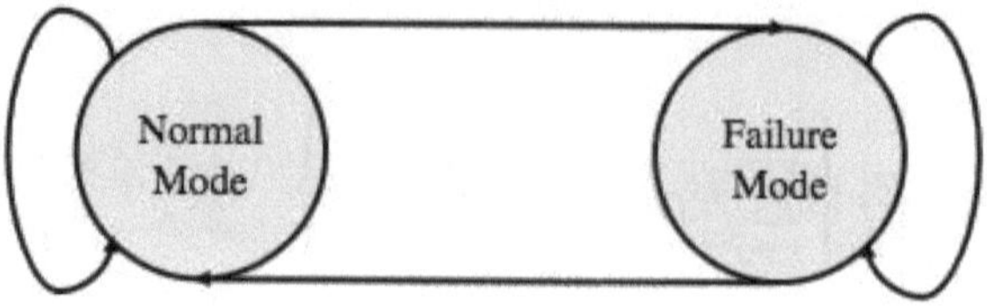

Fig. 2. System mode transitions from normal mode to failure mode. Such mode switch is dictated by cyber defense mechanisms, and it has impact on evolution of the physical state of the system.

5 A Numerical Example

In this section, we use a numerical example to illustrate the solution outlined in the section above. In Fig. 2, we depict the transition of individual systems from their normal operation ($\theta_j = 1$) to failure mode ($\theta_j = 2$). In the two-system case, this leads to a total four aggregate modes, i.e., $\Theta = \{\theta_1, \theta_2, \theta_3, \theta_4\}$ with $\theta_1 = (1,1), \theta_2 = (1,2), \theta_3 = (2,1), \theta_4 = (2,2)$. We let θ_4 be the absorbing mode, where no cyber recovery mechanisms are available at the same time scale of the state evolution at the physical layer. The rate matrix Λ is chosen as follows:

$$\Lambda = \begin{bmatrix} -\frac{3}{4} & \frac{1}{4} & \frac{1}{4} & \frac{1}{4} \\ 0 & -\frac{1}{2} & 0 & \frac{1}{2} \\ 0 & 0 & -\frac{1}{2} & \frac{1}{2} \\ 0 & 0 & 0 & 0 \end{bmatrix}. \tag{21}$$

Fig. 3 illustrates the transition between different modes.

The transition from one mode to another corresponds to the system failure due to cyber attacks. As a result, the aggregate system $\mathcal{S}$ has different system dynamics at each mode. In this example, we let x_j be scalars, and let $A^l, l \in \Theta$, be the only quantities that are mode-dependent and take the following values:

$$\tilde{A}^1 = \begin{bmatrix} 0 & 0 \\ 0 & 0 \end{bmatrix}, \tilde{A}^2 = \begin{bmatrix} 1 & 0 \\ 0 & 0 \end{bmatrix}, \tilde{A}^3 = \begin{bmatrix} 0 & 0 \\ 0 & 1 \end{bmatrix}, \tilde{A}^4 = \begin{bmatrix} 1 & 0 \\ 0 & 1 \end{bmatrix}. \tag{22}$$

It is easy to see that each failure causes the corresponding system "less stable" by switching the diagonal entry from 0 to 1. We let other parameters in the

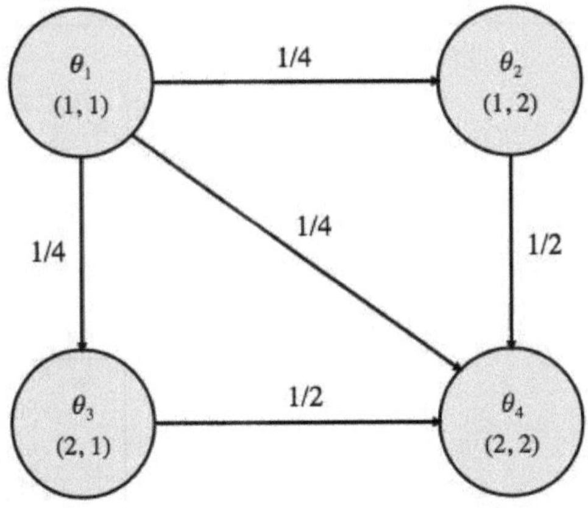

Fig. 3. Rate matrix Λ: θ_1 is the normal operating state. The entire system can fail and transition to another mode at equal rates to $\theta_1, \theta_2, \theta_3$. The intermediate failure modes θ_2, θ_3 switch to θ_4 when cyber attack occurs. θ_4 is the absorbing state. The system can not recover immediately once damaged.

system to be independent of modes, i.e., for all $l \in \Theta$, $j = 1, 2$, $B_j^l = 1$, $R_j^l = 1$, $\alpha = 1/2$, and

$$\tilde{Q}_1 = \tilde{Q}_1^f = \begin{bmatrix} 1 & -1 \\ -1 & 1 \end{bmatrix}, \quad \tilde{Q}_2^f = \begin{bmatrix} 0 & 0 \\ 0 & 1 \end{bmatrix}, \quad \tilde{Q}_2 = \begin{bmatrix} \alpha & -\alpha \\ -\alpha & 1 \end{bmatrix}. \tag{23}$$

We obtain the FBNE solution in the form of

$$u_1^*(t) = K_1^1(t, \theta(t))x_1 + K_1^2(t, \theta(t))x_2 + \xi_1(t, \theta(t)) \tag{24}$$
$$u_2^*(t) = K_2^1(t, \theta(t))x_1 + K_2^2(t, \theta(t))x_2 + \xi_2(t, \theta(t)) \tag{25}$$

and the resulting system dynamics are given by

$$\dot{x}(t) = (\tilde{A}(t, \theta(t)) + K_1(t, \theta(t)) + K_2(t, \theta(t)))x(t) + \xi(t, \theta(t)), \tag{26}$$

where $K_j = [K_j^1, K_j^2]^T, j = 1, 2$, and $\xi = [\xi_1, \xi_2]^T$. We set the initial condition as $\theta(0) = \theta_1$ and $x(0) = [0, 1/2]^T$. In Figs. 4 and 5, we show the feedback control gains K_1^1 and K_2^2 for systems S_1 and S_2 for the time interval $[0, 2.5]$. We can see that the feedback gain K_1^1 at modes θ_1, θ_2 are close and its behavior at modes θ_3, θ_4 are similar to each other. It is easy to see that in modes θ_1 and θ_2, the system S_1 is in normal operation mode. The difference in the gains $K_1^1(t, 1), K_1^1(t, 2)$ is due to the coupling from the malfunctioning of system S_2. We see here how the security mode of one system leads to different behavior of another system. In addition, the numerical solutions for the affine terms ξ_1, ξ_2 are illustrated in Figs. 6 and 7. These feedforward terms allow the system to

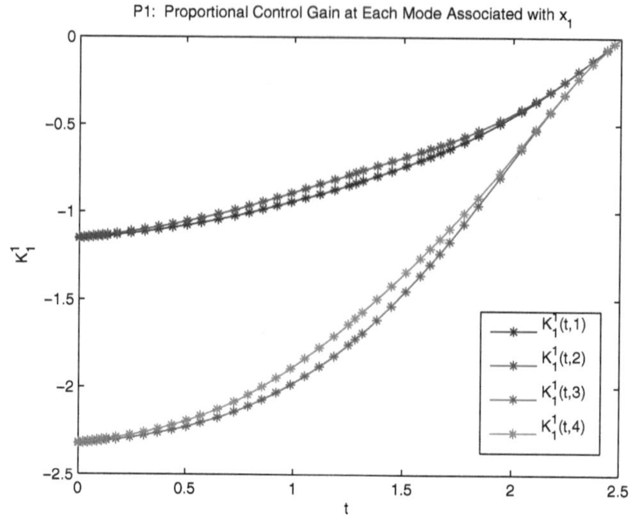

Fig. 4. Individual feedback control term of P_1 with respect to state x_1

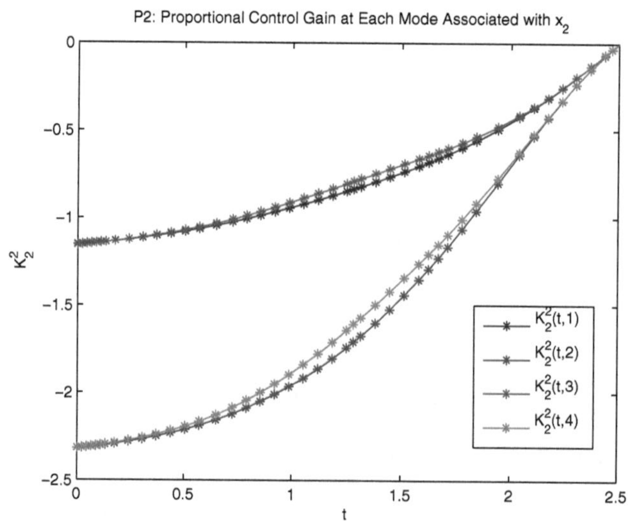

Fig. 5. Individual feedback control term of P_2 with respect to state x_2

track the desired trajectory. In Figs. 8 and 9, we show the sample state and mode trajectories. We see that the system mode eventually goes to the failure state θ_4. In Fig. 8, we observe that the malicious system S_2 first attracts S_1 for synchronization, and then misleads it to reach a value $\bar{x}_2 = 1$.

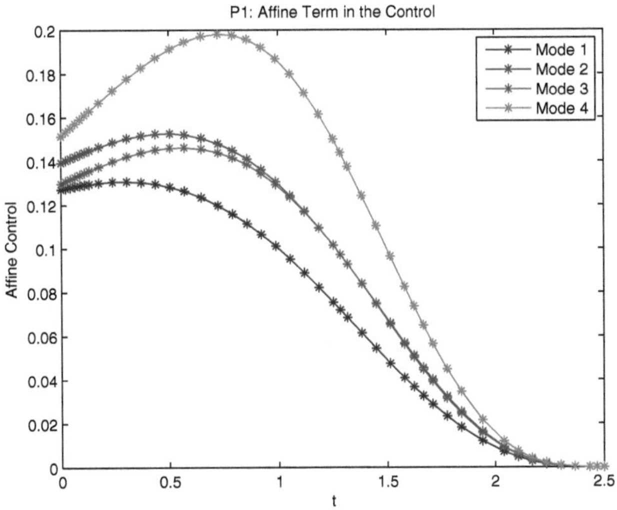

Fig. 6. Feedforward control term of P_1

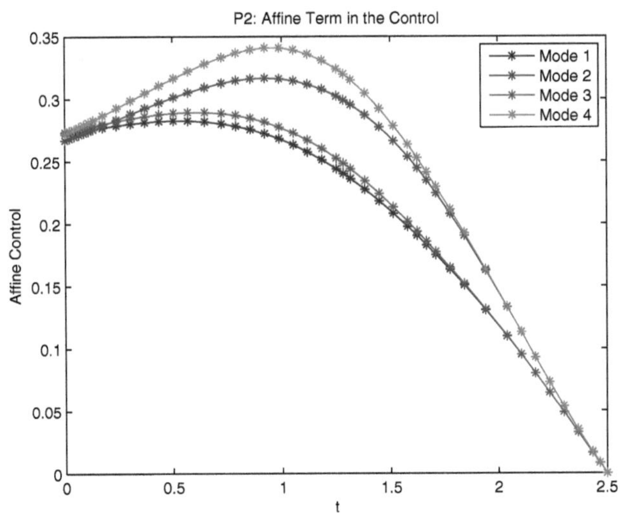

Fig. 7. Feedforward control term of P_2

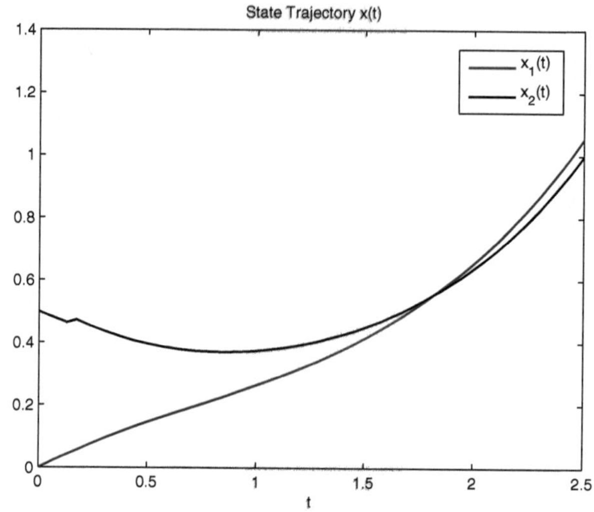

Fig. 8. Sample state trajectory

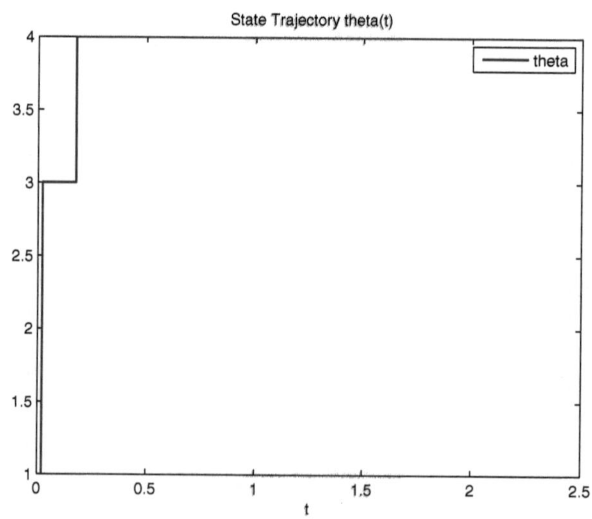

Fig. 9. Sample mode trajectory

6 Conclusion

Modern systems are increasingly complex due to cyber and physical system integrations as well as distributed interactions among different subsystems. This paper investigates resilient control design for multi-agent cyber-physical systems (CPSs). We have established a general system framework for describing the interactions between cyber and physical components within a CPS, as well as the interdependencies among multiple CPSs. We have focused on the physical layer control design and have studied a two-system problem with one malicious agent who intends to mislead and compromise the physical behaviors of the systems. We have designed distributed controllers based on feedback Nash equilibrium solutions. From the numerical example, we have observed that the performance of the systems are coupled at both physical and cyber layers. As for future work, we would extend this work to nonlinear and stochastic systems with additive noise. In addition, it would be interesting to study distributed cyber defense mechanisms based on this framework and employ tools from multi-resolution games to provide interface for cyber and physical decision problems in order to achieve required specifications for security and resilience.

References

1. Başar, T., Olsder, G.J.: Dynamic Noncooperative Game Theory. SIAM Series in Classics in Applied Mathematics (January 1999)
2. Alpcan, T., Başar, T.: Network Security: A Decision and Game Theoretic Approach. Cambridge University Press (January 2011)
3. Zhu, Q., Bushnell, L., Başar, T.: Game-theoretic analysis of node capture and cloning attack with multiple attackers in wireless sensor networks. In: Proc. 51st IEEE Conference on Decision and Control, CDC 2012, Maui, Hawaii, December 10-13 (2012)
4. Zhu, Q., Başar, T.: Robust and resilient control design for cyber-physical systems with an application to power systems. In: Proc. of 50th IEEE Conference on Decision and Control and European Control Conference (CDC/ECC), Orlando, Florida, December 12-15, pp. 4066–4071 (2011)
5. Zhu, Q., Başar, T.: A dynamic game-theoretic approach to resilient control system design for cascading failures. In: Proc. of International Conference on High Confidence Networked Systems (HiCoNS) at CPSWeek 2012, Beijing, China, pp. 41–46 (2012)
6. Başar, T., Bernhard, P.: H-infinity Optimal Control and Related Minimax Design Problems: A Dynamic Game Approach. Birkhäuser, Boston (August 1995)
7. Lye, K., Wing, J.M.: Game strategies in network security. International Journal of Information Security 4(1-2), 71–86 (2005)
8. Zhu, Q., Başar, T.: Dynamic policy-based IDS configuration. In: Proc. 48th IEEE Conference on Decision and Control, CDC 2009, Shanghai, China, December 16-18 (2009)
9. Zhu, Q., Başar, T.: Toward a theory of multi-resolution games. Submitted to SIAM Conference on Control and Its Applications (CT13)
10. Zhu, Q., Başar, T.: Multi-layer and multi-resolution large population stochastic games. In: 2012 SIAM Annual Meeting, Minneapolis, Minnesota, July 9-13 (2012)

11. Pan, Z., Başar, T.: H-infinity control of Markovian jump systems and solutions to associated piecewise-deterministic differential games. In: Olsder, G.J. (ed.) New Trends in Dynamic Games and Applications, pp. 61–94. Birkhäuser, Boston (1995)
12. Ji, Y., Chizeck, H.J.: Controllability, stabilizability, and continuous-time Markov jump linear quadratic control. IEEE Trans. on Automatic Control AC-35, 777–788 (1990)
13. Başar, T.: Minimax control of switching systems under sampling. Systems and Control Letters 25(5), 315–325 (1995)
14. Zhu, Q., Başar, T.: A hierarchical security architecture for smart grid. In: Hossain, E., Han, Z., Poor, H.V. (eds.) Smart Grid Communications and Networking. Cambridge University Press (2012)
15. Manshaei, M.H., Zhu, Q., Alpcan, T., Başar, T., Hubaux, J.-P.: Game theory meets network security and privacy. ACM Computing Survey 45(3) (2013, to appear)
16. Zhu, Q., Tembine, H., Başar, T.: Network security configuration: A nonzero-sum stochastic game approach. In: Proc. 2010 American Control Conference (ACC 2010), Baltimore, Maryland, June 30-July 2, pp. 1059–1064 (2010)
17. Kunreuther, H., Heal, G.: Interdependent security. Journal of Risk and Uncertainty 26(2), 231–249 (2003)
18. Liacco, T.D.: The adaptive reliability control system. IEEE Trans. on Power Apparatus & Systems PAS-86(5), 517–523 (1967)
19. Falliere, N., Murchu, L.O., Chien, E.: W32. Stuxnet Dossier. Symantec Reports (February 2011)
20. Başar, T., Haurie, A.: Feedback equilibria in differential games with structural and modal uncertainties. In: Cruz Jr., J.B. (ed.) Advances in Large Scale Systems, vol. 1, pp. 163–201. JAI Press Inc., Connecticut (1984)

Robust Team-Triggered Coordination of Networked Cyberphysical Systems

Cameron Nowzari and Jorge Cortés

University of California, San Diego
9500 Gilman Dr, La Jolla, California 92093
{cnowzari,cortes}@ucsd.edu

Abstract. This paper proposes a novel approach, termed team-triggered, to the real-time implementation of distributed controllers on networked cyberphysical systems. We build on the strengths of event- and self-triggered control to synthesize a single, unified approach that combines aspects of both and is implementable over distributed networked systems, while maintaining desired levels of performance. We establish provably correct guarantees of the distributed strategies resulting from the proposed approach and examine their robustness against multiple sources of errors including communication delays, packet drops, and communication noise. The results are illustrated in simulations of a multi-agent coverage control problem.

Keywords: cyberphysical systems, event-triggered control, self-triggered control, wireless robotic networks, set-valued dynamical systems, robust algorithms.

1 Introduction

The interest in the efficient and robust operation of cyberphysical systems has motivated a growing body of work that studies the distributed design and real-time implementation of controllers for networked sensors and actuators. In these systems, energy consumption is correlated with the rate at which sensors take samples, processors recompute control inputs, and actuator signals are transmitted. Performing these tasks periodically is costly and might be, at times, unnecessary or unfeasible due to physical constraints. Examples of unnecessary actions include sampling a part of the state that is already known or can be reconstructed with already available information, or recomputing a control signal that has not changed substantially. To address these issues, the goal of triggered control is to identify criteria that allow agents to tune the implementation of controllers and sampling schemes to the execution of the task at hand and the desired level of performance.

In event-triggered control, the focus is on detecting events during the network execution that are relevant from the point of view of task completion and should trigger specific agent actions. In self-triggered control, instead, the emphasis is on developing tests that rely only on the information available to individual agents

D.C. Tarraf (ed.), *Control of Cyber-Physical Systems*,
Lecture Notes in Control and Information Sciences 449,
DOI: 10.1007/978-3-319-01159-2_17,

to schedule future actions. Event-triggered control results in better performance but is costly to implement over fully distributed networked scenarios because of the need for continuous availability of information to check the triggers. Self-triggered control is more easily amenable to distributed implementation but results in conservative executions because of overapproximation by individual agents about the state of the environment and the network. This paper builds on the strengths of event- and self-triggered control to propose a single, unified approach for networked cyberphysical systems that combines the best of both worlds.

Literature Review. The need for systems integration and the importance of bridging the gap between computing, communication, and control in the study of cyberphysical systems cannot be overemphasized [1,2]. Real-time controller implementation is an area of extensive research, including periodic [3,4,5], event-triggered [6,7,8,9,10], and self-triggered [11,12,13,14] procedures. Our approach shares with these works the aim of trading computation and decision making for less communication, sensor, or actuator effort while still guaranteeing a desired level of performance.

Of particular relevance to this paper are works that study self- and event-triggered implementations of controllers for networked cyberphysical systems. The predominant paradigm is that of a single plant that is stabilized through a decentralized triggered controller over a sensor-actuator network, see e.g. [15,16,17]. Fewer works have considered scenarios where multiple plants or agents together are the subject of the overall control design, as is the case of this paper. Exceptions include consensus via event-triggered [18,19,15] or self-triggered control [18,20], model predictive control [21], and model-based event-triggered control [22,23]. The work [24] implements self-triggered communication schemes to perform distributed control where agents assume worst-case conditions occur for other agents when deciding when new information should be obtained. An idea to extend the event-triggered tools to decentralized systems with multiple plants is presented in [18]; however, agents require continuous information about each others' states in order to implement the resulting event-triggered controller. Distributed strategies based on event-triggered communication and control are explored in [25,27], where each agent has an a priori computed local error tolerance and once it violates it, the agent broadcasts its updated state to its neighbors. The same event-triggered approach is taken in [26] to implement gradient control laws that achieve distributed optimization. The work [22], closer in spirit to the ideas presented here, considers an interconnected system in which each subsystem helps neighboring subsystems by monitoring their estimates and ensuring that they stay within some performance bounds. The approach, however, requires different subsystems to have synchronized estimates of one another even though they do not communicate at all times. A main novelty of this manuscript with respect to the works listed above is the combination of elements of both event- and self-triggered strategies into a single unified approach.

Statement of Contributions. We propose a novel scheme for networked cyberphysical systems that combines ideas from event- and self-triggered control. The basic concept is for agents to make promises to one another about their future states and being responsible for warning each other if they later decide to break them (event-triggering). Such promises can be broad, from very tight state trajectories to loose descriptions of reachability sets. The information provided by these promises allows individual agents to autonomously determine when fresh information is needed (self-triggering). We refer to the approach as team-triggered because of the fact that agents need to make sure that their neighbors are operating with correct information about them. The team-triggered approach incorporates the reactive nature of event-triggered implementations while at the same time endows individual agents with autonomous tools and criteria to determine when and what information is needed, as self-triggered implementations do. The benefits of the proposed scheme are threefold. First, because of the availability of the promises, agents do not require continuous state information about neighbors, in contrast to event-triggered strategies implemented over distributed systems that assume continuous information is available in order to be able to detect the relevant triggers. Second, because of the extra information provided by promises about what other agents plan to do, agents can operate more efficiently and less conservatively than if only worst-case scenarios are assumed, as is done in self-triggered control. Lastly, we show that the networked system is guaranteed to maintain desired levels of performance while being robust to multiple physical sources of error such as communication delays, packet drops, and communication noise. We illustrate our results through simulations in a multi-agent coverage control problem. The journal version [28] of this work contains all proofs and additional results.

Organization. Section 2 lays out the problem of interest. Section 3 presents the team-triggered approach and Section 4 discusses correctness and robustness guarantees. Simulations illustrate our results in Section 5. Section 6 gathers our conclusions and ideas for future work.

Notation. We let $\mathbb{R}$, $\mathbb{R}_{\geq 0}$, and $\mathbb{Z}_{\geq 0}$ denote the sets of real, nonnegative real, and nonnegative integer numbers, respectively. The two-norm of a vector is denoted by $\|\cdot\|_2$. Given $x \in \mathbb{R}^d$ and $\delta \in \mathbb{R}_{\geq 0}$, $\overline{B}(x,\delta)$ denotes the closed ball centered at x with radius δ. For $A_i \subset \mathbb{R}^{m_i \times n_i}$ with $i \in \{1,\dots,N\}$, we denote by $\operatorname{diag}(A_1,\dots,A_N) \in \mathbb{R}^{m \times n}$ the block-diagonal matrix with A_1 through A_N on the diagonal, where $m = \sum_{i=1}^N m_i$ and $n = \sum_{i=1}^N n_i$.

Given a set S, we denote by $|S|$ its cardinality. We let $\mathbb{P}^{cc}(S)$ denote the collection of all closed and connected subsets of S. Similarly, we let $\mathbb{P}^{c}(S)$ denote the collection of all closed subsets of S. Given $S \subset \mathbb{R}^d$, we denote the distance from a point x to S as $\operatorname{dist}(x,S) = \inf_{y\in S}\|x-y\|_2$. Given $S_1, S_2 \subset \mathbb{R}^d$, the Hausdorff distance between S_1 and S_2 is

$$d_H(S_1,S_2) = \max\{\sup_{x\in S_1}\inf_{y\in S_2}\|x-y\|_2, \sup_{y\in S_2}\inf_{x\in S_1}\|x-y\|_2\}.$$

The Hausdorff distance is a metric on the set of all non-empty compact subsets of $\mathbb{R}^d$. Given two set-valued functions $C_1, C_2 \in \mathcal{C}^0(I \subset \mathbb{R}; \mathbb{P}^c(\mathbb{R}^d))$, we define the distance between the set-valued functions as

$$d_{\text{func}}(C_1, C_2) = \sup_{t \in I} d_H(C_1(t), C_2(t)). \tag{1}$$

An undirected graph $\mathcal{G} = (V, E)$ is a pair consisting of a set of vertices $V = \{1, \ldots, N\}$ and a set of edges $E \subset V \times V$ such that if $(i,j) \in E$, then $(j,i) \in E$ as well. The set of neighbors of a vertex i is given by $\mathcal{N}(i) = \{j \in V \mid (i,j) \in E\}$. Given $v \in \prod_{i=1}^N \mathbb{R}^{n_i}$, we let $v_{\mathcal{N}}^i = (v_i, \{v_j\}_{j \in \mathcal{N}(i)})$ denote the components of v that correspond to vertex i and its neighbors in $\mathcal{G}$.

2 Problem Statement

We consider a distributed control problem carried out over a realistic, unreliable wireless network. Consider N agents whose communication topology is described by an undirected graph $\mathcal{G}$. The fact that (i,j) belongs to E models the ability of agents i and j to communicate with one another. The set of all agents that i can communicate with is then given by its set of neighbors $\mathcal{N}(i)$ in the graph $\mathcal{G}$. The state of agent $i \in \{1, \ldots, N\}$, denoted x_i, belongs to a closed set $\mathcal{X}_i \subset \mathbb{R}^{n_i}$. The network state $x = (x_1, \ldots, x_N)$ therefore belongs to $\mathcal{X} = \prod_{i=1}^N \mathcal{X}_i$. According to the discussion above, agent i can access $x_{\mathcal{N}}^i$ when it communicates with its neighbors. We assume that each agent has access to its own state at all times.

We consider linear dynamics for each $i \in \{1, \ldots, N\}$,

$$\dot{x}_i = f_i(x_i, u_i) = A_i x_i + B_i u_i, \tag{2}$$

with $A_i \in \mathbb{R}^{n_i \times n_i}$, $B_i \in \mathbb{R}^{n_i \times m_i}$, and $u_i \in \mathcal{U}_i$. Here, $\mathcal{U}_i \subset \mathbb{R}^{m_i}$ is a closed set of allowable controls for agent i. We assume that the pair (A_i, B_i) is controllable with controls taking values in $\mathcal{U}_i$. Further, we assume there exists a *safe-mode* controller $u_i^{\text{sf}} : \mathcal{X}_i \to \mathcal{U}_i$ such that

$$A_i x_i + B_i u_i^{\text{sf}}(x_i) = 0, \quad \text{for all } x_i \in \mathcal{X}_i, \tag{3}$$

i.e., a controller able to keep agent i's state fixed. Note that this means that the null space of B_i should be contained in the null space of A_i.

2.1 Controller and System Certification

The goal of the network is to drive the agents' states to some desired set of configurations $D \subset \mathcal{X}$ and ensure that it stays there. Depending on how the set D is defined, this objective can capture different coordination tasks including deployment, rendezvous, or formation control. The scope of the paper is not to design the controller that achieves this, but rather synthesize efficient strategies for the real-time implementation of a given controller in the presence of communication delays, packet drops, and communication noise that still certifies convergence to the desired set D.

Given the agent dynamics, the communication graph $\mathcal{G}$, and the set D, our starting point is the availability of a continuous control law that drives the system asymptotically to D. Formally, we assume that a continuous map $u^* : \mathcal{X} \to \mathbb{R}^m$ and a continuously differentiable function $V : \mathcal{X} \to \mathbb{R}$, bounded from below, exist such that for all $x \notin D$,

$$\nabla_i V(x) \left(A_i x_i + B_i u_i^*(x)\right) \leq 0, \quad i \in \{1, \ldots, N\}, \tag{4a}$$

$$\sum_{i=1}^{N} \nabla_i V(x) \left(A_i x_i + B_i u_i^*(x)\right) < 0. \tag{4b}$$

We assume that both the control law u^* and the gradient ∇V are distributed over $\mathcal{G}$. By this we mean that, for each $i \in \{1, \ldots, N\}$, the ith component of each of these objects only depends on $x_{\mathcal{N}}^i$, rather than on the full network state x. For simplicity, and with a slight abuse of notation, we write $u_i^*(x_{\mathcal{N}}^i)$ and $\nabla_i V(x_{\mathcal{N}}^i)$ to emphasize this fact when convenient. This property has the important consequence that agent i can compute u_i^* and $\nabla_i V$ with the exact information it can obtain through communication on $\mathcal{G}$.

The controller u^* requires continuous agent-to-agent communication in order to be implemented. However, this is unrealistic and inefficient in practical networked scenarios, especially in the presence of communication delays and packet drops. Our goal is to relax this continuous information requirement to provide robust and efficient real-time controller implementations over networked cyber-physical systems.

2.2 Physical Sources of Error

We are ultimately interested in scenarios with unreliable communication among agents. More specifically, we look at three things. The first is communication noise, which we assume is bounded as follows: given a state x_i, a message y_i can be sent such that $\|y_i - x_i\|_2 \leq \bar{\omega}$ for some known $\bar{\omega} \in \mathbb{R}_{\geq 0}$. Second is the possibility of packet drops in the network. For any given message an agent sends to another agent, there is an unknown probability $0 \leq p < 1$ that the packet is dropped, and the message is never received. Lastly, we also consider the possibility that, at any time t, there is an unknown (possibly time-varying) communication delay $\Delta(t) \leq \bar{\Delta}$ in the network, where $\bar{\Delta} \in \mathbb{R}_{\geq 0}$ is known. In other words, if agent j sends agent i a message at time t, agent i will not receive it with probability p or receive it at time $t + \Delta(t)$ with probability $1-p$. We assume that small messages (i.e., 1-bit messages) can be sent reliably with negligible delay. This assumption is similar to the "acknowledgments" and "permission" messages used in other works, see [25,29] and references therein.

3 Team-Triggered Coordination for Real-Time Networked Control

Here we present a novel communication strategy to implement distributed controllers in real time on cyberphysical systems. Our strategy, termed team-triggered, combines ideas from event- and self-triggered approaches. Agents make promises to their neighbors about their future states and inform them later if these promises are violated (hence the connection with event-triggered control). With the information available to them, each agent computes the next time that an update is required based on the evolution of the Lyapunov function (hence the connection with self-triggered control).

3.1 Promises

A *promise* can be either a time-varying set of states (state promise) or controls (control promise) that an agent sends to another agent. Specifically, a state promise that agent j makes to agent i at time t is a set-valued, continuous (with respect to the Hausdorff distance) function $X_j^i[t] \in \mathcal{C}^0([t,\infty);\mathbb{P}^{cc}(\mathcal{X}_j))$. This means that agent j promises to agent i that its state at any time $t' \geq t$ will satisfy $x_j(t') \in X_j^i[t](t')$. Similarly, a control promise can be conveyed by a set-valued, continuous function $U_j^i[t] \in \mathcal{C}^0([t,\infty);\mathbb{P}^{c}(\mathcal{U}_j))$. This means that agent j promises to only use controls $u_j(t') \in U_j^i[t](t')$ for all $t' \geq t$. With this information and knowledge of the dynamics of agent j, agent i can compute the state promise

$$X_j^i[t](t') = \{z \in \mathcal{X}_j \mid \exists\, u_j : [t,t'] \to \mathcal{U}_j \text{ with } u_j(s) \in U_j^i[t](s) \text{ for } s \in [t,t'] \\ \text{such that } z = e^{A_j(t'-t)}x_j(t) + \int_t^{t'} e^{A_j(t'-\tau)}B_j u_j(\tau)d\tau\}. \quad (5)$$

For simplicity, when the time at which the promise is received is not relevant, we use the notation $X_j^i[\cdot]$ and $U_j^i[\cdot]$ or simply X_j^i and U_j^i, respectively. All promise information available to agent $i \in \{1,\dots,N\}$ at some time t is given by $X_{\mathcal{N}}^i[\cdot]|_{[t,\infty)} = (x_i|_{[t,\infty)}, \{X_j^i[\cdot]|_{[t,\infty)}\}_{j\in\mathcal{N}(i)}) \in \mathcal{C}^0\left([t,\infty);\prod_{j\in\mathcal{N}(i)\cup\{i\}}\mathbb{P}^{cc}(\mathcal{X}_j)\right)$. To extract information from this about a specific time t', we use $X_{\mathcal{N}}^i[\cdot](t')$ or simply $X_{\mathcal{N}}^i(t') = (x_i(t'), \{X_j^i[\cdot](t')\}_{j\in\mathcal{N}(i)}) \in \prod_{j\in\mathcal{N}(i)\cup\{i\}}\mathbb{P}^{cc}(\mathcal{X}_j)$. The generality of the above definitions allow promise sets to be arbitrarily complex. Here, we restrict ourselves to promise sets that can be described with a finite number of parameters so that these promises can be conveyed to one another in a realistic manner, i.e., without requiring an infinite number of bits.

The method of generating promises is not unique and can be done in a number of ways. A *promise rule* is a method to create promises. Formally, a state promise rule for agent $j \in \{1,\dots,N\}$ generated at time t is a continuous (with respect to the distance d_{func} between set-valued functions, cf. (1)) map $R_j^s : \mathcal{C}^0\left([t,\infty);\prod_{i\in\mathcal{N}(j)\cup\{j\}}\mathbb{P}^{cc}(\mathcal{X}_i)\right) \to \mathcal{C}^0([t,\infty);\mathbb{P}^{cc}(\mathcal{X}_j))$. This means that if agent j must send information to agent i at time t, it sends the state promise $X_j^i[t] = R_j^s(X_{\mathcal{N}}^j[\cdot]|_{[t,\infty)})$. A control promise rule for agent $j \in \{1,\dots,N\}$ generated at time t is a continuous map $R_j^c : \mathcal{C}^0\left([t,\infty);\prod_{i\in\mathcal{N}(j)\cup\{j\}}\mathbb{P}^{cc}(\mathcal{X}_i)\right) \to \mathcal{C}^0([t,\infty);\mathbb{P}^{c}(\mathcal{U}_j))$. This means that when agent j must send information to agent i at time t, it sends the control promise $U_j^i[t] = R_j^c(X_{\mathcal{N}}^j[\cdot]|_{[t,\infty)})$, from which the state promise is computed by (5). We make the assumption that, in the absence of communication delays or noise in the state measurements, the promises generated by these rules have the property that $X^i_j[t](t) = \{x_j(t)\}$. Note that because of this fact, it is unnecessary to send the current state $x_j(t)$ in addition to a state promise, since this information is already contained in the promise $X^i_j[t]$. However, when a control promise $U^i_j[t]$ is sent, the current state $x_j(t)$ should also be sent.

Example 1 (Static ball-radius promise rule). Here we describe one simple control promise rule, termed the static ball-radius rule, to create promises that can be described with a finite number of parameters. Given $j \in \{1, \ldots, N\}$, a continuous control law $u_j : \prod_{i \in \mathcal{N}(j) \cup \{j\}} \mathbb{P}^{cc}(\mathcal{X}_i) \to \mathbb{R}^{m_j}$, and $\delta > 0$, the static ball-radius control promise rule for agent j generated at time t is

$$R_j^{\mathrm{c,sb}}(X^j_{\mathcal{N}}[\cdot]|_{[t,\infty)})(t') = \overline{B}(u_j(X^j_{\mathcal{N}}(t)), \delta) \cap \mathcal{U}_j \qquad t' \geq t. \tag{6}$$

Note that this promise is a fixed ball of radius δ in the control space $\mathcal{U}_j$ centered at the control signal used at time t. This promise can be sent with two parameters (assuming δ is known by all agents), the state $x_j(t)$ when the promise was sent, and the control action $u_j(X^j_{\mathcal{N}}(t))$ at that time. •

Having introduced the notion of promise, several observations can be made. First, the availability of promises equips agents with set-valued information models about the state of other agents. This fact makes it necessary to address the definition of distributed controllers that operate on sets, rather than points. We discuss this point in Section 3.2. Then, based on the promises that agent i receives from its neighbors at a given time t, it is responsible for computing the next time it will require updated information. We discuss this in Section 3.3. On the other hand, if at any time agent j breaks its promise to agent i, i.e., its state no longer belongs to its promise set, this triggers an event requiring agent j to send updated information to agent i. We discuss this in Section 3.4.

3.2 Controllers on Set-Valued Information Models

Here we briefly discuss the type of controllers that the team-triggered approach relies on. The underlying idea is that since agents possess set-valued information about the state of other agents through promises, controllers themselves should be defined on sets, rather than on points. Our starting point is therefore the availability of a continuous controller of the form $u^{**} : \prod_{j \in \{1,\ldots,N\}} \mathbb{P}^{cc}(\mathcal{X}_j) \to \mathbb{R}^m$ that satisfies

$$\nabla_i V(x)\,(A_i x_i + B_i u_i^{**}(\{x\})) \leq 0, \quad i \in \{1, \ldots, N\}, \tag{7a}$$

$$\sum_{i=1}^{N} \nabla_i V(x)\,(A_i x_i + B_i u_i^{**}(\{x\})) < 0. \tag{7b}$$

In other words, if exact, singleton-valued information is available to the agents, then the controller u^{**} guarantees the monotonic evolution of the Lyapunov function V. We assume that u^{**} is distributed over the communication graph $\mathcal{G}$. As before, this means that for each $i \in \{1, \dots, N\}$, the ith component u_i^{**} can be computed with information in $\prod_{j \in \mathcal{N}(i) \cup \{i\}} \mathbb{P}^{cc}(\mathcal{X}_j)$ rather than in the full space $\prod_{j \in \{1,\dots,N\}} \mathbb{P}^{cc}(\mathcal{X}_j)$.

Controllers of the form described above can be obtained using a number of design methods. We do not enter into the specific details, but briefly mention how one such controller can be derived from the availability of the controller $u^* : \mathcal{X} \to \mathbb{R}^m$ introduced in Section 2. The intuitive idea is that, given the promise information $X_{\mathcal{N}}^i(t)$ that an agent i has about some time t, one can simply apply the controller u^* to any point $y_{\mathcal{N}} \in X_{\mathcal{N}}^i(t)$. This can be formalized as follows: let $E : \prod_{j=1}^{N} \mathbb{P}^{cc}(\mathcal{X}_j) \to \mathcal{X}$ be a continuous map that is distributed over the communication graph $\mathcal{G}$ and satisfies, for each $i \in \{1, \dots, N\}$, that $E_i(Y) \in Y_i$ for each $Y \in \prod_{j=1}^{N} \mathbb{P}^{cc}(\mathcal{X}_j)$ and $E_i(\{y\}) = y_i$ for each $y \in \mathcal{X}$. Now, define

$$u^{**}(Y) = u^*(E(Y)). \tag{8}$$

Note that this controller satisfies (7a) and (7b) because u^* satisfies (4a) and (4b).

3.3 Self-triggered Information Updates

Here, we discuss in detail how agents use the promises received from other agents to generate self-triggered information requests in the future. Let t_{last}^i be some time at which agent i receives updated information (i.e., promises) from its neighbors. Until the next time information is obtained, agent i has access to the collection of functions $X_{\mathcal{N}}^i$ describing its neighbors' states and can compute its own evolution under the controller u^{**} via

$$x_i(t) = e^{A_i(t - t_{\text{last}}^i)} x_i(t_{\text{last}}^i) + \int_{t_{\text{last}}}^{t} e^{A_i(t-\tau)} B_i u_i^{**}(X_{\mathcal{N}}^i(\tau)) d\tau, \quad t \geq t_{\text{last}}^i. \tag{9}$$

With this in place, agent i can schedule the next time t_{next}^i at which it will need updated information from its neighbors. To do so, we define, for any $Y_{\mathcal{N}} \in \prod_{j \in \mathcal{N}(i) \cup \{i\}} \mathbb{P}^{cc}(\mathcal{X}_j)$,

$$\mathcal{L}_i V^{\sup}(Y_{\mathcal{N}}) = \sup_{y_{\mathcal{N}} \in Y_{\mathcal{N}}} \nabla_i V(y_{\mathcal{N}}) \left(A_i y_i + B_i u_i^{**}(Y_{\mathcal{N}}) \right), \tag{10}$$

where y_i is the element of $y_{\mathcal{N}}$ corresponding to i. Using this, we create an implementable trigger that computes when an agent requires new information to guarantee the monotonically nonincreasing evolution of V. Specifically, the critical time at which information is requested is given by $t_{\text{next}}^i = \max\{t_{\text{last}}^i + T_{\text{d,self}}, t^*\}$, where $T_{\text{d,self}} > 0$ is an a priori chosen parameter that we discuss below and t^* is implicitly defined as the first time $t^* \geq t_{\text{last}}^i$ such that

$$\mathcal{L}_i V^{\sup}(X_{\mathcal{N}}^i(t^*)) = 0. \tag{11}$$

Note that as long as (11) has not yet occurred for all agents $i \in \{1, \ldots, N\}$ for some time t and the promises that agents $j \in \mathcal{N}(i)$ have made to i have not been broken, one can guarantee that

$$\frac{d}{dt}V(x(t)) \leq \sum_{i=1}^{N} \mathcal{L}_i V^{\text{sup}}(X^i_{\mathcal{N}}(t)) < 0.$$

This follows from assumptions (7a) and (7b) and the continuity of (10) when evaluated at the promise sets that each agent possesses.

Although an agent i sends a request REQ for new information from its neighbors $j \in \mathcal{N}(i)$ at time t^i_{next}, it does not mean that it arrives at this time. We assume that the REQ message can be sent reliably with negligible delay since it is a very small message. Given the model for delays and uncertainties, as long as agent i has still not received information from some neighbors j, it will continue to send j requests for new information every $\bar{\Delta}$ seconds. This is because if at time $t^i_{\text{next}} + \bar{\Delta}$ the message still has not arrived, it means the packet was dropped since $\bar{\Delta}$ is the maximum allowable delay.

The parameter $T_{\text{d,self}} > 0$ is known as the *self-triggered dwell time*. We introduce it because, in general, it is possible that $t^* = t^i_{\text{last}}$, implying that instantaneous communication is required. The dwell time is used to prevent this behavior as follows. Note that $\mathcal{L}_i V^{\text{sup}}(X^i_{\mathcal{N}}(t')) \leq 0$ is only guaranteed while $t' \in [t^i_{\text{last}}, t^*]$. Therefore, in case that $t^i_{\text{next}} = t^i_{\text{last}} + T_{\text{d,self}}$, i.e., if $t^* < t^i_{\text{last}} + T_{\text{d,self}}$, the agent uses the safe-mode control during $t' \in (t^*, t^i_{\text{last}} + T_{\text{d,self}}]$ to leave its state fixed. This design ensures the monotonicity of the evolution of the Lyapunov function V along the network execution. The team-triggered controller is therefore defined, for $t \in [t^i_{\text{last}}, t^i_{\text{next}})$, by

$$u_i^{\text{team}}(t) = \begin{cases} u_i^{**}(X^i_{\mathcal{N}}(t)), & \text{if } \mathcal{L}_i V^{\text{sup}}(X^i_{\mathcal{N}}(t)) \leq 0, \\ u_i^{\text{sf}}(x_i(t)), & \text{if } \mathcal{L}_i V^{\text{sup}}(X^i_{\mathcal{N}}(t)) > 0. \end{cases} \tag{12}$$

Next, we discuss how to ensure promises are still valid in the presence of communication noise. To deal with the communication noise, when an agent i receives an estimated promise $\widehat{X}^i_j$ from another agent j, it must be able to create a valid promise X^i_j that contains the promise that agent j intended to send. Note that the specific way of doing this depends on how promises are exchanged between agents. We refer to this action as making a promise set valid. The following example shows how it can be done for the static ball-radius promises described in Example 1.

Example 2 (Valid static ball-radius promise rule with communication noise). In the scenario with communication noise bounded by $\bar{\omega}$, when agent j attempts to send the control promise $\overline{B}(u_j(X^j_{\mathcal{N}}(t)), \delta)$ to agent i at time t as defined in Example 1, it will instead receive $\widehat{U}^i_j[t] = \overline{B}(\widehat{u}^i_j(X^j_{\mathcal{N}}(t)), \delta)$, where $u_j(X^j_{\mathcal{N}}(t)) \in \overline{B}(\widehat{u}^i_j(X^j_{\mathcal{N}}(t)), \bar{\omega})$. To ensure that the promise agent i operates with about agent j contains the true promise made by agent j, it can set

$$U^i_j[t](t') = \overline{B}(\widehat{u}^i_j(X^j_{\mathcal{N}}(t)), \delta + \bar{\omega}) \cap \mathcal{U}_j \qquad t' \geq t.$$

To create the state promise from this agent i would need the true state $x_j(t)$ of j at time t. However, since only the estimate $\widehat{x}^i_j(t)$ is available, we modify (5) by

$$X^i_j[t](t') = \cup_{x_j \in \overline{B}(\widehat{x}^i_j(t),\bar{\omega})} \{z \in \mathcal{X}_j \mid \exists u_j : [t,t'] \to \mathcal{U}_j \text{ with } u_j(s) \in U^i_j[t](s)$$
$$\text{for } s \in [t,t'] \text{ such that } z = e^{A_j(t'-t)}x_j + \int_t^{t'} e^{A_j(t'-\tau)}B_j u_j(\tau)d\tau\}. \quad \bullet$$

3.4 Event-Triggered Information Updates

Agent promises may need to be broken for a variety of reasons. For instance, an agent might receive new information from its neighbors and, based on it, decide to change its former plans. Disturbances in the agent dynamics or new requirements imposed by the level of completion of the network task are yet more reasons for why promises might be broken. In the team-triggered approach, promises get updated through events triggered by the agents that decide to break them. Specifically, consider an agent i that made a promise $X^j_i[t]$ to agent j at some time t that it is not able to keep after time $t' > t$, i.e., $x_i(t') \notin X^j_i[t](t')$. Then, agent i must send a new feasible promise $X^j_i[t']$ to agent j. This event-triggered mechanism requires each agent to keep track of the last promise it made to each one of its neighbors and constantly monitor them to detect when they are broken.

If the message arrived exactly at time t', there would be no issue because agent j will always be operating with correct information, namely that $x_i(t') \in X^j_i[t'](t')$. However, this message might arrive with a delay $\Delta(t')$ or even be dropped altogether. To deal with this, we require agent i to also send agent j a small message WARN at time t' that warns agent j that agent i has broken its promise at time t'. As before, we assume that the message WARN can be sent reliably with negligible delay. With this warning, agent j can make sure it is operating with correct information at all times as follows. Let us first define the notation of a reachable set. Given $y \in \mathcal{X}_i$, let $\mathcal{R}_i(s,y)$ be the reachable set of points under (2) starting from y in s seconds,

$$\mathcal{R}_i(s,y) = \{z \in \mathcal{X}_i \mid \exists u_i : [0,s] \to \mathcal{U}_i \text{ s.t. } z = e^{A_i s}y + \int_0^s e^{A_i(s-\tau)}B_i u_i(\tau)d\tau\}.$$

Assuming here that each agent has exact knowledge about the dynamics and control sets of its neighboring agents, each agent can construct, each time the WARN message is received, sets that are guaranteed to contain their neighbors' states. Formally, given the original promise $X^j_i[t]$ received by agent i at time t, and letting t^* be the time at which the WARN message was received from agent i, we then define the new promise for agent i as

$$X^j_i[t^*](t') = \cup_{x_i \in X^j_i[t](t^*)} \mathcal{R}_i(t'-t^*, x_i) \subset \mathcal{X}_i.$$

This new promise $X^j_i[t^*](t')$ is guaranteed to contain $x_i(t')$ for $t' \geq t^*$. As before, if the full message is not received after $\bar{\Delta}$ seconds, agent j will send requests REQ to agent i until the message is successfully transmitted.

Consider an agent $i \in \{1, \ldots, N\}$ that has sent a promise $X_i^j[t_{\text{last}}]$ to a neighboring agent j at some time t_{last}. If agent i ends up breaking its promise at time $t^* > t_{\text{last}}$, i.e., $x_i(t^*) \notin X_i^j[t_{\text{last}}](t^*)$, then it is responsible for sending a new promise $X_i^j[t_{\text{next}}]$ to agent j at time $t_{\text{next}} = \max\{t_{\text{last}} + T_{\text{d,event}}, t^*\}$, where $T_{\text{d,event}} > 0$ is an a priori chosen parameter that we discuss below. This implies that agent i must keep track of promises made to its neighbors and monitor them in case they are broken. Note that this mechanism is implementable because each agent only needs information about its own state and the promises it has made to determine whether the trigger is satisfied.

The parameter $T_{\text{d,event}} > 0$ is known as the *event-triggered dwell time*. We introduce it because, in general, the time $t^* - t_{\text{last}}$ between when agent i makes and breaks a promise to an agent j might be arbitrarily small. The issue, however, is that if $t^* < t_{\text{last}} + T_{\text{d,event}}$, agent j operates under incorrect information about agent i for $t \in [t^*, t_{\text{last}} + T_{\text{d,event}})$. We deal with this by introducing a warning message WARN that agent i must send to agent j when it breaks its promise at time $t^* < t_{\text{last}} + T_{\text{d,event}}$. If agent j receives such a warning message, it redefines the promise X_j^i as follows,

$$X_i^j[\cdot](t) = \bigcup_{x_i \in X_i^j[\cdot](t^*)} \mathcal{R}(t - t^*, x_i), \tag{13}$$

for $t \geq t^*$, until the new message arrives. By definition of the reachable set, the promise $X_i^j[\cdot](t)$ is guaranteed to contain $x_i(t)$ for $t \geq t^*$.

This algorithm ensures that promises are kept at all times. Even if promises are broken, the WARN message allows agents to redefine promises such that they always contain the true states of the relevant agents. The following remark points out minor modifications that would need to be made to the above discussion in the case of modeling uncertainties.

Remark 1 (Modeling uncertainties). In the case of modeling uncertainties or if each agent i does not know exactly the dynamics of its neighbors $j \in \mathcal{N}(i)$, the reachable sets $\mathcal{R}_j(s, y)$ can be replaced by any other set $\widehat{\mathcal{R}}_j(s, y)$ that contains $\mathcal{R}_j(s, y)$ for all $s \geq 0$. •

Combining the controller u_i^{team} with the event- and self-triggered methods of sharing information yields the ROBUST TEAM-TRIGGERED LAW, cf. Algorithm 1.

4 Analysis of the Robust Team-Triggered Law

In this section we analyze the convergence, performance, and robustness properties of the ROBUST TEAM-TRIGGERED LAW proposed in Section 3. We begin by noting the monotonic behavior of V with respect to the algorithm executions.

Proposition 1. *The function V is monotonically nonincreasing along the network dynamics* (2) *under the* ROBUST TEAM-TRIGGERED LAW, *with packet drops occurring with some unknown probability* $0 \leq p < 1$, *messages being delayed by some known maximum delay $\bar{\Delta}$, and communication noise bounded by $\bar{\omega}$.*

Algorithm 1 ROBUST TEAM-TRIGGERED LAW

(Self-triggered information update)

At any time t agent $i \in \{1, \dots, N\}$ receives new promise(s) $\widehat{X}_j^i[t]$ from neighbor(s) $j \in \mathcal{N}(i)$, agent i performs:

1: create valid promise $X_j^i[t]$ with respect to $\bar{\omega}$
2: compute own state evolution $x_i(t')$ for $t' \geq t$ using (9)
3: schedule information request to neighbors in $\max\{t^* - t, T_{\mathrm{d,self}}\}$ seconds
4: apply controller $u^{\mathrm{team}}(t')$ for $t' \geq t$
5: **while** message from j has not been received **do**
6:    **if** current time equals $t + \max\{t^* - t, T_{\mathrm{d,self}}\} + k\bar{\Delta}$ for $k \in \mathbb{Z}_{\geq 0}$ **then**
7:       send agent j a request REQ for new information
8:    **end if**
9: **end while**

(Respond to information request)

At any time t agent $j \in \mathcal{N}(i)$ requests information, agent i performs:

1: send new promise $Y_i^j[t] = R_i^{\mathrm{s}}(X_{\mathcal{N}}^i[\cdot]|_{[t,\infty)})$ to agent j

(Event-triggered information update)

At all times t, agent i performs:

1: **if** there exists $j \in \mathcal{N}(i)$ such that $x_i(t) \notin Y_i^j[\cdot](t)$ **then**
2:    send warning message WARN to agent j
3:    **if** agent i has sent a promise to j at some time $t_{\mathrm{last}} \in (t - T_{\mathrm{d,event}}, t]$ **then**
4:       schedule to send new promise $Y_i^j[t_{\mathrm{last}} + T_{\mathrm{d,event}}] = R_i^{\mathrm{s}}(X_{\mathcal{N}}^i[\cdot]|_{[t_{\mathrm{last}}+T_{\mathrm{d,event}},\infty)})$ to agent j in $t_{\mathrm{last}} + T_{\mathrm{d,event}} - t$ seconds
5:    **else**
6:       send new promise $Y_i^j[t] = R_i^{\mathrm{s}}(X_{\mathcal{N}}^i[\cdot]|_{[t,\infty)})$ to agent j
7:    **end if**
8: **end if**

(Respond to warning message)

At any time t agent $i \in \{1, \dots, N\}$ receives a warning message WARN from agent $j \in \mathcal{N}(i)$

1: redefine promise set $X_j^i[\cdot](t') = \cup_{x_j \in X_j^i[\cdot](t)} \mathcal{R}_j(t' - t, x_j)$ for $t' \geq t$
2: **while** message from j has not been received **do**
3:    **if** current time equals $t + k\bar{\Delta}$ for $k \in \mathbb{Z}_{\geq 0}$ **then**
4:       send agent j a request REQ for new information
5:    **end if**
6: **end while**

Proof. To prove the result, we write the time derivative of the Lyapunov function using the controller u_i^{team} for all agents $i \in \{1, \dots, N\}$,

$$\begin{aligned}\frac{d}{dt} V(x(t)) &= \sum_{i=1}^{N} \nabla_i V(x_{\mathcal{N}}^i(t)) \left(A_i x_i(t) + B_i u_i^{\mathrm{team}}(X_{\mathcal{N}}^i(t))\right) \qquad (14)\\ &\leq \sum_{i=1}^{N} \sup_{x_{\mathcal{N}}^i(t) \in X_{\mathcal{N}}^i(t)} \nabla_i V(x_{\mathcal{N}}^i(t)) \left(A_i x_i(t) + B_i u_i^{\mathrm{team}}(X_{\mathcal{N}}^i(t))\right) \leq 0.\end{aligned}$$

This is guaranteed by the design of the ROBUST TEAM-TRIGGERED LAW. In fact, the WARN messages allow agents to ensure that the information they are operating with is correct, i.e., that promises are valid at all times. When $\mathcal{L}_i V^{\sup}(X^i_{\mathcal{N}}(t)) \leq 0$, the control policy as defined in (12) is $u_i^{\text{team}}(X^i_{\mathcal{N}}(t)) = u_i^{**}(X^i_{\mathcal{N}}(t))$. In this case the summands of (14) are exactly $\mathcal{L}_i V^{\sup}(X^i_{\mathcal{N}}(t))$ as defined in (10). When $\mathcal{L}_i V^{\sup}(X^i_{\mathcal{N}}(t)) > 0$, the control $u_i^{\text{team}}(X^i_{\mathcal{N}}(t)) = u_i^{\text{sf}}(x_i)$, for which the summands of (14) are exactly 0. □

The following result states that, under the ROBUST TEAM-TRIGGERED LAW, Zeno behavior does not occur when using positive dwell times.

Lemma 1 (No Zeno behavior). *Consider a networked cyberphysical system as described in Section 2 executing the* ROBUST TEAM-TRIGGERED LAW, *cf. Algorithm 1, with dwell times $T_{d,self}, T_{d,event} > 0$. Then the network does not exhibit Zeno behavior.*

We are now ready to study the algorithm convergence. For simplicity, we first consider the case with no packet drops, communication delays, or communication noise and later extend the result.

Proposition 2. *Given the network dynamics* (2) *under the* ROBUST TEAM-TRIGGERED LAW *with no packet drops, communication delays, or communication noise, the system asymptotically approaches the set D.*

There are two main challenges in proving Proposition 2. The first is that we need a way to model the asynchronous executions of the agents' actions. The second is that because of the intermittent communications between agents, the memories of the agents evolve discontinuously in time, making it difficult to use standard stability methods of analyzing the trajectories of the system.

Proof Sketch. To model the asynchronism, let the time schedule of agent i be given by $\mathcal{T}^i = \{t_0^i, t_1^i, \dots\}$ where t_ℓ^i corresponds to the ℓth time that agent i receives information from one or more of its neighbors. Note that this information can be received because i requests it itself, or j sends it to i because an event is triggered. Analytic synchronization is a procedure of merging together the individual time schedules into a global one $\mathcal{T} = \{t_0, t_1, \dots\}$ by setting

$$\mathcal{T} = \cup_{i=1}^N \mathcal{T}^i.$$

This synchronization is done for analysis purposes only and the time schedules $\mathcal{T}^i$ are not known by the agents themselves. Note that more than one agent may receive information at any given time $t \in \mathcal{T}$.

The information possessed by any given agent are trajectories of sets for each of their neighbors, i.e., promises. For convenience, we denote by

$$S = \prod_{i=1}^N S_i, \quad \text{where}$$

$$S_i = \mathcal{C}^0\Big(\mathbb{R}; \mathbb{P}^{cc}(\mathcal{X}_1) \times \cdots \times \mathbb{P}^{cc}(\mathcal{X}_{i-1}) \times \mathcal{X}_i \times \mathbb{P}^{cc}(\mathcal{X}_{i+1}) \times \cdots \times \mathbb{P}^{cc}(\mathcal{X}_N)\Big),$$

the space that the state of the entire network lives in. Note that this set allows us to capture the fact that each agent i has perfect information about itself. Although agents only have information about their neighbors, the above space considers agents having promise information about all other agents to facilitate the analysis. This is only done to allow for a simpler technical presentation, and does not impact the validity of the arguments made here. The information possessed by all agents of the networks at some time t is collected in

$$(X^1[\cdot]|_{[t,\infty)}, \dots, X^N[\cdot]|_{[t,\infty)}) \in S,$$

where $X^i[\cdot]|_{[t,\infty)} = (X^i_1[\cdot]|_{[t,\infty)}, \dots, X^i_N[\cdot]|_{[t,\infty)}) \in S_i$. We can then formulate the ROBUST TEAM-TRIGGERED LAW as a discontinuous map of the form $S \times \mathbb{Z}_{\geq 0} \to S \times \mathbb{Z}_{\geq 0}$. This fact makes it difficult to use standard stability methods to analyze the convergence properties of the network. Our approach to this problem consists then of defining a set-valued map $M : S \times \mathbb{Z}_{\geq 0} \rightrightarrows S \times \mathbb{Z}_{\geq 0}$ whose trajectories contain the trajectories of the ROBUST TEAM-TRIGGERED LAW. Although this 'overapproximation procedure' enlarges the set of trajectories to consider, the gained benefit is that of having a set-valued map with suitable continuity properties that is amenable to set-valued stability analysis. We describe this in detail next.

We start by defining the set-valued map M. Let $(Z, \ell) \in S \times \mathbb{Z}_{\geq 0}$. We define the $(N+1)$th component of all the elements in $M(Z, \ell)$ to be $\ell + 1$. The ith component of the elements in $M(Z, \ell)$ is given by one of following possibilities. The first possibility is simply the ith component of Z,

$$\left(Z^i_1[\cdot]|_{[t_{\ell+1},\infty)}, \dots, Z^i_N[\cdot]|_{[t_{\ell+1},\infty)}\right), \tag{15}$$

which models the case when agent i does not receive any information from its neighbors. The second is

$$\left(Y^i_1[\cdot]|_{[t_{\ell+1},\infty)}, \dots, Y^i_N[\cdot]|_{[t_{\ell+1},\infty)}\right), \tag{16}$$

where for $j \neq i$

$$Y^i_j[\cdot]|_{[t_{\ell+1},\infty)} = \begin{cases} Z^i_j[\cdot]|_{[t_{\ell+1},\infty)}, & \text{if } i \text{ does not receive information from } j, \\ R^{\mathrm{s}}_j(Z^j_N[\cdot]|_{[t_{\ell+1},\infty)}), & \text{otherwise,} \end{cases} \tag{17a}$$

and

$$Y^i_i[\cdot](t) = e^{A_i(t-t_{\ell+1})} Z^i_i(t_{\ell+1}) + \int_{t_{\ell+1}}^t e^{A_i(t-\tau)} B_i u^{\mathrm{team}}_i(\tau) d\tau, \quad t \geq t_{\ell+1}, \tag{17b}$$

which models the case when the agent i has received updated information from at least one neighbor (here, with a slight abuse of notation, we use the notation u^{team} to denote the controller evaluated at the set $Y^i[\cdot]$).

Two properties regarding the set-valued map M are worth emphasizing. First, any trajectory of the ROBUST TEAM-TRIGGERED LAW is also a trajectory of the nondeterministic dynamical system defined by M,

$$(Z(t_{\ell+1}), \ell+1) \in M(Z(t_\ell), \ell).$$

Second, unlike the map defined by the ROBUST TEAM-TRIGGERED LAW, which is discontinuous, it can be shown that the set-valued map M is closed (a set-valued map $T : X \rightrightarrows Y$ is closed if $x_k \to x$, $y_k \to y$ and $y_k \in T(x_k)$ imply that $y \in T(x)$). Using this and the monotonicity of the Lyapunov function V, cf. Proposition 1, we can resort to a form of the LaSalle Invariance Principle for set-valued discrete-time dynamical systems to show that all trajectories of T converge to the largest weakly invariant set contained in

$$\begin{aligned} S^* &= \{(Z,\ell) \in S \times \mathbb{Z}_{\geq 0} \mid \exists (Z', \ell+1) \in M(Z,\ell) \text{ such that } V(Z') = V(Z)\}, \\ &= \{(Z,\ell) \in S \times \mathbb{Z}_{\geq 0} \mid \mathcal{L}_i V^{\sup}(Z^i_{\mathcal{N}}) \geq 0 \text{ for all } i \in \{1,\dots,N\}\}. \end{aligned} \quad (18)$$

Using this information, we are able to show that the omega-limit set of any trajectory of M specific to the ROBUST TEAM-TRIGGERED LAW corresponds to the set D. This concludes our proof sketch for Proposition 2.

In the scenario with possible packet drops, communication delays and sensor noise, we are able to state the following result.

Corollary 1. *Consider a networked cyberphysical system as described in Section 2 with packet drops occurring with some unknown probability* $0 \leq p < 1$, *messages being delayed by some known maximum delay* $\bar{\Delta}$, *and communication noise bounded by* $\bar{\omega}$, *executing the* ROBUST TEAM-TRIGGERED LAW *(cf. Algorithm 1) with dwell times* $T_{d,self}, T_{d,event} > 0$. *Then, any bounded network evolution with uniformly bounded promises asymptotically converges to the neighborhood of D given by*

$$D'(\bar{\Delta}, \bar{\omega}) = \{x \in \mathcal{X} \mid \inf_{x^i_{\mathcal{N}}{}' \in \overline{B}(x^i_{\mathcal{N}}, \bar{\omega})} \mathcal{L}_i V^{\sup}\Big(\{x_i\} \times \prod_{j \in \mathcal{N}(i)} \bigcup_{y_j \in \overline{B}(x^i_j{}', \bar{\omega})} \mathcal{R}_j(\bar{\Delta}, y_j)\Big) \geq 0, \text{ for all } i \in \{1,\dots,N\}\}, \quad (19)$$

with probability 1.

Note that by equation (7b), the definition (10), and the continuity of u^{**}, D precisely corresponds to $D'(0,0)$. We only provide a proof sketch of the result. Note that, under the hypotheses of the corollary, agents might never know the exact state of their neighbors at any time. The basic idea is the observation that all properties of M used in the proof of Proposition 2 still hold in the presence of packet drops, delays, and communication noise as long as the time schedule $\mathcal{T}^i$ remains unbounded for each agent $i \in \{1,\dots,N\}$. For this to happen, each agent i must receive an infinite number of messages, and $t^i_\ell \to \infty$. Since packet drops have probability $0 \leq p < 1$, the probability that there is a finite number of updates for any given agent i is 0. Thus, with probability 1, there are an infinite number of information updates for each agent. Using a similar argument to that in the proof of Proposition 2, one can establish that the bounded trajectories of M still converge to S^* as defined in (18). Finally, one can use this fact to conclude that the omega-limit set of any trajectory of M specific to the ROBUST TEAM-TRIGGERED LAW corresponds to the set $D'(\bar{\Delta}, \bar{\omega})$.

5 Simulations

This section presents simulations of a planar coverage control problem to illustrate the performance of the team-triggered approach and compare it with periodic and self-triggered approaches. Our starting point is the distributed coordination algorithm based on Voronoi partitioning introduced in [30]. The dynamics of each agent is a single integrator

$$\dot{x}_i = u_i, \qquad i \in \{1, \dots, N\}, \tag{20}$$

where $\|u_i\|_2 \leq u_{\max}$. Given a convex polygon $Q \subset \mathbb{R}^2$ and some known density function $\phi : Q \to \mathbb{R}_{\geq 0}$, consider the objective function

$$\mathcal{H}(x) = E_\phi \left[\min_{i \in \{1,\dots,N\}} \|q - x_i\|^2 \right] = \sum_{i=1}^{N} \int_{V_i} \|q - x_i\|^2 \phi(q) dq. \tag{21}$$

Here, $\{V_1, \dots, V_N\}$ denotes the Voronoi partition of Q, cf. [31]. Roughly speaking, the value $\mathcal{H}$ encodes the expected value of the minimum distance from some agent in the network to a generic point in the polygon, given the density function ϕ. The continuous control law $u^* = (u_1^*, \dots, u_N^*)$ is the gradient of $\mathcal{H}$,

$$u_i^* = -2M_{V_i}(p_i - C_{V_i}),$$

where M_{V_i} and C_{V_i} are the mass and centroid of V_i, respectively. Note that this control law is distributed on the Delaunay graph, i.e., where each agent's neighbors are its Voronoi neighbors. The system (20) under the control law u^* converges to the set of centroidal Voronoi configurations, i.e., configurations where each agent is at the centroid of its own Voronoi cell.

In the following simulations, we consider $N = 8$ agents operating in a square environment of side lengths 4 with $u_{\max} = 1$. The density function is given by $\phi(q) = e^{-\|q-p_1\|_2} + e^{-\|q-p_2\|_2}$, where $p_1 = (2, 3)$ and $p_2 = (3, 1)$. The promises among agents are generated using the static ball-radius rule described in Example 1 with $\delta = 0.5 u_{\max}$. The controller we use in the team-triggered approach is defined from u^* using the procedure described in Section 3.2, using $y_j^i = \mathrm{cc}(X_j^i) \in X_j^i$ for each $j \in \mathcal{N}(i)$ (here, $\mathrm{cc}(S)$ is the circumcenter of S). The dwell time in the team-triggered execution is $T_{\mathrm{d,self}} = 0.05$. According to Corollary 1, under communication delays bounded by $\bar{\Delta}$ and sensor noise bounded by $\bar{\omega}$, the system converges to a neighborhood of the set of centroidal Voronoi configurations. In this case, one can actually provide a characterization of this

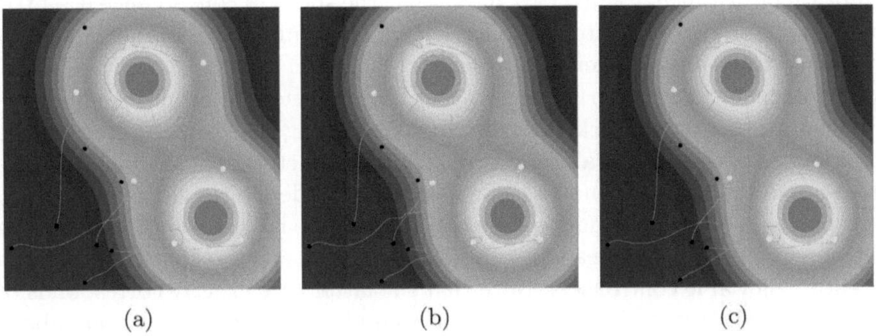

Fig. 1. Executions of the (a) periodic, (b) self-triggered, and (c) team-triggered implementations of the gradient-based continuous controller for optimal deployment in [30]. The black and gray dots correspond to initial and final conditions, respectively.

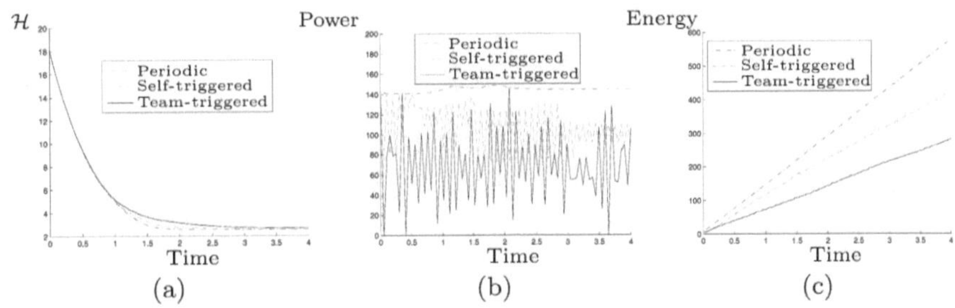

Fig. 2. (a) shows the evolution of the objective function (21), (b) shows the communication power (Watts) consumption over time and (b) shows the total transmission energy used (Joules) for the three executions in Figure 1

asymptotic behavior as follows: in the limit, each agent is within $2(\bar{\Delta} u_{\max} + \bar{\omega})$ of the centroid of its own Voronoi cell.

Figure 1 shows network executions under periodic, self-triggered, and team-triggered implementations of the controller. This figure also compares the evolution of the objective function (21). Figure 2 compares the total energy used to transmit messages by the entire network as the system evolves. For each agent $i \in \{1, \dots, N\}$, we quantify the power $\mathcal{P}_i$ used by i to communicate using [32],

$$\mathcal{P}_i = 10 \log_{10} \Big[\sum_{j \in \{1,\dots,n\}, i \neq j}^{n} \beta 10^{0.1 P_{i \to j} + \alpha \|x_i - x_j\|_2} \Big],$$

where $\alpha > 0$ and $\beta > 0$ depend on the characteristics of the wireless medium and $P_{i \to j}$ is the power received by j of the signal transmitted by i. In our simulations, all these values are set to 1. We can see from Figure 2(b) that the total amount of transmission energy used with the team-triggered approach is significantly less than those of the periodic and self-triggered approaches. Remarkably, this comes without compromising the stability of the system, cf.

Figure 1. For instance, Figure 1(b) shows that the speed of convergence is a little slower in the triggered strategies, but yields a large amount of savings in terms of message transmission energy.

We conclude this section by illustrating how tightness of promises affect the performance of the network. We do this by varying the parameter δ in the definition (6) of the static-ball radius rule. This parameter captures how large the promise sets that agents send to each other are. We define $\lambda = \frac{\delta}{2}$, so that $\lambda = 0$ corresponds to exact information (the control promise is a point in the allowable control set) and $\lambda = 1$ corresponds to no promises at all (the control promise is the entire allowable control set). Note that the latter case exactly corresponds to a self-triggered strategy because agents are simply using reachability sets about their neighbors as the promises. Figure 3 shows the average power consumption and the time to converge to 99% of the final value of the objective function for varying tightness on the promises. Note that for small λ, the amount of energy required for sending messages is minimal while the time to convergence only increases slightly.

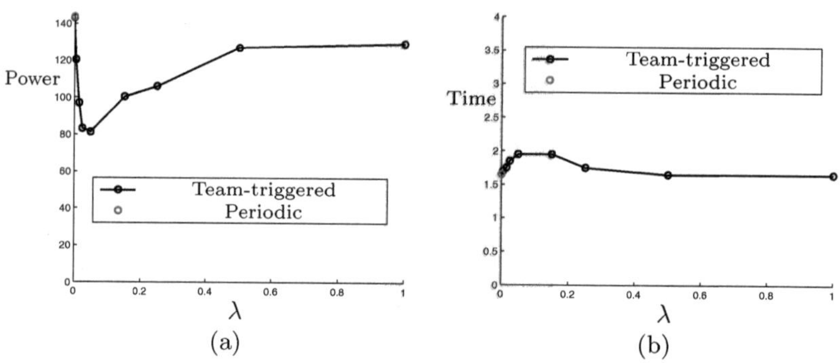

Fig. 3. Implementation of the team-triggered strategy with varying tightness of promises. Plot (a) shows average communication power consumption (Watts) by the whole network and (b) shows time to converge to 99% of the final value (seconds). The parameter λ captures tightness of promises, with $\lambda = 0$ corresponding to exact information and $\lambda = 1$ corresponding to the self-triggered case (no promises at all, just the description of the reachability set).

6 Conclusions

We have proposed a novel approach, termed team-triggered, for the real-time control of networked cyberphysical systems. When information between subsystems is obtained through wireless communication, event-triggered strategies may be costly to implement because they need continuous availability of information to check the triggers, and self-triggered strategies are conservative because they tend to generate more communications than strictly necessary. The ROBUST TEAM-TRIGGERED LAW combines ideas from both event- and self-triggered control into a unified paradigm that incorporates their strengths while maintaining desired levels of performance. The backbone of the team-triggered approach is the quality of promises that agents make to one another. Future work will be devoted to tuning the generation of promises to optimize system properties such as communication energy and time to convergence, analyzing the algorithm performance under other sources of errors such as disturbances in the dynamics, and relaxing our assumptions to make the approach more generally applicable.

Acknowledgments. This research was partially supported by NSF award CCF-0917166.

References

1. Kim, K.D., Kumar, P.R.: Cyber physical systems: A perspective at the centennial. Proceedings of the IEEE 100(Special Centennial Issue), 1287–1308 (2012)
2. Sztipanovits, J., Koutsoukos, X., Karsai, G., Kottenstette, N., Antsaklis, P., Gupta, V., Goodwine, B., Baras, J., Wang, S.: Toward a science of cyberphysical system integration. Proceedings of the IEEE 100(1), 29–44 (2012)
3. Hristu, D., Levine, W.: Handbook of Networked and Embedded Control Systems. Birkhäuser, Boston (2005)
4. Åström, K.J., Wittenmark, B.: Computer Controlled Systems: Theory and Design, 3rd edn. Prentice Hall, Englewood Cliffs (1996)
5. Laila, D.S., Nesic, D., Astolfi, A.: Sampled-data control of nonlinear systems. In: Loria, A., Lamnabhi-Lagarrigue, F., Panteley, E. (eds.) Advanced Topics in Control Systems Theory: Lecture Notes from FAP, vol. 328, pp. 91–137. Springer, New York (2005)
6. Wan, P., Lemmon, M.D.: Event-triggered distributed optimization in sensor networks. In: Symposium on Information Processing of Sensor Networks, San Francisco, CA, pp. 49–60 (2009)
7. Eqtami, A., Dimarogonas, D.V., Kyriakopoulos, K.J.: Event-triggered control for discrete-time systems. In: American Control Conference, Baltimore, MD, pp. 4719–4724 (July 2010)
8. Åström, K.J., Bernhardsson, B.M.: Comparison of Riemann and Lebesgue sampling for first order stochastic systems. In: IEEE Conf. on Decision and Control, Las Vegas, NV, pp. 2011–2016 (December 2002)
9. Tabuada, P.: Event-triggered real-time scheduling of stabilizing control tasks. IEEE Transactions on Automatic Control 52(9), 1680–1685 (2007)
10. Heemels, W.P.M.H., Sandee, J.H., van den Bosch, P.P.J.: Analysis of event-driven controllers for linear systems. International Journal of Control 81(4), 571–590 (2008)
11. Velasco, M., Marti, P., Fuertes, J.M.: The self triggered task model for real-time control systems. In: Proceedings of the 24th IEEE Real-Time Systems Symposium, pp. 67–70 (2003)
12. Subramanian, R., Fekri, F.: Sleep scheduling and lifetime maximization in sensor networks. In: Symposium on Information Processing of Sensor Networks, New York, NY, pp. 218–225 (2006)
13. Wang, X., Lemmon, M.D.: Self-triggered feedback control systems with finite-gain L_2 stability. IEEE Transactions on Automatic Control 54(3), 452–467 (2009)

14. Anta, A., Tabuada, P.: To sample or not to sample: self-triggered control for nonlinear systems. IEEE Transactions on Automatic Control 55(9), 2030–2042 (2010)
15. Mazo Jr., M., Tabuada, P.: Decentralized event-triggered control over wireless sensor/actuator networks. IEEE Transactions on Automatic Control 56(10), 2456–2461 (2011)
16. Wang, X., Hovakimyan, N.: L_1 adaptive control of event-triggered networked systems. In: American Control Conference, Baltimore, MD, pp. 2458–2463 (2010)
17. Donkers, M.C.F., Heemels, W.P.M.H.: Output-based event-triggered control with guaranteed L_∞-gain and improved and decentralised event-triggering. IEEE Transactions on Automatic Control 57(6), 1362–1376 (2012)
18. Dimarogonas, D.V., Frazzoli, E., Johansson, K.H.: Distributed event-triggered control for multi-agent systems. IEEE Transactions on Automatic Control 57(5), 1291–1297 (2012)
19. Shi, G., Johansson, K.H.: Multi-agent robust consensus-part II: application to event-triggered coordination. In: IEEE Conf. on Decision and Control, Orlando, FL, pp. 5738–5743 (December 2011)
20. Mazo Jr., M., Tabuada, P.: On event-triggered and self-triggered control over sensor/actuator networks. In: IEEE Conf. on Decision and Control, Cancun, Mexico, pp. 435–440 (2008)
21. Eqtami, A., Dimarogonas, D.V., Kyriakopoulos, K.J.: Event-triggered strategies for decentralized model predictive controllers. In: IFAC World Congress, Milano, Italy (August 2011)
22. Garcia, E., Antsaklis, P.J.: Model-based event-triggered control for systems with quantization and time-varying network delays. IEEE Transactions on Automatic Control 58(2), 422–434 (2013)
23. Heemels, W.P.M.H., Donkers, M.C.F.: Model-based periodic event-triggered control for linear systems. Automatica 49(3), 698–711 (2013)
24. Nowzari, C., Cortés, J.: Self-triggered coordination of robotic networks for optimal deployment. Automatica 48(6), 1077–1087 (2012)
25. Wang, X., Lemmon, M.D.: Event-triggering in distributed networked control systems. IEEE Transactions on Automatic Control 56(3), 586–601 (2011)
26. Zhong, M., Cassandras, C.G.: Asynchronous distributed optimization with event-driven communication. IEEE Transactions on Automatic Control 55(12), 2735–2750 (2010)
27. Wang, X., Sun, Y., Hovakimyan, N.: Relaxing the consistency condition in distributed event-triggered networked control systems. In: IEEE Conf. on Decision and Control, Atlanta, GA, pp. 4727–4732 (December 2010)
28. Nowzari, C., Cortés, J.: Team-triggered coordination for real-time control of networked cyberphysical systems. IEEE Transactions on Automatic Control (submitted, 2013)
29. Guinaldo, M., Lehmann, D., Moreno, J.S., Dormido, S., Johansson, K.H.: Distributed event-triggered control with network delays and packet losses. In: IEEE Conf. on Decision and Control, Hawaii, USA, pp. 1–6 (December 2012)
30. Cortés, J., Martínez, S., Karatas, T., Bullo, F.: Coverage control for mobile sensing networks. IEEE Transactions on Robotics and Automation 20(2), 243–255 (2004)
31. Okabe, A., Boots, B., Sugihara, K., Chiu, S.N.: Spatial Tessellations: Concepts and Applications of Voronoi Diagrams, 2nd edn. Wiley Series in Probability and Statistics. Wiley (2000)
32. Firouzabadi, S.: Jointly optimal placement and power allocation in wireless networks. Master's thesis, University of Maryland at College Park (2007)

Part VII
Networked Control Systems

Arbitrated Network Control Systems: A Co-Design of Control and Platform for Cyber-Physical Systems

Anuradha M. Annaswamy[1], Damoon Soudbakhsh[1], Reinhard Schneider[2], Dip Goswami[2], and Samarjit Chakraborty[2]

[1] Active-adaptive Control Laboratory, MIT, Cambridge, MA, USA
{aanna,damoon}@mit.edu
http://aaclab.mit.edu

[2] Institute for Real-time Systems, TU Munich, Germany
reinhard.schneider@rcs.ei.tum.de, {dip,samarjit}@tum.de

Abstract. The domain of networked control systems (NCS) has traditionally been concerned with modeling and designing distributed controllers in the presence of control message loss, varying delay and jitter. Here, the characteristics of the network are assumed to be given and the focus has largely been on the controller. In several cyber physical systems, it is possible to not only design distributed controllers, but also design both the scheduling parameters of the resident processors as well as those for the communication buses. We refer to such systems as *arbitrated networked control systems* (ANCS), where the parameters of the arbitration policies in the network are co-designed with the controller. Analytical methods from real-time calculus are used to design the former, and delay aware design procedures are used to determine the controller. A case study was performed on the co-design of platform and control of three quadrotors with delays.

Keywords: Co-Design, Network Control Systems, Cyber Physical Systems, Embedded Systems, Stability, Lyapunov Functions, Delay, LMI.

1 Introduction

Automotive, avionics, and industrial automation systems are often distributed embedded systems (DES) with a large number of processing units (PUs), sensors and actuators that communicate via shared buses such as CAN, LIN and FlexRay. Such architectures are used to run distributed control applications, often with multiple quality-of-control constraints. In such settings, the problem therefore is that of designing both – the network and the controller – with the choice of the network parameters depending on the plant, the quality-of-control requirements, and the choice of the controller, and similarly the choice of the controller parameters depending on the characteristics (and hence design) of the network (e.g., delay, jitter, etc). We refer to such settings as *arbitrated networked control systems* (ANCS) in order to emphasize that the control systems in question are to be designed for networks that are scheduled or arbitrated, i.e., we now have a handle on their characteristics.

Connections between control performance and architecture design have been explored in a number of papers (see, for example, [1–9]). The idea in most of these papers

D.C. Tarraf (ed.), *Control of Cyber-Physical Systems*,
Lecture Notes in Control and Information Sciences 449,
DOI: 10.1007/978-3-319-01159-2_18,

is that better control performance is achieved by redesigning the architecture in a suitable manner. In [1], the scheduling is proposed to be redesigned so as to allow more CPU time and therefore realize a better control performance. In [2, 7], a closed-loop state-based strategy is used to determine an aperiodic sampling sequence and therefore a corresponding scheduling strategy that ensures the desired control performance. Other event-based strategies are used in [3, 4, 8] in order to determine the sampling sequence. In [5, 9], automaton-based approaches are utilized for the design of online schedulers which in turn guarantee robust control performance with LQG controllers.

Other works such as [10–13] have addressed the problem of co-design of controllers and implementation architectures. In [10, 11], a simulation tool is used to estimate the distribution of message delay values and this distribution is used as an input to the Jitterbug toolbox to compute control performance. A good survey of such co-designs can be found in [12]. Reference [13] addresses analysis and implementation of a distributed control system on a network of communicating control unit via control system analysis in terms of sampling times and delays, mapping of control loops to computation/communication hardware components, and scheduling analysis. In this paper, we go one step further and tie in the areas of control systems and architecture design closer together, and provide a co-design procedure.

Our proposed co-design starts with the analytical computation of an upper bound for the delay τ between the sensed output and the delivered input to the actuator. This delay depends on a number of factors including the structure of the embedded platform, its scheduling properties, and the applications that the platform is required to service. By using analytical real-time system tools such as Real-Time Calculus (RTC) [14], this worst case delay is computed. The presence of a non-negligible delay introduces significant changes in the underlying sampled-data model of the plant [15] and the corresponding control performance [16]. The complexity of this model increases further depending on whether or not this delay value is smaller or larger than the sampling period; the requisite control design becomes correspondingly more complex as well. We consider both cases in this paper and propose a control design that explicitly makes use of the delay value, thereby ensuring a stable closed-loop system.

In this paper, the delay τ is assumed to be a constant for all time, which in general need not be true. The problem of varying delay is non-trivial as it can lead to dropped messages and switching between systems of varying dimensions. The latter makes a direct application of results in switched systems infeasible. Recent papers [16–20] have dealt with this problem leading to different co-designs. In these papers, it is shown that when the delay varies, by using a state-feedback controller for all τ less than a threshold τ_{th} and no control for $\tau > \tau_{th}$, it is possible to guarantee stability of the system under certain conditions on the plant. For example, in [19], the problem of varying delay is shown to lead to the phenomenon of dropped messages and that for a certain ratio of dropped to total messages in a given interval, suitable co-design can be carried out that ensures a stable performance. In all these papers, however, it is assumed that the plant model is stabilizable with the feedback of the plant state only, even with the delay. In this paper, we relax this assumption and derive stability conditions, under the assumption of a constant delay.

In §2, we consider a general embedded system architecture and show how rules of real time calculus can be used to derive an upper bound on the end-to-end delay between the sensor and the actuator. In § 3 we derive the control design using a nominal value for the delay using its worst case value. A detailed case study is presented in § 4 pertaining to the control of multiple quadrotors using an embedded platform. It is shown that our ANCS approach predicated on the available information about the platform delay results in improved performance compared to a control design that assumes that the delay is negligible.

2 A Co-Design Using the ANCS Approach

In order to illustrate an ANCS, we consider a typical distributed embedded architecture (see Figure 1), where an electronic control unit (ECU) collects sensor data (denoted as a task T_s). A communication bus (e.g., FlexRay) then transmits the data as message m_1 to a second ECU (marked as ECU_3), where the resident control algorithm is implemented (denoted as task T_c). The output of the controller is then sent as a message m_2 to the actuator in ECU_2, which activates the actuator task T_a. Here, the sampling period is h. The total end-to-end delay between the time the signal is measured and the time the actuator inputs the computed signal from the controller into the plant is indicated by τ (see Figure 1b). We define an arbitrated network control system as one where the implementation structure can be utilized to obtain prior information about τ and suitably exploit it in the control design. The specific problem that we consider is stated in the following section.

2.1 Problem Statement

We consider a platform architecture with multiple distributed control applications (see Figure 2). The control applications are partitioned into a number of tasks that are mapped onto different PUs. The PUs communicate via a shared communication bus and run different tasks from one or more control applications. Scheduling on the processors and arbitration on the shared bus introduces delays in various control signals. These delays are dependent on the arbitration/scheduling policy on the buses and on the PUs. For ease of illustration, this setup will form the basis of the techniques we propose. However, they hold true for other setups as well.

Hardware/Software Architecture: We consider three control applications: *controller 1, 2 and 3* shown in Figure 3, 4 and 5. Each controller is connected to an actuator and a sensor. The aim of each control application is to compute the control input to the plant such that the closed-loop system meets desired performance. Towards this, the controllers read sensors values to determine the state of the closed-loop system and compute commands based on the deviation from the desired performance. Control applications are partitioned into a number of tasks, which are mapped onto different PUs connected via the communication bus. In our architecture (see Figure 2) PU1 hosts the tasks responsible for reading the reference command from the user, PU2 hosts all tasks that compute control commands, PU3 hosts the tasks responsible for reading sensors, and the tasks responsible for providing plant commands are mapped onto PU4.

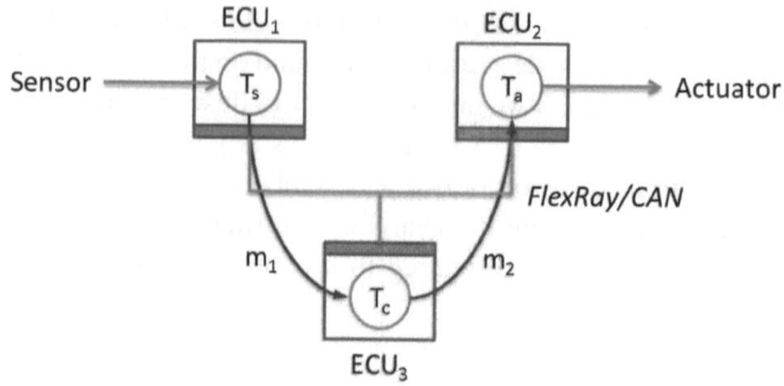

(a) Task allocation

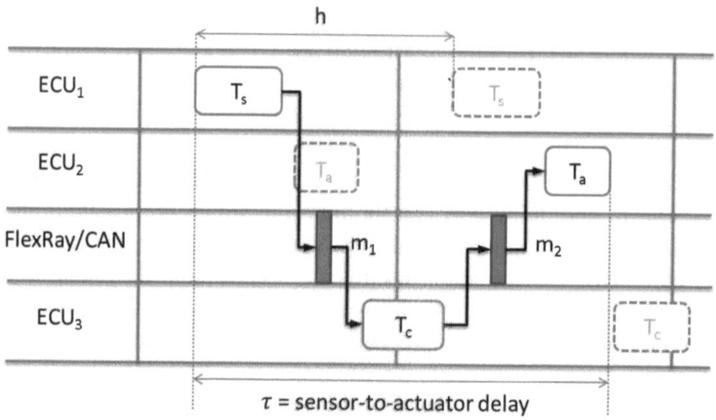

(b) Time line

Fig. 1. A typical distributed embedded system (DES)

Control Applications: *Control application 1* (Figure 3) is partitioned into four tasks: T_1, T_2, T_3 and T_4. Task T_1 reads sensor S_1 and sends sensor signal m_1 via the communication bus to the task T_3. Similarly, task T_2 reads the reference command from the user and sends the reference command m_2 via the communication bus to task T_3. The control input is computed in T_3 using the messages m_1 and m_2. The processed output of task T_3 is sent via the communication bus to task T_4. The plant P_1 receives the control input from the task T_4. Figure 3 (a) and (b) show the *task graph* and the *closed-loop block diagram* for *control application 1*.

Control application 2 (Figure 4) is partitioned into three tasks: T_5, T_6 and T_7. Task T_5 reads the sensor S_2 and sends the sensor signal m_5 via the communication bus to the task T_6. Controller task T_6 computes the control command based on the sensor feedback signal and the constant predefined reference command. The task T_6 sends the control command to task T_7 via the communication bus. The plant P_2 receives input signal from the task T_7. Figure 4 (a) and (b) show the *task graph* and the *closed-loop block diagram* for *control application 2*.

Control application 3 (Figure 5) is partitioned into four tasks: T_8, T_9, T_{10} and T_{11}. Task T_8 reads the sensor S_3 and sends the sensor signal m_8 via the communication bus

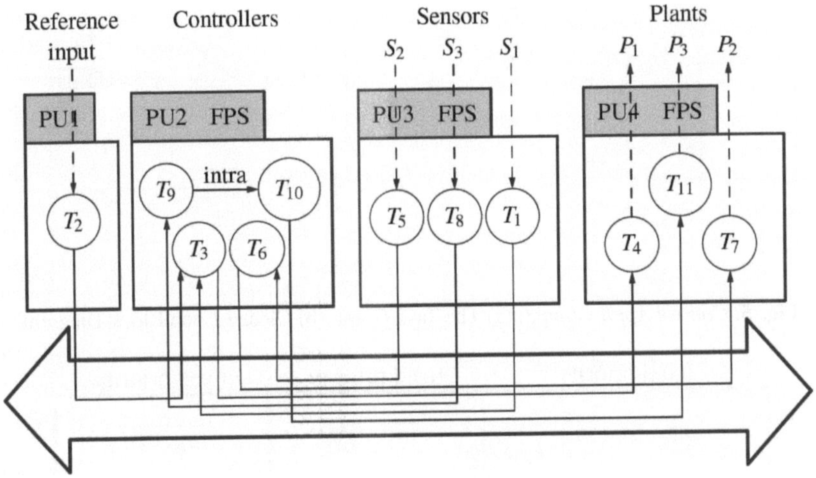

Fig. 2. System Architecture With Three Distributed Control Applications

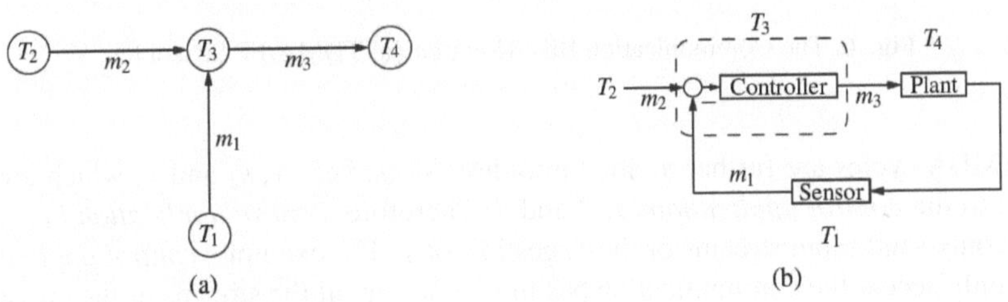

Fig. 3. *Control Application 1*: (a) The Task Graph (b) Closed-Loop Block Diagram

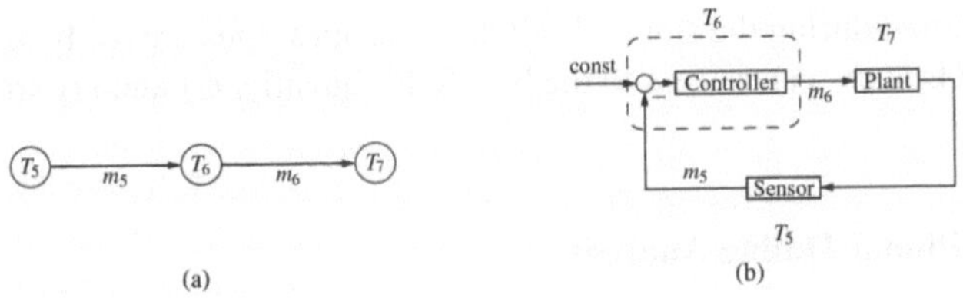

Fig. 4. *Control Application 2*: (a) The Task Graph (b) Closed-Loop Block Diagram

to the task T_9. Task T_9 computes one part of the control signal and sends the resulting command to the task T_{10}. Task T_{10} computes the control command m_{10} and sends it via the communication bus to the task T_{11} which sends the control command to the plant P_3. Figure 5 (a) and (b) show the *task graph* and the *closed-loop block diagram* for *control application 3*.

Communication Scheduling: The *shared communication bus* follows a hierarchical TDMA/FP scheduling policy (Figure 6). At the top-level of the scheduler runs a TDMA scheduler, i.e., the communication bandwidth is divided into equal cycles of length c.

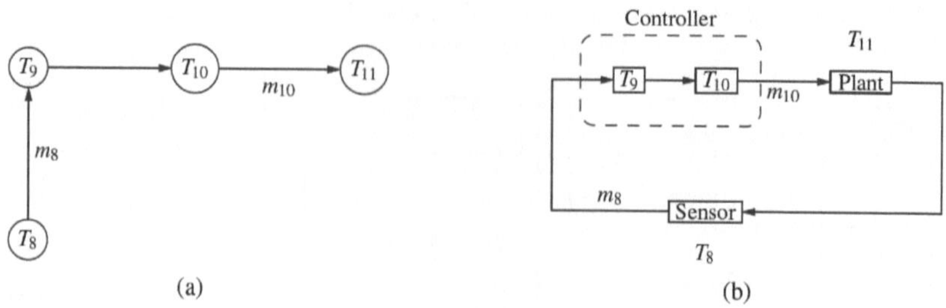

Fig. 5. *Control Application 3*: (a) The Task Graph (b) Closed-Loop Block Diagram

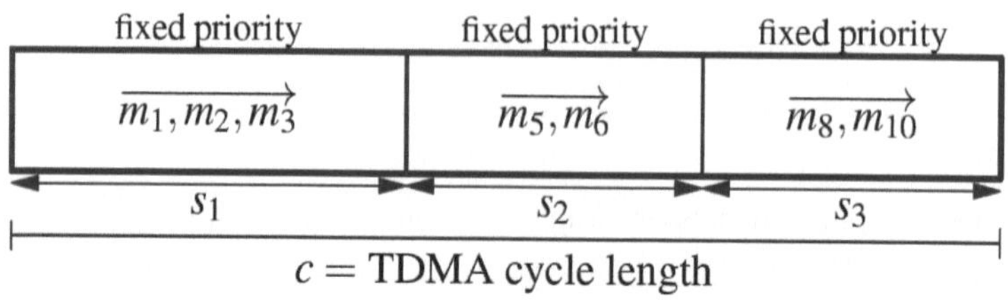

Fig. 6. The Communication Bus: Hierarchical TDMA/FP Scheduler

The TMDA cycles are further divided into three slots, i.e., s_1, s_2 and s_3 which are assigned to the *control applications 1, 2* and *3*. Therefore, *control application i (i=1, 2, 3)* can only send input streams or messages in slot s_i. For example, *control application 1* can only access the communication bus in s_1. Further, all the streams or the messages coming into/from the *control application* which are transmitted via the communication bus follow a fixed priority scheduler (FPS). For example, *control application 1* exchanges three messages among its various tasks: m_1, m_2 and m_3. These messages can only be transmitted during the slot s_1. If all the three messages are ready to be transmitted via s_1 then first m_1 gets access to the bus. Subsequently, m_2 and m_3 are transmitted respectively.

2.2 Compositional Timing Analysis

In this section we will give an overview of the *real-time calculus* (RTC) framework [14], which is an analytical method for analyzing performance properties of distributed embedded real-time systems. Given several distributed applications that are mapped onto different processing and communication resources, e.g., in the architecture depicted in Figure 2, we are now interested in computing the maximum end-to-end delay experienced by each of the *control applications*.

The input of a task can be represented by an arrival pattern $A(t)$ of a data stream, i.e., sensor readings. The cumulative function $A(t)$ denotes the total number of events that arrive during time interval $(0,t]$. However, the number of events that may arrive in *any* time interval of length Δ that trigger a task execution or a message transmission can be upper- and lower-bounded by the pair of *arrival curves* $\alpha = (\alpha^u, \alpha^l)$. Formally, we obtain the following mathematical inequality:

$$\forall \Delta \geq 0, \forall t \geq 0 : \alpha^l(\Delta) \leq A(\Delta + t) - A(t) \leq \alpha^u(\Delta) \tag{1}$$

In other words, $\alpha^u(\Delta)$ and $\alpha^l(\Delta)$ denote the maximum and minimum number of events that can arrive within *any* time interval of length Δ. Thus, using this event model, we can represent the timing properties of standard event models - like *periodic, periodic with jitter* and *sporadic* - as well as arbitrary arrival patterns by an appropriate choice of $\alpha^u(\Delta)$ and $\alpha^l(\Delta)$. Figure 8 (a) illustrates an example for the upper and lower arrival curves representing a strictly periodic event stream with period h. Figure 8 (b) depicts a pair of arrival curves with period h and jitter j, i.e., in a periodic sensor stream with jitter the sensor samples arrive at an average time interval of h time units, but can deviate from the ideal periodic arrival.

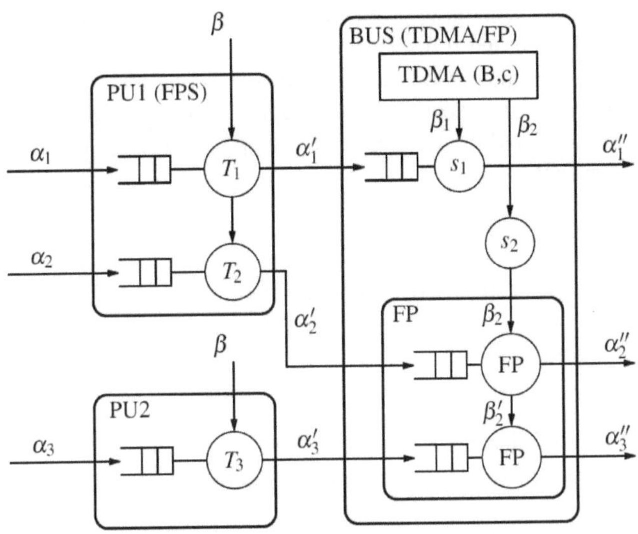

Fig. 7. System With Hierarchical Bus Scheduling

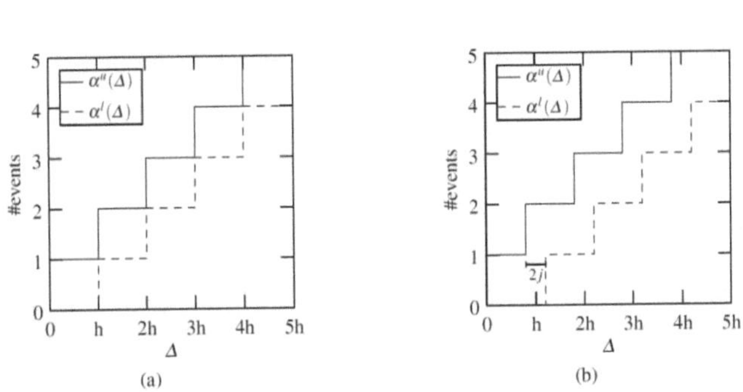

Fig. 8. (a) Periodic Arrival Curves (b) Periodic with jitter j

Similarly, we can model the available resource capacities on a processor or a bus to process a task or transmit a message. The cumulative function $C(t)$ captures the number of events that can be processed in $(0,t]$. The service available to a task or message is upper- and lower-bounded by the pair of *service curves* $\beta = (\beta^u, \beta^l)$. Thus, the following mathematical inequality hold true:

$$\forall \Delta \geq 0,\, \forall t \geq 0 : \beta^l(\Delta) \leq C(\Delta + t) - C(t) \leq \beta^u(\Delta) \tag{2}$$

Hence, $\beta^u(\Delta)$ and $\beta^l(\Delta)$ denote the maximum and minimum number of events that can be processed within *any* time interval of length Δ. The service curves can also be expressed in terms of the maximum and minimum number of resource units (e.g., processor cycles, execution times) that are required to process an event within any Δ. For this purpose, $\beta^u(\Delta)$ and $\beta^l(\Delta)$ are scaled with the execution requirement demanded by a task or message due to each activation. When running several tasks on a shared resource, e.g., T_1 and T_2 on $PU1$, the bounds on the service available to the tasks using this resource depend on the scheduling policy being used. On $PU1$ the tasks are executed according to FPS. Hence, the full service β is available to the highest priority task T_1 to process the input stream α_1. Consequently, the remaining service β' after processing α_1 serves as an input for the execution of T_2 processing α_2. Further, the processed data streams $\alpha' = (\alpha^{u'}, \alpha^{l'})$ and the remaining service $\beta' = (\beta^{u'}, \beta^{l'})$ can be computed by the following equations:

$$\alpha^{u'} = \min\{(\alpha^u \otimes \beta^u) \oslash \beta^l,\, \beta^u\} \tag{3}$$
$$\alpha^{l'} = \min\{(\alpha^l \oslash \beta^u) \otimes \beta^l,\, \beta^l\} \tag{4}$$
$$\beta^{u'} = (\beta^u - \alpha^l)\ \overline{\oslash}\ 0 \tag{5}$$
$$\beta^{l'} = (\beta^l - \alpha^u)\ \overline{\otimes}\ 0 \tag{6}$$

Let $\overline{\mathbb{R}} := \mathbb{R} \cup \{+\infty, -\infty\}$ and $\mathscr{F} = \{f : \mathbb{R}_+ \to \overline{\mathbb{R}} \mid \forall s < t,\, 0 \leq f(s) \leq f(t)\}$. The (min,+) convolution $\otimes$ and deconvolution $\oslash$ operators are defined by: $\forall f, g \in \mathscr{F}$ and $\forall t \in \mathbb{R}_+$,

$$(f \otimes g)(t) = \inf\{f(s) + g(t-s) \mid 0 \leq s \leq t\}$$
$$(f \oslash g)(t) = \sup\{f(t+u) - g(u) \mid u \geq 0\}$$

The (max,+) convolution $\overline{\otimes}$ and deconvolution $\overline{\oslash}$ operators are defined by: $\forall f, g \in \mathscr{F}$ and $\forall t \in \mathbb{R}_+$,

$$(f \overline{\otimes} g)(t) = \sup\{f(s) + g(t-s) \mid 0 \leq s \leq t\}$$
$$(f \overline{\oslash} g)(t) = \inf\{f(t+u) - g(u) \mid u \geq 0\}$$

Considering a set $S \subseteq \mathscr{F}$, the *supremum* (sup) is defined by the smallest $U \in \mathscr{F}$ such that $x \leq U$, $\forall x \in S$. Similarly, the *infimum* (inf) of S is the largest $L \in \mathscr{F}$ such that $x \geq L$, $\forall x \in S$. Given $\alpha^u(\Delta)$ and $\beta^l(\Delta)$ the maximum backlog b at the input buffer, and the maximum delay d that is experienced by the event stream α can be computed as follows:

$$b = \sup\{\alpha^u(\Delta) - \beta^l(\Delta) \mid \Delta \geq 0\} \tag{7}$$

$$d = \sup\{\inf\{\tau \geq 0 \mid \alpha^u(s) \leq \beta^l(s+\tau)\} \mid s \geq 0\} \tag{8}$$

Note that (7) and (8) denote the maximum vertical and horizontal deviations between $\alpha^u(\Delta)$ and $\beta^l(\Delta)$.

The processed data streams α_1', α_2' and α_3' trigger messages to be transmitted on the shared bus according to a *hierarchical* TDMA/FP scheduling policy. The top-level scheduler is modeled as a TDMA resource with a total bandwidth of B and a cycle length of c. Within every TDMA cycle the time slot s_i is assigned to one *control application*. In Figure 7, slot s_1 is assigned to the data stream α_1' whereas slot s_2 is assigned to α_2' and α_3'. The service bounds for a TDMA resource can be modeled as follows:

$$\beta_i^l = B\max\left\{\left\lfloor\frac{\Delta}{c}\right\rfloor s_i, \Delta - \left\lceil\frac{\Delta}{c}\right\rceil (c - s_i)\right\} \tag{9}$$

$$\beta_i^u = B\min\left\{\left\lceil\frac{\Delta}{c}\right\rceil s_i, \Delta - \left\lfloor\frac{\Delta}{c}\right\rfloor (c - s_i)\right\} \tag{10}$$

During time interval $\Delta = (c - s_i)$ no service is guaranteed to any data stream that is assigned to any slot s_i. According to the *hierarchical* scheduling policy the service available to the data streams assigned to any time slot s_i is determined according to the FPS. The full service β_1 is available for transmitting the data stream α_1'. The input streams α_2' and α_3' share slot s_2 providing the service β_2 where α_2' is assigned a higher priority than α_3'. Thus, the full service β_2 is available to α_2' while the remaining service β_2' is available for transmitting α_3' on the bus. Now, the output event streams α_1'', α_2'' and α_3'' can serve again as an input to further processing units, e.g., executing actuator tasks.

Considering the system in Figure 7, the end-to-end delay can be computed as the sum of the individual delays at the different processing components (due to task processing and bus communication) experienced by an input data stream, e.g., the data stream α_1 experiences a delay d_1 due to task processing of T_1 and another delay d_2 due to the message transmission on the bus. Thus, using (8), the total delay experienced by α_1 is computed as $\tau = d_1 + d_2$. Hence, using the compositional performance model presented in this section, we are able to compute the maximum end-to-end delay τ experienced by any message along the path from the sensor to the actuator, of any distributed *control application*. For example, in Figure 2 the worst case end-to-end delay experienced by *control application 3* is computed by

$$\tau_{worst} = d_{T_8} + d_{m_8} + d_{T_9} + d_{T_{10}} + d_{m_{10}} + d_{T_{11}}. \tag{11}$$

2.3 Estimation of End-to-End Delay for Control Design

The RTC tool described above provides a known estimate, albeit a conservative one, of the resident time-delay due to the implementation process in the embedded platform. When it comes to a cyber physical system where these platforms are used to control a physical application, the presence of a time-delay has a direct consequence in the performance of the control application. A knowledge of the time-delay and its incorporation in the control design can lead to a much better performance rather than ignoring its presence and appealing to robustness properties of the control design. We therefore use the information about time-delay using τ_{worst} in the following manner. The underlying sampled-data model of a system to be controlled depends not only on its physical properties but also the sampling period h. For ease of exposition, we assume in this paper that the best case delay, τ_{best}, is equal to nh, when $nh < \tau_{worst} < (n+1)h$. In particular, if τ_{act} is defined as the actual end to end delay between a sensor and an actuator, the underlying model structure varies significantly if $\tau_{act} < h$ or if $\tau_{act} > h$. Denoting τ_{best} as the minimum end to end delay between a sensor and an actuator, the compositional timing analysis presented in § 2.2 may be used to calculate τ_{best} by using a lower bound on the message arrival process (α^l) and the upper bound on the service (β^u). Instead, for the purposes of this paper, we assume τ_{best} to be

$$\tau_{best} = \begin{cases} 0 & \text{if} \quad \tau_{worst} < h \\ nh & \text{if} \quad nh \leq \tau_{worst} < (n+1)h, \quad n = 1,2,\cdots \end{cases} \tag{12}$$

With these bounds τ_{best} and τ_{worst} for the actual delay, we choose the the nominal time delay as

$$\tau_{nom} = \begin{cases} \frac{\tau_{worst}}{2} & \text{if} \quad \tau_{worst} < h \\ \frac{\tau_{worst}+nh}{2} & \text{if} \quad nh \leq \tau_{worst} < (n+1)h, \quad n = 1,2,\cdots \end{cases} \tag{13}$$

We therefore see that in this case the time-delay variation $\delta = \tau_{act} - \tau_{nom}$ has the bounds

$$\delta = \begin{cases} \frac{\tau_{worst}}{2} & \text{if} \quad \tau_{worst} < h \\ \frac{\tau_{worst}-nh}{2} & \text{if} \quad nh \leq \tau_{worst} < (n+1)h, \quad n = 1,2,\cdots \end{cases} \tag{14}$$

In principle, one can choose τ_{nom} and the control design such that the closed-loop system is stable for τ_{nom} and robust for all delay values τ_{act} such that $\tau_{best} \leq \tau_{act} \leq \tau_{worst}$.

3 Control Design

We first derive the plant model for the case when $\tau_{act} < h$ and when $\tau_{act} > h$, starting from a continuous-time description of the plant.

3.1 A Sampled Data Model of the Physical Plant

We assume that its model is linear and has the form

$$\dot{x}(t) = A_c x(t) + B_c u(t - \tau_{act}) \tag{15}$$

where $x(t) \in \Re^p$ and $u(t) \in \Re^q$ are states and inputs, respectively. The goal is to design u so that $x(t)$ tends to zero asymptotically with the closed-loop system remaining bounded. The delay of the system is $\tau_{act} > 0$, which represents the combined effect of lags due to communication and computation in the embedded system. When $\tau_{act} < h$, a sampled-data model of (15) leads to

$$x[k+1] = A_1 x[k] + B_{11} u[k] + B_{12} u[k-1]. \tag{16}$$

where

$$A_1 \stackrel{\text{def}}{=} e^{A_c h}, \quad B_{11} \stackrel{\text{def}}{=} (\int_0^{h-\tau_{act}} e^{A_c \nu} d\nu) B_c, \quad \text{and } B_{12} \stackrel{\text{def}}{=} (\int_{h-\tau_{act}}^{h} e^{A_c \nu} d\nu) B_c$$

if $\tau_{act} > h$, by defining $d_1 = \lfloor \frac{\tau_{act}}{h} \rfloor$, which is the largest integer smaller than $\frac{\tau_{act}}{h}$, we obtain that

$$x[k+1] = A_2 x[k] + B_{21} u[k-d_1] + B_{22} u[k-d_1-1] \tag{17}$$

where

$$A_2 \stackrel{\text{def}}{=} e^{A_c h}, \; B_{21} \stackrel{\text{def}}{=} (\int_0^{h-\tau'} e^{A_c \nu} d\nu) B_c, \; B_{22} \stackrel{\text{def}}{=} (\int_{h-\tau'}^{h} e^{A_c \nu} d\nu) B_c, \; \text{and } \tau' = \tau_{act} - h \cdot d_1.$$

We note that $d_1 = n$ if $nh \leq \tau_{act} < (n+1)h$. Equation (17) can be simplified further as

$$x[k+1] = \quad A_2^{d_1+1} x[k-d_1] + \sum_{\ell=0}^{d_1} \left(A_2^{\ell} B_{21} u[k-d_1-\ell] + A_2^{\ell} B_{22} u[k-d_1-1-\ell] \right) \tag{18}$$

The structure of the dynamic systems in (16) and (17) implies that the underlying state will vary in dimension and as such, two control designs with entirely different states may be needed in order to stabilize the underlying system. Since τ_{act} is unknown, the controllers will be designed using τ_{nom}. Certain off-line iterations can be done, however, by choosing the initial values of τ_{nom} as in (13), and iterating it until the closed-loop system is stable for all $\tau_{act} \in [\tau_{best}, \tau_{worst}]$. As mentioned in the previous section, τ_{best} is assumed to be of the form of (12). Any prior information about the task execution times leading to sharper bounds on τ_{best} can then be used in decreasing these iterations.

3.2 Case (i): $\tau_{act} < h$

Since the underlying plant model is of the form (16), noting that the system has an extended state $X(k) = [x^T[k], u^T[k-1]]^T$, a state feedback controller is chosen as

$$u[k] = Kx[k] + Gu[k-1]. \tag{19}$$

The closed-loop system is then given by

$$X[k+1] = \begin{bmatrix} A_1 + B_{11}K & B_{12} + B_{11}G \\ K & G \end{bmatrix} X[k] \stackrel{\text{def}}{=} \Gamma_1 X[k] \tag{20}$$

We discuss conditions under which Γ_1 is stable in Theorem 1. In what follows,

$$\begin{bmatrix} A & * \\ B & D \end{bmatrix} \stackrel{\text{def}}{=} \begin{bmatrix} A & B^T \\ B & D \end{bmatrix} \tag{21}$$

Theorem 1. *K and G exist such that (20) is stable if there exist positive definite matrices $Q_1 \succ 0$ and $Q_2 \succ 0$ such that the following Linear Matrix Inequalities (LMI) are satisfied for some matrices E and F:*

$$\begin{bmatrix} -Q_1 & \mathbf{0} & * & * \\ \mathbf{0} & -Q_2 & * & * \\ A_1Q_1+B_{11}E & B_{12}Q_2+B_{11}F & -Q_1 & \mathbf{0} \\ E & F & \mathbf{0} & -Q_2 \end{bmatrix} \prec \mathbf{0}, \qquad (22)$$

with

$$K = EQ_1^{-1} \qquad G = FQ_2^{-1}. \qquad (23)$$

Proof. Inequality (22) implies that the following inequalities are satisfied as well, with the Schur complement:

$$\begin{bmatrix} A_1Q_1+B_{11}E & B_{12}Q_2+B_{11}F \\ E & F \end{bmatrix}^T \begin{bmatrix} Q_1^{-1} & \mathbf{0} \\ \mathbf{0} & Q_2^{-1} \end{bmatrix} \begin{bmatrix} A_1Q_1+B_{11}E & B_{12}Q_2+B_{11}F \\ E & F \end{bmatrix} - \begin{bmatrix} Q_1 & \mathbf{0} \\ \mathbf{0} & Q_2 \end{bmatrix} \prec \mathbf{0} \qquad (24)$$

From (23) and defining a matrix Q as

$$Q = \begin{bmatrix} Q_1 & \mathbf{0} \\ \mathbf{0} & Q_2 \end{bmatrix},$$

we obtain that

$$Q\Gamma_1^T Q^{-1} \Gamma_1 Q - Q \prec \mathbf{0}. \qquad (25)$$

Multiplying (25) from left and right by Q^{-1}, and defining a new positive definite matrix $P \stackrel{\text{def}}{=} Q^{-1}$, we arrive at the following inequality

$$\Gamma_1^T P \Gamma_1 - P \prec \mathbf{0}, \qquad (26)$$

which implies that a Lyapunov function in the form of $V = X[k]^T PX[k]$ exists for system (20), proving Theorem 1. □

3.3 Case (ii): $\tau_{act} > h$

In this case, the governing plant model is given by (18).which has an extended state in the form of

$$X[k] = \left[x[k-d_1]^T \; u[k-2d_1-1]^T \cdots u[k-d_1-1]^T\right]^T,$$

Therefore, a state feedback control is proposed, of the form

$$u[k] = \kappa X[k] \qquad (27)$$

where

$$\kappa \stackrel{\text{def}}{=} \left[K_1 \; K_2 \cdots K_p\right], \qquad p \stackrel{\text{def}}{=} d_1 + 2 \qquad (28)$$

Using (27), (18) can be written as

$$X[k+d_1+1] = \begin{bmatrix} A_2^{d_2}+B_{21}K_1 & A_2^{d_1}B_{22}+B_{21}K_2 & \cdots & A_2B_{21}+B_{22}+B_{21}K_p \\ K_1 & K_2 & \cdots & K_p \\ \mathbf{0} & \mathbf{0} & \cdots & \mathbf{0} \\ \vdots & \vdots & \ddots & \vdots \\ \mathbf{0} & \mathbf{0} & \cdots & \mathbf{0} \end{bmatrix} X[k] \stackrel{\text{def}}{=} \Gamma_2 X[k] \tag{29}$$

Conditions under which Γ_2 is stable are presented in Theorem 2:

Theorem 2. *System (29) is stable with control law $u[k-d_1] = \kappa X$ as given in (27), if there exist positive definite matrices $Q_1, Q_2, \cdots, Q_{d_2}$, and matrices $F_1, \cdots, F_p$ such that the following Linear Matrix Inequalities (LMI) are satisfied.*

$$\begin{bmatrix} -Q & * \\ \mathscr{L} & -Q \end{bmatrix} \prec \mathbf{0} \tag{30}$$

where Q is a block diagonal matrix with $Q_1, \cdots, Q_p$ in the diagonals, and the control gains are given by

$$K_i = F_i Q_i^{-1}. \tag{31}$$

and $\mathscr{L}$ is defined to be

$$\mathscr{L} \stackrel{\text{def}}{=} \begin{bmatrix} A_2^{d_2}Q_1+B_{21}F_1 & A_2^{d_1}B_{22}Q_2+B_{21}F_2 & \cdots & (A_2B_{21}+B_{22})Q_p+B_{21}F_p \\ F_1 & F_2 & \cdots & F_p \\ \mathbf{0} & \mathbf{0} & \cdots & \mathbf{0} \\ \vdots & \vdots & \ddots & \vdots \\ \mathbf{0} & \mathbf{0} & \cdots & \mathbf{0} \end{bmatrix} \tag{32}$$

Proof. Inequality (30) implies that the following inequalities are satisfied as well, with the Schur complement:

$$\mathscr{L}^T Q^{-1} \mathscr{L} - Q \prec \mathbf{0} \tag{33}$$

From (31)-(33) and definition of matrix Q as

$$Q = \begin{bmatrix} Q_1 & & \mathbf{0} \\ & \ddots & \\ \mathbf{0} & & Q_p \end{bmatrix},$$

we obtain that

$$Q\Gamma_2^T Q^{-1} \Gamma_2 Q - Q \prec \mathbf{0}. \tag{34}$$

Multiplying (34) from left and right by Q^{-1}, and defining a new positive definite matrix $P \stackrel{\text{def}}{=} Q^{-1}$, we arrive at the following inequality

$$\Gamma_2^T P \Gamma_2 - P \prec \mathbf{0}, \tag{35}$$

which implies that a Lyapunov function in the form of $V = X[k]^T PX[k]$ exists for system (29), proving Theorem 2. □

Remark 1. Noting that Theorem 2 holds for $nh \leq \tau < (n+1)$ for any n, it follows that Theorem 1 is a special case of Theorem 2 for $n = 0$. This can be verified by setting $d_1 = 0$ in (27) to obtain Γ_1 as in (20). Then the solution of the LMI in (30), K_1 and K_2 are the K and G gains in Theorem 1, respectively.

3.4 Co-Design with a Varying Delay τ

The results presented in this paper assume that τ is constant, with $nh \leq \tau < (n+1)h$, and can either satisfy case (i) where $n = 0$ or case (ii), where $n \geq 1$, but fixed. In order to relax the restriction that this imposes on τ, the next question to be investigated is when τ varies between one sampling period to the next , i.e. when n changes with time. While at first glance, it may appear that the underlying problem in this case is simply one of a switching system with the dynamics switching between (16) and (18), the actual problem is significantly more complex. The main reason is that an increase in the delay τ from one period to another implies that at any given instant k, the inputs that affect $x[k+1]$ can vary and belong to one of several cases depending on the specific sequence of past delays over the interval $[k-nh,k]$. And if n itself varies with k, the underlying problem becomes significantly more complex. For example, in the simplest case where n can vary only between 0 and 1, $x[k+1]$ can be affected by (i) $u[k-1]$ and $u[k]$, (ii) $u[k-1]$ alone, (iii) $u[k-1]$ and $u[k-2]$, or (iv) $u[k]$ and $u[k-2]$. If case (iv) occurs, this implies that $u[k-1]$ is dropped. Also, in each of these cases, the underlying dynamics is different, with (18) reflecting case (iii). One therefore has to consider all possible switches between (17) and any of the above four cases, and address the underlying control problem. Obviously, the underlying complexity increases further as n varies between 0 and an arbitrary integer. This paper therefore should be viewed as a first step in laying the foundation for a co-design where the platform introduces a fixed delay. The next step is to address the problem outlined above for varying delay and consider all associated combinatorics.

4 Case Study

The problem we consider is the simultaneous control of multiple UAVs (see Figure 9), a problem of significant interest to the aerospace industry. The specific problem objective is to accomplish hover. Defining origin to be three distinct points in their respective workspace, we assume that they are perturbed from this point. The goal is to design closed-loop controllers such that they are all brought to their respective hover positions. We start with linearized open-loop models of the form [21].

We consider a linearized model of quadrotors to bring the UAV from a position (100, 50, 20) cm to the origin. Dynamics of the quadrotor can be described by

$$\ddot{x} = g\theta, \quad \ddot{y} = -g\phi, \quad \ddot{z} = \frac{\Delta U_1}{m}$$
$$\ddot{\phi} = \frac{L}{I_x}U_2, \quad \ddot{\theta} = \frac{L}{I_y}U_3, \quad \ddot{\psi} = \frac{1}{I_z}U_4. \tag{36}$$

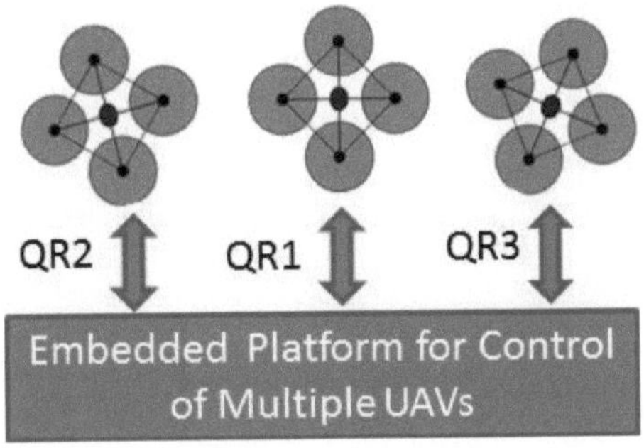

Fig. 9. Multiple UAVs

where x, y, and z describe the position of the center of mass of UAV in the inertial coordinate system; ϕ, θ, and ϕ are the Euler angles that describe the orientation of the local coordinate system with respect to the inertial frame; m is the mass and I_x, I_y, and I_z are the moments of inertia of the quadrotor with respect to x, y, and z axes, respectively; L is the distance from the center of mass to the rotors; and J_R and Ω_R are the moments of inertia and angular velocity of the propeller blades. Inputs U_1, U_2, U_3, and U_4 are the collective, roll, pitch, and yaw forces generated by the four propellers [21].

We consider 3 quadrotors QR_1, QR_2, and QR_3 that have to be controlled over the network with sampling times 15, 30, and 40*ms*[1], respectively. An embedded platform as in Figure 2 was chosen. The communication between the application and the platform is assumed to occur through a protected network. The scheduler runs a TDMA scheduler and the communication bandwidth is divided into equal cycles of lengths $c = 10ms$. The TMDA cycles are further divided into three slots, $s1 = 6$, $s2 = 2$, and $s3 = 2ms$, which are assigned to the control applications QR_1, QR_2, and QR_3. Using this information as well as service curves and arrival curves, using (8) we obtained the maximum delay of each task of control applications. The worst case end-to-end delay is the sum of individual component delays (see (11)), and they were computed as a τ_{worst} of 14.6, 65.1, and 43.2 for $QR_i, i = 1,2,3$, respectively.

The initial values for τ_{nom} were selected using (13) resulting in 7.3, 62.6, and 41.6 for QR_i, $i = 1,2,3$. These were then fine-tuned so as to result in a τ_{nom} such that the closed-loop controller was stable for all $\tau_{act} \in [\tau_{best}, \tau_{worst}]$. This resulted in the values 6, 61, and 43 for QR_1, QR_2, and QR_3, respectively. With these values of τ_{nom} and the corresponding control designs as in (19) and (27), closed-loop control of QR_i, i=1,2,3, was implemented in MATLAB using sampled data models (16) and (17). The resulting performances of the position outputs x, y, z and the corresponding inputs U_1, U_2, U_3, and U_4 are shown in Figures 10-11, for QR_1, Figures 12-13 for QR_2, and 12-15 for QR_3.

For the first quadrotor, QR_1, we designed the controller with nominal delay $\tau_{nom} = 6ms$. Figures 10 and 11 show the resulting closed-loop performance of QR_1 with $\tau_{act} = 10ms$. It was also observed that this controller was able to bring the UAV from its original position with any delay in the range of $0-15ms$. On the other hand, designing

[1] All units of time are in *ms* unless otherwise specified.

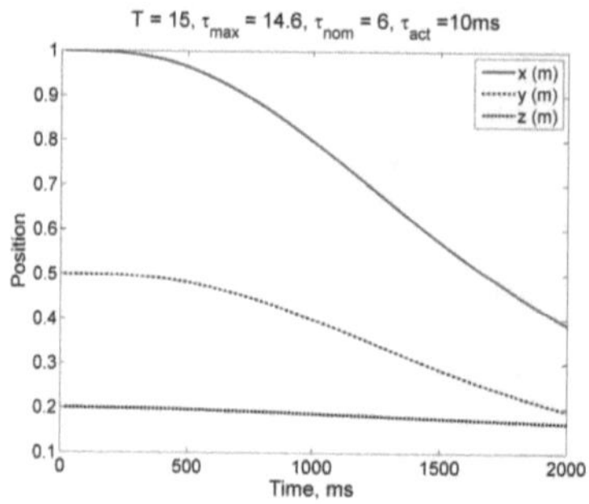

Fig. 10. Position of QR_1 with $\tau_{act} = 10ms$

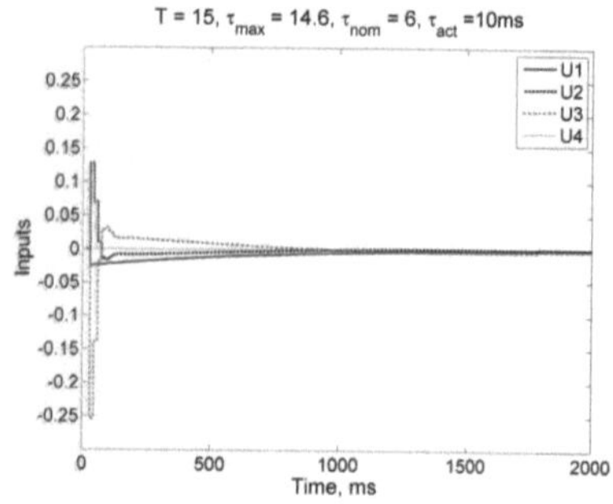

Fig. 11. Inputs to QR_1

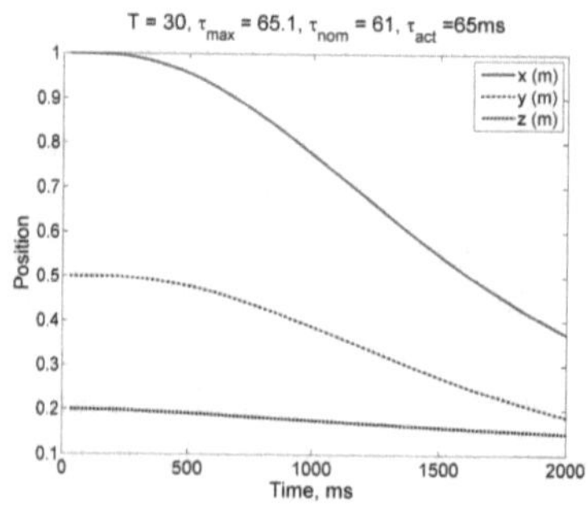

Fig. 12. Position of QR_2 with $\tau_{act} = 65ms$

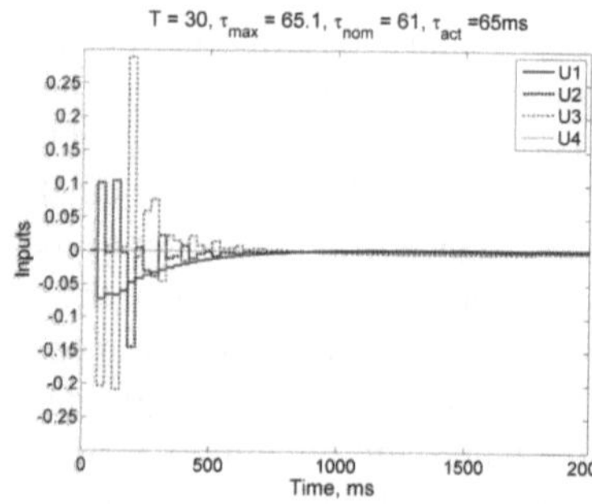

Fig. 13. Inputs to QR_2

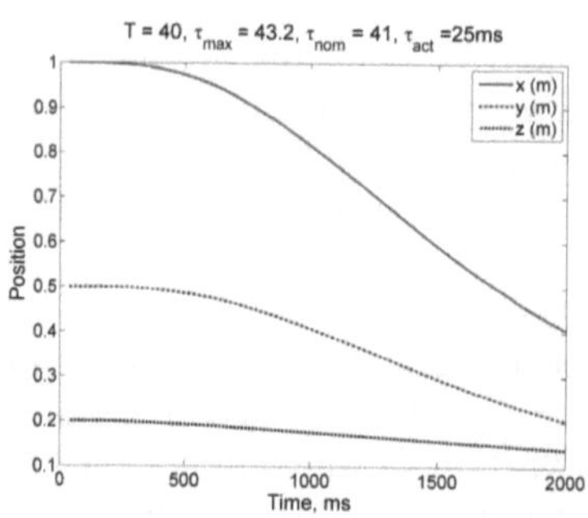

Fig. 14. Position of QR_3 with $\tau_{act} = 25ms$

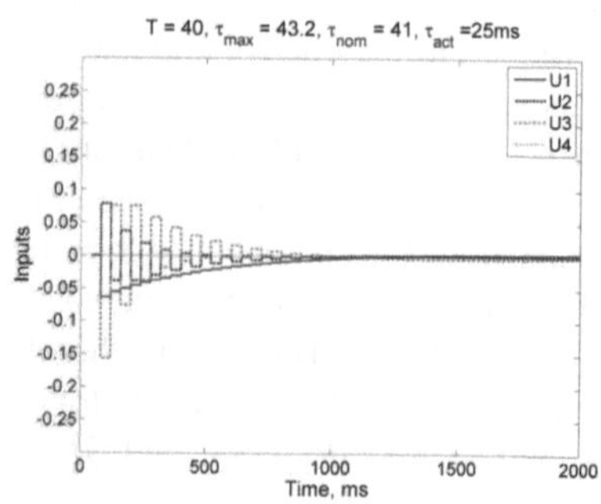

Fig. 15. Inputs to QR_3

the controller with nominal delay $\tau_{nom} = 0$ resulted in stable system only up to $13.2ms$, implying instability for the worst case delay of $14.6ms$. The second UAV has a sampling period of $30ms$ and suffers a worst case delay of $65.1ms$. The controller is designed for $\tau_{nom} = 61ms$. Figure 12 shows the closed-loop performance of QR_2, and the corresponding input is shown in Figure 13. Figure 14 shows resulting closed-loop system performance of QR_3. Similarly, the outputs and inputs corresponding to QR_3 for a τ_{nom}

of 25 are shown in Figures 15 and 14 respectively. It was observed that this controller stabilizes the system for delays from $21 - 58ms$. In all these cases, it was observed that the controller based on a $\tau_{nom} = 0$ led to a much poorer closed-loop performance.

5 Summary

In this paper, we considered arbitrated networked control systems (ANCS) that denote classes of problems where networks contain features of arbitration and allow flexibility in their design as well in the underlying controllers. We showed that an upper bound on the worst case delay between the sensor and actuator can be designed using tools from real time analysis. This upper bound in turn was used to design controllers that are delay-aware, for a range of delays and sampling periods. The overall co-design of platform and control was evaluated for the control of multiple UAVs and their implementation using a distributed implemented platform using simulation studies. The results showed improved performance compared to a classical control design that assumed that the implementation delays were negligible. These results show that there are a number of significant opportunities in ANCS for the development of novel tools for analysis and synthesis of cyber physical systems that arise in the context of embedded control systems.

References

1. Seto, D., Lehoczky, J., Sha, L., Shin, K.: On task schedulability in real-time control systems. In: 17th IEEE Real-Time Systems Symposium, pp. 13–21 (December 1996)
2. Tabuada, P.: Event-triggered real-time scheduling of stabilizing control tasks. IEEE Transactions on Automatic Control 52(9), 1680–1685 (2007)
3. Wang, X., Lemmon, M.D.: State based self-triggered feedback control systems with L2 stability. In: 17th IFAC World Congress (2008)
4. Bini, E., Cervin, A.: Delay-aware period assignment in control systems. In: Real-Time Systems Symposium, RTSS 2008, November 30- December 3, pp. 291–300 (2008)
5. Alur, R., Weiss, G.: Regular specifications of resource requirements for embedded control software. In: IEEE Real-Time and Embedded Technology and Applications Symposium, RTAS 2008, pp. 159–168 (April 2008)
6. Palopoli, L., Pinello, C., Bicchi, A., Sangiovanni-Vincentelli, A.: Maximizing the stability radius of a set of systems under real-time scheduling constraints. IEEE Transactions on Automatic Control 50(11), 1790–1795 (2005)
7. Mazo, M., Tabuada, P.: On event-triggered and self-triggered control over sensor/actuator networks. In: 47th IEEE Conference on Decision and Control, pp. 435–440 (December 2008)
8. Henningsson, T., Johannesson, E., Cervin, A.: Sporadic event-based control of first-order linear stochastic systems. Automatica 44(11), 2890–2895 (2008)
9. Weiss, G., Alur, R.: Automata based interfaces for control and scheduling. In: Bemporad, A., Bicchi, A., Buttazzo, G. (eds.) HSCC 2007. LNCS, vol. 4416, pp. 601–613. Springer, Heidelberg (2007)
10. Samii, S., Cervin, A., Eles, P., Peng, Z.: Integrated scheduling and synthesis of control applications on distributed embedded systems. In: DATE (2009)
11. Cervin, A., Eker, J.: Control-scheduling codesign of real-time systems: The control server approach. Journal of Embedded Computing 1(2), 209–224 (2005)

12. Xia, F., Sun, Y.: Control-scheduling codesign: A perspective on integrating control and computing. Dynamics of Continuous, Discrete and Impulsive Systems - Series B 13(S1), 1352–1358 (2006)
13. Naghshtabrizi, P., Hespanha, J.: Analysis of distributed control systems with shared communication and computation resources. In: American Control Conference, pp. 3384–3389 (June 2009)
14. Chakraborty, S., Künzli, S., Thiele, L.: A general framework for analysing system properties in platform-based embedded system designs. In: DATE (2003)
15. Aström, K., Wittenmark, B.: Computer-controlled systems: theory and design, vol. 3. Prentice-Hall, Englewood Cliffs (1984)
16. Bhave, A., Krogh, B.: Performance bounds on state-feedback controllers with network delay. In: 47th IEEE Conference on Decision and Control, pp. 4608–4613 (2008)
17. Annaswamy, A., Chakraborty, S., Soudbakhsh, D., Goswami, D., Voit, H.: The arbitrated networked control systems approach to designing cyber-physical systems. In: Proc. of the 3rd IFAC Workshop on Distributed Estimation and Control in Networked Systems (2012)
18. Kumar, P., Goswami, D., Chakraborty, S., Annaswamy, A., Lampka, K., Thiele, L.: A hybrid approach to cyber-physical systems verification. In: DAC (2012)
19. Soudbakhsh, D., Phan, L., Sokolsky, O., Lee, I., Annaswamy, A.: Co-design of control and platform with dropped signals. In: The 4th ACM/IEEE International Conference on Cyber-Physical Systems, ICCPS 2013 (April 2013)
20. Yu, M., Wang, L., Chu, T., Xie, G.: Stabilization of networked control systems with data packet dropout and network delays via switching system approach. In: 43rd IEEE Conference on Decision and Control, vol. 4, pp. 3539–3544. IEEE (2004)
21. Dydek, Z.T., Annaswamy, A.M., Lavretsky, E.: Adaptive control of quadrotor UAVs: A design trade study with flight evaluations. IEEE Transactions on Control Systems Technology PP(99), 1 (2012)

Mean Square Limitations of Spatially Invariant Networked Systems

Nicola Elia, Jing Wang, and Xu Ma*

Dept. of Electrical and Computer Engineering,
Iowa State University, U.S.A.
{nelia,maxu}@iastate.edu, jingwang@alumni.iastate.edu

Abstract. In this chapter, we summarize some of our recent results on spatially invariant networked systems. We study the effects of unreliable communication on these systems' stability and performance. Because of their special structure, the quantities that characterize the limitations of these systems can be computed more easily. In particular, we focus on Mean Square stability and performance, and investigate network architectures, which are more robust and better performing. We consider multi-agent networked systems where the communication links are unreliable and stochastically dropout. Spatial invariance leads to a simplified computation of the MS stability limitation, and allows us to derive an uncertainty conservation law enjoyed by such systems. We then focus on distributed averaging systems, for which the loss of Mean Square stability leads to the emergence of certain complex behavior related to Lévy flights. We present closed form formulae characterizing Mean Square stability and performance in the presence of unreliable communication among the nodes. Finally, we study Mean square performance in the presence of unreliable links. Our results allow to characterize the interplay between Mean Square performance and stability of torus networks of different dimensions.

Keywords: Mean Square stability, Mean Square performance, spatially invariant systems, packet-drop networks, complex systems.

1 Introduction

Many applications of cyber-physical systems involve multi-agent systems connected over a communication network that need to accomplish certain tasks cooperatively. Typical tasks require the ability for the agent to cooperatively compute. In recent years, interesting distributed computing systems based on a dynamical system view have been proposed, [1,2,3,4,5]. These early results often depend on the communication network being perfect. More generally, it is important to understand the effects of limited or unreliable communication on the computational capabilities and on the dynamic properties, of such networked systems. In this chapter, we summarize some of our recent results on spatially invariant networked systems. In particular, we study the effects of

* This research has been partially supported by NSF grants ECS-0524689, ECS-0901846 and ECS-1239319.

D.C. Tarraf (ed.), *Control of Cyber-Physical Systems*,
Lecture Notes in Control and Information Sciences 449,
DOI: 10.1007/978-3-319-01159-2_19, © Springer International Publishing Switzerland 2013

unreliable communication on these systems' stability and performance. We first review the main MS stability result that we use throughout the paper. We then present a condition to check MS performance for networked systems by extending the result of [31]. Analogously to the certain robust performance problems in robust control, which are equivalent to augmented robust stability problems, the MS performance problem is equivalent to a MS stability of an augmented system.

Although checking MS stability has polynomial time complexity, it can become cumbersome when the network is large. For spatially invariant systems, the MS stability condition can be more easily computed [26,29] and we can obtain more insights. We show that a spatially invariant networked system with SISO agents has a maximal amount of stochastic uncertainty it can tolerate before losing the MS stability. Moreover, the uncertainty budget can be linearly distributed among the random links.

We then turn our attention to the analysis of a distributed averaging system, when the communication links are unreliable. In our recent work, [27], we have shown that simple networked distributed averaging systems can produce complex behavior when subject to communication intermittency or fading and additive noise. The loss of Mean Square (MS) stability triggers for the emergence of Lévy flight like random process in the nodes' behavior and destroys the ability of the system to compute averages. In the spatial invariant case, we can derive closed form formulae for the cases of zero and large but fixed delays in the inter-agent communication and for large number of agents [29]. The derivations are made possible by the frequency domain characterization of the MS stability condition. The state-space conditions are of limited use when the dimensions of the system or the communication delays become large. In the limit of large delays, we find that networks with only one neighbor are the most robust. Moreover the larger the number of neighbors the more fragile the network system is to the combine effects of noise and channel fading. From our analysis, it follows that packet-drop networks are more robust for poor quality of service. This indicates that gossiping has a stabilizing and robustifying effect, although at the expenses of the convergence rate.

These results appear to contradict the result of [33], which however considers only additive noise effects but otherwise perfect channels. In this setting, spatially invariant torus networks have increasingly better performance with larger dimension. In particular, the 1-d torus has performance becoming linearly worse with the size of the network; for a 2-d torus the performance degrades logarithmically with the network size; finally for dimension larger than two, the performance degradation is instead bounded with respect to the network size. To study the performance in the presence random links, we use our MS performance result. We derive computable formulae characterizing the steady state covariance of the deviation from the collective average, and the MS stability of the averaging system. It turns out that the results of [33] are valid when the systems are all in MS stable regimes. However, when in settings that can lead to MS instability, the 3-d torus is the least resilient and becomes MS unstable before a 2-d torus. Similarly, the 2-d torus becomes unstable before the 1-d torus, which is the most resilient although the least performing. Thus, the result points to a tradeoff between robustness and performance.

The chapter is organized as follows. In Section 2, we introduce basic notation and terminology. In Section 3, we introduce the networked systems model we consider in the chapter. In Section 4, we recall the results on MS stability and performance of networked systems over fading networks. In Section 5, we present an uncertainty conservation formula for a class of spatially invariant systems. The result says that there is a maximal amount of uncertainty that can be tolerated by the system no matter how the uncertainty is distributed over the fading links. In Section 6, we consider distributed averaging systems, and report several results on spatially invariant robust network organizations. Finally, in Section 7, we apply the our methodology to evaluate MS performance in the presence of unreliable communication links to the spatially invariant averaging systems and obtain a closed-form formula.

2 Notation and Preliminaries

2.1 Basic Graph Theory Notation

A *directed* graph is represented as $\mathcal{G} := (\mathcal{V}, \mathcal{E})$, where $\mathcal{V} = \{1, \ldots, n\}$ is the set of nodes and $\mathcal{E} \subseteq \mathcal{V} \times \mathcal{V}$ is the set of edges. An edge of $\mathcal{G}$ is denoted by (i, j) implying there exists a directed link from node j to node i.

The first vertex i in the edge (i, j) is called its *tail* and the second vertex j is its *head*. Given a directed graph $\mathcal{G} = (\mathcal{V}, \mathcal{E})$, define the adjacency matrix $\mathcal{A}(\mathcal{G})$ to be a binary matrix such that

$$[\mathcal{A}(\mathcal{G})]_{ij} = 1 \text{ if } (j, i) \in \mathcal{E}, \ = 0 \text{ otherwise.} \tag{1}$$

Define directed neighborhoods around each node i, the in-neighbors $\mathcal{N}_i^- = \{j | (j, i) \in \mathcal{E}\}$ and the out-neighbors $\mathcal{N}_i^+ = \{j | (i, j) \in \mathcal{E}\}$, which are the sets of nodes that have edges to and from node i. The in-degree of each node is defined as $d_{\text{in}}^i = |\mathcal{N}_i^-|$ and the out-degree of each node is defined as $d_{\text{out}}^i = |\mathcal{N}_i^+|$. We denote the graph Laplacian matrix of $\mathcal{G}$ as $\mathcal{L} = -\mathcal{D} + \mathcal{A}(\mathcal{G})$, where $\mathcal{D} = \mathbf{Diag}[d_{\text{in}}^i]$. Thus $\mathcal{L}\mathbf{1} = 0$, where $\mathbf{1}$ is the vector consisting of all ones. The Laplacian is said to be *balanced* if $d_{\text{in}}^i = d_{\text{out}}^i$ for all $i \in \mathcal{V}$. Therefore, the balanced Laplacian has the property that $\mathbf{1}'\mathcal{L} = 0$. The incidence matrix $\mathcal{I}n$ of a directed graph $\mathcal{G}$ is a $n \times q$ matrix where n and q are the number of vertices and edges respectively, such that $\mathcal{I}n_{ik} = 1$ if the edge $k = (*, i)$ leaves vertex i, $\mathcal{I}n_{ik} = -1$ if edge $k = (i, *)$ enters vertex i and 0 otherwise. A graph is undirected if for any $(i, j) \in \mathcal{E}$, the edge $(j, i) \in \mathcal{E}$. The graph is said to be *strongly connected* if every node can reach every other node through a sequence of elements in $\mathcal{E}$. If the graph is strongly connected, $\mathcal{L}$ has only one eigenvalue at zero and all other eigenvalues are in open right half plane, and the left eigenvector associated with the zero eigenvalue has its components all positive.

2.2 Circulant Matrices

Given an n-dimensional real vector $c = [c_0, \ldots, c_{n-1}]$, the associated circulant matrix is denoted by $\text{Cir}(c)$ and is given by

$$\text{Cir}(c) = \begin{bmatrix} c_0 & c_1 & \cdots & c_{n-1} \\ c_{n-1} & c_0 & \cdots & c_{n-2} \\ \vdots & \ddots & \ddots & \vdots \\ c_1 & \cdots & c_{n-1} & c_0 \end{bmatrix}.$$

Proposition 1. *Circulant matrices have the following properties [12]:*

1 *The eigenvalues of the above circulant matrix are given by* $\lambda_k = \sum_{i=0}^{n-1} c_i \rho_k^i$ *for* $k = 1, 2, \ldots, n$, *where* $\rho_k = e^{j2\pi k/n}$ *is one of the* n *roots of* -1.

2 *As* $n \to \infty$ *under the assumption* $\sum_{i=0}^{\infty} |c_i| < \infty$, *the eigenvalues of the matrix are given by the discrete time fourier transform of its first row,* i.e., $\lambda(\omega) = \sum_{i=0}^{\infty} c_i e^{ji\omega}$.

3 *The* $n\times n$ *Fourier matrix* V, *whose* k*-th column is* $v_k = n^{-1/2}.[1, \rho_k, \rho_k^2, \ldots, \rho_k^{n-1}]'$ *diagonalizes any circulant matrix of size* n. *Furthermore,* $VV^* = V^*V = I$ *and* $\det(V) = 1$, *where* $\det(\cdot)$ *denotes the determinant of a square matrix.*

$\mathrm{Cir}(B_1, \ldots, B_m)$ is a generalized circulant matrix if it is a block-circulant matrix, whose blocks are (block-)circulant. Generalized circulant matrices can be diagonalized by multi-dimensional Fourier matrices V, $V^*V = I$. Like circulant matrices, the rows of a generalized circulant matrix are permutations of the first row.

3 Networked Systems

In this section, we describe systems that are built over communication networks.

Given a graph $\mathcal{G} = (\mathcal{V}, \mathcal{E})$ with n nodes, we associate a sub-system $\{P_i\}_{i\in\{1,\ldots,n\}}$ to each node. Each sub-system P_i is a Discrete-Time (DT) system with the following state-space description (assumed to be minimal)

$$P_i: \begin{bmatrix} x_i(k+1) \\ y_i(k) \\ \eta_i(k) \end{bmatrix} = \begin{bmatrix} A_i & B_i^u & B_i^\nu \\ C_i^y & D_i^{zu} & D_i^{y\nu} \\ C_i^\eta & D_i^{\eta u} & 0 \end{bmatrix} \begin{bmatrix} x_i(k) \\ u_i(k) \\ \nu_i(k) \end{bmatrix} \tag{2}$$

where $x_i(k)$ is the local state, $u_i(k)$ is the vector of local inputs, $y_i(k)$ represents the local outputs, while $\nu_i(k)$ and $\eta_i(k)$ represent the signals received from the network and those sent over the network, respectively. They correspond to the overall set of messages received and transmitted by P_i.

Define $P = \mathbf{Diag}[P_i]_{i=1:n}$ as a system with a state-space representation given by

$$\begin{bmatrix} x(k+1) \\ y(k) \\ \eta(k) \end{bmatrix} = \begin{bmatrix} \hat{A} & \hat{B}_u & \hat{B}_\nu \\ \hat{C}_y & \hat{D}_{yu} & \hat{D}_{y\nu} \\ \hat{C}_\eta & \hat{D}_{\eta u} & 0 \end{bmatrix} \begin{bmatrix} x(k) \\ u(k) \\ \nu(k) \end{bmatrix} \tag{3}$$

where, all the matrices are block diagonal consistent with the dimension of the corresponding state-space matrices of P_i's. The state, input and output vectors are given by $x(k)$, $u(k)$, $y(k)$ $\eta(k)$, and $\nu(k)$, which are the stacking of the corresponding local signals of each agent.

3.1 Systems over Noiseless and Delay Free Networks

The case of perfect, delay-free, network is important to study limitations of networked systems induced solely by the network topology. In this case, the network is a static relation between η and ν: $\nu = L_{\nu\eta}\eta$, where $L_{\nu\eta}$ partitioned consistently with P and structured according to $\mathcal{A}(\mathcal{G}) + I$.

From the above notation, a networked system is obtained by the feedback interconnection of P with $L_{\nu\eta}$, as shown in Fig. 1.

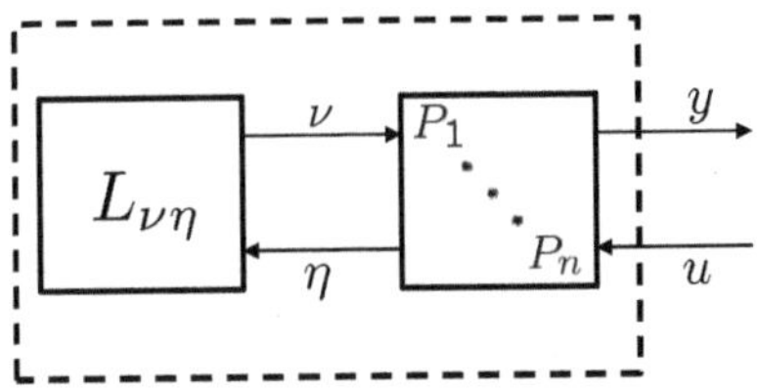

Fig. 1. The networked system over a noiseless and delay-free network

3.2 Fading Networks

More generally, the network is unreliable, subject to noise and intermittency or fading. To model systems over such networks, we introduce the following definitions adapted from [18]. The idea is to describe the random variables, which are part of the link model as the source of uncertainty in an otherwise purely deterministic model, the Mean Network. This view fits naturally within the traditional robust control framework, with a deterministic model in feedback with an uncertainty parameterized by the variance of the random variables.

Definition 1. *An analog Fading Network, N_F, is composed of two parts:*

1. *The Mean Network N.*
2. *The stochastic perturbation Δ.*

The Mean Network is a deterministic LTI system described by the following state space realization:

$$N: \quad \begin{bmatrix} \zeta^+ \\ \nu \\ z_\Delta \end{bmatrix} = \left[\begin{array}{c|c|c|c} G_\zeta & G_\upsilon & G_\eta & G_{w_\Delta} \\ \hline H_\nu & L_{\nu\upsilon} & L_{\nu\eta} & L_{\nu w_\Delta} \\ \hline H_{z_\Delta} & L_{z_\Delta \upsilon} & L_{z_\Delta \eta} & L_{z_\Delta w_\Delta} \end{array}\right] \begin{bmatrix} \xi \\ \upsilon \\ \eta \\ w_\Delta \end{bmatrix}, \tag{4}$$

where ζ is the state of the network, η is the network input vector coming from P, ν is the network output vector going to P. $w_\Delta \in \mathbb{R}^p$ and $z_\Delta \in \mathbb{R}^p$.
Δ maps $z_\Delta \to w_\Delta$ and is defined as $\Delta = \mathbf{Diag}[\Delta_i]_{i=1:p}$. For each $i = 1, \dots, p$, $\Delta_i(k)$ are IID random variables with

$$\mathbf{E}\{\Delta_i(k)\} = 0 \textit{ and } \mathbf{E}\{(\Delta_i(k))^2\} = \sigma_i^2 \ \forall k \geq 0.$$

Moreover, $\Delta_1(k), \dots, \Delta_p(k)$ are independent for each k, although not necessarily identically distributed. Δ acts as a multiplication operator on z to provide w; i.e., $w_{\Delta,i}(k) = \Delta_i(k) z_{\Delta,i}(k)$ for $i = 1, \dots, p$, $\forall k \geq 0$. Finally, υ is a vector of white noise signals independent from each other and independent from Δ.

Several fading channel models can be represented in the framework. In particular, a simple model of a packet-drop link can be represented as a Bernoulli switch $\xi(k)$ (neglecting quantization effects). $\xi(k)$ can be in turn re-parameterized as $\mu + \Delta(k)$, where

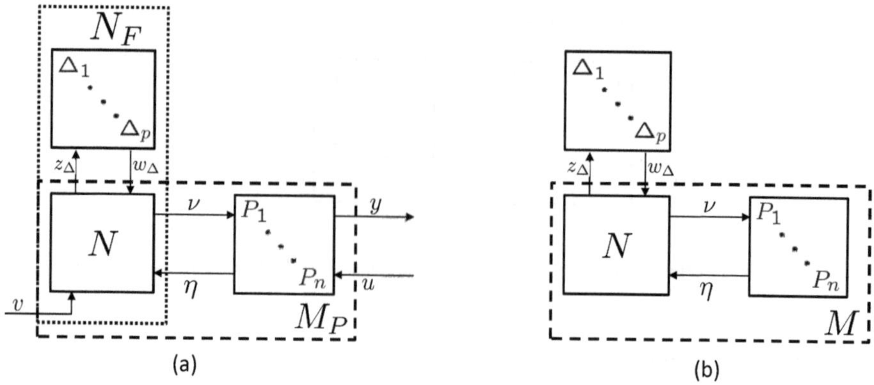

Fig. 2. Networked system over a Fading Network

μ is the mean of $\xi(k)$ and $\Delta(k)$ is a zero mean random variable with variance equal to the variance of $\xi(k)$. This extreme link fading model can also capture the randomness of network connections that is of interest in the networked control systems literature.

When there is no uncertainty and additive noise, then the fading network model reduces to the noiseless and delay-free network of the previous section, with $N = L_{\nu\eta}$ partitioned consistently with P and structured according to $\mathcal{A}(\mathcal{G}) + I$.

4 Systems over Fading Networks

A networked system is then given by the interconnection of the nodes, P, with the Fading Network as described in Figure 2(a).

Once the random perturbation Δ has been identified, it can be separated from the rest of the networked system.

Define

$$M_P: \quad \begin{array}{ll} \chi^+ &= \mathcal{A}\chi + \mathcal{B}w_\Delta + \mathcal{B}_w w \\ z_\Delta &= \mathcal{C}\chi + \mathcal{D}w_\Delta + \mathcal{D}_{z_\Delta w} w \\ z &= \mathcal{C}_z\chi + \mathcal{D}_{zw_\Delta} w_\Delta + \mathcal{D}_{zw} w \end{array} . \tag{5}$$

w denotes all the exogenous inputs to the networked system (u, v), and z denoted all the outputs of the networked system (y and possibly others form the network). w_Δ and z_Δ are respectively the input and output from and to Δ.

M_P represents the networked system with the exclusion of Δ. One important component of M_P is the system in feedback with Δ, which we denote by M (see Figure 2(b)).

M is the closed-loop system from w_Δ to z_Δ and it has the following state-space equations

$$M: \quad \begin{array}{ll} \chi^+ &= \mathcal{A}\chi + \mathcal{B}w_\Delta \\ z_\Delta &= \mathcal{C}\chi + \mathcal{D}w_\Delta \end{array} . \tag{6}$$

The linear time-invariant discrete-time system M has p-inputs and p-outputs and is in feedback with the diagonal uncertainty Δ described in Definition 1.

4.1 Mean Square Closed-Loop Stability

Let $H = \mathrm{F}_l(\Delta, M_P)$ denote the feedback interconnection of Δ and M_P, with $\chi_0 = \chi(0)$ independent from Δ, and w. w is a vector of IID random variables zero mean and variance equal to I, w.l.o.g.

$$H: \quad \begin{array}{ll} \chi^+ & = \mathcal{A}\chi + \mathcal{B}w_\Delta + \mathcal{B}_w w \\ z_\Delta & = \mathcal{C}\chi + \mathcal{D}w_\Delta + \mathcal{D}_{z_\Delta w} w \\ z & = \mathcal{C}_z\chi + \mathcal{D}_{zw_\Delta} w_\Delta + \mathcal{D}_{zw} w \\ w_\Delta & = \Delta z_\Delta \end{array} . \tag{7}$$

Assume that the feedback interconnection of Δ and M_P is well posed, namely that the solution to system (7) exists for any realization of Δ, and let $Q(k) = \mathbf{E}\{\chi(k)\chi(k)'\}$.

Definition 2. *System (7) is Mean Square stable if*
$\lim_{k\to\infty} Q(k) < \infty$, *if* $w(k) \neq 0$
$\lim_{k\to\infty} Q(k) = 0$, *if* $w(k) = 0, \forall k > 0$.
Otherwise the system is MS unstable.

The MS stability of H can be checked in terms of M and the variance of Δ.

Theorem 1. *[18] Assume that $M = \mathrm{F}_l(P, N) = (\mathcal{A}, \mathcal{B}, \mathcal{C}, \mathcal{D})$ is stable and that $\mathcal{D}$ is either strictly lower triangular or strictly upper triangular. Let $\Sigma^2 = \mathbf{Diag}[\sigma_i^2]_{i=1:p}$. H in (7) is Mean Square Stable if and only if $\rho\left(\Sigma^2 \hat{M}\right) < 1$, where $\rho(\cdot)$ denotes the spectral radius and* $\hat{M} = \begin{bmatrix} \|M_{11}\|_2^2 & \cdots & \|M_{1p}\|_2^2 \\ \vdots & \cdots & \vdots \\ \|M_{p1}\|_2^2 & \cdots & \|M_{pp}\|_2^2 \end{bmatrix}$.

See [18] for a more extended version of the theorem and for references to related work. Although rooted into the rich literature on stochastic systems [20,21,22,23], the approach provides new insights on the role of the channel generated uncertainty in affecting the stability of the closed loop.

4.2 Mean Square Stability and Performance

So far, we have considered the problem of checking that whether the networked system is MS stable in the presence of unreliable communication links. In this section, we study their effect on the system's performance. We assume that M_P is strictly proper for simplicity.

$$M_P: \quad \begin{array}{ll} \chi^+ & = \mathcal{A}\chi + \mathcal{B}w_\Delta + \mathcal{B}_w w \\ z_\Delta & = \mathcal{C}\chi \\ z & = \mathcal{C}_z\chi \end{array} . \tag{8}$$

Definition 3. *(Mean Square Performance) $H = \mathrm{F}_l(\Delta, M_P)$ is said to have Mean Square Performance less than γ^2, from w to z, if H is MS stable and*
$\lim_{k\to\infty} \mathbf{Tr}\{\mathbf{E}\{z(k)z(k)'\}\} < \gamma^2$.

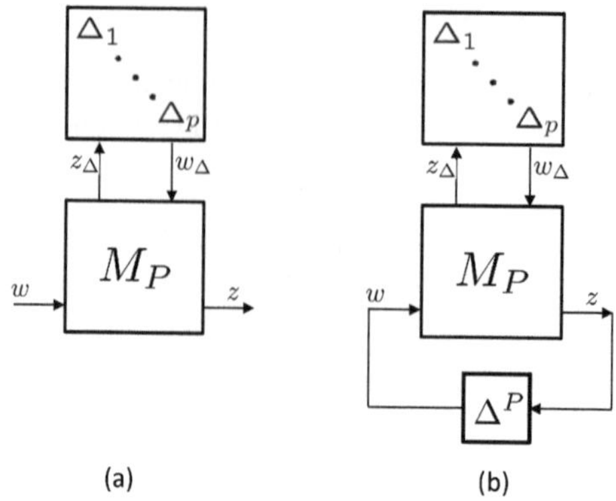

Fig. 3. (a) Robust Performance setup. (b) Equivalent Robust Stability problem.

4.3 Equivalence between Mean Square Stability and Mean Square Performance

Define

$$\hat{M}_P = \begin{bmatrix} \|M_{11}\|_2^2 \cdots \|M_{1p}\|_2^2 & \|M_{1w}\|_2^2 \\ \vdots \quad \cdots \quad \vdots & \vdots \\ \|M_{p1}\|_2^2 \cdots \|M_{pp}\|_2^2 & \|M_{pw}\|_2^2 \\ \|M_{z1}\|_2^2 \cdots \|M_{zp}\|_2^2 & \|M_{zw}\|_2^2 \end{bmatrix} = \begin{bmatrix} \hat{M} & \hat{M}_w \\ \hat{M}_z & \hat{M}_{zw} \end{bmatrix}$$

and

$$\tilde{\Sigma}^2 = \begin{bmatrix} \Sigma^2 & 0 \\ 0 & \frac{1}{\gamma^2} \end{bmatrix}.$$

Note that $\hat{M}_P$ includes $\hat{M}$ plus the vectors of $\mathcal{H}_2$ norm square of the transfer functions from w to z_Δ and z and form w_Δ to z. These add one extra column and row to$\hat{M}$. In particular, $\|M_{zw}\|_2^2$ is the performance of the nominal mean system.

Theorem 2. *[31,34] Under the current assumptions, H has Mean Square Performance strictly less than γ^2, if and only if $\rho(\tilde{\Sigma}^2\hat{M}_P) < 1$.*

Remark 1. *$\rho(\tilde{\Sigma}^2\hat{M}_P) < 1$ is equivalent to the Mean Square stability of the system in Figure 3(b) where an extra stochastic perturbation Δ^P has been added. Δ^P is a full random matrix for each k with independent elements. Each element has zero mean and variance $\frac{1}{\gamma^2}$ and is independent from $\Delta_i(k)$, $i = 1, ..., p$, $k > 0$.*

The condition of Theorem 1 (and thus of Theorem 2) for checking MS stability becomes computationally cumbersome when the system has large size and many fading links, although it is equivalent to solving a convex problem. Moreover, it is not clear as of yet, if it can be computed in a distributed fashion, allowing each agent to assess it. This motivates us to look for special network architectures where the condition is easier to check or to guarantee. In particular we consider spatially invariant networked systems.

5 Uncertainty Conservation Law for Spatially Invariant Networked Systems

In [30], we have derived Bode integral limitation of spatially invariant networked systems, which describes performance limitations when the communication links are ideal. In this section, we show that there is a maximum amount of uncertainty tolerable by a spatially invariant system with random links, and that this uncertainty can be linearly distributed over the various links.

Consider a set of n (SISO) agents in feedback over a fading network. Let $P = \mathbf{Diag}\,[P_i]_{1:n}$, where P_i represents the agent's dynamics with state-space model $P_i = P_A = (A, b, c, 0)$, i.e., all the agents have the same dynamics and are assumed to be stable for simplicity.

P sends η to the fading network N_F and receives ν from it at each k. For simplicity we restrict our presentation to systems with only one discrete spatial dimension, [13]. The results extend to the multi-spatial dimensional case naturally.

Assumption 1. *$\mathcal{A}(\mathcal{G})$ is circulant.*

This assumption implies that the neighborhood of each node has the same structure, thus each $\mathcal{N}_i^-$ contains the same number of vertices, d. We can order the vertices in $\mathcal{N}_1^-$, with $j = 1, \ldots, d$, where j represents the jth neighbor of node 1. Due to the spatial invariance, we can use the same ordering to identify the corresponding neighbors of node i, in $\mathcal{N}_i^-$. Thus, $j = 1$ corresponds to the first neighbor for each node.

For each node i, $i = 1, \ldots, n$, we have that

$$\begin{aligned} x_i(k+1) &= Ax_i(k) + b\nu_i(k) \\ \eta_i(k) &= cx_i(k) \end{aligned}$$

$$\nu_i(k) = \sum_{j=1}^{d} \xi_{i\mathcal{N}_i^-(j)}(k)\eta_{\mathcal{N}_i^-(j)}(k)$$

or

$$\begin{aligned} x_i(k+1) &= Ax_i(k) + b\sum_{j=1}^{d} \xi_{i\mathcal{N}_i^-(j)}(k)y_{\mathcal{N}_i^-(j)}(k) \\ y_i(k) &= cx_i(k). \end{aligned}$$

We potentially allow the gain $\xi_{i\mathcal{N}_i^-(j)}(k)$ to randomly vary with time and location i and $\mathcal{N}_i^-(j)$ Let $\mathcal{J} \subseteq \{1, \ldots, d\}$ denote a subset with cardinality $\kappa \leq d$ corresponding to a set of random gains.

Assumption 2. *For $j \notin \mathcal{J}$, $\xi_{i\mathcal{N}_i^-(j)}(k) = \mu_j$ for all $k \geq 0$ and all $i = 1, \ldots n$. For $j \in \mathcal{J}$, $\xi_{i\mathcal{N}_i^-(j)}(k)$ is a random variable IID in time k, independent in $i = 1, \ldots, n$ and in $j \in \mathcal{J}$, with mean μ_j and variance σ_j^2.*

With this assumption, the networked systems has a spatially invariant Mean Network with spatially invariant uncertainty structure. This is an important class of systems for which the testing of the Mean Square stability scales linearly with the number of nodes. Note that the actual network interconnection is not spatially invariant for each k. For example, consider the case where $d = n$ for simplicity of exposition. Then

$$L_{\nu\eta} = \text{Cir}([0, \mu_2, \ldots, \mu_n]).$$

For $j \in \mathcal{J}$, let $\tilde{\Delta}_{ij}(k) = \xi_{i\mathcal{N}_i^-(j)}(k) - \mu_j$, for $i = 1, \ldots, n$. $\tilde{\Delta}_{ij}(k)$ is zero mean and has variance σ_j^2. Let $\tilde{\Delta}_j = \textbf{Diag}\{\Delta_{1j}, \cdots, \Delta_{n,j}\}$, and $\tilde{\Delta} = \textbf{Diag}\{\tilde{\Delta}_1, \cdots, \tilde{\Delta}_\kappa\}$ where κ is the cardinality of $\mathcal{J}$. Finally, let $\Pi_j = \text{Cir}([0, \ldots, 0, 1, 0, \ldots 0])$ with the element at location $\mathcal{N}_1^-(j)$ equal to 1. Note that Π_j is a permutation matrix. Then

$$N_F(k) = L_{\nu\eta} + \sum_{j=1}^{\kappa} \Delta_j(k)\sigma_j \Pi_j$$

where $\Delta_j(k)$ is the re-scaling of $\tilde{\Delta}_j(k)$ so that $\textbf{var}(\Delta_{ij}(k)) = 1$. Then the Mean Network has then the following structure

$$N: \quad \begin{bmatrix} \nu \\ z_\Delta \end{bmatrix} = \left[\begin{array}{c|c} L_{\nu\eta} & L_{\nu w_\Delta} \\ \hline L_{z_\Delta \eta} & 0 \end{array}\right] \begin{bmatrix} \eta \\ w_\Delta \end{bmatrix} \tag{9}$$

with

$$L_{\nu w_\Delta} = \underbrace{[I, \ldots, I]}_{\kappa}$$

where I is the $n \times n$ identity matrix and

$$L_{z_\Delta \eta} = \begin{bmatrix} \sigma_1 \Pi_1 \\ \vdots \\ \sigma_\kappa \Pi_\kappa \end{bmatrix}.$$

Note that N has a block circulant structure. We leave to the reader to verify that $M(z)$ is given by the following expression,

$$M(z) = L_{z_\Delta \eta} P(z)(I - L_{\nu\eta} P(z))^{-1} L_{\nu w_\Delta} = L_{z_\Delta \eta} \tilde{M}(z) L_{\nu w_\Delta}.$$

Lemma 1. *Assume M is stable, then*

$$\rho(\hat{M}) = \frac{1}{n\kappa} \|M\|_2^2.$$

Proof. It is easy to see that $\tilde{M}(z)$ circulant transfer function matrix and each block of $M(z)$ is obtained from $\tilde{M}(z)$ by rows permutation and scaling by σ_i. Thus each column block of $M(z)$ has the same set of transfer matrices where the set is specified as $\{\sigma_1 \Pi_1(z)\tilde{M}(z), \ldots, \sigma_\kappa \Pi_\kappa \tilde{M}(z)\}$. Since Π_i and $\tilde{M}(z)$ are circulant with consistent dimensions, their product is also circulant. Therefore, each column of $M(z)$ has the same set of transfer functions. This implies that each column of $\hat{M}$ has the same column sum. From Theorem 8.1.22 in [25], it follows that $\rho(\hat{M})$ is equal to $\sum_{i=1}^{n\kappa} \|M_{ij}\|_2^2$. However, $\|M\|_2^2 = \sum_i^{n\kappa} \sum_{j=1}^{n\kappa} \|M_{ij}\|_2^2$ and the result follows.

Theorem 3. *Assume M is stable. Then the networked system is MS stable if and only if*

$$\sum_{j=1}^{\kappa} \sigma_j^2 < \frac{1}{\frac{1}{n}\sum_{i=1}^{n}\left\|\frac{P_A(z)}{I-\lambda_i(L_{\nu\eta})P_A(z)}\right\|_2^2}$$

where λ_i's are the eigenvalues of nominal mean network $L_{\nu\eta}$.

Proof.

$$\begin{aligned}
\|M\|_2^2 &= \frac{1}{2\pi}\int_0^{2\pi} \mathbf{Tr}\{M^*(e^{j\theta})M(e^{j\theta})\}d\theta \\
&= \frac{1}{2\pi}\int_0^{2\pi} \mathbf{Tr}\{L'_{\nu w_\Delta}\tilde{M}^*(e^{j\theta})L'_{z_\Delta \eta}L_{z_\Delta \eta}\tilde{M}(e^{j\theta})L_{\nu w_\Delta}\}d\theta \\
&= \left(\sum_{j=1}^{\kappa}\sigma_j^2\right)\frac{1}{2\pi}\int_0^{2\pi} \mathbf{Tr}\{L'_{\nu w_\Delta}\tilde{M}^*(e^{j\theta})\tilde{M}(e^{j\theta})L_{\nu w_\Delta}\}d\theta \\
&= \kappa\left(\sum_{j=1}^{\kappa}\sigma_j^2\right)\frac{1}{2\pi}\int_0^{2\pi} \mathbf{Tr}\{\tilde{M}^*(e^{j\theta})\tilde{M}(e^{j\theta})\}d\theta \\
&= \kappa\left(\sum_{j=1}^{\kappa}\sigma_j^2\right)\left\|\tilde{M}\right\|_2^2 .
\end{aligned} \tag{10}$$

Circulant matrices are diagonalized by the Fourier Matrix V. In particular $V'L_{\nu\eta}V = \Lambda_{\nu\eta}$. Also, since P has a diagonal transfer function, $V'P(z)V = P(z)$. Thus,

$$\tilde{M}(z) = V'P(z)(I-\Lambda_{\nu\eta}P(z))^{-1}V.$$

Therefore,

$$\begin{aligned}
\left\|\tilde{M}(z)\right\|_2^2 &= \left\|P(z)(I-\Lambda_{\nu\eta}P(z))^{-1}V\right\|_2^2 \\
&= \left\|P(z)(I-\Lambda_{\nu\eta}P(z))^{-1}\right\|_2^2 \\
&= \sum_{i=1}^{n}\left\|P_A(z)(I-\lambda_i(L_{\nu\eta})P_A(z))^{-1}\right\|_2^2 .
\end{aligned} \tag{11}$$

Finally, using the result of Lemma 1,

$$\rho(\hat{M}) = \left(\sum_{j=1}^{\kappa}\sigma_j^2\right)\frac{1}{n}\sum_{i=1}^{n}\left\|\frac{P_A(z)}{I-\lambda_i(L_{\nu\eta})P_A(z)}\right\|_2^2 .$$

Corollary 1. *In the limit of large n, if $\lim_{n\to\infty}\sum_{j=0}^{n}|\mu_j| < \infty$ and M is stable, then the networked system is MS stable if and only if*

$$\sum_{j=1}^{\kappa}\sigma_j^2 < \frac{1}{\frac{1}{4\pi^2}\int_0^{2\pi}\int_0^{2\pi}\left|\frac{P_A(e^{j\theta})}{I-\lambda(\omega)P_A(e^{j\theta})}\right|^2 d\theta d\omega}$$

where $\lambda(\omega) = \sum_{\ell=0}^{\infty}\mu_\ell e^{j\,\ell\omega}$.

Remark 2. *The result of the above theorem and corollary has an interesting interpretation in term of conservation of the total amount (variance) of the uncertainty that can be tolerated by a spatially invariant system. The conditions for Mean Square stability say that the there is a maximum amount of link unreliability/uncertainty that can be tolerated by the interconnection no matter how it is distributed along the spatial dimension. Links with large variance must be balanced by enough reliable links so that the sum stays within this bound. The robustness bound is given by the inverse of the $\mathcal{H}_2$ norm of the Mean system. This quantity is of interest in the literature [32] and here we are showing that it does have a well-defined robust stability measure interpretation.*

6 Distributed Averaging over Unreliable Channels

The fading network framework applies to the MS stability analysis of a simple but classical distributed averaging algorithm when the communication links are unreliable. Consider the classical consensus algorithm with the addition of noise, switching, and (fixed) delays, i.e.,

$$x_i(k+1) = x_i(k) + \beta\sum_{j\in\mathcal{N}_i^-}\xi_{ij}(k)[x_j(k-\tau_{ij}) - x_i(k)] + v_i(k), \quad (12)$$

where $\tau_{ij} \geq 0$ denotes the *time-invariant* and *bounded* delay in the channel e_{ij}, and $\xi_{ij}(k)'s$ are IID a Bernoulli random variables with probability μ_{ij} of being 1 characterizing the fading property of the channel at time k. $v_i(k)$'s (the communication noise) are Gaussian random variables independently identically distributed over both k and i with zero mean and unit variance.

Assumption 3. *In the rest of the paper all the graphs are assumed to be strongly connected.*

One important characteristic of many distributed averaging systems, including those in this chapter, is their lack of minimality. More precisely, the presence of marginally stable poles corresponding to eigenvalues on the boundary of the stability region that are unobservable and/or uncontrollable. For example, consider the system in (12) in the case of $\tau_{ij} = 0$ for simplicity. Assume that all $\xi_{ij}(k)$ independent but with the same mean μ and variance σ^2. The agents/nodes are discrete-time integrators (adders), $P_i = \frac{\beta}{z-1}$ $i = 1, \ldots, n$ and the Mean Network is given by

$$N: \quad \begin{bmatrix} \nu \\ z_\Delta \end{bmatrix} = \left[\begin{array}{c|c|c} I & -\mu L & B \\ \hline 0 & C & 0 \end{array}\right] \begin{bmatrix} \upsilon \\ \eta \\ w_\Delta \end{bmatrix}, \tag{13}$$

where L is the graph Laplacian with $0-1$ weights, C is the transpose of incidence matrix of the graph, and B is such that $L = BC$.

It follows that $M_P = \mathrm{F}_l(\mathbf{Diag}[P_A]_{1:N}, N)$ has state-space

$$\begin{aligned} x^+ &= (I - \beta\mu L)x + Bw_\Delta + \upsilon \\ z_\Delta &= Cx. \end{aligned}$$

The Laplacian of a strongly connected graph has one eigenvalue at zero with eigenvector $\mathbf{1}$, i.e., $L\mathbf{1} = 0$. One the other hand, $C\mathbf{1} = 0$ too. Therefore, M_P (or M) has one eigenvalue at 1 (marginally stable) with eigenvector $\mathbf{1}$, but this mode is not observable from C. Thus, the feedback interconnection of P and N is not detectable as it has one eigenvalues on the unit circle, (not strictly stable). The impulse response of M, and therefore its transfer function matrix $M(z)$, will not show the marginal internal instability. We see that the distributed averaging system cannot be (strictly) stable, thus Theorem 1 will not directly apply and the system is not MS stable.

However, since the mode at 1 in the direction $\mathbf{1}$ is not observable from Δ, we could study how Δ affect the MS stability of the rest of the modes which can be made stable by β small enough. A simple approach would be to obtain a minimal state-space representation for M and to study that. However, this will destroy the networked system structure and reduce our insights. Alternatively, we use a natural decomposition of the state. Let γ be the left eigenvector of $A = I - \beta\mu L$ associated with this eigenvalue, and by normalization such that $\gamma^T A = \gamma^T$ and $\gamma^T \mathbf{1} = 1$. It is natural to define operators

$$\mathcal{P} := I - \mathbf{1}\gamma', \;\; \mathcal{F} := \mathbf{1}\gamma'. \tag{14}$$

Let the *deviation state* and *conserved state* of the system be defined as $x_d(k) := \mathcal{P}x(k)$ and $x_c(k) := \mathcal{F}x(k)$ respectively, then their evolutions are governed by

$$x_d(k+1) = \mathcal{P}A\mathcal{P}x_d(k) + \mathcal{P}B\Delta(k)Cx_d(k) + \mathcal{P}\upsilon(k) \tag{15}$$

$$x_c(k+1) = x_c(k) + \mathcal{F}B\Delta(k)Cx_d(k) + \mathcal{F}\upsilon(k) \tag{16}$$

where we have used the facts that $\mathcal{P}A = A\mathcal{P}$ and $\mathcal{P}^2 = \mathcal{P}$, which are easy to verify from the definitions (14). In the sequel, we refer to (15) as the *deviation system* and (16) as the *conserved system*. Although x_c is in $\mathbb{R}^n$, it evolves on the subspace spanned by $\mathbf{1}$. Thus, essentially x_c is one-dimensional. For future reference, it is convenient to define any component of x_c as $\hat{x}_c$, or equivalently

$$\hat{x}_c(k) = \gamma' x_c(k). \tag{17}$$

In simple terms, $\hat{x}_c$ is the result of the integration (adder) of a process that depends on x_d through the coupling provided by $\Delta(k)$ and $\mathcal{F}B$. Instead of reducing the system M,

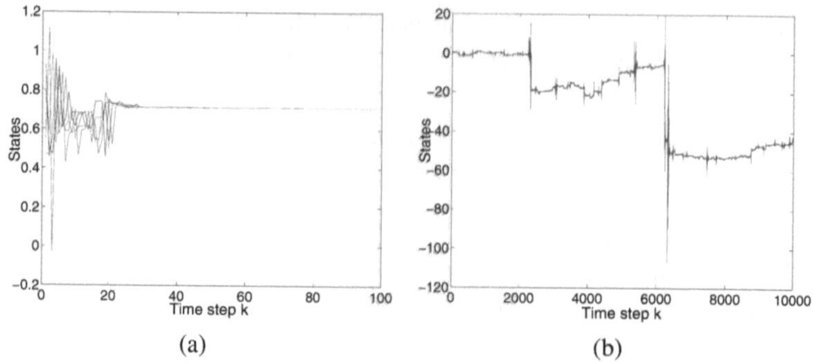

Fig. 4. (a) Convergence without additive noise. (b) Hypersensitivity to additive noise that can happen when the system is MS unstable.

with this decomposition we are augmenting it by one dimension preserving most of the structure of the system. In the deviation system, the eigenvalue at 1 is replaced with one at 0. This eigenvalue at 0 is still unobservable and does not impact the behavior of the network systems but, makes the deviation system stabilizable and stable for small enough β. The marginal internal instability is now captured in x_c by the conserved system.

6.1 Emergence of Lévy Flights

One can now apply Theorem 1 to the deviation system (15) and study how its MS (in)-stability impacts the behavior of the networked system. In [28,27], we have shown that the MS instability of the deviation system is responsible for the emergence of a complex behavior of the networked system when additive noise is present. In particular, we have characterized $\hat{x}_c$ as an hype-jump diffusion process and as a uncorrelated Lévy flight in a special two agents case. We refer the reader to the cited references for more details and in depth analysis. Thus, the MS instability of the deviation system is an indication of a fragility of the system to the combined effect of additive and multiplicative noises, which are typically both present in networked systems. While the state of system (12) may converges to a common value when noise in not present, although the deviation system is MS unstable (Figure 4(a)). The system is in a critical state (for appropriate values of the parameters) and is hypersensitive to additive noise. Figure 4(b) shows a collective behavior of the system when a small noise is added to the communications. There are abrupt jumps in the agents states, and between jumps, periods where the agents' states are reasonably close to each other. Lévy flights have been observed in many natural sciences as well as economics and many other fields, [7,8,9,10,11]. [27] provides the first example of a simple multi-agent model that exhibits such a collective diffusion behavior and shows that limited communication can have serious effects on the behavior of computing networked systems. However, we now have analysis tools to study and design such systems and to predict, prevent or mitigate the emergence of critical phenomena when we consider distributed optimization systems.

6.2 Robust Organizations

As we have already seen, spatially invariant architectures are amenable to a simplified analysis. Here we assume the following,

Assumption 4. *i) $\mathcal{A}(\mathcal{G})$ is circulant.*

ii) All the links fade with the same distribution. For Bernoulli fading each link has the same $0 < e < 1$, $\mu = (1 - e)$, and $\sigma_B^2 = e(1 - e)$. For Gaussian fading, each link is distributed as $N(\mu, \sigma_G^2)$ with $\mu = 1$. For simplicity, we use σ^2 to denote the variance of the fading, and σ^2 could be either σ_B^2 or σ_G^2.

iii) The delays are spatially invariant.

With these assumptions, System (12) and the Mean System become spatially invariant with the mean connectivity graph strongly connected.

As done in Section 5, the spatially invariant structure leads to simpler computation of the spectral radius condition, see [29] for details.

Because of the spatial invariance, each $\mathcal{N}_i^-$ contains the same number of vertices, d. We can order the vertices in $\mathcal{N}_1^-$, with $j = 1, \ldots, d$, where j represents the jth neighbor of node 1. Due to the spatial invariance, we can use the same ordering to identify the corresponding neighbors of node i, in $\mathcal{N}_i^-$. Thus, $j = 1$ corresponds to the first neighbor for each node.

We take the Z-transform of (12). Consequently, L and C now become functions of the complex variable z. We have

$$\begin{aligned} zX_i(z) &= X_i(z) + \beta\mu \sum_{j=1}^{d} [z^{-\tau_{ij}} X_{\mathcal{N}_i^-(j)}(z) - X_i(z)] \\ &\quad + \beta \sum_{j=1}^{d} W_{\Delta, i\mathcal{N}_i^-(j)}(z), \\ Z_{\Delta,ij}(z) &= [z^{-\tau_{ij}} X_{\mathcal{N}_i^-(j)}(z) - X_i(z)], \qquad \forall i \in \mathcal{V}. \end{aligned}$$

We order the inputs and outputs according to the sequence of nodes, and the sequence of neighbors, i.e., $\Delta_\ell = \Delta_{i,j}$ where $\ell = i + (j-1)n$, for $i = 1, \ldots, n$ and $j = 1, \ldots, d$. We can write the transformed equations compactly as

$$\begin{aligned} zX(z) &= (I - \beta\mu L(z))\, X(z) + \beta B W_\Delta(z), \\ Z_\Delta(z) &= -C(z)X(z). \end{aligned} \tag{18}$$

The structure of the matrices $C(z)$, B and $L(z)$ in (18) is specified as follows. $C(z)$ is the vertical stacking of d matrices $L^j(z)$ each generalized circulant. Each row, k, of L_j has only two nonzero elements, a 1 at location k, and $-z^{-\tau_j}$ at the index corresponding to the jth neighbor node in the graph. $B = [I, \ldots, I]$[1], and $L(z) = \sum_{j=1}^{d} L_j(z) = BC(z)$ is generalized circulant. Note that $B \in \mathbb{C}^{n \times p}$ and $C(z) \in \mathbb{C}^{p \times n}$, where $p = nd$. Define $H(z) := (z-1)I + \beta\mu L(z)$, note that as $L(z)$ is generalized circulant, $H(z)$ is

[1] Here, we separate the coupling gain β from B in writing (18).

also generalized circulant with circulant blocks. We can derive the transfer matrix from input z to output w (which corresponds to system G) as

$$M(z) = -\beta C(z)H^{-1}(z)B. \tag{19}$$

To see the structure of $M(z)$ more clearly, we expand it as follows

$$M(z) = -\beta \begin{bmatrix} L_1(z)H^{-1}(z) \cdots L_1(z)H^{-1}(z) \\ L_2(z)H^{-1}(z) \cdots L_2(z)H^{-1}(z) \\ \vdots \quad\quad \vdots \quad\quad \vdots \\ L_d(z)H^{-1}(z) \cdots L_d(z)H^{-1}(z) \end{bmatrix}. \tag{20}$$

As we use the condition of Theorem 1, we also need the following assumption, which holds for the rest of the chapter.

Assumption 5. *The Mean System M is input-output stable.*

Note that this assumption is not difficult to verify given state-space realization of M. The structure of M leads to an important simplification in the computation of $\rho(\hat{M})$.

Lemma 2. *Under the current assumptions*

$$\rho(\hat{M}) = \frac{1}{p}\,\|M\|_2^2\,,$$

where $p = nd$.

This follows the same line of proof as Lemma 1.

Corollary 2. *Suppose $\tau = 0$, then the deviation system (15) is MS stable if an only if*

$$\sigma^2\rho(\hat{M}) = \sigma^2\beta^2 \cdot \frac{1}{n}\sum_{k=2}^{n} \frac{2Re\{\lambda_k(L)\}}{1-|1-\beta\mu\lambda_k(L)|^2} < 1, \tag{21}$$

where $\lambda_k(L)$ is the k-th eigenvalue of the Laplacian matrix associated with the connectivity graph $\mathcal{G}$, $\lambda_1(L) = 0$, and $Re(\cdot)$ denotes the real part of a complex number.

6.3 Topology-Independent MS Stability Limitations for Large Delays

In this section, we summarize some results for the distributed averaging system (12) in the limit of large but uniform delays. The reader is referred to [29] for a more complete analysis. We summarize conclusions that are general and do not depend on specific network interconnections besides the in-degree, *d* [29].

The results in this section are under Assumptions 3, 4, 5 and uniform delays, τ, and are independent from the location of the eigenvalues of the graph.

Corollary 3. *In the limit of large delays $\tau \to \infty$,*

$$\frac{\sigma^2\beta}{\mu\left(1-\beta\mu\dfrac{d}{2}\right)} < \rho(\sigma^2\hat{M}) \le \frac{\sigma^2\beta}{\mu\,(1-\beta\mu d)} \tag{22}$$

and the upper bound is tight if and only if $d = 1$.

Note that both bounds are independent of the size of the network n, the specific connection among the nodes, but only affected by the number of in-neighbors d.

We can infer interesting properties of the model from those bounds.

Robustness of Single Neighbor Networks. The next result says that the MS stability index is constant, independent from the size of the network, in single neighbor topologies, in the limit of large delays.

Corollary 4. *In the limit of large delays $\tau \to \infty$,*

$$\sigma^2 \rho(\hat{M}) = \frac{\sigma^2 \beta}{\mu(1 - \beta\mu)}$$

for all the spatially invariant networks where each node has only one neighbor, $d = 1$, and any size $n \geq 2$.

It is natural to investigate the network topology most robust to link uncertainties when the inter-agent coupling gain β and QoS μ and σ^2 is fixed. The next result shows that having more than one in-neighbor increases the fragility of the consensus system under the same β, μ, σ^2.

Let $\rho(\sigma^2 \hat{M})(\beta, \mu, d, n)$ denote the dependence of the MS stability index, in the limit of $\tau \to \infty$, on the system parameters, where d denotes the number of neighbors and n denotes the number of nodes.

Corollary 5. *Under the current assumptions, in the limit of large delays, $\tau \to \infty$,*

$$\rho(\sigma^2 \hat{M})(\beta, \mu, d, m) \geq \rho(\sigma^2 \hat{M})(\beta, \mu, 1, n)$$

for any m and $n \geq 2$ and any $d \geq 2$.

Gain Robustness Tradeoff

Corollary 6. *In the limit of large delays, the deviation system is MS stable if $\beta < \dfrac{\mu}{\sigma^2 + \mu^2 d}$ and MS unstable if $\beta > \dfrac{\mu}{\sigma^2 + \mu^2 \frac{d}{2}}$.*

Corollary 6 suggests that β cannot be too large, and that the complex behavior could be avoided by scaling the coupling between agents appropriately, if d is known. From another prospective, networks with nodes that are not aware of d go through phase transitions when d increases. Of course, if the agents do not have a way to know d, or to use the information, they can only play safe and have a small β. On the other hand, the agents may naturally tend to increase β in the attempt of increasing convergence speed. Thus, there is a natural tradeoff between speed and robustness.

Robustness of Gossiping. In the case of packet-drop networks with Bernoulli fading, let us assume that $\beta < 1$ is fixed. Then, the deviation system is MS stable if

$d < \frac{1-\beta(1-\mu)}{\beta\mu}$ and MS unstable if $d > 2\frac{1-\beta(1-\mu)}{\beta\mu}$. These bounds indicate that there is a maximum number of neighbors supportable by a certain QoS (QoS$=\mu = (1-e)$). However, the quality of service needs to decrease in order to increase the number of neighbors d. This follows from the fact that $g(\mu) = \frac{1-\beta(1-\mu)}{\beta\mu}$ is a monotonically decreasing function of μ for fixed $\beta < 1$. Thus, to accommodate more neighbors we need to decrease μ, i.e., reduce the quality of service.

When μ is very small, the network is usually disconnected and only few couples of nodes talk at any time. This setting is similar to the random gossip algorithm [36], and our analysis explains why such algorithms are found to be robust.

7 MS Stability and Performance of Distributed Averaging Systems

In this section, we apply Theorem 2 to quantifying the performance of the distributed averaging algorithm in the presence of fading links and additive noise (white Gaussian zero mean, unit variance). Because of the internal marginally stable pole, the states of the agents will undergo a random walk even when the links are perfect. To evaluate performance, it is then natural to consider the variance of the deviation state as suggested by [33]

$$\begin{aligned} x_i(k+1) &= x_i(k) + \beta \sum_{j\in\mathcal{N}_i^-} \xi_{ij}(k)[x_j(k-\tau_{ij}) - x_i(k)] + \beta w_i(k) \\ z_i(k) &= x_i(k) - \tfrac{1}{n}\sum_{\ell=1}^{n} x_\ell(k) \end{aligned} \tag{23}$$

In the case where the communication links are perfect, we have that the $\mathcal{H}_2$ norm of the transfer function from w to z, which is given by

$$C_z(zI - (I-\beta L))^{-1}B_w$$

where $C_z = I - \frac{1}{n}\mathbf{1}\mathbf{1}'$, and $B_w = \beta I$.

C_z is circulant and it has all its eigenvalues at 1 except one a at zero corresponding to the eigenvector $\mathbf{1}$. Thus,

$$\left\|C_z(zI - (I-\beta L))^{-1}B_w\right\|_2^2 = \beta^2 \cdot \sum_{k=2}^{n} \frac{1}{1-|1-\beta\lambda_k(L)|^2}. \tag{24}$$

Figure 5(a) shows the results for torus networks of dimension 1, 2, 3. This is the discrete-time version of the results of [33]. They showed, that the performance per agent grows linearly with n in the case $d = 1$, logarithmically with n in the case $d = 2$, while is bounded above in the case $d \geq 3$. This suggests that the three dimensional torus leads to a more cohesive behavior, where each node behaves more similarly to the collective average.

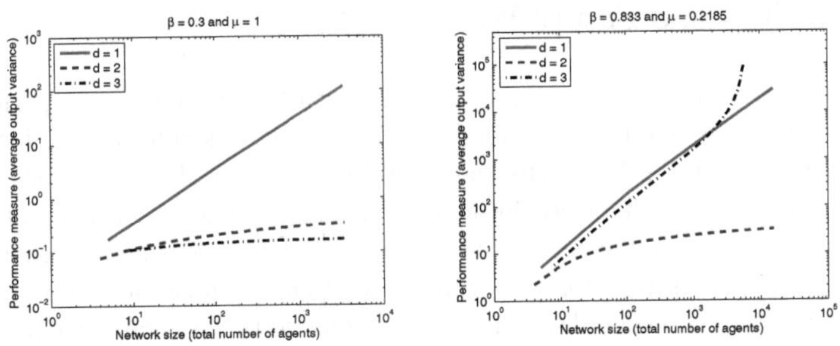

(a) (b)

Fig. 5. Mean Square performance for $d-$ torus networks for different sizes and dropout probabilities. (a) $\beta = 0.3$ with perfect links $\mu = 1$. (b) $\beta = 0.833$ with packet drop links with $\mu = 0.22$.

In the general case, we use the theory of Section 4.2. From Theorem 2. The MS performance problem is equivalent to a the Mean Square stability problem involving M_P given by

$$M_p: \quad \begin{aligned} x^+ &= (I - \beta\mu L)x + Bw_\Delta + B_w w \\ z_\Delta &= Cx \\ z &= C_z x. \end{aligned}$$

Theorem 4. *[34] System (23) has MS performance γ^2, if it is MS stable, and*

$$\gamma^2 > \frac{\beta^2 \cdot \sum_{k=2}^{n} \frac{1}{1 - |1 - \beta\mu\lambda_k(L)|^2}}{1 - \sigma^2\beta^2 \cdot \frac{1}{n} \sum_{k=2}^{n} \frac{2Re\{\lambda_k(L)\}}{1 - |1 - \beta\mu\lambda_k(L)|^2}}.$$

The above formula can be used to compare the performance of different networked systems under various conditions. For example, Figure 5(b) shows the results for d-torus networks, for $d = 1, 2, 3$. In the example, we set $\beta = 0.833$ and quality of services: $\mu = 0.22$. We recall that small μ, more sporadic communication, tends to stabilize.

When $\mu = 0.2185$, we see that the 3-d torus becomes MS unstable after certain size and the performance curve reflect the grater fragility of this structure for the given parameters. The performance for the 2-d torus is however better and grows slower with the size of the system than in the 1-d case. If we set $\mu = 0.4$, (not shown) the 3-d torus is MS unstable for all sizes. Moreover, the 2-d torus now becomes MS unstable for a certain size and its performance is worst than the 1-d torus. Thus, in the settings of this example, we see that higher dimensional torus have better performance under normal regimes, but they are more fragile and tend to collapse in regimes that can lead to MS instability. The 1-d torus instead has the worst performance in normal settings but it is most resilient to packet drops.

8 Conclusions and Future Work

In this chapter, we have summarized some recent results on stability and performance of networked systems. We have extended the MS stability result to include MS performance and shown that the MS performance problem is equivalent to an extended MS stability problem. Although the condition is computable efficiently, its computation becomes cumbersome when the networked system has many nodes and unreliable links. To gain more insights into network organization which might be more robust or better performing, we have considered spatially invariant networked systems. We have shown that, given the nodes' dynamics and the network topology, there is a maximal amount of stochastic uncertainty that the networked system can tolerate in its links. Each uncertain link consumes a part of the total budget of the total available uncertainty. Thus, highly uncertain links need to be compensated by very reliable other links in order to maintain MS stability. We have also reviewed topology independent bounds on the spectral radius in the limit of large communication delays for a classical distributed averaging algorithm. The bounds imply that the single neighbor networks are the most robust to link uncertainty. We have then obtained a formula to compute the MS performance of spatially invariant networked systems. The formula extends the existing result on performance to the case of unreliable communication links, and captures the interplay between MS performance and robustness to unreliable links. Future research should identify and consider other network structures, besides spatially invariant ones, for which checking MS stability is easy. Another related direction is to whether the spectral radius condition could be computed in a distributed way in certain cases.

References

1. Tsitsiklis, J.N.: Problems in decentralized decision making and computation, Ph.D. Thesis, Massachusetts Institute of Technology (1984)
2. Tsitsiklis, J.N., Bertsekas, D.P., Athans, M.: Distributed asynchronous deterministic and stochastic gradient optimization algorithms. IEEE TAC 31(9), 803–812 (1986)
3. Jadbabaie, A., Lin, J., Morse, A.S.: Coordination of groups of mobile autonomous agents using nearest neighbor rules. IEEE Trans. Automat. Contr. 48(6), 988–1001 (2003)
4. Olfati-Saber, R., Murray, R.M.: Consensus problems in networks of agents with switching topology and time-delays. IEEE Transactions on Automatic Control 49(9), 1520–1533 (2004)
5. Fax, J.A., Murray, R.M.: Information flow and cooperative control of vehicle formations. IEEE Transactions on Automatic Control Special Issue on Networked Control 49(9), 1465–1476 (2004)
6. Spanos, D.P., Olfati-Saber, R., Murray, R.M.: Dynamic consensus on mobile networks. Presented at the 16th IFAC World Congr., Prague, Czech Republic (July 2005)
7. Reynolds, A.M., Smith, A.D., Menzel, R., Greggers, U., Reynolds, D.R., Riley, J.R.: Displaced honey bees perform optimal scale-free search flights. Ecology 88(8), 1955–1961 (2007)
8. Bartumeus, F., Daluz, M.G.E., Viswanathan, G.M., Catalan, J.: Animal search strategies: A quantitative random-walk analysis. Ecology 86(11), 3078–3087 (2005)

9. Brockmann, D., Hufnagel, L., Geisel, T.: The scaling laws of human travel. Nature 439, 462–465 (2006)
10. Mandelbrot, B.: The variation of certain speculative prices. Journal of Business 36, 394–419 (1963)
11. Mantegna, R., Stanley, H.: Scaling behavior in the dynamics of an economic index. Nature 376, 46–49 (1995)
12. Davis, P.J.: Circulant Matrices. A Wiley-Interscience Publication, Pure and Applied Mathematics. John Wiley & Sons, New York (1979)
13. Bamieh, B., Paganini, F., Dahleh, M.: Distributed control of spatially invariant systems. IEEE Trans. Automat. Contr. 47, 1091–1118 (2002)
14. Wang, J., Elia, N.: Distributed agreement in the presence of noise. In: 47th Annual Allerton Conference, September 30-October 2. Allerton House, UIUC, Illinois (2009)
15. Wang, J., Elia, N.: Dynamic average consensus over random networks with additive noise. In: IEEE Conference on Decision and Control, pp. 4789–4794 (2010)
16. Wang, J., Elia, N.: Control approach to distributed optimization. In: Allerton Conference on Communication, Control and Computing, September 29-October 1, pp. 557–561 (2010)
17. Wang, J., Elia, N.: A Control perspective to centralized and distributed optimization. Finalist for Best Student Paper Award in Proc. 50th IEEE CDC-ECC (2011)
18. Elia, N.: Remote stabilization over fading channels. Systems and Control Letters 54(3), 237–249 (2005)
19. Elia, N., Eisenbeis, J.N.: Limitations of linear control over packet drop networks
20. Arnold, L.: Stochastic differential equations: theory and applications. Krieger, Malabar (1992)
21. Kushner, H.J.: Stochastic Stability and Control. Academic Press, New York (1967)
22. Meyn, S.P., Tweedie, R.L.: Markov chains and stochastic stability. Springer, London (1993)
23. Costa, O.L.V., Fragoso, M.D.: Stability Results for Discrete-Time Linear Systems with Markovian Jumping Parameters. Journal of Mathematic Analysis Applications 179, 154–178 (1993)
24. Costa, O.L.V., Fragoso, M.D., Marques, R.P.: Discrete-Time Markov Jump Linear Systems. Springer (2005)
25. Horn, R.A., Johnson, C.R.: Matrix Analysis. Cambridge Univ. Press, Cambridge (1987)
26. Elia, N.: Emergence of Power Laws in Networked Control Systems. In: Proc. IEEE CDC 2006, pp. 490–495 (2006)
27. Wang, J., Elia, N.: Distributed averaging under constraints on information exchange: emergence of Lévy flights. IEEE Trans. on Automat. Control 57(10), 2435–2449 (2012)
28. Wang, J., Elia, N.: Mean square stability of consensus over fading networks with nonhomogeneous communication delays. In: Proc. of the 47th IEEE Conf. on Decision and Control, pp. 4614–4619 (December 2008)
29. Wang, J., Elia, N.: Mitigation of Complex Behavior over Networked Systems: Analysis of Spatially Invariant Structures. To appear in Automatica (2013)
30. Padmasola, P., Elia, N.: Bode Integral Limitations of Spatially Invariant Systems. In: IEEE CDC 2006, pp. 4327–4332 (2006)
31. Padmasola, P., Elia, N.: Mean Square Stabilization of LTI Systems Over Packet-Drop Networks with Imperfect Side Information. In: Proc. American Control Conference, pp. 5550–5555 (2006)
32. Napoli, M., Bamieh, B., Dahleh, M.: H_2 Norm Minimization for Distributed Continuous Time Systems: an Input/ Output Approach. In: Proc. 1998 Mathematical Theory of Networks and Systems, Padova, Italy (July 1998)

33. Bamieh, B., Jovanovic, M.R., Mitra, P., Paterson, S.: Coherence in large-scale networks: dimension dependent limitations of local feedback. IEEE Trans. on Automatic Control 57(9), 2235–2249 (2012)
34. Ma, X., Elia, N.: Mean Square Performance Analysis for Torus-Networked Distributed Consensus under Stochastic Disturbances. Submitted to IEEE CDC (2013)
35. Zhou, K., Doyle, J.C., Glover, K.: Robust and Optimal Control. Prentice Hall, New Jersey (1996)
36. Boyd, S., Ghosh, A., Prabhakar, B., Shah, D.: Randomized gossip algorithms. IEEE Trans. Inform. Theory 52(6), 2508–2530 (2006)

Author Index

MIX
Papier aus verantwortungsvollen Quellen
Paper from responsible sources
FSC® C105338

If you have any concerns about our products,
you can contact us on
ProductSafety@springernature.com

In case Publisher is established outside the EU,
the EU authorized representative is:
Springer Nature Customer Service Center GmbH
Europaplatz 3, 69115 Heidelberg, Germany

Printed by Libri Plureos GmbH
in Hamburg, Germany